Vibration with Control

Vibration with Control

Second Edition

Daniel John Inman
University of Michigan, USA

This edition first published 2017

Registered Offices
John Wiley & Sons, Inc., 111 River Street, Hoboken, NJ 07030, USA
John Wiley & Sons Ltd, The Atrium, Southern Gate, Chichester, West Sussex, PO19 8SQ, UK

Editorial Office
The Atrium, Southern Gate, Chichester, West Sussex, PO19 8SQ, UK

For details of our global editorial offices, customer services, and more information about Wiley products visit us at www.wiley.com.

Wiley also publishes its books in a variety of electronic formats and by print-on-demand. Some content that appears in standard print versions of this book may not be available in other formats.

Library of Congress Cataloging-in-Publication Data

Names: Inman, D. J., author.
Title: Vibration with control / Daniel John Inman.
Description: Second edition. | Chichester, West Sussex, UK ; Hoboken, NJ, USA: John Wiley & Sons Inc., [2017] | Includes bibliographical references and index.
Identifiers: LCCN 2016048220 | ISBN 9781119108214 (cloth) | ISBN 9781119108238 (Adobe PDF) | ISBN 9781119108221 (epub)
Subjects: LCSH: Damping (Mechanics). | Vibration.
Classification: LCC TA355 .I523 2017 | DDC 620.3–dc23 LC record available at https://lccn.loc.gov/2016048220

Cover Images: (background) © jm1366/Getty Images, Inc.; (left to right) © Teun van den Dries / Shutterstock.com; © simonlong/Getty Images, Inc.; © Edi_Eco/Getty Images, Inc.; © karnaval2018 / Shutterstock.com; © Stefano Borsani / iStockphoto
Cover Design: Wiley

Set in 10/12pt WarnockPro by Aptara Inc., New Delhi, India

V B

10 9 8 7 6 5 4 3 2 1

This book is dedicated to our grand children

Griffin Amherst Pitre
Benjamin Lafe Scamacca
Lauren Cathwren Scamacca
Conor James Scamacca
Jacob Carlin Scamacca

And our children

Daniel John Inman, II
Angela Wynne Pitre
Jennifer Wren Scamacca

Contents

Preface

Advance level vibration topics are presented here, including lumped mass and distributed mass systems in the context of the appropriate mathematics along with topics from control that are useful in vibration analysis, testing and design. This text is intended for use in a second course in vibration, or a combined course in vibration and control. It is also intended as a reference for the field of structural control and could be used as a text in structural control. The control topics are introduced at the beginners' level with no prerequisite knowledge in controls needed to read the book.

The text is an attempt to place vibration and control on a firm mathematical basis and connect the disciplines of vibration, linear algebra, matrix computations, control and applied functional analysis. Each chapter ends with notes on further references and suggests where more detailed accounts can be found. In this way I hope to capture a "bigger picture" approach without producing an overly large book. The first chapter presents a quick introduction using single degree of freedom systems (second-order ordinary differential equations) to the following chapters, which extend these concepts to multiple degree of freedom systems (matrix theory and systems of ordinary differential equations) and distributed parameter systems (partial differential equations and boundary value problems). Numerical simulations and matrix computations are also presented through the use of MATLAB™.

New In This Edition – The book chapters have been reorganized (there are now 12 instead of 13 chapters) with the former chapter on design removed and combined with the former chapter on control to form a new chapter titled *Vibration Suppression*. Some older, no longer used material, has been deleted in an attempt to keep the book limited in size as new material has been added.

The new material consists of adding several modeling sections to the text, including corresponding problems and examples. Many figures have been redrawn throughout to add clarity with more descriptive captions. In addition, a number of new figures have been added. New problems and examples have been added and some old ones removed. In total, seven new sections have been added to introduce modeling, coupled systems, the use of piezoelectric materials, metastructures, and validation and verification.

Instructor Support – Power Point slides are available for presentation of the material, along with a complete solutions manual. These materials are available from the publisher for those who have adapted the book. The author is pleased to answer questions via the email listed below.

Student Support – The best place to get help is your instructor and others in your peer group through discussion of the material. There are also many excellent texts as referenced throughout the book and of course Internet searches can provide lots of help. In addition, feel free to email the author at the address below (but don't ask me to do your homework!).

Acknowledgements – I would like to thank two of my current PhD students, Katie Reichl and Brittany Essink, for checking some of the homework and providing some plots. I would like to thank all of my former and current PhD students for 36 years of wonderful research and discussions. Thanks are owed to the instructors and students of the previous edition who have sent suggestions and comments. Last, thanks to my lovely wife Catherine Ann Little for putting up with me.

Leland, Michigan *Daniel J. Inman*
daninman@umich.edu

About the Companion Website

Vibration with Control, Second Edition is accompanied by a companion website:

www.wiley.com/go/inmanvibrationcontrol2e

The website includes:

- Powerpoint slides
- Solutions manual

1

Single Degree of Freedom Systems

1.1 Introduction

In this chapter, the vibration of a single degree of freedom system (SDOF) will be analyzed and reviewed. Analysis, measurement, design and control of SDOF systems are discussed. The concepts developed in this chapter constitute a review of introductory vibrations and serve as an introduction for extending these concepts to more complex systems in later chapters. In addition, basic ideas relating to measurement and control of vibrations are introduced that will later be extended to multiple degree of freedom systems and distributed parameter systems. This chapter is intended to be a review of vibration basics and an introduction to a more formal and general analysis for more complicated models in the following chapters.

Vibration technology has grown and taken on a more interdisciplinary nature. This has been caused by more demanding performance criteria and design specifications of all types of machines and structures. Hence, in addition to the standard material usually found in introductory chapters of vibration and structural dynamics texts, several topics from control theory are presented. This material is included not to train the reader in control methods (the interested student should study control and system theory texts), but rather to point out some useful connections between vibration and control as related disciplines. In addition, structural control has become an important discipline requiring the coalescence of vibration and control topics. A brief introduction to nonlinear SDOF systems and numerical simulation is also presented.

1.2 Spring-Mass System

Simple harmonic motion, or oscillation, is exhibited by structures that have elastic restoring forces. Such systems can be modeled, in some situations, by a spring-mass schematic (Figure 1.1). This constitutes the most basic vibration model of a structure and can be used successfully to describe a surprising number of devices, machines and structures. The methods presented here for solving such a simple mathematical model may seem to be more sophisticated than the problem

Vibration with Control, Second Edition. Daniel John Inman.

Companion Website: www.wiley.com/go/inmanvibrationcontrol2e

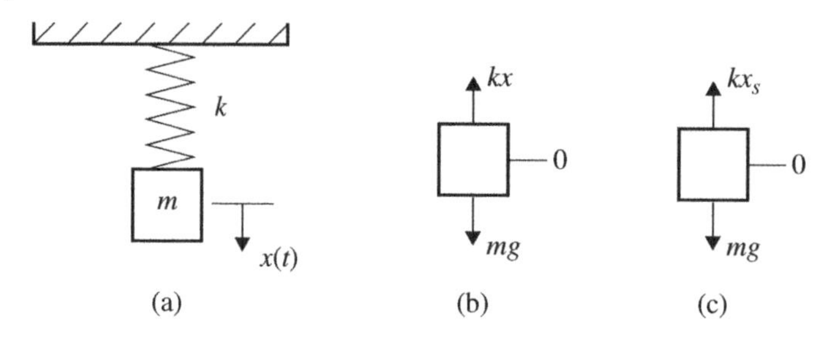

Figure 1.1 (a) A spring-mass schematic, (b) a free body diagram, and (c) a free body diagram of the static spring mass system.

requires. However, the purpose of this analysis is to lay the groundwork for solving more complex systems discussed in the following chapters.

If $x = x(t)$ denotes the displacement (in meters) of the mass m (in kg) from its equilibrium position as a function of time, t (in sec), the equation of motion for this system becomes (upon summing the forces in Figure 1.1b)

$$m\ddot{x} + k(x + x_s) - mg = 0$$

where k is the stiffness of the spring (N/m), x_s is the static deflection (m) of the spring under gravity load, g is the acceleration due to gravity (m/s^2) and the over dots denote differentiation with respect to time. A discussion of dimensions appears in Appendix A and it is assumed here that the reader understands the importance of using consistent units. From summing forces in the free body diagram for the static deflection of the spring (Figure 1.1c), $mg = kx_s$ and the above equation of motion becomes

$$m\ddot{x}(t) + kx(t) = 0 \tag{1.1}$$

This last expression is the equation of motion of an SDOF system and is a linear, second-order, ordinary differential equation with constant coefficients.

Figure 1.2 indicates a simple experiment for determining the spring stiffness by adding known amounts of mass to a spring and measuring the resulting static deflection, x_s. The results of this static experiment can be plotted as force (mass times acceleration) versus x_s, the slope yielding the value of k for the linear portion of the plot. This is illustrated in Figure 1.3.

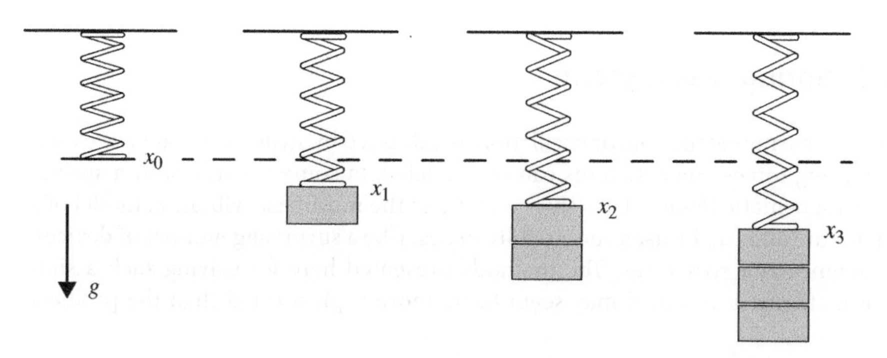

Figure 1.2 Measurement of spring constant using static deflection caused by added mass.

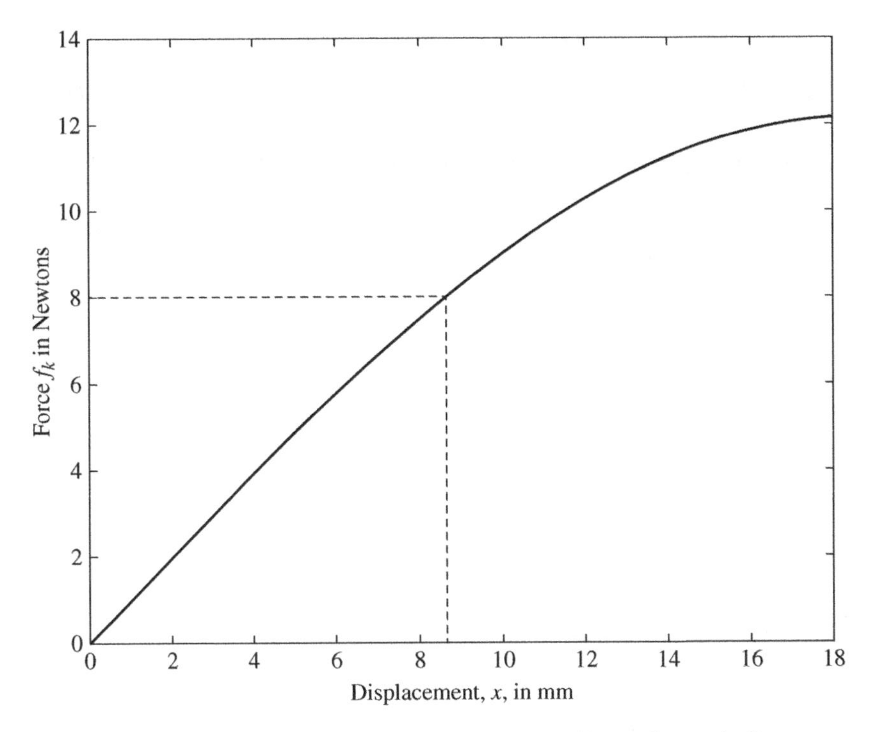

Figure 1.3 Determination of the spring constant. The dashed box indicates the linear range of the spring.

Once m and k are determined from static experiments, Equation (1.1) can be solved to yield the time history of the position of the mass m, given the initial position and velocity of the mass. The form of the solution of Equation (1.1) is found from substitution of an assumed periodic motion (from experience watching vibrating systems) of the form

$$x(t) = A\sin(\omega_n t + \phi) \tag{1.2}$$

where $\omega_n = \sqrt{k/m}$ is called the *natural frequency* in radians per second (rad/s). Here A, the *amplitude*, and ϕ, the *phase shift*, are constants of integration determined by the initial conditions.

The existence of a *unique* solution for Equation (1.1) with two specific initial conditions is well known and is given in Boyce and DiPrima (2012). Hence, if a solution of the form of Equation (1.2) is guessed and it works, then it is *the* solution. Fortunately, in this case, the mathematics, physics and observation all agree.

To proceed, if x_0 is the specified initial displacement from equilibrium of mass m, and v_0 is its specified initial velocity, simple substitution allows the constants of integration A and ϕ to be evaluated. The unique solution is

$$x(t) = \sqrt{\frac{\omega_n^2 x_0^2 + v_0^2}{\omega_n^2}} \sin\left[\omega_n t + \tan^{-1}\left(\frac{\omega_n x_0}{v_0}\right)\right] \tag{1.3}$$

Alternately, $x(t)$ can be written as

$$x(t) = \frac{v_0}{\omega_n} \sin \omega_n t + x_0 \cos \omega_n t \tag{1.4}$$

by using a simple trigonometric identity or by direct substitution of the initial conditions (Example 1.2.1).

A purely mathematical approach to the solution of Equation (1.1) is to assume a solution of the form $x(t) = Ae^{\lambda t}$ and solve for λ, i.e.

$$m\lambda^2 e^{\lambda t} + ke^{\lambda t} = 0$$

This implies that (because $e^{\lambda t} \neq 0$ and $A \neq 0$)

$$\lambda^2 + \left(\frac{k}{m}\right) = 0$$

or that

$$\lambda = \pm j \left(\frac{k}{m}\right)^{1/2} = \pm \omega_n j$$

where $j = (-1)^{1/2}$. Then the general solution becomes

$$x(t) = A_1 e^{-\omega_n jt} + A_2 e^{\omega_n jt} \tag{1.5}$$

where A_1 and A_2 are arbitrary complex conjugate constants of integration to be determined by the initial conditions. Use of Euler's formulas then yields Equations (1.2) and (1.4) (Inman, 2014). For more complicated systems, the exponential approach is often more appropriate than first guessing the form (sinusoid) of the solution from watching the motion.

Another mathematical comment is in order. Equation (1.1) and its solution are valid only as long as the spring is linear. If the spring is stretched too far or too much force is applied to it, the curve in Figure 1.3 will no longer be linear. Then Equation (1.1) will be nonlinear (Section 1.10). For now, it suffices to point out that initial conditions and springs should always be checked to make sure that they fall into the linear region, if linear analysis methods are going to be used.

Example 1.2.1

Assume a solution of Equation (1.1) of the form

$$x(t) = A_1 \sin \omega_n t + A_2 \cos \omega_n t$$

and calculate the values of the constants of integration A_1 and A_2 given arbitrary initial conditions x_0 and v_0, thus verifying Equation (1.4).

Solution: The displacement at time $t = 0$ is

$$x(0) = x_0 = A_1 \sin(0) + A_2 \cos(0)$$

or $A_2 = x_0$. The velocity at time $t = 0$ is

$$\dot{x}(0) = v_0 = \omega_n A_1 \cos(0) - \omega_n x_0 \sin(0)$$

Solving this last expression for A_1 yields $A_1 = v_0/x_0$, so that Equation (1.4) results in

$$x(t) = \frac{v_0}{x_0}\sin\omega_n t + x_0\cos\omega_n t$$

Example 1.2.2
Compute and plot the time response of a linear spring-mass system to initial conditions of $x_0 = 0.5$ mm and $v_0 = 2\sqrt{2}$ mm/s, if the mass is 100 kg and the stiffness is 400 N/m.

Solution: The frequency is

$$\omega_n = \sqrt{k/m} = \sqrt{400/100} = 2 \text{ rad/s}$$

Next compute the amplitude from Equation (1.3):

$$A = \sqrt{\frac{\omega_n^2 x_0^2 + v_0^2}{\omega_n^2}} = \sqrt{\frac{2^2(0.5)^2 + (2\sqrt{2})^2}{2^2}} = 1.5 \text{ mm}$$

From Equation (1.3) the phase is

$$\phi = \tan^{-1}\left(\frac{\omega_n x_0}{v_0}\right) = \tan^{-1}\left(\frac{2(0.5)}{2\sqrt{2}}\right) \approx 10 \text{ rad}$$

Thus the response has the form

$$x(t) = 1.5\sin(2t + 10)$$

and this is plotted in Figure 1.4.

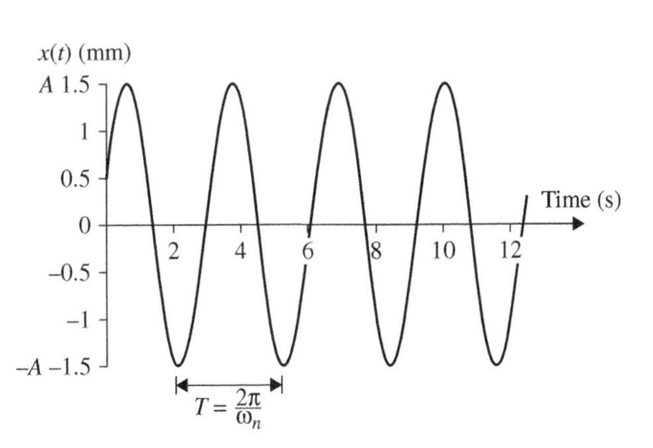

Figure 1.4 The response of a simple spring-mass system to an initial displacement of $x_0 =$ 0.5 mm and an initial velocity of $v_0 = 2\sqrt{2}$ mm/s. The period, defined as the time it takes to complete one cycle off oscillation, $T = 2\pi/\omega_n$, becomes $T = 2\pi/2 = \pi$ s.

1.3 Spring-Mass-Damper System

Most systems will not oscillate indefinitely when disturbed, as indicated by the solution in Equation (1.3). Typically, the periodic motion dies down after some time. The easiest way to treat this mathematically is to introduce a velocity term, $c\dot{x}$, into Equation (1.1) and examine the equation

$$m\ddot{x} + c\dot{x} + kx = 0 \tag{1.6}$$

This also happens physically with the addition of a *dashpot* or *damper* to dissipate energy, as illustrated in Figure 1.5.

Equation (1.6) agrees with summing forces in Figure 1.5 if the dashpot exerts a dissipative force proportional to velocity on the mass m. Unfortunately, the constant of proportionality, c, cannot be measured by static methods as m and k are. In addition, many structures dissipate energy in forms not proportional to velocity. The constant of proportionality c is given in Newton-second per meter (Ns/m) or kilograms per second (kg/s) in terms of fundamental units.

Again, the unique solution of Equation (1.6) can be found for specified initial conditions by assuming that $x(t)$ is of the form

$$x(t) = Ae^{\lambda t}$$

and substituting this into Equation (1.6) to yield

$$A\left(\lambda^2 + \frac{c}{m}\lambda + \frac{k}{m}\right)e^{\lambda t} = 0 \tag{1.7}$$

Since a trivial solution is not desired, $A \neq 0$, and since $e^{\lambda t}$ is never zero, Equation (1.7) yields

$$\lambda^2 + \frac{c}{m}\lambda + \frac{k}{m} = 0 \tag{1.8}$$

Equation (1.8) is called the *characteristic equation* of Equation (1.6). Using simple algebra, the two solutions for λ are

$$\lambda_{1,2} = -\frac{c}{2m} \pm \frac{1}{2}\sqrt{\frac{c^2}{m^2} - 4\frac{k}{m}} \tag{1.9}$$

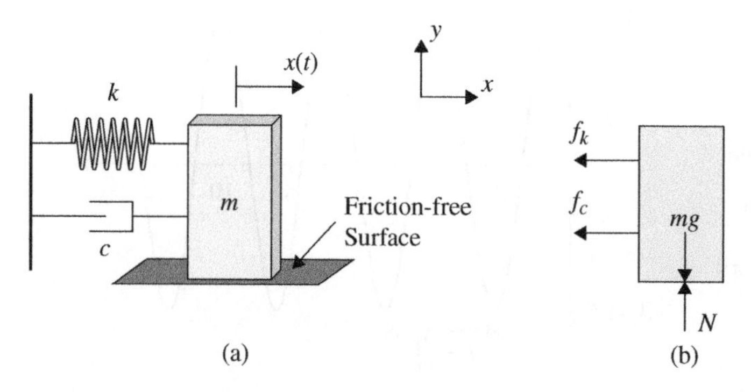

Figure 1.5 (a) Schematic of spring-mass-damper system. (b) A free-body diagram of the system in part (a).

The quantity under the radical is called the *discriminant* and together with the sign of m, c and k determines whether or not the roots are complex or real. Physically, m, c and k are all positive in this case, so the value of the discriminant determines the nature of the roots of Equation (1.8).

It is convenient to define the dimensionless *damping ratio*, ζ, as

$$\zeta = \frac{c}{2\sqrt{km}}$$

In addition, let the *damped natural frequency*, ω_d, be defined by (for $0 < \zeta < 1$)

$$\omega_d = \omega_n\sqrt{1-\zeta^2} \tag{1.10}$$

Then Equation (1.6) becomes

$$\ddot{x} + 2\zeta\omega_n\dot{x} + \omega_n^2 x = 0 \tag{1.11}$$

and Equation (1.9) becomes

$$\lambda_{1,2} = -\zeta\omega_n \pm \omega_n\sqrt{\zeta^2-1} = -\zeta\omega_n \pm \omega_d j, \quad 0 < \zeta < 1 \tag{1.12}$$

Clearly the value of the damping ratio, ζ, determines the nature of the solution of Equation (1.6). There are three cases of interest. The derivation of each case is left as an exercise and can be found in almost any introductory text on vibrations (Inman, 2014; Meirovitch, 1986).

Underdamping occurs if the system's parameters are such that

$$0 < \zeta < 1$$

so that the discriminant in Equation (1.12) is negative and the roots form a complex conjugate pair of values. The solution of Equation (1.11) then becomes

$$x(t) = e^{-\zeta\omega_n t}(A\cos\omega_d t + B\sin\omega_d t) \tag{1.13}$$

or

$$x(t) = Ce^{-\zeta\omega_n t}\sin(\omega_d t + \phi)$$

where A, B, C and ϕ are constants determined by the specified initial velocity, v_0 and position, x_0

$$\begin{aligned} A &= x_0 & C &= \frac{\sqrt{(v_0+\zeta\omega_n x_0)^2 + (x_0\omega_d)^2}}{\omega_d} \\ B &= \frac{(v_0+\zeta\omega_n x_0)}{\omega_d} & \phi &= \tan^{-1}\left[\frac{x_0\omega_d}{(v_0+\zeta\omega_n x_0)}\right] \end{aligned} \tag{1.14}$$

The underdamped response has the form given in Figure 1.6 and consists of a decaying oscillation of frequency ω_d.

Overdamping occurs if the system's parameters are such that

$$\zeta > 1$$

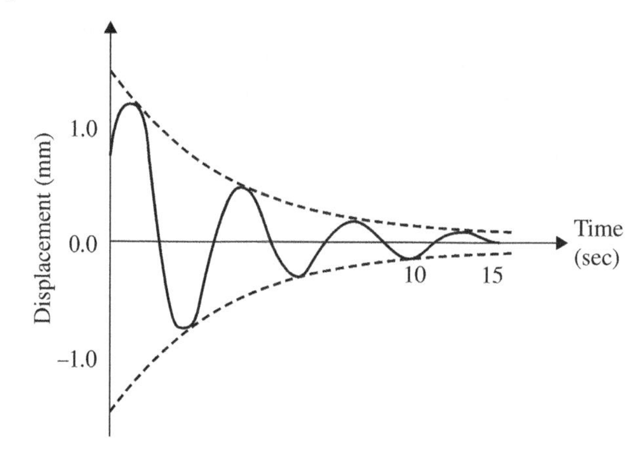

Figure 1.6 Response of an underdamped system illustrating oscillation with exponential decay.

so that the discriminant in Equation (1.12) is positive and the roots are a pair of negative real numbers. The solution of Equation (1.11) then becomes

$$x(t) = Ae^{\left(-\zeta+\sqrt{\zeta^2-1}\right)\omega_n t} + Be^{\left(-\zeta-\sqrt{\zeta^2-1}\right)\omega_n t} \tag{1.15}$$

where A and B are again constants determined by v_0 and x_0. They are

$$A = \frac{v_0 + \left(\zeta + \sqrt{\zeta^2-1}\right)\omega_n x_0}{2\omega_n\sqrt{\zeta^2-1}} \text{ and } B = -\frac{v_0 + \left(\zeta - \sqrt{\zeta^2-1}\right)\omega_n x_0}{2\omega_n\sqrt{\zeta^2-1}} \tag{1.16}$$

The overdamped response has the form given in Figure 1.7. An overdamped system does not oscillate, but rather returns to its rest position exponentially.

Critical Damping occurs if the system's parameters are such that $\zeta = 1$, so that the discriminant in Equation (1.12) is zero and the roots are a pair of negative real repeated numbers. The solution of Equation (1.11) then becomes

$$x(t) = e^{-\omega_n t}[(v_0 + \omega_n x_0)t + x_0] \tag{1.17}$$

The critically damped response is plotted in Figure 1.8 for values of the initial velocity v_0 of different signs and $x_0 = 0.25$ mm.

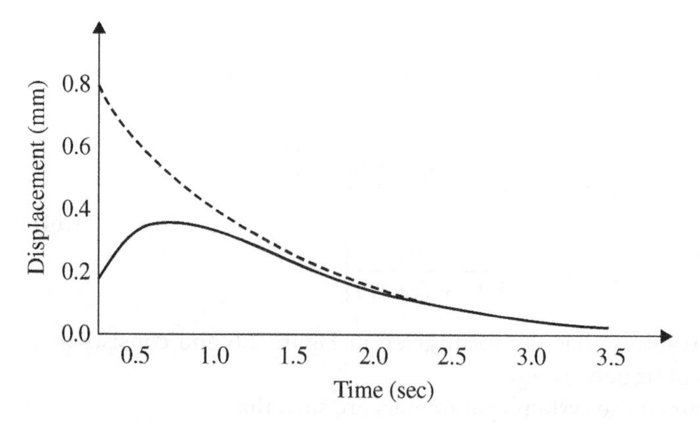

Figure 1.7 Response of an overdamped system illustrating exponential decay without oscillation.

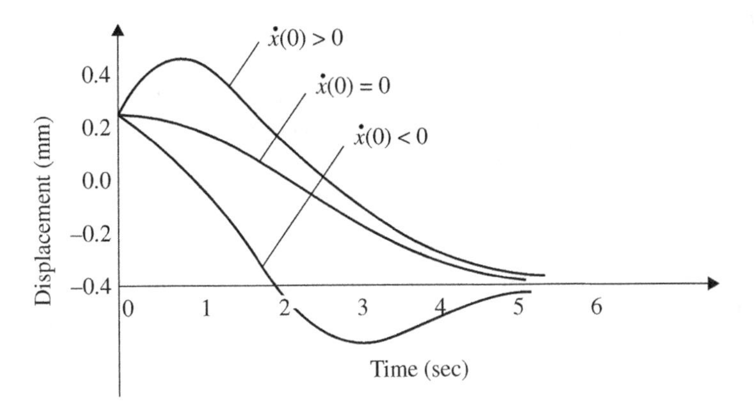

Figure 1.8 Response of critically damped system to an initial displacement and three different initial velocities indicating no oscillation.

It should be noted that critically damped systems can be thought of in several ways. First, they represent systems with the minimum value of damping rate that yields a non-oscillating system (Exercise 1.5). Critical damping can also be thought of as the case that separates non-oscillation from oscillation.

Example 1.3.1

Derive the constants A and B of integration for the overdamped case of Equation (1.15).

Solution: Substitution of $x(0) = x_0$ into Equation (1.15) yields

$$x(0) = Ae^0 + Be^0 \text{ or } x_0 = A + B \tag{1.18}$$

Differentiating Equation (1.15) and setting $t = 0$ in the result yields

$$\dot{x}(0) = A\lambda_1 e^0 + B\lambda_2 e^0 \text{ or } v_0 = \lambda_1 A + \lambda_1 B \tag{1.19}$$

where λ_1 and λ_2 are defined in Equation (1.12). These two initial conditions result in two independent equations in two unknowns, A and B, which can be solved in many ways. Writing Equations (1.17) and (1.18) as a single matrix equation yields

$$\begin{bmatrix} x_0 \\ v_0 \end{bmatrix} = \begin{bmatrix} 1 & 1 \\ \lambda_1 & \lambda_2 \end{bmatrix} \begin{bmatrix} A \\ B \end{bmatrix} \text{ or } \begin{bmatrix} A \\ B \end{bmatrix} = \begin{bmatrix} 1 & 1 \\ \lambda_1 & \lambda_2 \end{bmatrix}^{-1} \begin{bmatrix} x_0 \\ v_0 \end{bmatrix}$$

Solving by computing matrix inverse (see Appendix B for details on computing a matrix inverse) yields

$$\begin{bmatrix} A \\ B \end{bmatrix} = \frac{1}{\lambda_2 - \lambda_1} \begin{bmatrix} \lambda_2 & -1 \\ -\lambda_1 & 1 \end{bmatrix} \begin{bmatrix} x_0 \\ v_0 \end{bmatrix}$$

Expanding, substituting in the values for λ_1 and λ_2, recalling that they are real numbers (i.e. $\zeta^2 > 1$) and writing as two separate equations results in

$$A = \frac{-v_0 + (-\zeta - \sqrt{\zeta^2 - 1})\omega_n}{-2\omega_n\sqrt{\zeta^2 - 1}} \text{ and } B = \frac{v_0 + (\zeta - \sqrt{\zeta^2 - 1})\omega_n}{-2\omega_n\sqrt{\zeta^2 - 1}}$$

Factoring out the minus sign in the denominator results in Equations (1.16).

1.4 Forced Response

The preceding analysis considers the vibration of a device or structure due to some initial disturbance (nonzero v_0 and x_0). In this section, the vibration of a spring-mass-damper system subjected to an external force is considered. In particular, the response to harmonic excitations, impulses and step forcing functions is examined.

In many environments, rotating machinery, motors, etc., cause periodic motions of structures to induce vibrations into other mechanical devices and structures nearby. It is common to approximate the driving forces, $F(t)$, as periodic of the form

$$F(t) = F_0 \sin \omega t \tag{1.20}$$

where F_0 represents the amplitude of the applied force and ω denotes the frequency of the applied force, or the driving frequency, in rad/s. On summing forces, the equation for the forced vibration of the system in Figure 1.9 becomes

$$m\ddot{x} + c\dot{x} + kx = F_0 \sin \omega t \tag{1.21}$$

Recall from the discipline of differential equations (Boyce and DiPrima, 2012), that the solution of Equation (1.21) consists of the sum of the homogeneous solution Equation (1.5) and a particular solution. These are usually referred to as the *transient response* and the *steady-state response*, respectively. Physically, there is motivation to assume that the steady state response will follow the forcing function. Hence, it is tempting to assume that the particular solution has the form

$$x_p(t) = X \sin(\omega t - \theta) \tag{1.22}$$

where X is the steady-state amplitude and θ is the phase shift at steady state. Mathematically, the method is referred to as the method of undetermined coefficients. Substitution of Equation (1.22) into Equation (1.21) yields

$$X = \frac{F_0/k}{\sqrt{(1 - m\omega^2/k)^2 + (c\omega/k)^2}}$$

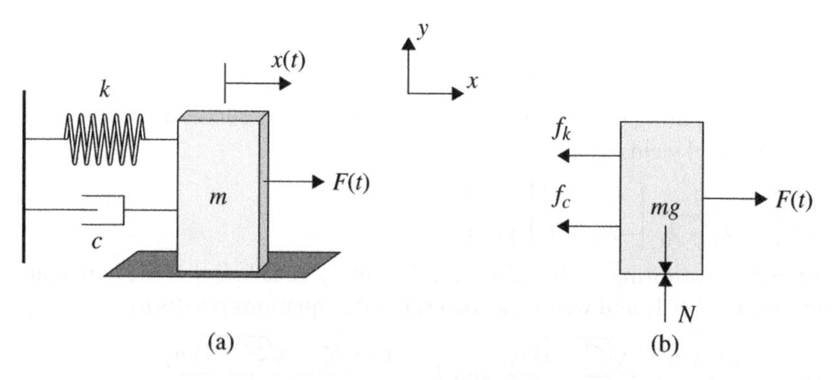

Figure 1.9 **(a)** The schematic of the forced spring-mass-damper system, assuming no friction on the surface. **(b)** The free-body diagram of the system of part (a).

or

$$\frac{Xk}{F_0} = \frac{1}{\sqrt{\left[1 - (\omega/\omega_n)^2\right]^2 + [2\zeta(\omega/\omega_n)]^2}} \tag{1.23}$$

and

$$\tan\theta = \frac{(c\omega/k)}{1 - m\omega^2/k} = \frac{2\zeta(\omega/\omega_n)}{1 - (\omega/\omega_n)^2} \tag{1.24}$$

where $\omega_n = \sqrt{k/m}$ as before. Since the system is linear, the sum of two solutions is a solution, and the total time response for the system in Figure 1.9 for the case $0 < \zeta < 1$ becomes

$$x(t) = e^{-\zeta\omega_n t}(A\sin\omega_d t + B\cos\omega_d t) + X\sin(\omega t - \theta) \tag{1.25}$$

Here A and B are constants of integration determined by the initial conditions and the forcing function (and in general will be different than the values of A and B determined for the free response). See Examples 1.4.2 and 1.5.1 for the case where the driving force is a cosine function.

Examining Equation (1.25), two features are important and immediately obvious. First, as t gets larger, the transient response (the first term) becomes very small – hence the term steady-state response is assigned to the particular solution (the second term). The second observation is that the coefficient of the steady state response, or particular solution, becomes large when the excitation frequency is close to the undamped natural frequency, i.e. $\omega \approx \omega_n$. This phenomenon is known as *resonance* and is extremely important in design, vibration analysis and testing.

Example 1.4.1

Compute the response of the following system (assuming consistent units)

$$\ddot{x}(t) + 0.4\dot{x}(t) + 4x(t) = \frac{1}{\sqrt{2}}\sin 3t, \quad x(0) = \frac{-3}{\sqrt{2}}, \quad \dot{x}(0) = 0$$

Solution: First solve for the particular solution by using the more convenient form of

$$x_p(t) = X_1 \sin 3t + X_2 \cos 3t$$

rather than the magnitude and phase form, where X_1 and X_2 are the constants to be determined. Differentiating x_p yields

$$\dot{x}_p(t) = 3X_1 \cos 3t - 3X_2 \sin 3t$$
$$\ddot{x}_p(t) = -9X_1 \sin 3t - 9X_2 \cos 3t$$

Substitution of x_p and its derivatives into the equation of motion and collecting like terms yields

$$\left(-9X_1 - 1.2X_2 + 4X_1 - \frac{1}{\sqrt{2}}\right)\sin 3t + (-9X_2 + 1.2X_1 + 4X_2)\cos 3t = 0$$

Since the sine and cosine are independent, the two coefficients in parenthesis must vanish, resulting in two equations in the two unknowns, X_1 and X_2. This solution yields

$$x_p(t) = -0.134 \sin 3t - 0.032 \cos 3t$$

Next consider adding the free response to this. From the problem statement

$$\omega_n = 2 \text{ rad/s}, \quad \zeta = \frac{0.4}{2\omega_n} = 0.1 < 1, \quad \omega_d = \omega_n \sqrt{1-\zeta^2} = 1.99 \text{ rad/s}$$

Thus, the system is underdamped, and the total solution is of the form

$$x(t) = e^{-\zeta\omega_n t}(A \sin \omega_d t + B \cos \omega_d t) + X_1 \sin \omega t + X_2 \cos \omega t$$

Applying the initial conditions requires the derivative

$$\begin{aligned} \dot{x}(t) &= e^{-\zeta\omega_n t}(\omega_d A \cos \omega_d t - \omega_d B \sin \omega_d t) + \omega X_1 \cos \omega t \\ &\quad - \omega X_2 \sin \omega t - \zeta\omega_n e^{-\zeta\omega_n t}(A \sin \omega_d t + B \cos \omega_d t) \end{aligned}$$

The initial conditions yield the constants A and B

$$x(0) = B + X_2 = \frac{-3}{\sqrt{2}} \Rightarrow B = -X_2 - \frac{3}{\sqrt{2}} = -2.089$$

$$\dot{x}(0) = \omega_d A + \omega X_1 - \zeta\omega_n B = 0 \Rightarrow A = \frac{1}{\omega_d}(\zeta\omega_n B - \omega X_1) = -0.008$$

Thus the total solution is

$$x(t) = -e^{-0.2t}(0.008 \sin 1.99t + 2.089 \cos 1.99t) - 0.134 \sin 3t - 0.032 \cos 3t$$

Example 1.4.2

Calculate the form of the forced response if, instead of a sinusoidal driving force, the applied force is given by

$$F(t) = F_0 \cos \omega t.$$

Solution: In this case, assume that the response is also a cosine function out of phase or

$$x_p(t) = X \cos(\omega t - \theta)$$

To make the computations easy to follow, this is written in the equivalent form using a basic trig identity

$$x_p(t) = A_s \cos \omega t + B_s \sin \omega t$$

where the constants $A_s = X \cos \theta$ and $B_s = X \sin \theta$ satisfying

$$X = \sqrt{A_s^2 + B_s^2} \quad \text{and} \quad \theta = \tan^{-1} \frac{B_s}{A_s}$$

are undetermined constant coefficients. Taking derivatives of the assumed form of the solution and substitution of these into the equation of motion yields

$$\left(-\omega^2 A_s + 2\zeta\omega_n\omega B_s + \omega_n^2 A_s - f_0 \right) \cos \omega t$$
$$+ \left(-\omega^2 B_s - 2\zeta\omega_n\omega A_s + \omega_n^2 B_s \right) \sin \omega t = 0$$

This equation must hold for all time, in particular for $t = \pi/2\omega$, so that the coefficient of sin ωt must vanish. Similarly, for $t = 0$, the coefficient of cos ωt must vanish. This yields the two equations

$$\left(\omega_n^2 - \omega^2\right) A_s + \left(2\zeta\omega_n\omega\right) B_s = f_0$$

and

$$(-2\zeta\omega_n\omega)A_s + \left(\omega_n^2 - \omega^2\right) + B_s = 0$$

in the two undetermined coefficients A_s and B_s. Solving yields

$$A_s = \frac{(\omega_n^2 - \omega^2)f_0}{\left(\omega_n^2 - \omega^2\right)^2 + (2\zeta\omega_n\omega)^2}$$

$$B_s = \frac{2\zeta\omega_n\omega f_0}{\left(\omega_n^2 - \omega^2\right)^2 + (2\zeta\omega_n\omega)^2}$$

Substitution of these expressions into the equations for X and θ yields the particular solution

$$x_p(t) = \overbrace{\frac{f_0}{\sqrt{\left(\omega_n^2 - \omega^2\right)^2 + (2\zeta\omega_n\omega)^2}}}^{X} \cos\left(\omega t - \overbrace{\tan^{-1}\frac{2\zeta\omega_n\omega}{\omega_n^2 - \omega^2}}^{\theta}\right)$$

Resonance is generally to be avoided in designing structures, since it means large amplitude vibrations, which can cause fatigue failure, discomfort, loud noises, etc. Occasionally, the effects of resonance are catastrophic. However, the concept of resonance is also very useful in testing structures and in certain applications such as energy harvesting (Section 7.10). In fact, the process of modal testing (Chapter 12) is based on resonance. Figure 1.10 illustrates how ω_n and ζ affect the amplitude at resonance. The dimensionless quantity Xk/F_0 is called the *magnification factor* and Figure 1.10 is called a *magnification curve* or *magnitude plot*. The maximum value at resonance, called the *peak resonance*, and denoted by M_p, can be shown (Inman, 2014) to be related to the damping ratio by

$$M_p = \frac{1}{2\zeta\sqrt{1 - \zeta^2}} \tag{1.26}$$

Also, Figure 1.10 can be used to define the *bandwidth* of the structure, denoted by BW, as the value of the driving frequency at which the magnitude drops below 70.7% of its zero frequency value (also said to be the 3-dB down point from the zero

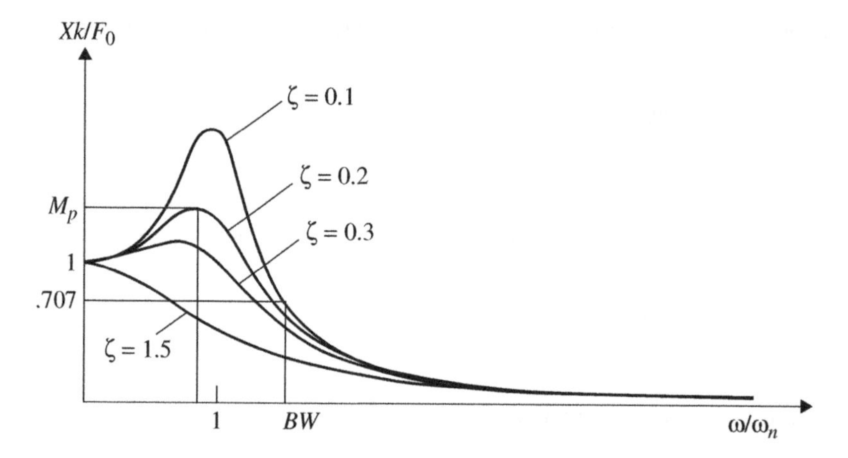

Figure 1.10 Magnification curves (dimensionless) for an SDOF system showing the normalized amplitude of vibration versus the ratio of driving frequency to natural frequency ($r = \omega/\omega_n$).

frequency point). The bandwidth can be calculated (Kuo and Golnaraghi, 2009: p. 359) in terms of the damping ratio by

$$BW = \omega_n\sqrt{(1-2\zeta^2)+\sqrt{4\zeta^4-4\zeta^2+2}} \tag{1.27}$$

Two other quantities are used in discussing the vibration of underdamped structures. They are the *loss factor* defined at resonance (only) to be

$$\eta = 2\zeta \tag{1.28}$$

and the *Q value*, or *resonance sharpness* factor, given by

$$Q = \frac{1}{2\zeta} = \frac{1}{\eta} \tag{1.29}$$

Another common situation focuses on the transient nature of the response, namely, the response of Equation (1.6) to an impulse, to a step function, or to initial conditions. Many mechanical systems are excited by loads, which act for a very brief time. Such situations are usually modeled by introducing a fictitious function called the *unit impulse function*, or the *Dirac delta function*. This delta function, denoted δ, is defined by the two properties

$$\begin{aligned} &\delta(t-a) = 0 \qquad t \neq a \\ &\int_{-\infty}^{\infty} \delta(t-a)\,dt = 1 \end{aligned} \tag{1.30}$$

where a is the instant of time at which the impulse is applied. Strictly speaking, the quantity $\delta(t)$ is not a function; however, it is very useful in quantifying important physical phenomena of an impulse.

The response of the system of Figure 1.9 for the underdamped case (with $a = x_0 = v_0 = 0$) can be given by

$$x(t) = \begin{Bmatrix} 0 & t < a \\ \dfrac{1}{m\omega_d} e^{-\zeta\omega_n t} \sin \omega_d t & t \geq a \end{Bmatrix} \tag{1.31}$$

Note from Equation (1.13) that this corresponds to the transient response of the system to the initial conditions $x_0 = 0$ and $v_0 = 1/m$. Hence, the impulse response is equivalent to giving a system at rest an initial velocity of $(1/m)$. This makes the impulse response, $x(t)$, important in discussing the transient response of more complicated systems. The impulse is also very useful in making vibration measurements, as described in Chapter 12.

A physical impact applied to a structure can be modeled by using the Dirac delta function with a magnitude representing the size of the impact. In this case, the impulse applied to the structure is modeled as having a magnitude F applied over a short time period Δt so that the effective change in momentum is $mv_0 - 0 = F\,\Delta t$, assuming the structure is initially at rest. This is equivalent to imparting an initial velocity of $v_0 = F\,\Delta t/m$. Thus, for an impulse of magnitude F applied over time Δt, the response becomes

$$x(t) = \begin{Bmatrix} 0 & t < a \\ \dfrac{F\Delta t}{m\omega_d} e^{-\zeta\omega_n t} \sin \omega_d t & t \geq a \end{Bmatrix} \tag{1.32}$$

Often design problems are stated in terms of certain specifications based on the response of the system to step function excitation. The response of the system in Figure 1.9 to a step function (of magnitude $m\omega_n^2$ for convenience), with initial conditions both set to zero, is calculated for underdamped systems from

$$m\ddot{x} + c\dot{x} + kx = m\omega_n^2 \mu(t), \qquad \mu(t) = \begin{cases} 0 & t < 0 \\ 1 & t \geq 0 \end{cases} \tag{1.33}$$

to be

$$x(t) = 1 - \frac{e^{-\zeta\omega_n t} \sin(\omega_d t + \phi)}{\sqrt{1-\zeta^2}} \tag{1.34}$$

where

$$\phi = \arctan\left[\frac{\sqrt{1-\zeta^2}}{\zeta}\right] \tag{1.35}$$

A sketch of the response is given in Figure 1.11, along with the labeling of several significant specifications for the case $m = 1$, $\omega_n = 2$ and $\zeta = 0.2$.

In some situations, the steady-state response of a structure may be at an acceptable level, but the transient response may exceed acceptable limits. Hence, one important measure is the *overshoot*, labeled O.S. in Figure 1.11 and defined to be

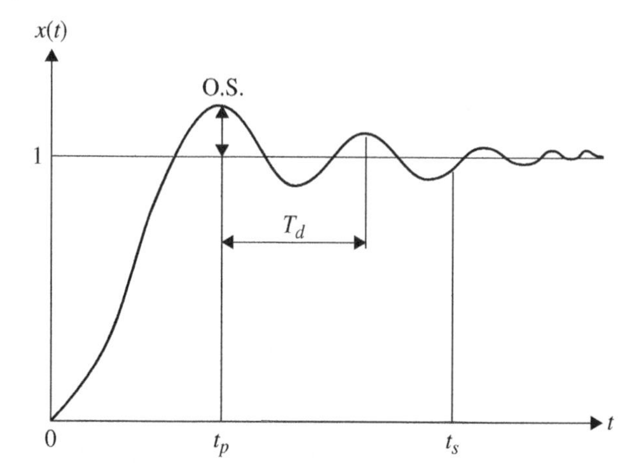

Figure 1.11 Step response of an SDOF system.

the maximum value of the response minus the steady-state value of the response. From Equation (1.34) it can be shown that

$$\text{overshoot} = \text{O.S.} = x_{\max}(t) - 1 = e^{-\zeta\pi/\sqrt{1-\zeta^2}} \tag{1.36}$$

This occurs at the *peak time*, t_p, which can be shown to be

$$t_p = \frac{\pi}{\omega_n\sqrt{1-\zeta^2}} \tag{1.37}$$

In addition, the period of oscillation, T_d, is given by

$$T_d = \frac{2\pi}{\omega_n\sqrt{1-\zeta^2}} = 2t_p \tag{1.38}$$

Another useful quantity, which indicates the behavior of the transient response, is the *settling time*, t_s. This is the time it takes the response to get within ±5% of the steady-state response and remain within ±5%. One approximation of t_s is given by Kuo and Golnaraghi (2009: p. 263)

$$t_s = \frac{3.2}{\omega_n\zeta} \tag{1.39}$$

The preceding definitions allow designers and vibration analysts to specify and classify precisely the nature of the transient response of an underdamped system. These definitions also give some indication of how to adjust the physical parameters of the system so that the response has a desired shape.

The response of a system to an impulse may be used to determine the response of an underdamped system to any input $F(t)$ by defining the *impulse response function* by

$$h(t) = \frac{1}{m\omega_d}e^{-\zeta\omega_n t}\sin\omega_d t \tag{1.40}$$

Then the solution of

$$m\ddot{x}(t) + c\dot{x}(t) + kx(t) = F(t)$$

can be shown to be

$$x(t) = \int_0^t F(\tau)h(t-\tau)d\tau = \frac{1}{m\omega_d}e^{-\zeta\omega_n t}\int_0^t F(\tau)e^{\zeta\omega_n \tau}\sin\omega_d(t-\tau)d\tau \qquad (1.41)$$

for the case of zero initial conditions. This last expression gives an analytical representation for the response to any driving force that has an integral.

Example 1.4.3

Consider a spring-mass-damper system with $m = 1$ kg, $c = 2$ kg/s and $k = 2000$ N/m, with an impulsive force applied to it of 10,000 N for 0.01 s. Compute the resulting response.

Solution: A 10,000 N force acting over 0.01 s provides (area under the curve) a value of $F\Delta t = 10000 \times 0.01 = 100$ N · s Using the values given, the equation of motion is

$$\ddot{x}(t) + 2\dot{x}(t) + 2000x(t) = 100\delta(t)$$

Thus the natural frequency, damping ratio and damped natural frequency are

$$\omega_n = \sqrt{\frac{2000}{1}} = 44.721 \text{ rad/s}, \zeta = \frac{2}{2\sqrt{1 \times 2000}} = 0.022,$$

$$\omega_d = 44.721\sqrt{1-0.022^2} = 44.71 \text{ rad/s}$$

Using Equation (1.32), the response becomes

$$x(t) = \frac{\hat{F}e^{-\zeta\omega_n t}}{m\omega_d}\sin\omega_d t = 2.237e^{-0.1t}\sin(44.71t)$$

1.5 Transfer Functions and Frequency Methods

The preceding analysis of the response was carried out in the time domain. Current vibration measurement methodology (Ewins, 2000), as well as much control analysis (Kuo and Golnaraghi, 2009), often takes place in the frequency domain. Hence, it is worth the effort to reexamine these calculations using frequency domain methods (a phrase usually associated with linear control theory). The frequency domain approach arises naturally from mathematics (ordinary differential equations) via an alternative method of solving differential equations, such as Equations (1.21) and (1.33), using the Laplace transform (Boyce and DiPrima, 2012; Chapter 6).

Taking the Laplace transform of Equation (1.33), assuming both initial conditions to be zero, yields

$$X(s) = \left[\frac{1}{ms^2 + cs + k}\right]\mu(s) \qquad (1.42)$$

where $X(s)$ denotes the Laplace transform of $x(t)$, and $\mu(s)$ is the Laplace transform on the right-hand side of Equation (1.33). If the same procedure is applied to Equation (1.21), the result is

$$X(s) = \left[\frac{1}{ms^2 + cs + k}\right] F_0(s) \tag{1.43}$$

where $F_0(s)$ denotes the Laplace transform of $F_0 \sin \omega t$. Note that

$$G(s) = \frac{X(s)}{\mu(s)} = \frac{X(s)}{F_0(s)} = \frac{1}{ms^2 + cs + k} \tag{1.44}$$

Thus, it appears that the quantity $G(s) = [1/(ms^2 + cs + k)]$, the ratio of the Laplace transform of the output (response) to the Laplace transform of the input (applied force) to the system characterizes the system (structure) under consideration. This characterization is independent of the input or driving function. This ratio, $G(s)$, is defined as the *transfer function* of this system in control analysis (or of this structure in vibration analysis). The transfer function can be used to provide analysis of the vibrational properties of the structure, as well as to provide a means of measuring the structure's dynamic response.

In control theory, the transfer function of a system is defined in terms of an output to input ratio, but the use of a transfer function in structural dynamics and vibration testing implies certain physical properties, depending on whether position, velocity or acceleration is considered as the response (output). It is common, for instance, to measure the response of a structure by using an accelerometer. The transfer function resulting is then $s^2X(s)/U(s)$, where $U(s)$ is the Laplace transform of the input and $s^2X(s)$ is the Laplace transform of the acceleration. This transfer function is called the *inertance* and its reciprocal is referred to as the *apparent mass*. Table 1.1 lists the nomenclature of various transfer functions. The physical basis for these names can be seen from their graphical representation.

The transfer function representation of a structure is very useful in control theory as well as in vibration testing. It also forms the basis of impedance methods discussed in the next section. The variable s in the Laplace transform is a complex variable, which can be further denoted by

$$s = \sigma + j\omega_d$$

where the real numbers σ and ω_d denote the real and imaginary parts of s, respectively ($j = \sqrt{-1}$). Thus, the various transfer functions are also complex-valued.

Table 1.1 Various transfer functions.

Response Measurement	Transfer Function	Inverse Transfer Function
Acceleration	Inertance	Apparent mass
Velocity	Mobility	Impedance
Displacement	Compliance	Dynamic stiffness

Figure 1.12 Complex s-plane of the poles (roots of the characteristics of Equation (1.39).

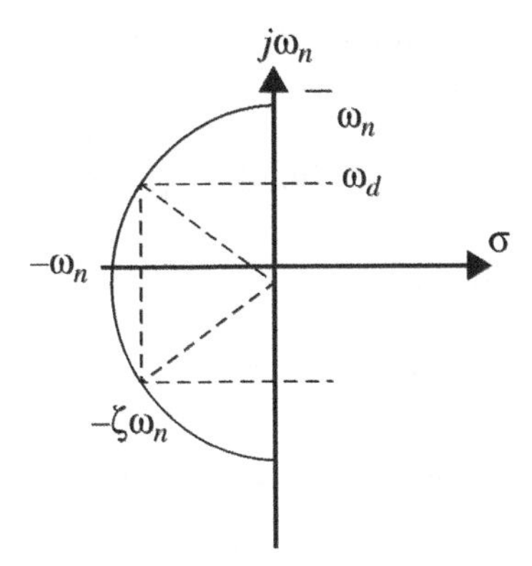

In control theory, the values of s where the denominator of the transfer function $G(s)$ vanishes are called the *poles* of the transfer function. A plot of the poles of the compliance (also called receptance) transfer function for Equation (1.44) in the complex s-plane is given in Figure 1.12. The points on the semi-circle occur where the denominator of the transfer function is zero. These values of s ($s = -\zeta\omega_n \pm \omega_d j$) are exactly the roots of the characteristic equation for the structure. The values of the physical parameters m, c and k determine the two quantities ζ and ω_n, which in turn determine the position of the poles in Figure 1.12.

Another graphical representation of a transfer function useful in control is the *block diagram* illustrated in Figure 1.13a. This diagram is an icon for the definition of a transfer function. The control terminology for the physical device represented by the transfer function is the *plant*, whereas in vibration analysis the plant is usually referred to as the structure. The block diagram of Figure 1.13b is meant to imply the formula

$$\frac{X(s)}{U(s)} = \frac{1}{(ms^2 + cs + k)} \tag{1.45}$$

exactly.

The response of Equation (1.21) to a sinusoidal input (forcing function) motivates a second description of a structure's transfer function called the *frequency response function* (often denoted by FRF). The FRF is defined as the transfer function evaluated at $s = j\omega$, i.e. $G(j\omega)$. The significance of the FRF follows from Equation (1.22), namely, that the steady-state response of a system driven sinusoidally is a sinusoid of the same frequency with different amplitude and phase. In fact,

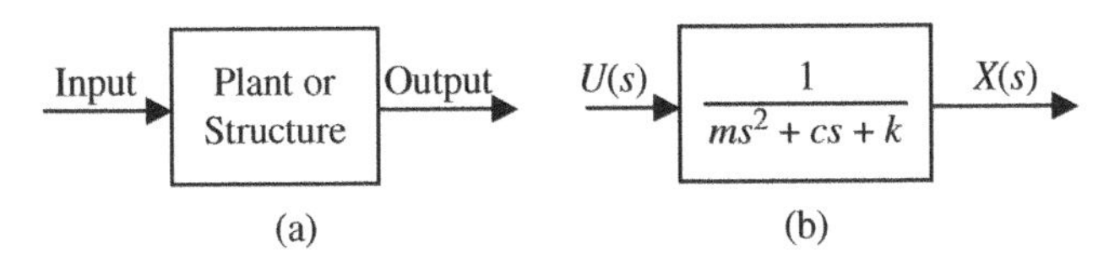

Figure 1.13 Block diagram representation of an SDOF system.

substitution of $j\omega$ into Equation (1.45) yields exactly Equations (1.23) and (1.24) from

$$\frac{X}{F_0} = |G(j\omega)| = \sqrt{x^2(\omega) + y^2(\omega)} \tag{1.46}$$

where $|G(j\omega)|$ indicates the magnitude of the complex FRF

$$\phi = \tan^{-1} G(j\omega) = \tan^{-1}\left[\frac{y(\omega)}{x(\omega)}\right] \tag{1.47}$$

indicates the phase of the FRF, and

$$G(j\omega) = x(\omega) + y(\omega)j \tag{1.48}$$

This mathematically expresses two ways to represent a complex function, as the sum of its real part (Re $G(j\omega) = x(\omega)$) and its imaginary part (Im $(G(j\omega)) = y(\omega)$), or by its magnitude ($|G(j\omega)|$) and phase (ϕ). In more physical terms, the FRF of a structure represents the magnitude and phase shift of its steady-state response under sinusoidal excitation. While Equations (1.23), (1.24), (1.46) and (1.47) verify this for an SDOF viscously damped structure, it can be shown in general for any linear time invariant plant (Melsa and Schultz, 1969: p. 187)).

It should also be noted that the FRF of a linear system can be obtained from the transfer function of the system and vice versa. Hence, the FRF uniquely determines the time response of the structure to any known input.

Graphical representations of the FRF form an extensive part of control analysis and also form the backbone of vibration measurement analysis. Next, three sets of FRF plots that are useful in testing vibrating structures are examined. The first set of plots consists simply of plotting the imaginary part of the FRF versus the driving frequency and the real part of the FRF versus the driving frequency. These are shown for the damped SDOF system in Figure 1.14 (the compliance FRF for $\zeta = 0.01$ and $\omega_n = 20$ rad/s).

The second representation consists of a single plot of the imaginary part of the FRF versus the real part of the FRF. This type of plot is called a *Nyquist plot* (also called an *Argand plane plot*) and is used for measuring the natural frequency and damping in testing methods and for stability analysis in control system design. The

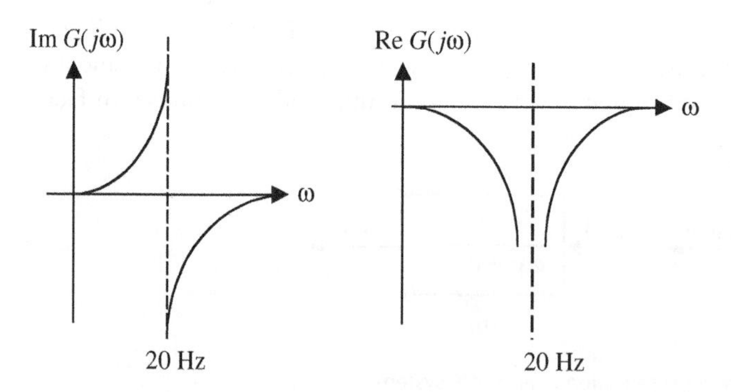

Figure 1.14 Plots of the real part and the imaginary part of the FRF.

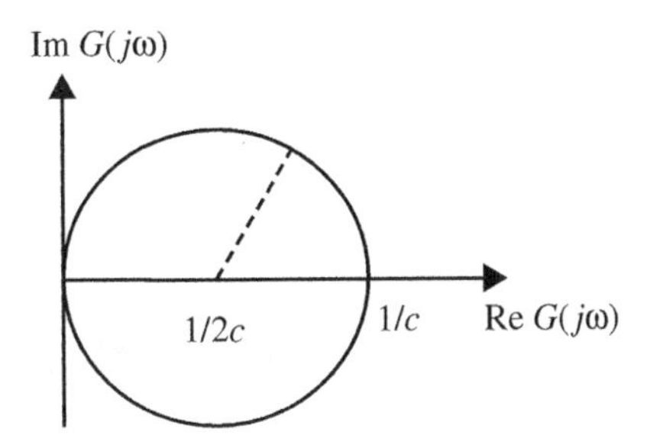

Figure 1.15 Nyquist plot for Equation 1.44.

Nyquist plot of the mobility FRF of a structure modeled by Equation (1.44) is given in Figure 1.15.

The last plots considered for representing the FRF are called *Bode plots* and consist of a plot of the magnitude of the FRF versus the driving frequency and the phase of the FRF versus the driving frequency (a complex number requires *two* real numbers to describe it completely). Bode plots have long been used in control system design and analysis as well as for determining the plant transfer function of a system. More recently, Bode plots have been used in analyzing vibration test results and in determining the physical parameters of the structure.

In order to represent the complete Bode plots in a reasonable space, $\log_{10}$ scales are often used to plot $|G(j\omega)|$. This has given rise to the use of the decibel and decades in discussing the magnitude response in the frequency domain. The magnitude and phase plots (for the compliance transfer function) for the system in Equation (1.21) are shown in Figures 1.16 and 1.17 for different values of ζ. Note the phase change at resonance (90°), as this is important in interpreting measurement data.

Note that Figures 1.10 and 1.17 show the same physical phenomenon and are both plots of the compliance transfer function. However, the magnitude in Figure 1.10 is dimensionless versus dimensionless frequency, while Figure 1.17 is usually the magnitude in decibels versus frequency on a semi-log scale.

Example 1.5.1

Solve the following system using the Laplace Transform method and using a Table of Laplace Transform pairs (from the Internet)

$$m\ddot{x}(t) + kx(t) = F_0 \cos \omega(t),\ x(0) = x_0,\ \dot{x}(0) = v_0$$

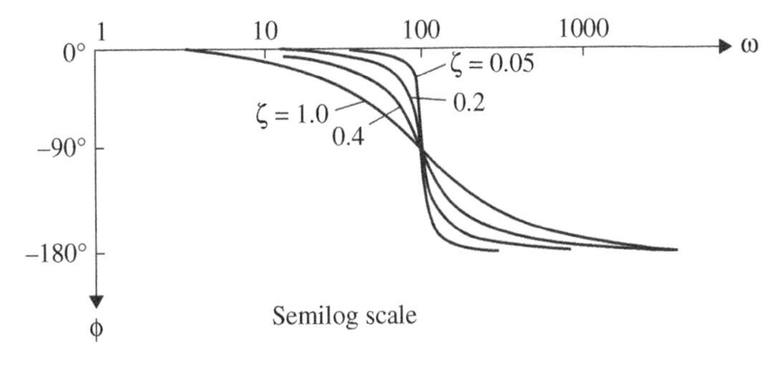

Figure 1.16 Bode phase plot for Equation (1.39) showing resonance at −90°.

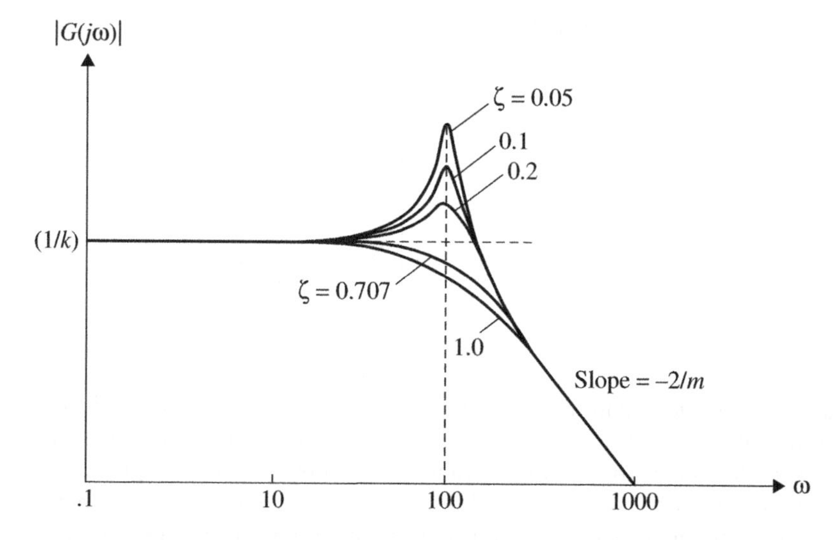

Figure 1.17 Bode magnitude plot for Equation (1.39) showing resonance and values of mass and stiffness.

Solution: First divide through by the mass to get

$$\ddot{x}(t) + \omega_n^2 x(t) = f_0 \cos \omega t, \; x(0) = x_0, \; \dot{x}(0) = v_0$$

Here $f_0 = F_0/m$. Taking the Laplace Transform (see the Table of Laplace Transforms: from the Internet) of the equation of motion considering the initial conditions yields

$$s^2 X(s) - s x_0 - v_0 + \omega_n^2 X(s) = \frac{s f_0}{s^2 + \omega^2}$$

$$\Rightarrow \left(s^2 + \omega_n^2\right) X(s) = s x_0 + v_0 + \frac{s f_0}{s^2 + \omega^2}$$

Solving this for $X(s)$ yields

$$X(s) = \frac{s x_0 + v_0}{s^2 + \omega_n^2} + \frac{s f_0}{\left(s^2 + \omega_n^2\right)\left(s^2 + \omega^2\right)}$$

$$= (x_0) \frac{s}{s^2 + \omega_n^2} + \left(\frac{v_0}{\omega_n}\right) \frac{\omega_n}{s^2 + \omega_n^2} + \frac{s f_0}{\left(s^2 + \omega_n^2\right)\left(s^2 + \omega^2\right)}$$

Taking the Inverse Laplace Transform using an online table of each term yields

$$x(t) = x_0 \cos \omega_n t + \frac{v_0}{\omega_n} \sin \omega_n t + \frac{f_0}{\omega_n^2 - \omega^2} (\cos \omega t - \cos \omega_n t)$$

$$= \frac{v_0}{\omega_n} \sin \omega_n t + \left(x_0 - \frac{f_0}{\omega_n^2 - \omega^2}\right) \cos \omega_n t + \frac{f_0}{\omega_n^2 - \omega^2} \cos \omega t$$

In comparing this with the solution given in Equation (1.25) for zero damping, note that Equation (1.25) is the solution for the case where the driving force is a sine function instead of a cosine as solved here.

1.6 Complex Representation and Impedance

Table 1.1 formally defines impedance as the ratio of a sinusoidal driving force, F, acting on the system to the resulting velocity, v, of the system. Impedance is usually denoted by the symbol Z and is a measure of a structure's resistance to motion. In working with impedance methods it is common to use the complex exponential notation to represent harmonic quantities. Using the exponential notation, the sinusoidal force in Equation (1.21) can be written as

$$F(t) = F_0 e^{j\omega t} \tag{1.49}$$

Here, ω is the driving frequency as before. The impedance approach offers an alternative way to examine systems vibrating harmonically based on using complex functions to represent the response.

A useful way to visualize harmonic motion is to think of the response $x(t)$ as a vector rotating in the complex plane, as illustrated in Figure 1.18. Here the vector has magnitude A and rotates an angle ωt in the complex plane. From Euler's formula for the complex exponential function

$$x(t) = Ae^{j\omega t} = A\cos\omega t + Aj\sin\omega t \tag{1.50}$$

which agrees with representation in Figure 1.18. Differentiation of the complex exponential yields simply

$$\begin{aligned} \frac{d}{dt}(Ae^{j\omega t}) &= j\omega Ae^{j\omega t} = j\omega x(t) \\ \frac{d^2}{dt^2}(Ae^{j\omega t}) &= j^2\omega^2 Ae^{j\omega t} = -\omega^2 x(t) \end{aligned} \tag{1.51}$$

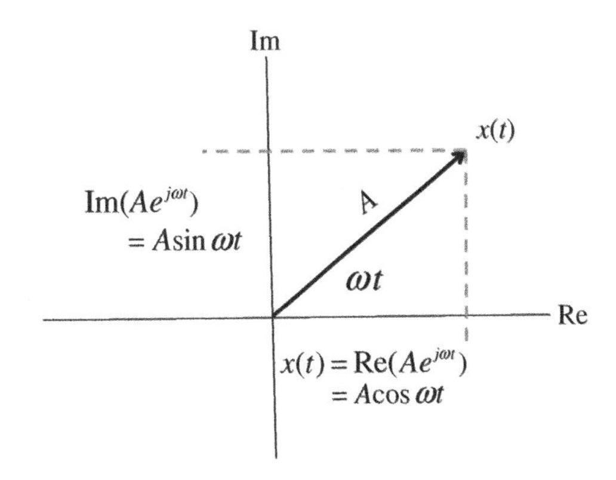

Figure 1.18 Graphic illustration of Euler's formula of the complex exponential.

Thus, each differentiation of the complex exponential results in simply multiplying by $j\omega$, similar to multiplying by s in the Laplace domain.

From the Figure 1.18, the physical displacement is interpreted from the complex exponential as just the real part of Equation (1.50). Thus the velocity becomes the real part of the derivative of the complex exponential and the acceleration is the real part of the derivative of that or

$$\begin{aligned} x(t) &= \mathrm{Re}(Ae^{j\omega t}) = A\cos(\omega t) \\ \dot{x}(t) &= \mathrm{Re}(j\omega Ae^{j\omega t}) = -\omega A\sin(\omega t) \\ \ddot{x}(t) &= \mathrm{Re}(j^2\omega^2 Ae^{j\omega t}) = -\omega^2 A\cos(\omega t) \end{aligned} \tag{1.52}$$

If the displacement is thought to be a sine function, then the physical motion variables become the imaginary parts of the complex exponential. Using the complex notation equation for the forced response of an SDOF system becomes

$$m\ddot{x}(t) + c\dot{x}(t) + kx(t) = F_0 e^{jwt} \tag{1.53}$$

Assuming the resulting displacement is of the form

$$x(t) = A\sin(\omega t - \theta)$$

its complex form is the corresponding velocity as

$$v(t) = Aj\omega e^{j(\omega t+\theta)} \tag{1.54}$$

Here ω and θ are the driving frequency and phase shift between the applied force and the resulting response respectively. Substituting the complex form of $x(t)$ into Equation (1.48) yields

$$[-\omega^2 m + j\omega c + k]Ae^{j\omega - j\theta} = F(t) \tag{1.55}$$

Solving for the complex value A yields

$$A = \frac{F_0 e^{j\theta}}{[-\omega^2 m + j\omega c + k]} \tag{1.56}$$

which has magnitude and phase given by

$$|A| = \frac{F}{\sqrt{(k-\omega^2 m)^2 + (\omega c)^2}} \quad \text{and} \quad \theta = \tan\frac{\omega c}{k - \omega^2 m} \tag{1.57}$$

These values are of course the same as those derived in the previous section in Equations (1.23 and 1.24).

Examination of the force/velocity expressions for each element reveals the impedance of each, and these are given in Table 1.2.

Table 1.2 Impedance values for mass, damping and stiffness.

Mass	$Z = j\omega m$
Damping	$Z = c$
Stiffness	$Z = {}^{-jk}/_{\omega}$

Example 1.6.1
Compute the mechanical impedance of the spring-mass-damper system of Figure 1.9.

Solution: Dividing Equation (1.55) by (1.54) and simplifying yields that directly the mechanical impedance of the spring-mass-damper system becomes

$$\begin{aligned} Z &= \frac{F}{v} = \frac{[k - \omega^2 m + j\omega c]Ae^{j\omega t - j\theta}}{Aj\omega e^{j\omega t - j\theta}} = \frac{1}{j\omega}(k - \omega^2 m + j\omega c) \\ &= \omega jm + c - \frac{k}{j\omega} \end{aligned} \tag{1.58}$$

Comparing this expression to the terms in Table 1.2 reveals that the mechanical impedance of the system is just the sum of the impedance expressions for each element. The use of the impedance method is essentially the existence of following rules developed in electrical engineering for combining deferent circuit elements by adding their impedances (e.g. series and parallel combinations) and making the analogy to electrical components of capacitance (reciprocal of stiffness), inductance (mass) and resistance (damping). The units of mechanical impedance are kg/s, the same as the viscous damping coefficient.

1.7 Measurement and Testing

One can also use the quantities defined in the previous sections to measure the physical properties of a structure. As mentioned before, resonance can be used to determine a system's natural frequency. Methods based on resonance are referred to as resonance testing (or modal analysis techniques) (Bishop and Gladwell, 1963) and are briefly introduced here and discussed in more detail in Chapter 8.

As mentioned earlier, the mass and stiffness of a structure can often be determined by making simple static measurements. However, damping rates require a dynamic measurement and hence are more difficult to determine. For underdamped systems one approach is to realize, from Figure 1.6, that the decay envelope is the function $e^{-\zeta\omega_n t}$. The points on the envelope illustrated in Figure 1.19

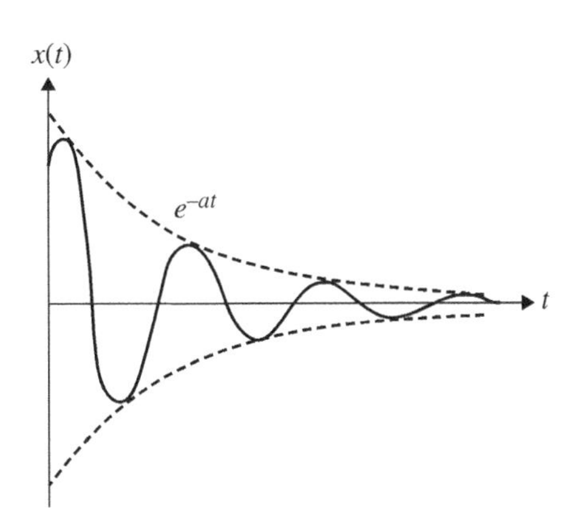

Figure 1.19 Free decay measurement method.

can be used to curve-fit the function e^{-at}, where a is the constant determined by the curve fit. The relation $a = \zeta\omega_n$ can next be used to calculate ζ and hence the damping rate c (assuming that m and k or ω_n are known).

A second approach is to use the concept of logarithmic decrement, denoted by δ (delta) and defined by

$$\delta = \ln\frac{x(t)}{x(t+T_d)} \tag{1.59}$$

where T_d is the period of oscillation. Using Equation (1.13) in the form

$$x(t) = Ae^{-\zeta\omega_n t}\sin(\omega_d t + \phi) \tag{1.60}$$

the value for δ becomes

$$\delta = \ln\left[\frac{e^{-\zeta\omega_n t}\sin(\omega_d t + \phi)}{e^{-\zeta\omega_n (t+T_d)}\sin(\omega_d t + \omega_d T_d + \phi)}\right] = \ln e^{\zeta\omega_n T_d} = \zeta\omega_n T_d \tag{1.61}$$

where the sine functions cancel because $\omega_d T_d$ is a one period shift by definition. Further evaluating δ yields

$$\delta = \zeta\omega_n T_d = \frac{2\pi\zeta}{\sqrt{1-\zeta^2}} \tag{1.62}$$

Equation (1.62) can be manipulated to yield the damping ratio in terms of the decrement, i.e.

$$\zeta = \frac{\delta}{\sqrt{4\pi + \delta^2}} \tag{1.63}$$

Hence, if the decrement is measured, Equation (1.63) yields the damping ratio.

The various plots of the previous section can also be used to measure ω_n, ζ, m, c and k. For instance, the Bode diagram of Figure 1.17 can be used to determine the natural frequency, stiffness and damping ratio. The stiffness is determined from the intercept of the FRF and the magnitude axis, since the value of the magnitude of the FRF for small ω is $\log(1/k)$. This can be seen by examining the function $\log_{10}|G(j\omega)|$ for small ω. Note that

$$\log|G(j\omega)| = \log\frac{1}{k} - \frac{1}{2}\log\left[\left(1-\frac{\omega^2}{\omega_n^2}\right)^2 + \left(\frac{2\zeta\omega}{\omega_n}\right)^2\right] = \log\left(\frac{1}{k}\right) \tag{1.64}$$

for very small values of ω. Also note that $|G(j\omega)|$ evaluated at ω_n yields

$$k|G(j\omega_n)| = \frac{1}{2\zeta} \tag{1.65}$$

which provides a measure of the damping ratio from the magnitude plot of the FRF.

Note that Equations (1.65) and (1.26) appear to contradict each other, since

$$\frac{1}{2\zeta\sqrt{1-\zeta^2}} = k\max|G(j\omega)| = M_p \neq k|G(j\omega_n)| = \frac{1}{2\zeta}$$

except in the case of very small ζ (i.e. the difference between M_p and $|G(j\omega_n)|$ goes to zero as ζ goes to zero). This indicates a subtle difference between using the damping ratio obtained by using resonance as the value of ω, where $|G(j\omega_n)|$ is a maximum, and using the point, where $\omega = \omega_n$, the undamped natural frequency. This point is also illustrated by noting that the damped natural frequency, Equation (1.8), is $\omega_d = \omega_n\sqrt{1-\zeta^2}$ and ω_p, the frequency at which $|G(j\omega_n)|$ is maximum, is

$$\omega_p = \omega_n\sqrt{1-2\zeta^2} \tag{1.66}$$

Also note that Equation (1.66) is valid only if $0 < \zeta < 0.707$.

Finally, the mass can be related to the slope of the magnitude plot for the inertance transfer function, denoted by $G_I(s)$, by noting that

$$G_I(s) = \frac{s^2}{(ms^2 + cs + k)} \tag{1.67}$$

and for large ω (i.e. $\omega_n \ll \omega$), the value of $|G_I(j\omega)|$ is

$$|G_I(j\omega)| \approx (1/m) \tag{1.68}$$

Plots of these values are referred to as straight-line approximations to the actual magnitude plot (Bode, 1945).

The preceding formulas relating the physical properties of the structure to the magnitude Bode diagrams suggest an experimental way to determine a structure's parameters: namely, if the structure can be driven by a sinusoid of varying frequency and if the magnitude and phase (needed to locate resonance) of the resulting response are measured, then the Bode plots and the preceding formulas can be used to obtain the desired physical parameters. This process is referred to as plant identification in the controls literature and can be extended to systems with more degrees of freedom (see Melsa and Schultz (1969), for a more complete account).

There are several other formulas for measuring the damping ratio and natural frequency from the results of such experiments, sine sweeps. For instance, if the Nyquist plot of the mobility transfer function is used, a circle of diameter $1/c$ results (Figure 1.15). Another approach is to plot the magnitude of the FRF on a linear scale near the region of resonance (Figure 1.20). If the damping is small enough so that the peak at resonance is sharp, the damping ratio can be determined by measuring the frequencies at 0.707 at the maximum value (also called

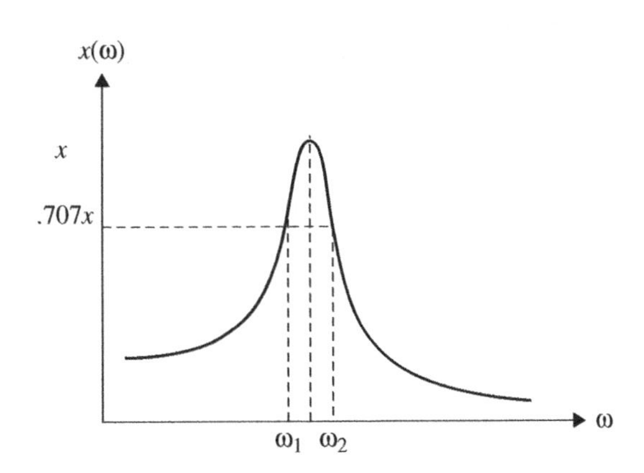

Figure 1.20 Quadrature peak picking method.

the 3-dB down point or half-power points), denoted by ω_1 and ω_2, respectively. Then, using the formula (Ewins, 2000)

$$\zeta = \frac{1}{2}\left[\frac{\omega_2 - \omega_1}{\omega_d}\right] \tag{1.69}$$

to compute the damping ratio. This method is referred to as *quadrature peak picking* and is illustrated in Figure 1.20.

1.8 Stability

In all the preceding analysis, the physical parameters m, c and k are, of course, positive quantities. There are physical situations, however, in which equations of the form of Equations (1.1) and (1.6) result but have one or more negative coefficients. Such systems are not well behaved and require some additional analysis.

Recalling that the solution to Equation (1.1) is of the form $A\sin(\omega t + \phi)$, where A is a constant, it is easy to see that the response, in this case $x(t)$, is bounded. That is to say that

$$|x(t)| \leq A \tag{1.70}$$

for all t where A is some finite constant and $|x(t)|$ denotes the absolute value of $x(t)$. In this case, the system is well behaved or *stable* (called marginally stable in the control's literature). In addition, note that the roots (also called *characteristic values* or eigenvalues) of

$$\lambda^2 m + k = 0$$

are purely complex numbers $\pm j\omega_n$ as long as m and k are positive (or have the same sign). If k happens to be negative and m is positive, the solution becomes

$$x(t) = A\sinh\omega_n t + B\cosh\omega_n t \tag{1.71}$$

which increases without bound as t does. Such solutions are called *divergent* or *unstable.*

If the solution of the damped system of Equation (1.6) with positive coefficients is examined, it is clear that $x(t)$ approaches zero as t becomes large, because of the exponential term. Such systems are considered to be *asymptotically stable* (called stable in the controls literature). Again, if one or two of the coefficients are negative, the motion grows without bound and becomes unstable as before. In this case, however, the motion may become unstable in one of two ways. Similar to overdamping and underdamping, the motion may grow without bound and not oscillate, or it may grow without bound and oscillate. The first case is referred to as *divergent instability* and the second case as *flutter instability*; together they fall under the topic of self-excited vibrations.

Apparently, the sign of the coefficient determines the stability behavior of the system. This concept is pursued in Chapter 4, where these stability concepts are formally defined. Figures 1.21 to 1.24 illustrate each of these concepts.

These stability definitions can also be stated in terms of the roots of the characteristic Equation (1.8) or in terms of the poles of the transfer function of the

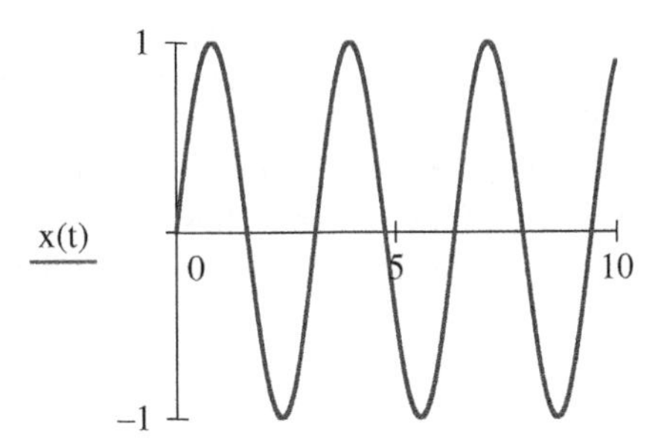

Figure 1.21 Response of a stable system.

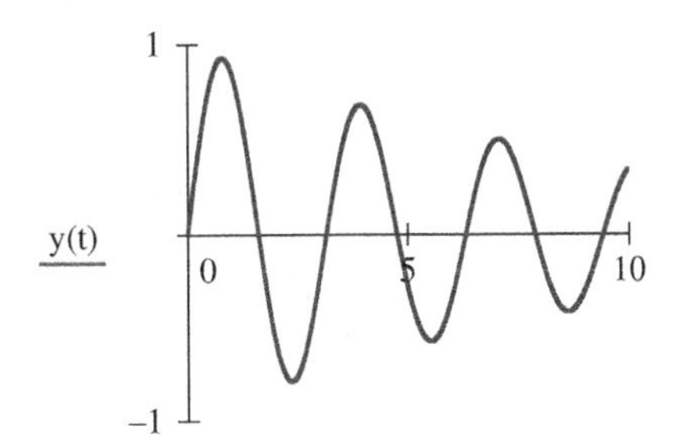

Figure 1.22 Response of an asymptotically stable system.

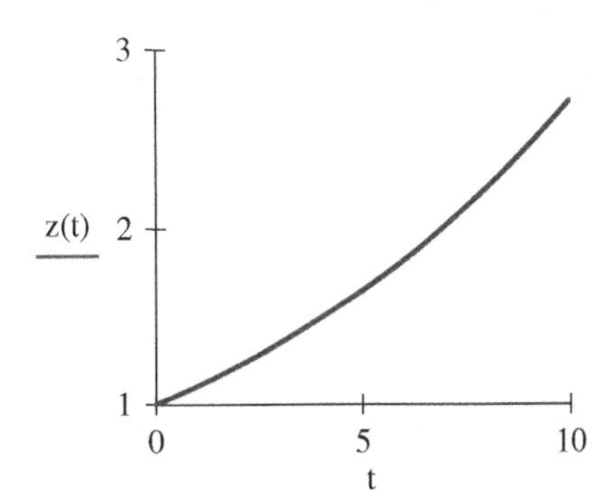

Figure 1.23 Response of a system with a divergent instability.

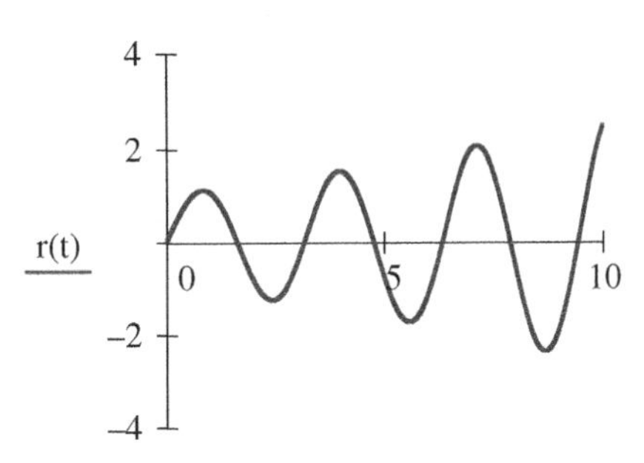

Figure 1.24 Response of a system with flutter instability.

system. In fact, referring to Figure 1.12, the system is stable if the poles of the structure lie along the imaginary axis (called the $j\omega$ axis), unstable if one or more poles are in the right half-plane, and asymptotically stable if all of the poles lie in the left half-plane. Flutter occurs when the poles are in the right half-plane and not on the real axis (complex conjugate pairs of roots with positive real part) and divergence occurs when the poles are in the right-half plane along the real axis. In the simple SDOF case considered here, the pole positions are entirely determined by the signs of m, c and k.

The preceding definitions and ideas about stability are stated for the free response of the system. These concepts of a well-behaved response can also be applied to the forced motion of a vibrating system. The stability of the forced response of a system can be defined by considering the nature of the applied force or input. The system is said to be *bounded-input, bounded-output stable* (or, simply, BIBO stable) if for *any* bounded input (driving force) the output (response) is bounded for any arbitrary set of initial conditions. Such systems are manageable at resonance.

It can be seen immediately that Equation (1.21) with $c = 0$, the undamped system, is not BIBO stable, since for $f(t) = \sin(\omega_n t)$, the response $x(t)$ goes to infinity (at resonance), whereas $f(t)$ is certainly bounded. However, the response of Equation (1.21) with $c > 0$ is bounded whenever $f(t)$ is. In fact, the maximum value of $x(t)$ at resonance M_p is illustrated in Figure 1.10. Thus, the system of Equation (1.21) with damping is said to be BIBO stable.

The fact that the response of an undamped structure is bounded when $f(t)$ is an impulse or step function suggests another, weaker, definition for the stability of the forced response. A system is said to be *bounded*, or *Lagrange stable*, with respect to a *given* input if the response is bounded for any set of initial conditions. Structures described by Equation (1.1) are Lagrange stable with respect to many inputs. This definition is useful when $f(t)$ is known completely or known to fall in some specified class of functions.

Stability can also be thought of in terms of whether or not the energy of the system is increasing (unstable), constant (stable) or decreasing (asymptotically stable), rather than in terms of the explicit response. Lyapunov stability, defined in Chapter 4, extends this idea. Another important view of stability is based on how sensitive a motion is to small perturbations in the system's parameters (m, c and k) and/or small perturbations in initial conditions. Unfortunately, there does not appear to be a universal definition of stability that fits all situations. The concept of stability becomes further complicated for nonlinear systems. The definitions and concepts mentioned here are extended and clarified in Chapter 4.

Example 1.8.1

Most structures are asymptotically stable (m, c, k are all positive) or at least stable (m, k positive $c = 0$). However, if other forces are present, such as flow through a pipe or over an airfoil, stability can be lost as the effective coefficients in the equation of motion could become negative. In addition, active control systems constructed to improve performance also add forces that can potentially destabilize a structure or machine. Discuss the stability properties of the following equation of motion

$$J\ddot{\theta} + (c - f_d)\dot{\theta} + k\theta = 0$$

Here, θ is the angle of rotation of the flap, J is the moment of inertia of the flap (assumed positive), k is the rotational stiffness of the device plus a control force and assumed positive, c is the internal damping of the device (positive) and f_d is the aerodynamic force applied to the flap (also positive). This is a crude representation of a control surface (flap or tab).

Solution: As long as k and J are positive, stability is controlled by the sign of the equivalent damping term, $c - f_d$. If $c = f_d$, then the system is stable, having the response of the form illustrated in Figure 1.21. If $c - f_d > 0$, then the solution is that of Equation (1.13) and exponential decay and the system is asymptotically stable, as illustrated in Figure 1.22. If, on the other hand, aerodynamic force overcomes the actuation force and internal damping so that $c - f_d < 0$, then the exponent in Equation (1.13) becomes positive and flutter instability occurs, as illustrated in Figure 1.23.

1.9 Design and Control of Vibrations

One can use the quantities defined in the previous sections to design structures and machines to have a desired transient and steady state response to some extent. For instance, it is a simple matter to choose m, c and k so that the overshoot is a specified value. However, if one needs to specify the overshoot, the settling time and the peak time, then there may not be a choice of m, c and k that will satisfy all three criteria. Hence, the response cannot always be completely shaped, as the formulas in Section 1.4 may seem to indicate.

Another consideration in designing structures is that each of the physical parameters m, c and k may already have design constraints that have to be satisfied. For instance, the material the structure is made of may fix the damping rate, c. Then, only the parameters m and k can be adjusted. In addition, the mass may have to be within 10% of a specified value, for instance, which further restricts the range of values of overshoot and settling time. The stiffness is often designed based on the static deflection limitation and strength.

For example, consider the system of Figure 1.11 and assume it is desired to choose values of m, c and k so that ζ and ω_n specify a response with a settling time $t_s = 3.2$ units and a time to peak, t_p, of 1 unit. Then Equations (1.37) and (1.39) imply that $\omega_n = 1/\zeta$ and $\zeta = 1/\sqrt{1+\pi^2}$. This, unfortunately, also specifies the overshoot, since

$$\text{O.S.} = \exp\left(\frac{-\zeta\pi}{\sqrt{(1+\zeta^2)}}\right)$$

Thus, all three performance criteria cannot be satisfied. This leads the designer to have to make compromises, to reconfigure the structure or to add additional components.

Hence, in order to meet vibration criteria such as avoiding resonance, it may be necessary in many instances to alter the structure by adding vibration absorbers or isolators (Machinante, 1984; Rivin, 2003). Another possibility is to use active vibration control and feedback methods. Both of these approaches are discussed in Chapters 6 and 7.

The choice of the physical parameters m, c and k determines the shape of the response of the system. The choice of these parameters can be considered as the design of the structure. Passive control can also be considered as a redesign process of changing these parameters on an already existing structure to produce a more desirable response. For instance, some mass could be added to a given structure to lower its natural frequency. Although passive control or redesign is generally the most efficient way to control or shape the response of a structure, the constraints on m, c and k are often such that the desired response cannot be obtained. Then the only alternative, short of starting over again, is to try active control.

There are many different types of active control methods, and only a few will be considered to give the reader a feel for the connection between the vibration and control disciplines. Again, the comments made in this text on control should not be considered as a substitute for studying standard control or linear systems texts. Output feedback control is briefly introduced here and discussed in more detail in Chapter 7.

First, a clarification of the difference between active and passive control is in order. Basically, an active control system uses some external adjustable or active (e.g. electronic) device, called an actuator, to provide a means of shaping or controlling the response. Passive control, on the other hand, depends only on a fixed (passive) change in the physical parameters of the structure. Passive control can also involve adding external devices to the structure; however, such devices are not powered. Active control often depends on current measurements of the response of the system and passive control does not. Active control requires an external energy source and passive control typically does not.

Feedback control consists of measuring the output, or response, of the structure and using that measurement to determine the force to apply to the structure to obtain a desired response. The device used to measure the response (sensor), the device used to apply the force (actuator) and any electronics required to transfer the sensor signal into an actuator command (control law) make up the *control hardware*. This is illustrated in Figure 1.25 by using a *block diagram*. Systems with feedback are referred to as closed-loop systems, while control systems without feedback are called open-loop systems, as illustrated in Figures 1.25 and

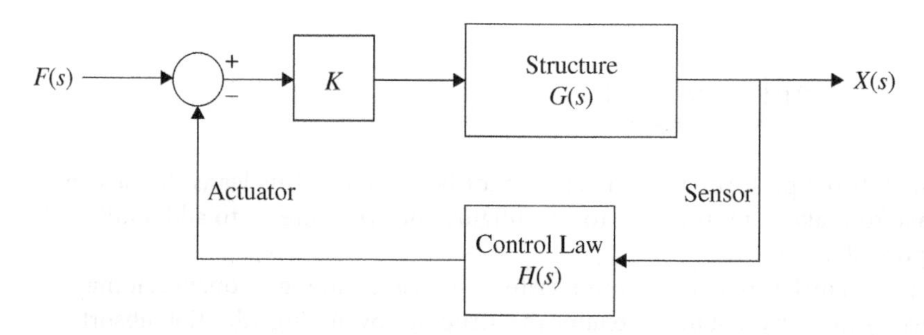

Figure 1.25 Block diagram of closed-loop system.

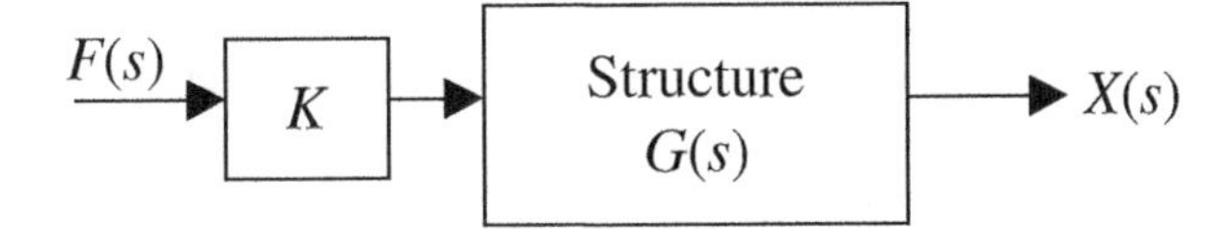

Figure 1.26 Block diagram of an open-loop system.

1.26, respectively. A major difference between open-loop and closed-loop control is simply that closed-loop control depends on information about the system's response and open-loop control does not.

The rule that defines how the measurement from the sensor is used to command the actuator to effect the system is called the *control law*, denoted $H(s)$ in Figure 1.25. Much of control theory focuses on clever ways to choose the control law to achieve a desired response.

A simple open-loop control law is to multiply (or amplify) the response of the system by a constant. This is referred to as *constant gain control*. The magnitude of the FRF for the system in Figure 1.25 is multiplied by the constant K, called the *gain*. The frequency domain equivalent of Figure 1.25 is

$$\frac{X(s)}{F(s)} = KG(s) = \frac{K}{(ms^2 + cs + k)} \tag{1.72}$$

where the plant is taken to be an SDOF model of structure. In the time domain, this becomes

$$m\ddot{x}(t) + c\dot{x}(t) + kx(t) = Kf(t) \tag{1.73}$$

The effect of this open-loop control is simply to multiply the steady-state response by K and to increase the value of the peak response, M_p.

On the other hand, the closed-loop control, illustrated in Figure 1.25, has the equivalent frequency domain representation given by

$$\frac{X(s)}{F(s)} = \frac{KG(s)}{(1 + KG(s)H(s))} \tag{1.74}$$

If the feedback control law is taken to be one that measures both the velocity and position, multiplies them by some constant gains g_1 and g_2, respectively, and adds the result, the control law $H(s)$ is given by

$$H(s) = g_1 s + g_2 \tag{1.75}$$

As the velocity and position are the *state variables* for this system, this control law is called *full state feedback*, or PD control (for position and derivative). In this case, Equation (1.74) becomes

$$\frac{X(s)}{F(s)} = \frac{K}{ms^2 + (kg_1 + c)s + (kg_2 + k)} \tag{1.76}$$

The time domain equivalent of this equation is (obtained by using the inverse Laplace Transform) is

$$m\ddot{x}(t) + (c + kg_1)\dot{x}(t) + (k + kg_2)x(t) = Kf(t) \tag{1.77}$$

By comparing Equations (1.73) and (1.77), the versatility of closed-loop control versus open-loop, or passive, control is evident. In many cases the choice of values of K, g_1 and g_2 can be made electronically. By using a closed-loop control, the designer has the choice of three more parameters to adjust than are available in the passive case to meet the desired specifications.

On the negative side, closed-loop control can cause some difficulties. If not carefully designed, a feedback control system can cause an otherwise stable structure to have an unstable response. For instance, suppose the goal of the control law is to reduce the stiffness of the structure so that the natural frequency is lower. From examining Equation (1.77), this would require that g_2 be a negative number. Then suppose that the value of k was over-estimated and g_2 calculated accordingly. This could result in the possibility that the coefficient of $x(t)$ becomes negative, causing instability. That is, the response of Equation (1.77) would be unstable if $(k + Kg_2) < 0$. This would amount to positive feedback and is not likely to arise by design on purpose, but can happen if the original parameters are not well known. On physical grounds, instability is possible because the control system is adding energy to the structure. One of the major concerns in designing high-performance control systems is to maintain stability. This introduces another constraint on the choice of the control gains and is discussed in more detail in Chapter 7. Of course, closed-loop control is also expensive because of the sensor, actuator and electronics required to make a closed-loop system. On the other hand, closed-loop control can always result in better performance provided the appropriate hardware is available.

Feedback control uses the measured response of the system to modify and add back into the input to provide an improved response. Another approach to improving the response consists of producing a second input to the system that effectively cancels the disturbance to the system. This approach, called *feedforward control*, uses knowledge of the response of a system at a point to design a control force that when subtracted from the uncontrolled response yields a new response with desired properties, usually a response of zero. Feedforward control is most commonly used for high frequency applications and in acoustics (for noise cancellation) and is not considered here. An excellent treatment of feedforward controllers can be found in Fuller et al. (1996).

Example 1.9.1

Consider the step response of an underdamped system (Figure 1.11). Calculate the value of the damping ratio ζ in terms of the performance measures t_p and t_s. Show that it is not possible to specify all three performance measures O.S., t_p and t_s in the design of a passive system.

Solution: Rearranging the definition of settling time given in Equation (1.37) yields

$$\omega_n = \frac{3.2}{t_s \zeta}$$

The time to peak is given in Equation (1.39) for underdamped systems as

$$t_p = \frac{\pi}{\omega_n \sqrt{1 - \zeta^2}}$$

Squaring this last expression and solving for ζ yields

$$\zeta = \frac{t_p}{\sqrt{t_p^2 + a^2 t_s^2}}, \quad \text{where} \quad a = \left(\frac{\pi}{3.2}\right)$$

Thus, specifying t_p and t_s completely determines both ζ and ω_n. Since O.S. is only a function of ζ, its value is also determined once t_p and t_s are specified. Recall that

$$\zeta = \frac{c}{2\sqrt{km}}, \quad \text{and} \quad \omega_n = \sqrt{\frac{k}{m}}$$

Thus, no passive adjustment of m, c and/or k can arbitrarily assign all three performance values of O.S., t_p and t_s.

1.10 Nonlinear Vibrations

The force versus displacement plot for a spring of Figure 1.3 curves off after the deflections and/or forces become large enough. Before enough force is applied to permanently deform or break the spring, the force deflection curve becomes nonlinear and curves away from a straight line, as indicated in Figure 1.27. So rather than the linear spring relationship $f_k = kx$, a model such as $f_k = \alpha x - \beta x^3$, called a *softening spring*, might better fit the curve. This nonlinear spring behavior greatly changes the physical nature of the vibratory response and complicates the mathematical description and analysis to the point that numerical integration usually has to be employed to obtain a solution. Stability analysis of nonlinear systems also becomes more complicated.

In Figure 1.27, the force-displacement curves for three springs are shown. Notice that the linear range for the two nonlinear springs is a good approximation until about 1.8 units of displacement or 2000 units of force. If the spring is to be used beyond that range, then the linear vibration analysis of the preceding sections no longer applies.

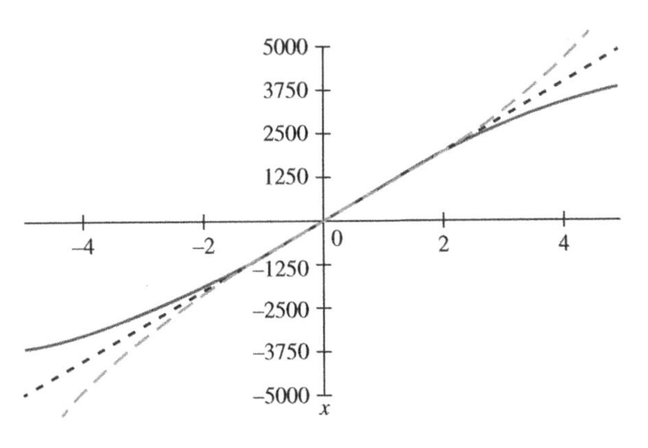

Figure 1.27 Force (vertical axis) deflection (horizontal axis) curves for three different springs in dimensionless terms, indicating their linear range. The curve $g(x) = kx$ is a linear spring (short dashed line), the curve $f(x) = kx - \beta x^3$ is called a softening spring (solid line) and the curve $h(x) = kx + \beta x^3$ is called a hardening spring (long dashed line), for the case $k = 1000$ and $\beta = 10$.

Consider then the equation of motion of a system with a nonlinear spring of the form

$$m\ddot{x}(t) + \alpha x(t) - \beta x^3(t) = 0 \tag{1.78}$$

which is subject to two initial conditions. In the linear system, there was only one equilibrium point to consider, $v(t) = x(t) = 0$. As will be shown in the following, the nonlinear system of Equation (1.78) has more than one equilibrium position. The equilibrium point of a system, or set of governing equations, may be defined best by first placing the equation of motion into *state space* form.

A general SDOF system may be written as

$$\ddot{x}(t) + f(x(t), \dot{x}(t)) = 0 \tag{1.79}$$

where the function f can take on any form, linear or nonlinear. For example, for a linear spring-mass-damper system, the function f is just $f(x, \dot{x}) = 2\zeta\omega_n\dot{x}(t) + \omega_n^2 x(t)$, which is a linear function of the state variables of position and velocity. For a nonlinear system, the function f will be some nonlinear function of the state variables. For instance, for the nonlinear spring of Equation (1.78), the function is

$$f(x, \dot{x}) = \alpha x - \beta x^3$$

The *state space* model of Equation (1.79) is written by defining the two *state variables* the position: $x_1 = x(t)$, and the velocity: $x_2 = \dot{x}(t)$. Then Equation (1.79) can be written as the first-order pair

$$\begin{aligned} \dot{x}_1(t) &= x_2(t) \\ \dot{x}_2(t) &= -f(x_1, x_2) \end{aligned} \tag{1.80}$$

This state space form of the equation of motion is used for numerical integration, in control analysis and for formally defining an equilibrium position. Define the state vector, $\mathbf{x}$, and a nonlinear vector function $\mathbf{F}$, as

$$\mathbf{x}(t) = \begin{bmatrix} x_1(t) \\ x_2(t) \end{bmatrix}, \quad \text{and} \quad \mathbf{F} = \begin{bmatrix} x_2(t) \\ -f(x_1, x_2) \end{bmatrix} \tag{1.81}$$

Then Equation (1.80) may be written in the simple form of a vector equation

$$\dot{\mathbf{x}} = \mathbf{F}(\mathbf{x}) \tag{1.82}$$

An *equilibrium point* of this system, denoted $\mathbf{x}_e$, is defined to be any value of the vector $\mathbf{x}$ for which $\mathbf{F}(\mathbf{x})$ is identically zero (called zero phase velocity). Thus the equilibrium point is any vector of constants, $\mathbf{x}_e$, that satisfies the relations

$$\mathbf{F}(\mathbf{x}_e) = \mathbf{0} \tag{1.83}$$

Placing the linear SDOF system into state space form then yields

$$\dot{\mathbf{x}} = \begin{bmatrix} x_2 \\ -2\zeta\omega_n x_2 - \omega_n^2 x_1 \end{bmatrix} \tag{1.84}$$

The equilibrium of a linear system is thus the solution of the vector equality

$$\begin{bmatrix} x_2 \\ -2\zeta\omega_n x_2 - \omega_n^2 x_1 \end{bmatrix} = \begin{bmatrix} 0 \\ 0 \end{bmatrix} \tag{1.85}$$

which has the single solution: $x_1 = x_2 = 0$. Thus, for any linear system, the equilibrium point is a single point consisting of the origin. On the other hand, the equilibrium condition of the soft spring system of Equation (1.78) requires that

$$\begin{aligned} x_2 &= 0 \\ -\alpha x_1 + \beta x_1^3 &= 0 \end{aligned} \tag{1.86}$$

Solving for x_1 and x_2, yields the *three* equilibrium points

$$\mathbf{x}_e = \begin{bmatrix} 0 \\ 0 \end{bmatrix}, \quad \begin{bmatrix} \sqrt{\dfrac{\alpha}{\beta}} \\ 0 \end{bmatrix}, \quad \begin{bmatrix} -\sqrt{\dfrac{\alpha}{\beta}} \\ 0 \end{bmatrix} \tag{1.87}$$

In principle, the soft spring system of Equation (1.78) could oscillate around any of these equilibrium points and which one will depend on the initial conditions (and the magnitude of any applied forcing function). Each of these equilibrium points may also have a different stability property.

Note that placing an nth order, ordinary differential equation into n first-order, ordinary differential equations can always be done (Boyce and DiPrima, 2012).

This existence of multiple equilibrium points also complicates the notion of stability introduced in Section 1.7. In particular, solutions near each equilibrium point could potentially have different stability behavior. Since the initial conditions may determine which equilibrium the solution centers around, the behavior of a nonlinear system will depend on the initial conditions. In contrast, for a linear system with fixed parameters, the solution form is the same regardless of the initial conditions. This represents another important difference to consider when working with nonlinear components.

Example 1.10.1

Sliding or Coulomb friction applied to a spring-mass system results in an equation of motion of the form

$$m\ddot{x}(t) + \mu mg\,\text{sgn}(\dot{x}) + kx(t) = 0$$

where m and k are the mass and stiffness values, μ is the coefficient of sliding friction and the function sgn denotes the signum function which is zero when the velocity is zero and is 1 the rest of the time, having the same sign of the velocity. This force reflects the fact that dry friction is always opposite to the direction of motion. Discuss the equilibrium positions of this nonlinear system.

Solution: First put the equation of motion into state space form resulting in

$$\dot{\mathbf{x}} = \begin{bmatrix} x(t) \\ \dot{x}(t) \end{bmatrix} = \frac{d}{dt}\begin{bmatrix} x_1 \\ x_2 \end{bmatrix} = \begin{bmatrix} x_2 \\ -\mu\text{sgn}(x_2) - kx_1 \end{bmatrix}$$

The equilibrium is thus defined as

$$\begin{bmatrix} x_2 \\ -\mu\text{sgn}(x_2) - kx_1 \end{bmatrix} = \begin{bmatrix} 0 \\ 0 \end{bmatrix}$$

This requires $x_2 = 0$ and $-\mu$ sgn $(x_2) - kx_1 = 0$. This last expression is a static condition, which is also satisfied for any value of x_1 that satisfies

$$-\frac{\mu_s mg}{k} < x_1 < \frac{\mu_s mg}{k}$$

Thus the equilibrium is not a series of points but rather a region of values. Basically the solution can only move out of this region if the spring force is large enough to overcome static friction. Hence, the response will end up in this equilibrium position depending on the initial condition, as typical of a nonlinear system.

1.11 Computing and Simulation in MATLAB™

Modern computer codes such as MATLAB™ make the visualization and computation of vibration problems available without much programming effort. Such codes can help enhance understanding through plotting responses, can help find solutions to complex problems lacking closed form solutions through numerical integration and can often help with symbolic computations. Plotting certain parametric relations or plotting solutions can often aid in visualizing the nature of relationships or the effect of parameter changes on the response. Most of the plots used in this text are constructed from simple MATLAB commands, as the following examples illustrate. If you are familiar with MATLAB, you may wish to skip this section.

MATLAB is a high-level code, with many built-in commands for numerical integration (simulation), control design, performing matrix computations, symbolic manipulation, etc. MATLAB has two areas to enter information. The first is the command window, which is an active area where the entered command is compiled as it is entered. Using the command window is somewhat like a calculator. The second area is called an m-file, which is a series of commands that are saved then called from the command window to execute. All of the plots in the figures in this chapter can be reproduced using these simple commands.

Example 1.11.1

Plot the free response of the underdamped system to initial conditions $x_0 = 0.01$ m, $v_0 = 0$ for values of $m = 100$ kg, $c = 25$ kg/s and $k = 1000$ N/m, using MATLAB and Equation (1.13).

Solution: To enter numbers in the command window, just type a symbol and use an equal sign after the blinking cursor. The following entries in the command window will produce the plot of Figure 1.28. Note that the prompt symbol "≫" is provided by MATLAB and the information following it is code typed in by the user. The symbol % is used to indicate comments so that anything following this symbol is ignored by the code and is included to help explain the situation. A semicolon typed after a command suppresses the command from displaying the output. MATLAB uses matrices and vectors so that numbers can be entered and computed in arrays. Thus, there are two types of multiplication. The notation `a*b` is a vector operation demanding that the number of rows of `a` be equal to

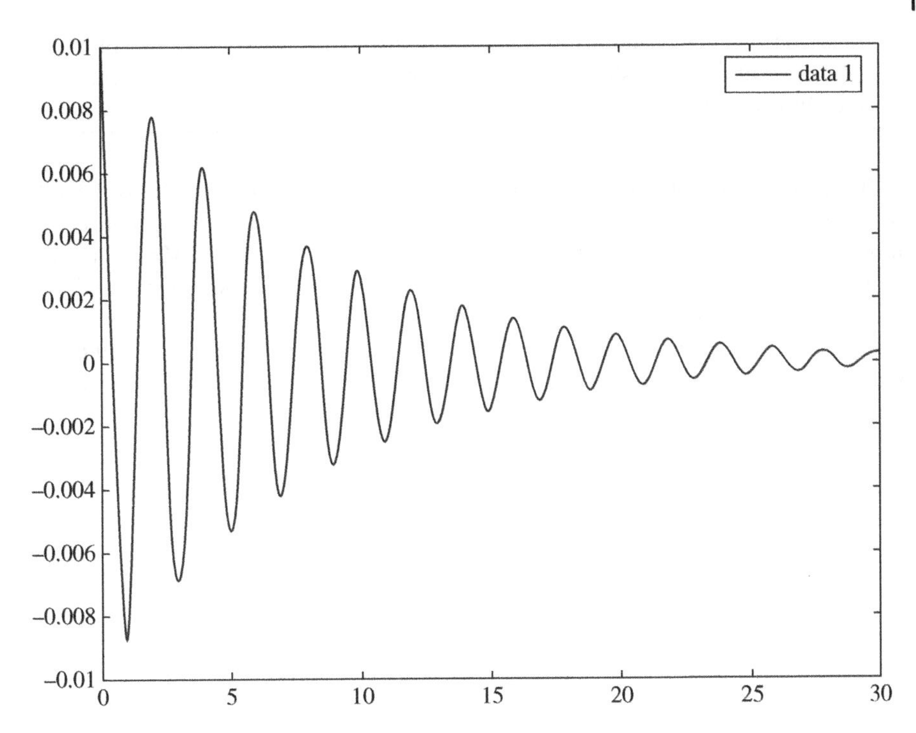

Figure 1.28 The response of an underdamped system ($m = 100$ kg, $c = 25$ kg/s and $k = 1000$ N/m) to the initial conditions $x_0 = 0.01$ m, $v_0 = 0$ plotted using MATLAB.

the number of columns of b. The product a*b, on the other hand, multiplies each element of a times the corresponding element in b.

```
>> clear % used to make sure no previous values are stored
>> %assign the initial conditions, mass, damping and stiffness
>> x0=0.01;v0=0.0;m=100;c=25;k=1000;
>> %compute omega and zeta, display zeta to check if under-
damped
>> wn=sqrt(k/m);z=c/(2*sqrt(k*m))

z =

  0.0395

>> %compute the damped natural frequency
>> wd=wn*sqrt(1-z^2);
>> t=(0:0.01:15*(2*pi/wn));%set the values of time from 0 in
increments of 0.01 up to 15 periods
>> x=exp(-z*wn*t).*(x0*cos(wd*t)+((v0+z*wn*x0)/wd)*sin(wd*t));
% computes x(t)
>> plot(t,x)%generates a plot of x(t) vs t
```

The MATLAB code used in this example is not the most efficient way to plot the response and does not show the detail of labeling the axis, etc. but is given as a quick introduction.

The next example illustrates the use of m-files in a numerical simulation. Instead of plotting the closed-form solution given in Equation (1.13), the equation of motion can be numerically integrated using the ode command in MATLAB. The ode45 command uses a fifth-order Runge-Kutta, automated time step method for numerically integrating the equation of motion (Pratap, 2002).

In order to use numerical integration, the equations of motion must first be placed in first order, or state space form, as in Equation (1.84). This state space form is used in MATLAB to enter the equations of motion.

Vectors are entered in MATLAB by using square brackets, spaces and semicolons. Spaces are used to separate columns and semicolons are used to separate rows. So that a row vector is entered by typing

```
>> u = [1 -1 2]
```

which returns the row

```
u =
      1   -1   2
```

and a column vector is entered by typing

```
>> u = [1; -1; 2]
```

which returns the column

```
u =
      1
     -1
      2
```

To create a list of formulas in an m-file, choose "New" from the file menu and select "m-file". This will display a text editor window, in which you can enter commands. The following example illustrates the creation of an m-file and how to call it from the command window to numerically integrate the equation of motion given in Example 1.11.1.

Example 1.11.2

Numerically integrate and plot the free response of the underdamped system to initial conditions $x_0 = 0.01$ m, $v_0 = 0$ for values of $m = 100$ kg, $c = 25$ kg/s and $k = 1000$ N/m, using MATLAB and Equation (1.13).

Solution: First create an m-file containing the equation of motion to be integrated and save it. This is done by selecting "New" and "M-File" from the File menu in MATLAB, then typing:

```
----------------------
function xdot=f2(t,x)
c=25; k = 1000; m = 100;
% set up a column vector with the state Equations
xdot=[x(2); -(c/m)*x(2)-(k/m)*x(1)];
----------------------
```

This file is now saved with the name `f2.m`. Note that the name of the file must agree with the name following the equal sign in the first line of the file. Now open the command window and enter the following:

```
>> ts=[0 30]; % this enters the initial and final time
>>x0 =[0.01 0]; % this enters the initial conditions
>>[t, x]=ode45('f2',ts,x0);
>>plot(t,x(:,1))
```

The third line of code calls the Runge-Kutta program ode45 and the state equations to be integrated contained in the file named `f2.m`. The last line plots the simulation of the first state variable $x_1(t)$, which is the displacement, denoted `x(:,1)` in MATLAB. The plot is given in Figure 1.29.

Note that the plots of Figures 1.28 and 1.29 look the same. However, Figure 1.28 was obtained by simply plotting the analytical solution, whereas the plot of Figure 1.29 was obtained by numerically integrating the equation of motion. The numerical approach can be used successfully to obtain the solution of a nonlinear state equation, such as Equation (1.78), just as easily.

The forced response can also be computed using numerical simulation and this is often more convenient then working through an analytical solution when the forcing functions are discontinuous or not made up of simple functions. Again the equations of motion (this time with the forcing function) must be placed in state space form. The equation of motion for damped system with general applied force is

$$m\ddot{x}(t) + c\dot{x}(t) + kx(t) = F(t)$$

In state space form, this expression becomes

$$\begin{bmatrix} \dot{x}_1(t) \\ \dot{x}_2(t) \end{bmatrix} = \begin{bmatrix} 0 & 1 \\ -\dfrac{k}{m} & -\dfrac{c}{m} \end{bmatrix} \begin{bmatrix} x_1(t) \\ x_2(t) \end{bmatrix} + \begin{bmatrix} 0 \\ f(t) \end{bmatrix}, \quad \begin{bmatrix} x_1(0) \\ x_2(0) \end{bmatrix} = \begin{bmatrix} x_0 \\ v_0 \end{bmatrix} \tag{1.88}$$

where $f(t) = F(t)/m$ and $F(t)$ is any function that can be integrated. The following example illustrates the procedure in MATLAB.

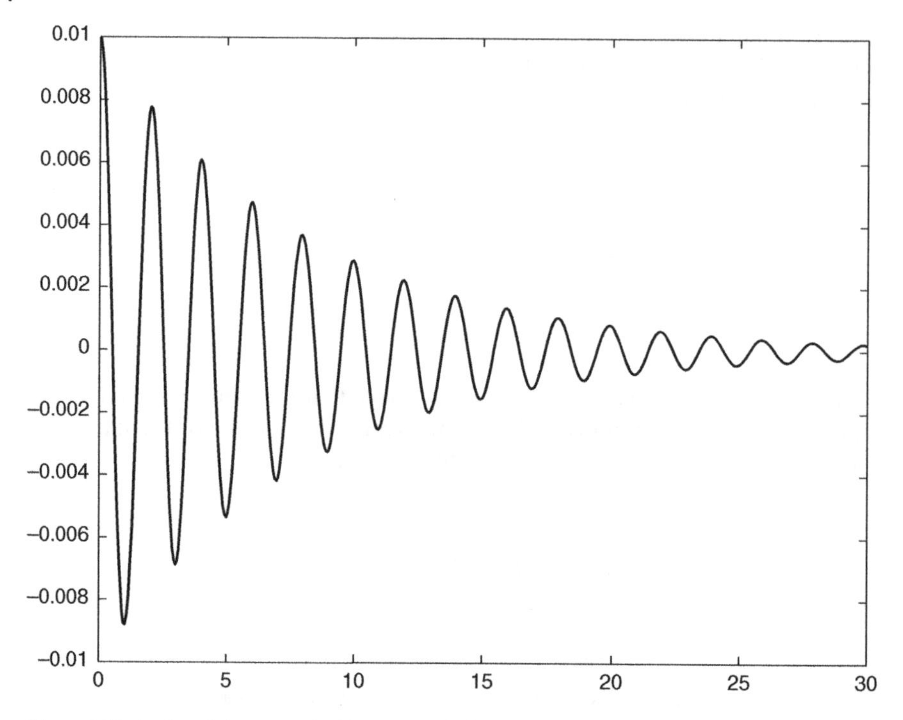

Figure 1.29 A plot of the numerical integration of the underdamped system of Example 1.10.1 resulting from the MATLAB code given in Example 1.10.2.

Example 1.11.3

Use MATLAB to compute and plot the response of the following system

$$100\ddot{x}(t) + 10\dot{x}(t) + 500x(t) = 150\cos 5t,\ x_0 = 0.01, v_0 = 0.5.$$

Solution: The Matlab code for computing these plots is given. First an m-file is created with the equation of motion given in first-order form.

```
-------------------------------------------
function v=f(t,x)
m=100; k=500; c=10; Fo=150; w=5;
v=[x(2); x(1)*-k/m+x(2)*-c/m + Fo/m*cos(w*t)];
-------------------------------------------
```

Then the following is typed in the command window:

```
>>clear all
>>xo=[0.01; 0.5]; %enters the initial conditions
>>ts=[0 40]; %enters the initial and final times
>>[t,x]=ode45('f',ts,xo); %calls the dynamics and integrates
>>plot(t,x(:,1)) %plots the result
```

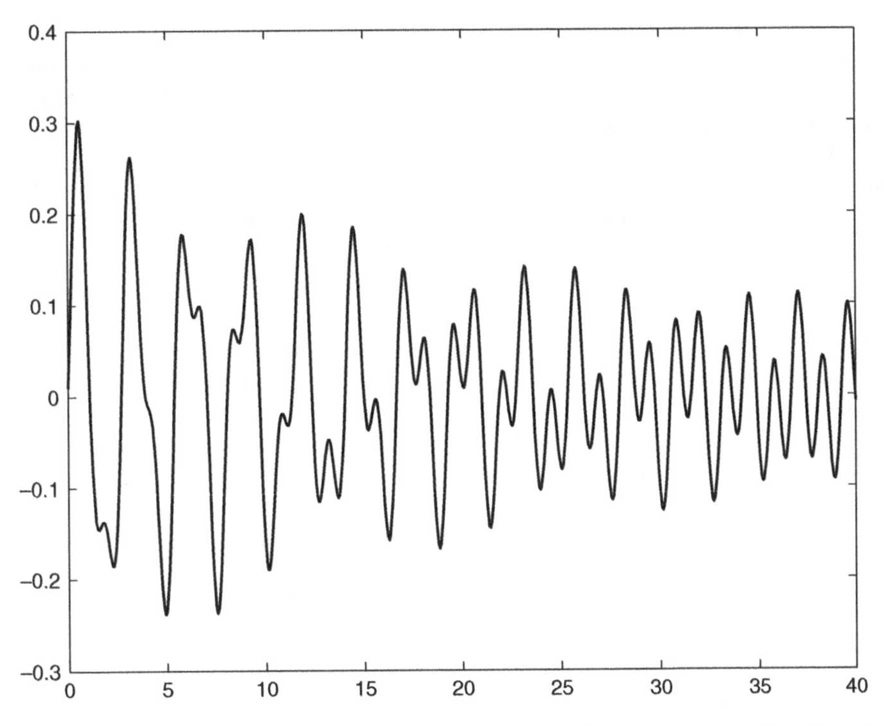

Figure 1.30 A plot of the numerical integration of the damped forced system resulting from the MATLAB code given in Example 11.1.3

This code produces the plot given in Figure 1.30. Note that the influence of the transient dynamics dies off due to the damping after about 20 sec.

Such numerical integration methods can also be used to simulate the nonlinear systems discussed in the previous section. Use of high-level codes in vibration analysis such as MATLAB is now commonplace and has changed the way vibration quantities are computed. More detailed codes for vibration analysis can be found in Inman (2014). In addition, there are many books written on using MATLAB (Pratap, 2002) as well as available online help.

Chapter Notes

This chapter attempts to provide a review of introductory vibrations and to expand the discipline of vibration analysis and design, by intertwining elementary vibration topics with the disciplines of design, control, stability and testing. An early attempt to relate vibrations and control at an introductory level was written by Vernon (1967). More recent attempts are by Meirovitch (1985, 1990) and Inman (1989), which is the predecessor or first edition of this text. Leipholz and Abdel-Rohman (1986) give the civil engineering approach to structural control. The latest attempt to combine vibration and control is by Preumont (2011) and

Benaroya (2004), who also provides excellent treatment of uncertainty in vibrations. Pruemont and Seto (2008) presents control of structures slanted towards civil structures. Moheimani *et al.* (2003) focuses on vibration control of flexible structures. The information contained in Sections 1.2, 1.3 and part of 1.4 can be found in every introductory vibrations text, such as my own (Inman, 2014) and such as the standards by Thomson and Dahleh (1993), Rao (2012) and Meirovitch (1986). A complete summary of most vibration related topics can be found in Braun et al. (2002) and Harris and Piersol (2002).

A good reference for vibration measurement is McConnell (1995). The reader is encouraged to consult a basic text on control, such as the older text of Melsa and Schultz (1969), which contains some topics dropped in modern texts, or Kuo and Golnaraghi (2009), which contains more modern topics integrated with MATLAB. These two texts also give background on specifications and transfer functions given in Sections 1.4 and 1.5, as well as feedback control discussed in Section 1.9. A complete discussion of plant identification, as presented in Section 1.7, can be found in Melsa and Schultz (1969). The book by Neubert (1987) was issued by the Naval Sea Systems Command and provides a treatise on impedance methods (Section 1.6).

There are many introductory controls texts with various slants and focus on sub-topics, as addressed by Davison et al. (2007). The excellent text by Fuller et al. (1996) examines controlling high frequency vibration. Control is introduced here, not as a discipline by itself, but rather as a design technique for vibration engineers. A standard reference on stability is Hahn (1967), which provided the basic ideas of Section 1.8. The topic of flutter and self-excited vibrations is discussed in Den Hartog (1985). Nice introductions to nonlinear vibration can be found in Virgin (2000), Worden and Tomlinson (2001) and the standards by Nayfeh and Mook (1978) and Nayfeh and Balachandra (1995). While there are many excellent texts introducing how to use MATLAB, the website of the MathWorks contains excellent tutorials for using their code.

References

Benaroya, H. (2004) *Mechanical Vibration: Analysis, Uncertainties and Control*, 2nd Edition. Marcel Dekker, Inc., New York.

Bishop, R. E. D. and Gladwell, G. M. L. (1963) An investigation into the theory of resonance testing. *Proceedings of the Royal Society Philosophical Transactions* 255(A): 241.

Bode, H. W. (1945) *Network Analysis and Feedback Amplifier Design*. D. Van Nostrand, New York.

Boyce, W. E. and DiPrima, R. C. (2012) *Elementary Differential Equations and Boundary Value Problems*, 10th Edition. John Wiley & Sons, Hoboken, NJ.

Braun, S. G., Ewins, D. J. and Rao, S. S. (eds) (2002) *Encyclopedia of Vibration*. Academic Press, London.

Davison, D. E., Chen, J., Ploen, S. R. and Bernstein, D. S. (2007) What is your favorite book on classical control? *IEEE Control Systems Magazine*, 27, 89–94.

Den Hartog, J. P. (1985) *Mechanical Vibrations*. Dover Publications, Mineola, NY.

Ewins, D. J. (2000) *Modal Testing: Theory and Practice*, 2nd Edition. Research Studies Press, Hertfordshire, UK.
Fuller, C. R., Elliot, S. J. and Nelson, P. A. (1996) *Active Control of Vibration*. Academic Press, London.
Hahn, W. (1967) *Stability of Motion*. Springer Verlag, New York.
Harris, C. M. and Piersol, A. G. (eds) (2002) *Harris' Shock and Vibration Handbook*, 5th Edition. McGraw Hill, New York
Inman, D. J. (1989) *Vibrations: Control, Measurement and Stability*. Prentice Hall, New Jersey.
Inman, D. J. (2014) *Engineering Vibration*, 4th Edition. Pearson Education, Upper Saddle River, NJ.
Kuo, B. C. and Golnaraghi, F. (2009) *Automatic Control Systems*, 9th Edition. John Wiley & Sons, New York.
Leipholz, H. H. and Abdel-Rohman, M. (1986) *Control of Structures*. Martinus Nijhoff, Boston, MA.
McConnell, K. G. (1995) *Vibration Testing; Theory and Practice*. John Wiley & Sons, New York.
Machinante, J. A. (1984) *Seismic Mountings for Vibration Isolation*. John Wiley & Sons, New York.
Meirovitch, L. (1985) *Introduction to Dynamics and Control*. John Wiley & Sons, New York.
Meirovitch, L. (1986) *Elements of Vibration Analysis*, 2nd Edition. McGraw-Hill, New York.
Meirovitch, L. (1990) *Dynamics and Control of Structures*. John Wiley & Sons, New York.
Melsa, J. L. and Schultz, D. G. (1969) *Linear Control System*. McGraw-Hill, New York.
Moheimani, S.O.R., Halim, D. and Fleming, A. J. (2003) *Spatial Control of Vibration Theory and Experiments*. World Scientific, Singapore.
Nayfeh, A. H. and Balachandra, B. (1995) *Applied Nonlinear Dynamics*. John Wiley & Sons, New York.
Nayfeh, A. H. and Mook, D. T. (1978) *Nonlinear Oscillations*. John Wiley & Sons, New York.
Neubert, V. H. (1987) *Mechanical Impedance: Modelling/Analysis of Structures*. Jostens Pringint and Publishing Co, State College, PA.
Pratap, R. (2002) *Getting Started with MATLAB: A Quick Introduction for Scientists and Engineers*. Oxford University Press, New York.
Preumont, A. and Seto, K. (2008) *Active Control of Structures*. John Wiley & Sons, Chichester, UK.
Preumont, A. (2011) *Vibration Control of Active Structures: An Introduction*, 3rd Edition. Springer-Verlag, Berlin.
Rao, S. S. (2012) *Mechanical Vibrations*, 5th Edition. Pearson, New Jersey.
Rivin, E. I. (2003) *Passive Vibration Isolation*. ASME Press, New York.
Thomson, W. T. and Dahleh, M. D. (1993) *Theory of Vibration with Applications*, 5th Edition. Prentice Hall, Englewood Cliffs, NJ.
Vernon, J. B. (1967) *Linear Vibrations and Control System Theory*. John Wiley & Sons, New York.

Virgin, L. N. (2000) *Introduction to Experimental Nonlinear Dynamics: A case study in mechanical vibrations*. Cambridge University Press, Cambridge, UK.
Wordon, K. and Tomlinson, G. T. (2001) *Nonlinearity in Structural Dynamics: Detection, Identification and Modeling*. Institute of Physics Publishing, Bristol, UK.

Problems

1.1 Derive the solution of $m\ddot{x} + kx = 0$ and sketch your result (for at least 2 periods) for the case.

1.2 Solve $m\ddot{x} - kx = 0$ for $x_0 = 1$, $v_0 = 0$, for $x(t)$ and sketch the solution.

1.3 Derive the solutions given in the text for $\zeta > 1$, $\zeta = 1$ and $0 < \zeta < 1$ with x_0 and v_0 as the initial conditions (i.e. derive Equations 1.14 to 1.16 and corresponding constants).

1.4 Solve $\ddot{x} - \dot{x} + x = 0$ with $x_0 = 1$ and $v_0 = 0$ for $x(t)$, and sketch the solution.

1.5 Prove that $\zeta = 1$ corresponds to the smallest value of c such that no oscillation occurs. (*Hint:* Let $\lambda = -b$, b be a positive real number, and differentiate the characteristic equation.)

1.6 Consider a small spring about 30 mm long, welded to a stationary table (ground) so that it is fixed at the point of contact, with a 12-mm bolt welded to the other end, which is free to move. The mass of this system is about 49.2×10^{-3} kg. The spring stiffness is $k = 857.8$ N/m. Calculate the natural frequency, period and the maximum amplitude of the response if the spring is initially deflected 10 mm.

1.7 A simple model of a vehicle wheel, tire and suspension assembly is just the basic spring-mass equation of motion. If its mass is measured to be about 30 kilograms (kg) and its frequency of oscillation is observed to be 10 Hz, what is the approximate stiffness of the suspension?

1.8 Calculate t_p, OS, T_d, M_p and BW for a system described by

$$2\ddot{x} + 0.8\dot{x} + 8x = f(t)$$

where $f(t)$ is either a unit step function or a sinusoidal as required.

1.9 Derive an expression for the forced response of the undamped system

$$m\ddot{x}(t) + kx(t) = F_0 \sin \omega t, \; x(0) = x_0, \; \dot{x}(0) = v_0$$

to a sinusoidal input and nonzero initial conditions. Compare your result to Equation (1.25) with $\zeta = 0$.

1.10 Compute the total response to the system

$$4\ddot{x}(t) + 16x(t) = 8 \sin 3t, \; x_0 = 1 \text{ mm}, \; v_0 = 2 \text{ mm/s}$$

1.11 Calculate the maximum value of the peak response (magnification factor) for the system of Figure 1.10 with $\zeta = 1/\sqrt{2}$.

1.12 Derive Equation (1.26).

1.13 Calculate the impulse response function for a critically damped system.

1.14 Solve for the forced response of an SDOF system to a harmonic excitation with $\zeta = 1.1$ and $\omega_n^2 = 4$. Plot the magnitude of the steady state response versus the driving frequency. For what value of ω_n is the response a maximum (resonance)?

1.15 Consider the forced vibration of a mass m connected to a spring of stiffness 2000 N/m being driven by a 20-N harmonic force at 10 Hz (20π rad/s). The maximum amplitude of vibration is measured to be 0.1 m and the motion is assumed to have started from rest ($x_0 = v_0 = 0$). Calculate the mass of the system.

1.16 Consider a spring-mass-damper system with $m = 100$ kg, $c = 20$ kg/s and $k = 2000$ N/m, with an impulsive force applied to it of 1000 N for 0.01 s. Compute the resulting response.

1.17 Calculate the compliance transfer function for the system described by the differential equation

$$a\ddddot{x} + b\dddot{x} + c\ddot{x} + d\dot{x} + ex = f(t)$$

where $f(t)$ is the input and $x(t)$ is a displacement. Also calculate the FRF for this system.

1.18 Use the frequency response approach to compute the amplitude of the particular solution for the undamped system of the form

$$m\ddot{x}(t) + kx(t) = F_0 \cos \omega t$$

1.19 Derive Equation (1.66).

1.20 Plot (using a computer) the unit step response of an SDOF system with $\omega_n^2 = 4$, $k = 1$ for several values of the damping ratio ($\zeta = 0.01, 0.1, 0.5$ and 1.0).

1.21 Plot ω_p/ω_n versus ζ and ω_d/ω_n versus ζ, and comment on the difference as a function of ζ.

1.22 For the system of Problem 1.8, construct the Bode plots for (a) the inertance transfer function, (b) the mobility transfer function, (c) the compliance transfer function, and (d) the Nyquist diagram for the compliance transfer function.

1.23 The free response of the damped SDOF system with a mass of 2 kg is observed to be underdamped. A static deflection test is performed and the stiffness is determined as 1.5×10^3 N/m. The displacements at two successive maximum amplitudes t_1 and t_2 are measured to be 9 and 1 mm, respectively. Calculate the damping coefficient.

1.24 Discuss the stability of the following system

$$2\ddot{x}(t) - 3\dot{x}(t) + 8x(t) = -3\dot{x}(t) + \sin 2t$$

1.25 An inverted pendulum has equation of motion

$$ml^2\ddot{\theta} + \left(\frac{kl^2}{2}\sin\theta\right)\cos\theta - mgl\sin\theta = 0$$

Linearize the equation and discuss the stability of the result.

1.26 Using the system of Problem 1.8, refer to Equation (1.77) and choose the gains K, g_1 and g_2 so that the resulting closed-loop system has a 5% overshoot and a settling time of less than 10.

1.27 Calculate an allowable range of values for the gains K, g_1 and g_2 for the system of Problem 1.8, such that the closed loop system is stable and the formulas overshoot and peak time of an underdamped system is valid.

1.28 Compute a feedback law with full state feedback (of the form given in Equation 1.77) that stabilizes (makes it asymptotically stable) the following system: $4\ddot{x}(t) + 16x(t) = 0$ and causes the closed loop settling time to be 1 second.

1.29 Compute the equilibrium positions of the pendulum equation

$$ml^2\ddot{\theta}(t) + mgl\sin\theta(t) = 0$$

1.30 Compute the equilibrium points for a system with Coulomb damping given by

$$m\ddot{x}(t) + \mu mg\mathrm{sgn}(\dot{x}) + kx(t) = 0$$

where μ is the coefficient of friction and g denotes the acceleration due to gravity. Here sgn denotes the signum function takes on a plus, minus or zero value, depending on whether the argument is plus, minus or zero.

1.31 Compute the equilibrium points for the system defined by

$$\ddot{x} + \beta\dot{x} + x + x^2 = 0$$

1.32 The linearized version of the pendulum equation is given by $\ddot{\theta}(t) + \frac{g}{l}\theta(t) = 0$. Use numerical integration to plot the solution of the nonlinear equation of Problem 1.29 and this linearized version for the case that

$$g/l = 0.01,\ \theta(0) = 0.1 \text{ rad},\ \dot{\theta}(0) = 0.1 \text{ rad/s}$$

Compare your two simulations.

2

Lumped Parameter Models

2.1 Introduction

Many physical systems cannot be modeled successfully by the single-degree-of-freedom (SDOF) model discussed in Chapter 1. That is, to describe the motion of the structure or machine, several coordinates may be required. Such systems are referred to as *lumped parameter systems* to distinguish them from the distributed parameter systems considered in Chapters 7 to 10. Chapter 11 considers approximations of distributed parameter systems as lumped parameter systems. Such systems are also called *lumped mass systems* and sometimes as *discrete systems* (referring to mass not time). Each lumped mass potentially corresponds to 6 degrees of freedom. Such systems are referred to as multiple degree of freedom systems (often abbreviated MDOF). In order to keep a record of each coordinate of the system, vectors are used. This is done both for ease of notation and to enable vibration theory to take advantage of the power of linear algebra. This section organizes the notation to be used throughout the rest of the text and introduces several common examples.

Before the motions of such systems are considered, it is important to recall the definition of a matrix and a vector as well as a few simple properties of each. Vectors were used in Chapter 1 to write the equation of motion in state space form, and are formalized here. If you are familiar with vector algebra, skip ahead to Equation (2.7). Let $\mathbf{q}$ denote a vector of dimension n defined by

$$\mathbf{q} = \begin{bmatrix} q_1 \\ q_2 \\ \vdots \\ q_n \end{bmatrix} \tag{2.1}$$

Here, q_i denotes the ith element of the vector $\mathbf{q}$. This is not to be confused with $\mathbf{q}_i$, which denotes the ith vector in a set of vectors. Two vectors $\mathbf{q}$ and $\mathbf{r}$ of the same dimension (n in this case) may be summed under the rule

$$\mathbf{q} + \mathbf{r} = \mathbf{s} = \begin{bmatrix} q_1 + r_1 \\ q_2 + r_2 \\ \vdots \\ q_n + r_n \end{bmatrix} \tag{2.2}$$

Vibration with Control, Second Edition. Daniel John Inman.

Companion Website: www.wiley.com/go/inmanvibrationcontrol2e

and multiplied under the rule (dot product or inner product)

$$\mathbf{q} \cdot \mathbf{r} = \sum_{i=1}^{n} q_i r_i = \mathbf{q}^T \mathbf{r} \tag{2.3}$$

where the superscript T denotes the transpose of the vector. Note that the inner product of two vectors, $\mathbf{q}^T\mathbf{q}$, yields a scalar. With $\mathbf{q}$ given in Equation (2.1) as a column vector, $\mathbf{q}^T$ is a row vector given by $\mathbf{q}^T = [q_1\ q_2\ \cdots\ q_n]$.

The product of a scalar, a, times a vector, $\mathbf{q}$, is a vector given by

$$a\mathbf{q} = [aq_1\ \ aq_2 \ldots aq_{\mathrm{n}}]^{\mathrm{T}} \tag{2.4}$$

If the zero vector is defined to be a vector of proper dimension whose entries are all zero, then rules Equations (2.2) and (2.4) define a linear vector space (Appendix B) of dimension n.

A matrix, A, is defined to be a rectangular array of numbers (scalars) of the form

$$A = \begin{bmatrix} a_{11} & a_{12} & \cdots & a_{1n} \\ a_{21} & a_{22} & \cdots & a_{2n} \\ \vdots & \vdots & & \vdots \\ a_{m1} & a_{m2} & \cdots & a_{mn} \end{bmatrix}$$

consisting of m rows and n columns. The matrix A is said to have dimension $m \times n$. For the most part, the equations of motion used in vibration theory result in real valued square matrices of dimension $n \times n$. Each of the individual elements of a matrix are labeled as a_{ik}, which denotes the element of the matrix in the position of the intersection of the ith row and k^{th} column.

Two matrices of the same dimension can be summed by adding the corresponding elements in each position, as illustrated by a 2×2 example

$$\begin{bmatrix} a_{11} & a_{12} \\ a_{21} & a_{22} \end{bmatrix} + \begin{bmatrix} b_{11} & b_{12} \\ b_{21} & b_{22} \end{bmatrix} = \begin{bmatrix} a_{11}+b_{11} & a_{12}+b_{12} \\ a_{21}+b_{21} & a_{22}+b_{22} \end{bmatrix} \tag{2.5}$$

The product of two matrices is more complicated and is given by the formula $C = AB$, where the resulting matrix C has elements given by

$$c_{ij} = \sum_{k=1}^{n} a_{ik} b_{kj} \tag{2.6}$$

and is only defined if the number of columns of A is the same as the number of rows of B, which is n in Equation (2.6). Note that the product BA is not defined for this case, unless A and B are of the same dimensions. In most cases, the product of two square matrices or the product of a square matrix and a vector are used in vibration analysis. Note that a vector is just a rectangular matrix with the smallest dimension being 1.

With this introduction, the equations of motion for lumped parameter systems can be discussed. All linear equations of motion that describe vibration can be put in the following form

$$A_1\ddot{\mathbf{q}} + A_2\dot{\mathbf{q}} + A_3\mathbf{q} = \mathbf{f}(t) \tag{2.7}$$

which is a vector differential equation with matrix coefficients. This form is sometimes referred to as the matrix form of the equations of motion, but really it is a vector equation. Here, $\mathbf{q} = \mathbf{q}(t)$ is an n vector of time-varying elements representing the displacements of the masses in the lumped mass model. The vectors $\ddot{\mathbf{q}}$ and $\dot{\mathbf{q}}$ represent the accelerations and velocities, respectively. The over dot implies that each element of $\mathbf{q}$ is differentiated with respect to time. The vector $\mathbf{q}$ could also represent a generalized coordinate that may not be an actual physical coordinate or position but is related, usually in a simple manner, to the physical displacement. The coefficients A_1, A_2 and A_3 are n square matrices of constant real elements representing the various physical parameters of the system. The n vector $\mathbf{f} = \mathbf{f}(t)$ represents applied external forces and is also time-varying. The system of Equation (2.7) is also subject to initial conditions on the initial displacement $\mathbf{q}(0) = \mathbf{q}_0$ and initial velocities $\dot{\mathbf{q}}(0) = \dot{\mathbf{q}}_0$.

The matrices A_i will have different properties, depending on the physics of the problem under consideration. As will become evident in the remaining chapters, the mathematical properties of these matrices will determine the physical nature of the solution $\mathbf{q}(t)$, just as the properties of the scalars m, c and k determined the nature of the solution $x(t)$ to the SDOF system of Equation (1.6) in Chapter 1.

In order to understand these properties, there is the need to classify square matrices further. The *transpose* of a matrix A, denoted by A^T, is the matrix formed by interchanging the rows of A with the columns of A. A square matrix is said to be *symmetric* if it is equal to its transpose, i.e. $A = A^T$. Otherwise it is said to be *asymmetric*. A square matrix is said to be *skew-symmetric* if it satisfies $A = -A^T$. It is useful to note that any real square matrix may be written as the sum of a symmetric matrix and a skew-symmetric matrix. To see this, notice that the symmetric part of A, denoted by A_s, is given by

$$A_s = \frac{(A^T + A)}{2} \tag{2.8}$$

and the skew-symmetric part of A, denoted by A_{ss}, is given by

$$A_{ss} = \frac{(A - A^T)}{2} \tag{2.9}$$

so that $A = A_s + A_{ss}$.

With these definitions in mind, Equation (2.7) can also be rewritten as

$$M\ddot{\mathbf{q}} + (C + G)\dot{\mathbf{q}} + (K + H)\mathbf{q} = \mathbf{f} \tag{2.10}$$

where $\mathbf{q}$ and $\mathbf{f}$ are as before, but

$M = M^T =$ *mass, or inertia matrix*
$C = C^T =$ *viscous damping matrix* (sometimes denoted by D)
$G = -G^T =$ *gyroscopic matrix*
$K = K^T =$ *stiffness matrix*
$H = -H^T =$ *circulatory matrix* (constraint damping or follower forces).

Some physical systems may also have asymmetric mass matrices (Soom and Kim, 1983).

In the following sections, each of these matrix cases will be illustrated by a physical example indicating how such forces may arise. The physical basis for the nomenclature is as expected. The mass matrix arises from the inertial forces in the system, the damping matrix arises from dissipative forces proportional to velocity, and the stiffness matrix arises from elastic forces proportional to displacement. The nature of the skew-symmetric matrices G and H is pointed out by the examples that follow.

Symmetric matrices and the physical systems described by Equation (2.10) can be further characterized by defining the concept of the definiteness of a matrix. A matrix differs from a scalar in many ways. One way in particular is the concept of sign, or ordering. In Chapter 1, it was pointed out that the sign of the coefficients in an SDOF system determined the stability of the resulting motion. Similar results will hold for Equation (2.10) when the "sign of a matrix" is interpreted as the *definiteness* of a matrix (this is discussed in Chapter 4). The definiteness of an $n \times n$ symmetric matrix is defined by examining the sign of the scalar $\mathbf{x}^T A\mathbf{x}$, called the *quadratic* form of A, where $\mathbf{x}$ is an arbitrary n-dimensional real vector. Note that if $A = I$, the identity matrix consisting of ones along the diagonal and all other elements 0, then

$$\mathbf{x}^T A\mathbf{x} = \mathbf{x}^T I\mathbf{x} = \mathbf{x}^T\mathbf{x} = x_1^2 + x_2^2 + x_3^2 + \cdots x_n^2$$

which is clearly quadratic.

In particular, the symmetric matrix A is said to be:

1) *Positive definite* if $\mathbf{x}^T A\mathbf{x} > 0$ for all nonzero real vectors $\mathbf{x}$ and $\mathbf{x}^T A\mathbf{x} = 0$, if and only if $\mathbf{x}$ is zero,
2) *Positive semidefinite* (or *nonnegative definite*), if $\mathbf{x}^T A\mathbf{x} \geq 0$ for all nonzero real vectors $\mathbf{x}$ (here $\mathbf{x}^T A\mathbf{x}$ could be zero for some nonzero vector $\mathbf{x}$),
3) *Indefinite* (or *sign variable*) if $(\mathbf{x}^T A\mathbf{x})(\mathbf{y}^T A\mathbf{y}) < 0$ for some pair of real vectors $\mathbf{x}$ and $\mathbf{y}$.

Definitions of *negative definite* and *negative semidefinite* should be obvious from (1) and (2).

In many cases, M, C and K will be positive definite, a condition that ensures stability (as illustrated in Chapter 4) and follows from physical considerations, as it does in the SDOF case.

2.2 Modeling

There are two basic approaches to writing down the equations of motion, or modeling: free body diagrams followed by force and moment balance (Newton's Laws, Euler's second law) and energy methods (Lagrange's Equations): Davison et al., 2007; Goldstein, 2002). For mechanical systems and structures coupled to other physical fields, linear graph theory (Rosenberg and Karnopp, 1983; Shearer et al., 1967) is also a useful modeling concept. For more complex systems, finite element modeling, boundary element modeling and spectral elements are useful and these are briefly introduced in Chapter 11. These analytical models can be further refined by comparison to experimental data (Friswell and Mottershead, 1995; Hughes, 2000).

Recall the basic interpretation of Newton's and Euler's laws is to isolate each body, write down all the forces and moments acting on that body and invoke (for constant mass systems) the following two sums:

$$\sum_{i}^{m} f_i(t) = m\ddot{x}(t) \text{ and } \sum_{i}^{m} m_{0i}(t) = J\ddot{\theta}(t) \tag{2.11}$$

where f_i denotes the ith force acting on the mass under consideration and the summation is over the number of such forces. Likewise, m_{0i} denotes the ith torques taken around point 0 acting on mass of moment of inertia J. The displacement of the mass is denoted $x(t)$ and its rotation by $\theta(t)$. The over dots are time derivatives as before. These equations for each body are combined into one vector differential equation of the form of Equation (2.7), where the coordinates in $\mathbf{q}$ might be angles of rotations, displacements or generalized coordinates. The following example provides a review of this approach.

Example 2.2.1

Consider the two degree of freedom, spring-mass system vibrating on a friction free horizontal surface, as indicated in Figure 2.1, and determine the equations of motion.

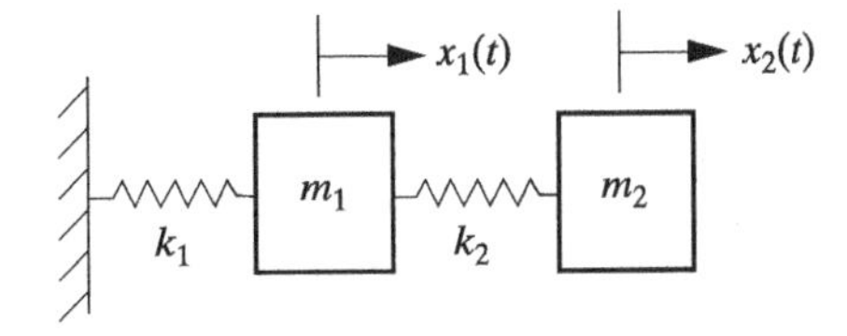

Figure 2.1 A two-degree of freedom spring-mass system on a frictionless horizontal surface.

Solution: Separating each mass and identifying the forces acting on each is given in Figure 2.2.

Examining the free body diagram of m_1 and summing forces in the y-direction yields static equilibrium in that direction: $N_1 = m_1 g$ where g is the acceleration due to gravity and N_1 is the normal force provided by the supporting plane (not pictured). In the x-direction, assuming that the masses are sliding on a friction-free surface, the sum of forces yields

$$\begin{aligned} \sum f_i = m_1\ddot{x}_1 &\Rightarrow m_1\ddot{x}_1 = -k_1 x_1 + k_2(x_2 - x_1) \\ &\Rightarrow m_1\ddot{x}_1 + (k_1 + k_2)x_1 - k_2 x_2 = 0 \end{aligned}$$

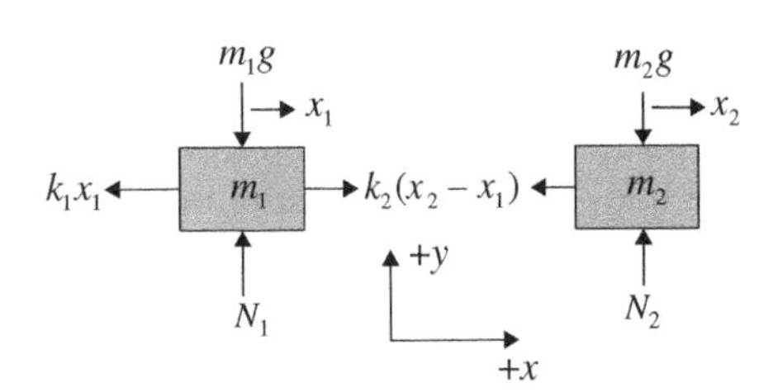

Figure 2.2 Free body diagrams of the two masses in Figure 2.1.

Likewise, the forces acting on mass m_2 result in static equilibrium in the y-direction and in the x-direction result in

$$\sum f_i = m_2\ddot{x}_2 \Rightarrow m_2\ddot{x}_2 = -k_2(x_2 - x_1) = k_2x_1 - k_2x_2$$
$$\Rightarrow m_2\ddot{x}_2 - k_2x_1 + k_2x_2 = 0$$

Combining these two equations of motion into a single matrix equation yields

$$\begin{bmatrix} m_1 & 0 \\ 0 & m_2 \end{bmatrix}\begin{bmatrix} \ddot{x}_1 \\ \ddot{x}_2 \end{bmatrix} + \begin{bmatrix} k_1 + k_2 & -k_2 \\ -k_2 & k_2 \end{bmatrix}\begin{bmatrix} x_1 \\ x_2 \end{bmatrix} = \begin{bmatrix} 0 \\ 0 \end{bmatrix}$$

The Lagrange formulation can be used as an alternative to summing forces and moments for those cases where the free-body diagram is not as obvious. The Lagrange formulation requires identification of the energy in the system, rather than the identification of forces and moments acting on the system, and requires the use of generalized coordinates. A brief working account of the Lagrange formulation is given here. A more precise and detailed account is given in Meirovitch (2001). The procedure begins by assigning a generalized coordinate to each moving part. The standard rectangular coordinate system is an example of a generalized coordinate, but any length, angle or other coordinate that *uniquely* defines the position of the part at any time, forms a generalized coordinate. It is usually desirable to choose coordinates that are independent.

Lagrange's method for conservative systems consists of defining the *Lagrangian*, L, of the system defined by $L = T - U$. Here T is the total kinetic energy of the system and U is the total potential energy in the system, both stated in terms of "generalized" coordinates, denoted "$q_i(t)$". Lagrange's method for conservative systems states that the equations of motion for the free response of an undamped system result from

$$\frac{d}{dt}\left(\frac{\partial L}{\partial \dot{q}_i}\right) - \frac{\partial L}{\partial q_i} = 0,\ i = 1,\ 2,\ 3\ \ldots n \tag{2.12}$$

where n corresponds to the number of degrees of freedom. Substitution of the expression for L into Equation (2.12) yields

$$\frac{d}{dt}\left(\frac{\partial T}{\partial \dot{q}_i}\right) - \frac{\partial T}{\partial q_i} + \frac{\partial U}{\partial q_i} = Q_i \qquad i = 1,\ 2,\ 3,\ \ldots n \tag{2.13}$$

where $\dot{q}_i = \partial q_i / \partial t$ is the generalized velocity and Q_i represents all the nonconservative forces corresponding to q_i. Here $\partial / \partial q_i$ denotes the partial derivative with respect to the coordinate q_i. For conservative systems, $Q_i = 0$ and Equation (2.13) becomes

$$\frac{d}{dt}\left(\frac{\partial T}{\partial \dot{q}_i}\right) - \frac{\partial T}{\partial q_i} + \frac{\partial U}{\partial q_i} = 0 \qquad i = 1,\ 2,\ 3,\ \ldots n \tag{2.14}$$

which represents one equation for each generalized coordinate. This expression allows the equations of motion of complicated systems to be derived without using free-body diagrams and summing forces and moments. The following example illustrates the use of the Lagrange approach to derive the equation of motion of the simple spring mass system of Example 2.2.1.

Example 2.2.2
Use the Lagrangian approach to derive the equations of motion for the system of Figure 2.1.

Solution: The total kinetic energy can be written in terms of the velocities of each mass and is

$$T = \frac{1}{2}m_1\dot{x}_1^2 + \frac{1}{2}m_2\dot{x}_2^2$$

The total potential energy is written in terms of the displacements, which are the generalized coordinates in this case and is

$$U = \frac{1}{2}k_1x_1^2 + \frac{1}{2}k_2(x_1 - x_2)^2$$

Taking the derivatives of T as indicated in Equation (2.14) yields

$$\frac{d}{dt}\left(\frac{\partial T}{\partial x_1}\right) = \frac{d}{dt}(m_1\dot{x}_1) = m_1\ddot{x}_1 \text{ and } \frac{d}{dt}\left(\frac{\partial T}{\partial x_2}\right) = m_2\ddot{x}_2$$

Note that the second term in Equation (2.14) is zero, because T does not depend on displacement in this case. Taking the derivatives of U as indicated in Equation (2.14) yields

$$\frac{\partial U}{\partial x_1} = k_1x_1 + k_2(x_1 - x_2) = (k_1 + k_2)x_1 - k_2x_2$$

and

$$\frac{\partial U}{\partial x_2} = k_2(x_1 - x_2)(-1) = -k_2x_1 + k_2x_2$$

Using these last four expressions to assemble Equation (2.14) for each coordinate x_1 and x_2 yields the two coupled equations

$$m_1\ddot{x}_1 + (k_1 + k_2)x_1 - k_2x_2 = 0 \text{ and } m_2\ddot{x}_2 - k_2x_1 + k_2x_2 = 0$$

These are of course the same set of equations obtained in Example 2.2.1 by summing forces.

Viscous damping is a nonconservative force and may be modeled using the Lagrangian approach by defining the Rayleigh dissipation function. This function assumes that the damping forces are proportional to the velocities. The Rayleigh dissipation function then takes the form

$$F = \frac{1}{2}\sum_{r=1}^{n}\sum_{s=1}^{n} c_{rs}\dot{q}_r\dot{q}_s \tag{2.15}$$

Here the damping coefficients are symmetric, $c_{rs} = c_{sr}$, and n is again the number of generalized coordinates. With this form, the generalized forces for viscous damping can be derived from

$$Q_j = -\frac{\partial F}{\partial \dot{q}_j}, \text{ for each } j = 1, 2, 3, \ldots n \tag{2.16}$$

To derive equations of motion with viscous damping, substitute Equation (2.15) into Equation (2.16) and Equation (2.16) into Equation (2.13).

2.3 Classifications of Systems

This section lists the various classifications of systems modeled by Equation (2.10) commonly found in the literature. Each particular engineering application may have a slightly different nomenclature and jargon. The definitions presented here are meant only to simplify discussion in this text and are intended to conform with most other references. These classifications are useful in verbal communication of the assumptions made when discussing a vibration problem. In the following, each word in italics is defined to imply the assumptions made in modeling the system under consideration.

The phrase *conservative system* usually refers to a system of the form

$$M\ddot{\mathbf{q}}(t) + K\mathbf{q}(t) = \mathbf{f}(t) \tag{2.17}$$

where M and K are both symmetric and positive definite. However, the system

$$M\ddot{\mathbf{q}}(t) + G\dot{\mathbf{q}}(t) + K\mathbf{q}(t) = \mathbf{f}(t) \tag{2.18}$$

where G is skew-symmetric is also conservative, in the sense of conserving energy, but is referred to as a *gyroscopic conservative system*, or an *undamped gyroscopic system*. Such systems arise naturally when spinning motions are present, such as in a gyroscope, rotating machine or spinning satellite.

Systems of the form

$$M\ddot{\mathbf{q}}(t) + C\dot{\mathbf{q}}(t) + K\mathbf{q}(t) = \mathbf{f}(t) \tag{2.19}$$

where M, C and K are all positive definite, are referred to as *damped nongyroscopic systems* and are also considered to be damped conservative systems in some instances, although they certainly do not conserve energy. Systems with symmetric and positive definite coefficient matrices are sometimes referred to as *passive* systems.

Classification of systems with asymmetric coefficients is not as straightforward, as the classification depends on more matrix theory than has yet been presented. However, systems of the form

$$M\ddot{\mathbf{q}}(t) + (K + H)\mathbf{q}(t) = \mathbf{f}(t) \tag{2.20}$$

are referred to as *circulatory systems* (Ziegler, 1968). In addition, systems of the more general form

$$M\ddot{\mathbf{q}}(t) + C\dot{\mathbf{q}}(t) + (K + H)\mathbf{q}(t) = \mathbf{f}(t) \tag{2.21}$$

result from dissipation referred to as *constraint damping* as well as external damping in rotating shafts. Combining all of these effects provides motivation to study the most general system of the form of Equation (2.10), i.e.

$$M\ddot{\mathbf{q}}(t) + (C + G)\dot{\mathbf{q}}(t) + (K + H)\mathbf{q}(t) = \mathbf{f}(t) \tag{2.22}$$

This expression is the most difficult model considered in the first half of the text. To be complete, however, it is appropriate to mention that this model of a structure does not account for time-varying coefficients or nonlinearities, which are sometimes present. Physically the expression represents the most general forces considered in the majority of linear vibration analysis, with the exception of certain external forces. Mathematically, Equation (2.22) will be further classified in terms of the properties of the coefficient matrices.

2.4 Feedback Control Systems

The vibrations of many structures and devices are controlled by sophisticated control methods. Examples of the use of feedback control to remove vibrations range from machine tools to tall buildings and large spacecraft. As discussed in Section 1.9, one popular way to control the vibrations of a structure is to measure the structure's position and velocity vectors and to use that information to drive the system in direct proportion to its positions and velocities. The measurement of the system's states (position and velocity) is denoted by the measurement equation (called output equation)

$$\mathbf{y}(t) = C_p\mathbf{q}(t) + C_v\dot{\mathbf{q}}(t) \tag{2.23}$$

Here, C_p is a matrix defining the location of sensors and associated gains that measure the position, C_v is a matrix defining the location of sensors and associated gains that measure the velocities, and $\mathbf{y}(t)$ is called the output vector. In order to perform active control, actuators are needed and the positions of the actuators are indicated by the matrix B_f. The structure of B_f is determined by actuator locations and the numerical values of the entries of the matrix are weighting factors (usually control gains). The vector $\mathbf{u}(t)$, called the input vector, denotes the control force input to the structure and for closed-loop control is a function of the measurements. In particular, the control force applied to a structure is modeled as $\mathbf{f}_f = B_f\mathbf{u}(t)$. However, $\mathbf{u}(t)$ is a function of the measurements so that $\mathbf{u}(t) = -G_f\mathbf{y}(t)$, where G_f consists of constant feedback gains manipulated to achieve the desired closed-loop response much like choosing the gains g_1 and g_2 in the SDOF case of Equation (1.77). The minus sign is used to help with stability, as the following equations will illustrate. The equations of motion, Equation (2.22), with applied force $\mathbf{f}(t)$ and control forces $\mathbf{f}_f(t)$ become

$$M\ddot{\mathbf{q}}(t) + (G + C)\dot{\mathbf{q}}(t) + (K + H)\mathbf{q}(t) = \mathbf{f}(t) + \mathbf{f}_f(t)$$

Substitution of the values of $\mathbf{f}_f$, $\mathbf{u}(t)$ and $\mathbf{y}(t)$ into this last expression yields

$$M\ddot{\mathbf{q}}(t) + (G + C)\dot{\mathbf{q}}(t) + (K + H)\mathbf{q}(t) = -B_fG_fC_p\mathbf{q}(t) - B_fG_fC_v\dot{\mathbf{q}}(t) + \mathbf{f}(t)$$

The coefficients of the position and velocity on the right-hand side can each be represented as a single matrix

$$M\ddot{\mathbf{q}}(t) + (G + C)\dot{\mathbf{q}}(t) + (K + H)\mathbf{q}(t) = -K_p\mathbf{q}(t) - K_v\dot{\mathbf{q}}(t) + \mathbf{f}(t) \tag{2.24}$$

where $K_p = B_f\, G_f\, C_p$ and $K_v = B_f\, G_f\, C_v$. Equation (2.24) is the vector version of Equation (1.77). Here K_p and K_v are called feedback gain matrices and in terms of analysis these values are determined to give the best response. However, when one considers the hardware, then it is important to understand B_f, G_f, C_p and C_v.

The control system of Equation (2.24) can be rewritten in the form of Equation (2.10) by moving the terms $K_p\mathbf{q}$ and $K_v\dot{\mathbf{q}}$ to the left-hand side of Equation (2.24). Thus, analysis performed on Equation (2.10) will also be useful for studying the vibrations of structures controlled by position and velocity feedback, also called *state feedback*.

Control and system theory (Davidson et al., 2007) are very well developed areas. Most of the work carried out in linear systems has been developed for systems in the *state space* form introduced in Section 1.10 to define equilibrium and in Section 1.11 for numerical integration (Rugh, 1996). The state space form is

$$\begin{aligned}\dot{\mathbf{x}}(t) &= A\mathbf{x}(t) + B\mathbf{u}(t)\\ \mathbf{u}(t) &= -G_c\mathbf{y}(t), \mathbf{y}(t) = C_c\mathbf{x}(t)\end{aligned} \tag{2.25}$$

where $\mathbf{x}(t)$ is called the *state vector*, A is the *state matrix*, B is the *input matrix*, $\mathbf{u}$ is the applied control force, or control vector, $\mathbf{y}(t)$ is the output matrix, C_c is a location and gain matrix and G_c is a gain matrix. Many software routines and theoretical developments exist for systems in the form of Equation (2.25). Equation (2.10) can be written in this form by several very simple transformations. To this end, let $\mathbf{x}_1 = \mathbf{q}$ and $\mathbf{x}_2 = \dot{\mathbf{q}}$, then Equation (2.22) can be written as the two coupled equations

$$\begin{aligned}\dot{\mathbf{x}}_1(t) &= \mathbf{x}_2(t)\\ M\dot{\mathbf{x}}_2(t) &= -(D+G)\mathbf{x}_2(t) - (K+H)\mathbf{x}_1(t) + \mathbf{f}(t)\end{aligned} \tag{2.26}$$

This form allows the theory of control and systems analysis to be directly applied to vibration problems.

Now suppose there exists a matrix, M^{-1}, called the inverse of M, such that $M^{-1}M = I$, the $n \times n$ identity matrix. Then Equation (2.26) can be written as

$$\dot{\mathbf{x}}(t) = \begin{bmatrix} 0 & I \\ -M^{-1}(K+H) & -M^{-1}(D+G)\end{bmatrix}\mathbf{x}(t) + \begin{bmatrix} 0 \\ M^{-1}\end{bmatrix}\mathbf{f}(t) \tag{2.27}$$

where the state matrix A and the input matrix B are

$$A = \begin{bmatrix} 0 & I \\ -M^{-1}(K+H) & -M^{-1}(D+G)\end{bmatrix}, \quad B = \begin{bmatrix} 0 \\ M^{-1}\end{bmatrix} \tag{2.28}$$

and where

$$\mathbf{x} = [\mathbf{x}_1 \quad \mathbf{x}_2]^T = [\mathbf{q} \quad \dot{\mathbf{q}}]^T.$$

Along with the output equation, $\mathbf{y}(t) = C_c\mathbf{x}(t)$ and control law $\mathbf{u}(t) = -G_c\mathbf{y}(t)$ forms the state space model of a closed-loop structure under state feedback indicated in Equation (2.24). The state space approach has made a big impact on the development of control theory and, to a lesser but still significant extent, on vibration theory. This state space representation also forms the approach used for numerical simulation and calculation for vibration analysis. The state space formulation is also used for the numerical solution of ordinary differential equations.

The matrix inverse M^{-1}can be calculated by a number of different numerical methods readily available in most mathematical software packages along with other factorizations. This is discussed in detail in Appendix B. The reader should show that for second-order matrices of the form

$$M = \begin{bmatrix} a & b \\ c & d \end{bmatrix}$$

the inverse is given by

$$M^{-1} = \frac{1}{\det(M)} \begin{bmatrix} d & -b \\ -c & a \end{bmatrix},$$

where $\det(M) = ad - cb$. This points out that if $ad = cb$, then M is called *singular* and M^{-1} does not exist. In general, it should be noted that if a matrix inverse exists, then it is unique. Furthermore, the inverse of a product of square matrices is given by $(AB)^{-1} = B^{-1}A^{-1}$.

The following section of examples illustrates the preceding ideas and notations. Additional useful matrix definitions and concepts are presented in the next chapter and as the need arises.

2.5 Examples

This section lists several examples to illustrate how the various symmetries and asymmetries arise from mechanical devices. The equations of motion have all been derived using either free body diagrams or energy methods. Complete derivations may be found in the references listed or in most texts on dynamics.

Example 2.5.1
The first example (Meirovitch, 1980: p. 37) consists of a rotating ring of negligible mass containing an object of mass m, that is free to move in the plane of rotation, as indicated in Figure 2.3. In the figure, k_1 and k_2 are both positive

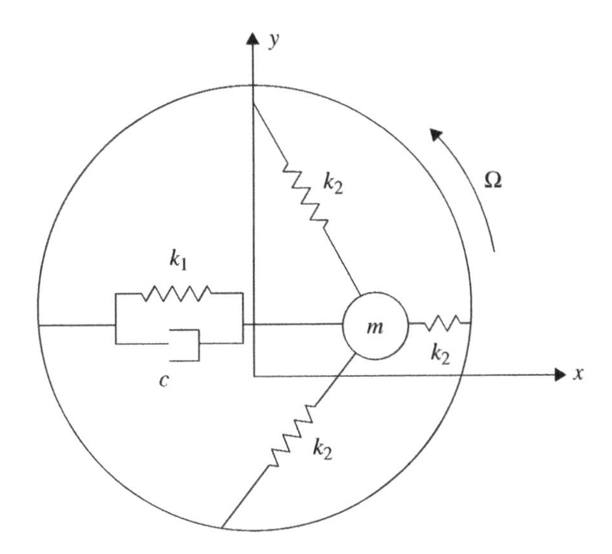

Figure 2.3 Schematic of a simplified model of a spinning satellite.

spring stiffness values, c is a damping rate (also positive), and Ω is the constant angular velocity of the disc. The linearized equations of motion are

$$\begin{bmatrix} m & 0 \\ 0 & m \end{bmatrix} \ddot{\mathbf{q}} + \left\{ \begin{bmatrix} c & 0 \\ 0 & 0 \end{bmatrix} + 2m\Omega \begin{bmatrix} 0 & -1 \\ 1 & 0 \end{bmatrix} \right\} \dot{\mathbf{q}}$$
$$+ \begin{bmatrix} k_1 + k_2 - m\Omega^2 & 0 \\ 0 & 2k_2 - m\Omega^2 \end{bmatrix} \mathbf{q} = \mathbf{0} \tag{2.29}$$

where $\mathbf{q} = [x(t) \;\; y(t)]^T$, the vector of displacements. Here M, C and K are symmetric, while G is skew symmetric, so the system is a damped gyroscopic system.

Note, that for any arbitrary nonzero vector $\mathbf{x}$, the quadratic form associated with M becomes

$$\mathbf{x}^T M \mathbf{x} = [x_1 \;\; x_2] \begin{bmatrix} m & 0 \\ 0 & m \end{bmatrix} \begin{bmatrix} x_1 \\ x_2 \end{bmatrix} = m\left(x_1^2 + x_2^2\right) > 0$$

So, $\mathbf{x}^T M \mathbf{x}$ is positive for all nonzero choices of $\mathbf{x}$, thus the matrix M is (symmetric) positive definite (and nonsingular, meaning that M has an inverse). Likewise, the quadratic form for the damping matrix becomes

$$\mathbf{x}^T \begin{bmatrix} c & 0 \\ 0 & 0 \end{bmatrix} \mathbf{x} = cx_1^2 \geq 0.$$

Note here that while this quadratic form will always be nonnegative, the quantity $\mathbf{x}^T C \mathbf{x} = cx_1^2 = 0$ for the nonzero vector $\mathbf{x} = [0 \;\; 1]^T$, so that C is only positive semidefinite (and singular). Other methods for checking the definiteness of a matrix are given in the next chapter.

The matrix G for the preceding system is obviously skew-symmetric. It is interesting to calculate its quadratic form and note that for any real value of $\mathbf{x}$

$$\mathbf{x}^T G \mathbf{x} = 2m\Omega(x_1 x_2 - x_2 x_1) = 0 \tag{2.30}$$

This is true in general. The quadratic form of any order real skew-symmetric matrix is always zero.

Example 2.5.2

A gyroscope is an instrument based on using gyroscopic forces to sense motion and is commonly used for navigation as well as in other applications. One model of a gyroscope is shown in Figure 2.4. This is a three-dimensional device

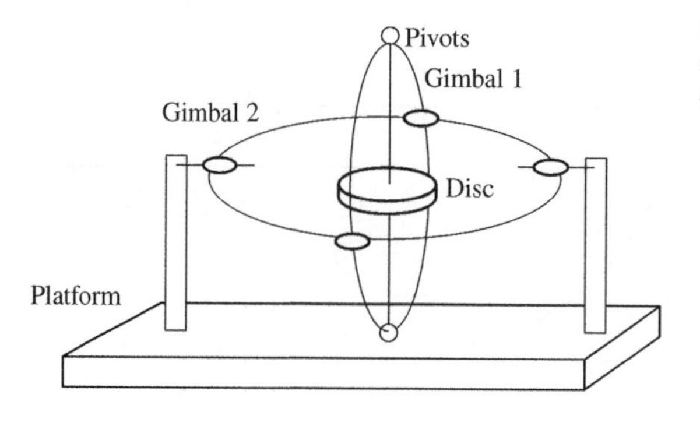

Figure 2.4 Schematic of a simplified model of a gyroscope.

consisting of a rotating disc (with electric motor), two gimbals (hoops) and a platform, all connected by pivots or joints. The disc and the two gimbals each have three moments of inertia – one around each of the principal axes of reference. There is also a stiffness associated with each pivot. Let:

- a, b and c be the moments of inertia of the disc (rotor);
- a_i, b_i and c_i, for $i = 1, 2$ be the principal moments of inertia of the two gimbals;
- k_{11} and k_{12} be the torsional stiffness elements connecting the drive shaft to the first and second gimbal, respectively,
- k_{21} and k_{22} be the torsional stiffness elements connecting the rotor to the first and second gimbal, respectively; and
- Ω denote the constant rotor speed.

The equations of motion are then given by Burdess and Fox (1978) to be

$$\begin{bmatrix} A+a_1 & 0 \\ 0 & B+b_1 \end{bmatrix} \ddot{\mathbf{q}} + \Omega(a+b-c) \begin{bmatrix} 0 & -1 \\ 1 & 0 \end{bmatrix} \dot{\mathbf{q}}$$
$$+ \begin{bmatrix} k_{11}+k_{22}+2\Omega^2(c-b+c_1-b_1) & 0 \\ 0 & k_{12}+k_{21}+2\Omega^2(c-a+c_2-b_2) \end{bmatrix} \mathbf{q} = \mathbf{0} \tag{2.31}$$

where $\mathbf{q}$ is the displacement vector of the rotor.

Here we note that M is symmetric and positive definite, C and H are zero, G is nonzero skew-symmetric, and the stiffness matrix K will be positive definite if $(c - b + c_1 - b_1)$ and $(c - a + c_2 - b_2)$ are both positive. This is a conservative gyroscopic system.

Example 2.5.3

A lumped parameter version of the rod illustrated in Figure 2.5 yields another example of a system with asymmetric matrix coefficients. The rod is called Pflüger's Rod, and its equation of motion and the lumped parameter version of it used here can be found in Huseyin (1978) or by using the methods of Chapter 11. The equations of motion are given by the vector equation

$$\frac{m}{2}\begin{bmatrix} 1 & 0 \\ 0 & 1 \end{bmatrix} \ddot{\mathbf{q}} + \left\{ \frac{EI\pi^4}{l^3} \begin{bmatrix} \frac{1}{2} & 0 \\ 0 & 8 \end{bmatrix} - \eta \begin{bmatrix} \frac{\pi^2}{4} & \frac{32}{9} \\ \frac{8}{9} & \pi^2 \end{bmatrix} \right\} \mathbf{q} = \mathbf{0} \tag{2.32}$$

where η is the magnitude of the applied force, EI is the flexural rigidity, m is the mass density, l is the length of the rod, and $\mathbf{q}(t) = [x_1(t)\ x_2(t)]^T$ represents the displacements of two points on the rod.

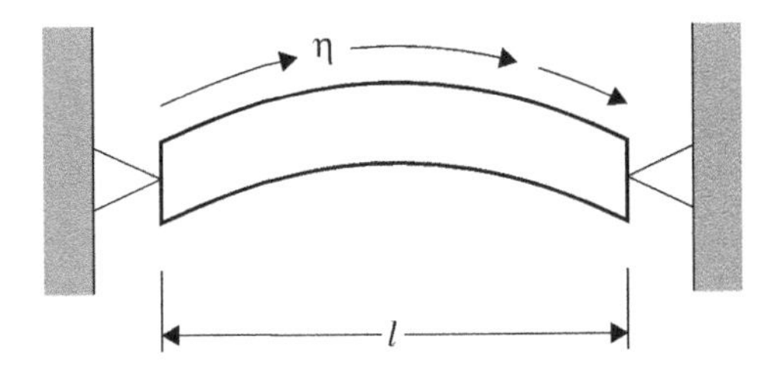

Figure 2.5 Pflüger's Rod: a simply supported bar subjected to uniformly distributed tangential forces.

Again note that the mass matrix is symmetric and positive definite. However, due to the presence of the so-called follower force, η, the coefficient of $\mathbf{q}(t)$ is not symmetric. Using Equations (2.8) and (2.9), the stiffness matrix K becomes

$$K = \begin{bmatrix} \dfrac{EI\pi^4}{2l^3} - \dfrac{\pi}{4}\eta & -\dfrac{20}{9}\eta \\ -\dfrac{20}{9}\eta & \dfrac{8EI\pi^4}{l^3} - \eta\pi^2 \end{bmatrix}$$

and the skew-symmetric matrix H becomes

$$H = \frac{12\eta}{9}\begin{bmatrix} 0 & -1 \\ 1 & 0 \end{bmatrix}$$

Example 2.5.4

As an example of the type of matrices that can result from feedback control systems, consider the two-degree of freedom system of Figure 2.6 and its free body diagram given in Figure 2.7. Here, f_2 indicates a control force applied to m_2. From Example 2.2.1, we know that the sum of forces in the y-direction yield static equilibrium, because the masses are assumed to be sliding on a friction free, horizontal surface.

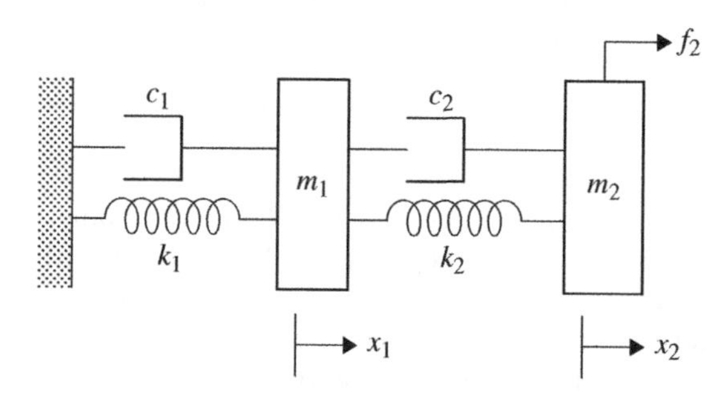

Figure 2.6 Schematic of a two degree of freedom system with applied control force f_2. As in Figure 2.1, the masses are assumed to be moving on a horizontal friction-free surface.

Solution: Summing forces in the x-direction on mass m_1 in Figure 2.7 yields

$$\begin{aligned} \sum f_i = m_1\ddot{x}_1 \Rightarrow m_1\ddot{x}_1 &= -k_1x_1 + k_2(x_2 - x_1) - c_1\dot{x}_1 + c_2(\dot{x}_2 - \dot{x}_1) \\ &\Rightarrow m_1\ddot{x}_1 + (c_1 + c_2)\dot{x}_1 - c_2\dot{x}_2 + (k_1 + k_2)x_1 - k_2x_2 = 0 \end{aligned} \tag{2.33}$$

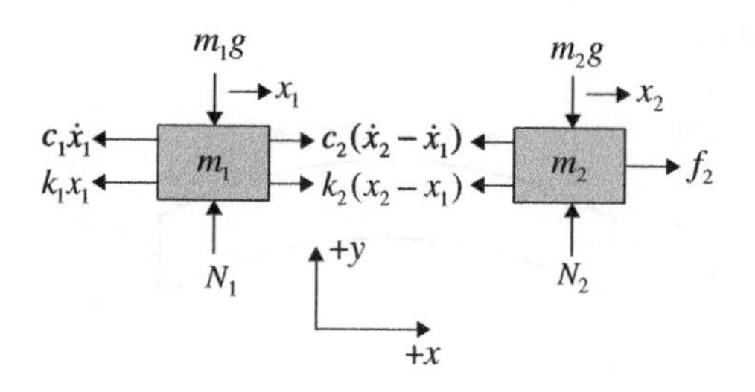

Figure 2.7 Free-body diagram of the system in Figure 2.6.

Summing forces in the x-direction on mass m_2 in Figure 2.7 yields

$$\begin{aligned}\sum f_i = m_2\ddot{x}_2 \Rightarrow m_2\ddot{x}_2 &= -c_2(\dot{x}_2 - \dot{x}_1) - k_2(x_2 - x_1)\\ &= c_2\dot{x}_1 - c_2\dot{x}_2 + k_2x_1 - k_2x_2\\ &\Rightarrow m_2\ddot{x}_2 - c_2\dot{x}_1 + c_2\dot{x}_2 - k_2x_1 + k_2x_2 = 0\end{aligned} \tag{2.34}$$

Writing Equations (2.33) and (2.34) together in matrix form yields

$$\begin{bmatrix} m_1 & 0 \\ 0 & m_2 \end{bmatrix}\ddot{\mathbf{q}} + \begin{bmatrix} c_1 + c_2 & -c_2 \\ -c_2 & c_2 \end{bmatrix}\dot{\mathbf{q}} + \begin{bmatrix} k_1 + k_2 & -k_2 \\ -k_2 & k_2 \end{bmatrix}\mathbf{q} = \begin{bmatrix} 0 \\ f_2 \end{bmatrix} \tag{2.35}$$

where

$$\mathbf{q} = [x_1(t) \quad x_2(t)]^T$$

Next consider the control force f_2 applied to the mass m_2. In order to represent feedback control, the control force must be a function of a measurement. Suppose sensors are places so that both positions x_1 and x_2 are measured. However, in this case, the control system electronics are such that the signal from m_1 is fed to m_2 so that C_p is fully populated. Then the output matrix of Equation (2.23) becomes

$$C_p = \begin{bmatrix} c_{p1} & c_{p2} \\ c_{p3} & c_{p4} \end{bmatrix}, \quad \text{so that } \mathbf{y}(t) = \begin{bmatrix} c_{p1} & c_{p2} \\ c_{p3} & c_{p4} \end{bmatrix}\begin{bmatrix} x_1 \\ x_2 \end{bmatrix}$$

If C_p is diagonal, the feedback from the sensor at m_1 is cancelled. Note that forming a fully populated C_p matrix requires some hardware to implement. Because no velocity is sensed, $C_v = 0$. Since a control force is only applied to mass m_2, the input matrix is defined by

$$B_f = \begin{bmatrix} 0 & 0 \\ 0 & b_2 \end{bmatrix}, \quad \text{so that } \mathbf{f}_f(t) = B_f\mathbf{u} = \begin{bmatrix} 0 & 0 \\ 0 & b_2 \end{bmatrix}\begin{bmatrix} u_1 \\ u_2 \end{bmatrix}$$

and $\mathbf{u}(t) = G_f\mathbf{y}(t)$ is defined by the gain matrix

$$G_f = \begin{bmatrix} g_1 & 0 \\ 0 & g_2 \end{bmatrix}$$

From Equation (2.24), for position feedback, the equations of motion are altered by the product

$$B_fG_pC_p = \begin{bmatrix} 0 & 0 \\ \underbrace{c_3b_2g_2}_{k_{p1}} & \underbrace{c_2b_2g_2}_{k_{p2}} \end{bmatrix}$$

where k_{p1} and k_{p2} are constant gains. The closed-loop Equation (2.24) becomes

$$\begin{bmatrix} m_1 & 0 \\ 0 & m_2 \end{bmatrix}\ddot{\mathbf{q}} + \begin{bmatrix} c_1 + c_2 & -c_2 \\ -c_2 & c_2 \end{bmatrix}\dot{\mathbf{q}} + \begin{bmatrix} k_1 + k_2 & -k_2 \\ -k_2 + k_{p1} & k_2 + k_{p2} \end{bmatrix}\mathbf{q} = \begin{bmatrix} 0 \\ 0 \end{bmatrix} \tag{2.36}$$

Equation (2.36) is analogous to Equation (1.77) for an SDOF system.

The displacement coefficient matrix is no longer symmetric due to the feedback gain constant k_{p1}. Since only x_1 and x_2 are used in the control, this is called *position feedback*. Velocity feedback could result in the damping matrix becoming asymmetric as well. Without the control this is a damped symmetric system or non-conservative system. With position and/or velocity feedback, the coefficient matrices may become asymmetric, greatly changing the nature of the response and, as discussed in Chapter 4, the system's stability.

These examples are referred to in the remaining chapters of this book, which develops theories to test, analyze and control such systems.

2.6 Experimental Models

Many structures are not configured in nice lumped arrangements, as in Examples 2.5.1, 2.5.2 and 2.5.4. Instead, they appear as distributed parameter arrangements (Chapter 7), such as the rod of Figure 2.5. However, lumped parameter MDOF models can be assigned to such structures on an experimental basis. As an example, a simple beam may be experimentally analyzed for the purpose of obtaining an analytical model of the structure by measuring the displacement at one end due to a harmonic excitation ($\sin \omega t$) at the other end and sweeping through a wide range of driving frequencies, ω. Using the ideas of Section 1.5, a magnitude-versus-frequency similar to Figure 2.8 may result. Because of the three very distinct peaks in Figure 2.8, one is tempted to model the structure as a three-degree of freedom system (corresponding to the three resonances). In fact, if each peak is thought of as an SDOF system, using the formulations of Section 1.7 yields a value for m_i, k_i and c_i (or ω_i and ζ_i) for each of the three peaks ($i = 1$, 2 and 3). A reasonable model for the system *might* then be

$$\begin{bmatrix} m_1 & 0 & 0 \\ 0 & m_2 & 0 \\ 0 & 0 & m_3 \end{bmatrix} \ddot{\mathbf{q}}(t) + \begin{bmatrix} c_1 & 0 & 0 \\ 0 & c_2 & 0 \\ 0 & 0 & c_3 \end{bmatrix} \dot{\mathbf{q}}(t) + \begin{bmatrix} k_1 & 0 & 0 \\ 0 & k_2 & 0 \\ 0 & 0 & k_3 \end{bmatrix} \mathbf{q}(t) = \mathbf{0} \quad (2.37)$$

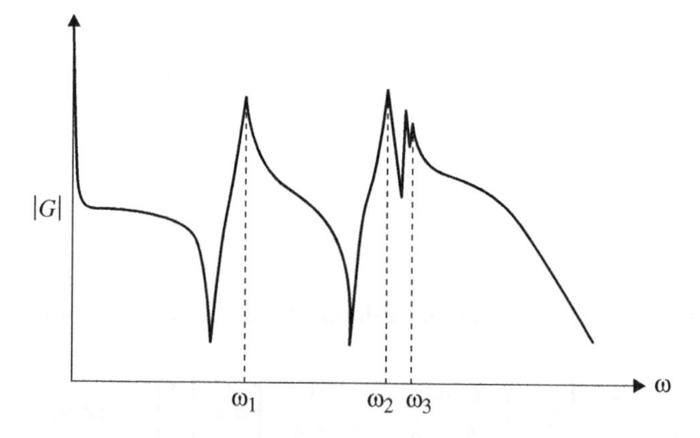

Figure 2.8 Experimentally obtained magnitude versus frequency plot for a simple beam-like structure.

which is referred to as a *physical model.* Alternatively, values of ω_i and ζ_i could be used to model the structure by the equation

$$\begin{bmatrix} 1 & 0 & 0 \\ 0 & 1 & 0 \\ 0 & 0 & 1 \end{bmatrix} \ddot{\mathbf{r}} + \begin{bmatrix} 2\zeta_1\omega_1 & 0 & 0 \\ 0 & 2\zeta_2\omega_2 & 0 \\ 0 & 0 & 2\zeta_3\omega_3 \end{bmatrix} \dot{\mathbf{r}} + \begin{bmatrix} \omega_1^2 & 0 & 0 \\ 0 & \omega_2^2 & 0 \\ 0 & 0 & \omega_3^2 \end{bmatrix} \mathbf{r} = \mathbf{0} \tag{2.38}$$

which is referred to as a *modal model.* The problem with each of these models is that it is not clear what physical motion to assign to the coordinates $q_i(t)$ or $r_i(t)$. In addition, as discussed in Chapter 12, it is not always clear that each peak corresponds to a single resonance (however, phase plots of the experimental transfer function can help).

Modal models, however, are useful for discussing the vibrational responses of the structure and will be considered in more detail in Chapter 12. The point in introducing this model here is to note that experimental methods can result in viable analytical models of structures directly and that these models are fundamentally based on the phenomenon of resonance.

2.7 Nonlinear Models and Equilibrium

If one of the springs in Equation (2.17) is stretched beyond its linear region, then Newton's law would result in an MDOF system with nonlinear terms. For such systems, the equations of motion become coupled nonlinear equations instead of coupled linear equations. The description of the nonlinear equations of motion can still be written in vector form, but this does not result in matrix coefficients so that linear algebra does not help. Rather the equations of motion are written in the state space form of Equation (1.82) repeated here

$$\dot{\mathbf{x}} = \mathbf{F}(\mathbf{x}) \tag{2.39}$$

As in Section 1.10, Equation (2.39) is used to define the equilibrium position of the system. Unlike the linear counter part, there will be multiple equilibrium positions defined by solutions to the nonlinear algebraic Equation (1.83)

$$\mathbf{F}(\mathbf{x}_e) = \mathbf{0} \tag{2.40}$$

The existence of these multiple equilibrium solutions forms the first basic difference between linear and nonlinear systems. In addition to being useful for defining equilibria, Equation (2.39) is also useful for numerically simulating the response of a nonlinear system with multiple degrees of freedom.

If one of the springs is nonlinear or if a damping element is nonlinear, the stiffness and/or damping terms can no longer be factored into a matrix times a vector, but must be left in state space form. Instead of Equation (2.19), the form of the equations of motion can only be written as

$$M\ddot{\mathbf{q}} + \mathbf{G}(\mathbf{q}, \dot{\mathbf{q}}) = \mathbf{0} \tag{2.41}$$

where $\mathbf{G}$ is some nonlinear vector function of the displacement and velocity vectors. As in the SDOF case discussed in Section 1.10, it is useful to place the system in Equation (2.41) into state space form by defining new coordinates corresponding to the position and velocity. To that end let $\mathbf{x}_1 = \mathbf{q}$ and $\mathbf{x}_2 = \dot{\mathbf{q}}$ and multiply the above by M^{-1}. Then the equation of motion for the nonlinear system of Equation (2.41) becomes

$$\dot{\mathbf{x}} = \mathbf{F}(\mathbf{x}). \tag{2.42}$$

Here

$$\mathbf{F}(\mathbf{x}) = \begin{bmatrix} \mathbf{x}_2 \\ -M^{-1}\mathbf{G}(\mathbf{x}_1, \mathbf{x}_2) \end{bmatrix} \tag{2.43}$$

and the $2n \times 1$ state vector $\mathbf{x}(t)$ is

$$\mathbf{x}(t) = \begin{bmatrix} \mathbf{q} \\ \dot{\mathbf{q}} \end{bmatrix} = \begin{bmatrix} \mathbf{x}_1 \\ \mathbf{x}_2 \end{bmatrix} \tag{2.44}$$

Applying the definition given in Equation (1.83), equilibrium is defined as the vector $\mathbf{x}_e$ satisfying $\mathbf{F}(\mathbf{x}_e) = \mathbf{0}$. The solution yields the various equilibrium points for the system.

Example 2.7.1
Compute the equilibrium positions for the linear system of Equation (2.25).

Solution: Equation (2.25) is of the form

$$\dot{\mathbf{x}} = A\mathbf{x} + \begin{bmatrix} 0 \\ M^{-1} \end{bmatrix} \mathbf{f}.$$

Equilibrium is concerned with the free response. Thus, set $\mathbf{f} = \mathbf{0}$ in this last expression and the equilibrium condition becomes $A\mathbf{x} = \mathbf{0}$. As long as the matrix A has an inverse, $A\mathbf{x} = \mathbf{0}$ implies that the equilibrium position is defined by $\mathbf{x}_e = \mathbf{0}$. This is the origin with zero velocity: $\mathbf{x}_1 = \mathbf{0}$ and $\mathbf{x}_2 = \mathbf{0}$, or $\mathbf{x} = \mathbf{0}$ and $\dot{\mathbf{x}} = \mathbf{0}$. Physically this condition is the rest position for each mass.

Much analysis and theory of nonlinear systems focuses on SDOF systems. Numerical simulation is used extensively in trying to understand the behavior of a multi-degree of freedom (MDOF) nonlinear systems. Here we present a simple example of a two degree of freedom nonlinear system and compute its equlibria. This is just a quick introduction to nonlinear MDOF and the references should be consulted for a more detailed understanding.

Example 2.7.2
Consider the two degree of freedom system of Figure 2.6, where spring $k_1(q_1)$ is driven into its nonlinear region, so that $k_1(q_1) = k_1 q_1 - \beta q_1^3$ and the force and dampers are set to zero. For convenience, let $m_1 = m_2 = 1$. Note the coordinates are relabeled q_i to be consistent with the state space coordinates. Determine the equilibrium points.

The equations of motion become

$$m_1\ddot{q}_1 = k_2(q_2 - q_1) - k_1 q_1 + \beta q_1^3$$
$$m_2\ddot{q}_2 = -k_2(q_2 - q_1)$$

Next define the state variables by $x_1 = q_1, x_2 = \dot{x}_1 = \dot{q}_1, x_3 = q_2$ and $x_4 = \dot{x}_3 = \dot{q}_2$. Then the equations of motion in first order can be written (for $m_1 = m_2 = 1$) as

$$\begin{aligned} \dot{x}_1 &= x_2 \\ \dot{x}_2 &= k_2(x_3 - x_1) - k_1 x_1 + \beta x_1^3 \\ \dot{x}_3 &= x_4 \\ \dot{x}_4 &= -k_2(x_3 - x_1) \end{aligned}$$

In vector form this becomes

$$\dot{\mathbf{x}} = \mathbf{F}(\mathbf{x}) = \begin{bmatrix} x_2 \\ k_2(x_3 - x_1) - k_1 x_1 + \beta x_1^3 \\ x_4 \\ -k_2(x_3 - x_1) \end{bmatrix}$$

Setting $\mathbf{F}(\mathbf{x}) = \mathbf{0}$ yields the equilibrium equations

$$\mathbf{F}(\mathbf{x}_e) = \begin{bmatrix} x_2 \\ k_2(x_3 - x_1) - k_1 x_1 + \beta x_1^3 \\ x_4 \\ -k_2(x_3 - x_1) \end{bmatrix} = \begin{bmatrix} 0 \\ 0 \\ 0 \\ 0 \end{bmatrix}$$

This is four algebraic equations in four unknowns. Solving yields the three equilibrium points

$$\mathbf{x}_e = \begin{bmatrix} 0 \\ 0 \\ 0 \\ 0 \end{bmatrix}, \begin{bmatrix} \sqrt{k_1/\beta} \\ 0 \\ \sqrt{k_1/\beta} \\ 0 \end{bmatrix}, \begin{bmatrix} -\sqrt{k_1/\beta} \\ 0 \\ -\sqrt{k_1/\beta} \\ 0 \end{bmatrix}$$

The first equilibrium vector corresponds to the linear system.

Chapter Notes

The material in Section 2.1 can be found in any text on matrices or linear algebra. The classification of vibrating systems is discussed in Huseyin (1978), which also contains an excellent introduction to matrices and vectors. The use of velocity and position feedback, as discussed in Section 2.4, is common in the literature but is usually not discussed for MDOF mechanical systems in control texts. The experimental models of Section 2.6 are discussed in Ewins (2000). Influence methods are discussed in more detail in texts on structural dynamics, such as Clough and Penzien (1975).

As mentioned, there are several approaches to deriving the equation of vibration of a mechanical structure, as indicated by the references in Section 2.1. Many texts on dynamics and modeling are devoted to the topic of deriving equations of motion (Meirovitch, 1986). The interest here is in analyzing these models.

References

Burdess, J. S. and Fox, S. H. J. (1978) The dynamic of a Multigimbal Hooke's joint gyroscope. *Journal of Mechanical Engineering Science*, 20(5) 254–262.

Clough, R. W. and Penzien, J. (1975) *Dynamics of Structures*. McGraw-Hill, New York.

Davison, D. E., Chen, J., Ploen, S. R. and Bernstein, D. S. (2007) What is your favorite book on classical control? *IEEE Control Systems Magazine*, 27, 89–94.

Ewins, D. J. (2000) *Modal Testing: Theory and Practice*, 2nd Edition. Research Studies Press, Hertfordshire, UK.

Friswell, M. I. and Motershead, J. E. (1995) *Finite Element Updating in Structural Dynamics*. Kluwer Academic Publications, Dordrecht, The Netherlands.

Goldstein, H. (2002) *Classical Mechanics*, 3rd Edition. Prentice Hall, Upper Saddle River, NJ.

Hughes, T. R. J. (2000) *The Finite Element Method, Linear Static and Dynamic Finite Element Analysis*. Dover Publications, Mineloa, New York.

Huseyin, K. (1978) *Vibrations and Stability of Multiple Parameter Systems*. Sijthoff and Noordhoff International Publishers, Alphen aan der Rijn.

Rugh, W. J. (1996) *Linear System Theory*, 2nd Edition. Prentice Hall, Upper Saddle River, NJ.

Meirovitch, L. (1980) *Computational Methods in Structural Dynamics*. Sijthoff and Noordhoff International Publishers, Alphen aan der Rijn.

Meirovitch, L. (1986) *Elements of Vibration Analysis*, 2nd Edition. McGraw-Hill, New York.

Meirovitch, L. (2001) *Fundamentals of Vibration*. McGraw-Hill, New York.

Rosenberg, R. C. and Karnopp, D. C. (1983) *Introduction to Physical System Dynamics*. McGraw-Hill, New York.

Shearer, J. L., Murphy, A. T. and Richardson, H. H. (1967) *Introduction to System Dynamics*. Addison-Wesley, Reading, MA.

Soom, A. and Kim, C. (1983) Roughness induced dynamic loading at dry and boundary-lubricated sliding contacts. *ASME Journal of Lubrication Technology*, 105(4), 514–517.

Ziegler, H. (1968) *Principles of Structural Stability*. Blaisdell Publishing Company, New York.

Problems

For Problems 2.1 to 2.5, consider the system described by

$$\begin{bmatrix} 3 & 0 \\ 0 & 1 \end{bmatrix} \ddot{\mathbf{q}} + \begin{bmatrix} 6 & 2 \\ 0 & 2 \end{bmatrix} \dot{\mathbf{q}} + \begin{bmatrix} 3 & -2 \\ 2 & -1 \end{bmatrix} \mathbf{q} = \mathbf{0}$$

2.1 Identify the matrices M, C, G, K and H.

2.2 Which of these matrices are positive definite and why?

2.3 Write the preceding equations in the form $\dot{\mathbf{x}} = A\mathbf{x}$, where $\mathbf{x}$ is a vector of four elements given by

$$\mathbf{x} = [\mathbf{q} \quad \dot{\mathbf{q}}]^T.$$

2.4 Calculate the definiteness of M, C and K from Problem 2.1, as well as the values of $\mathbf{x}^T G\mathbf{x}$ and $\mathbf{x}^T H\mathbf{x}$ for an arbitrary value of $\mathbf{x}$.

2.5 Calculate M^{-1}, C^{-1} and K^{-1}, as well as the inverse of $C + G$ and $K + H$ from Problem 2.1 and illustrate that they are, in fact, inverses.

2.6 Derive the equations of motion of the system of Figure 2.9 using the Lagrange's equation.

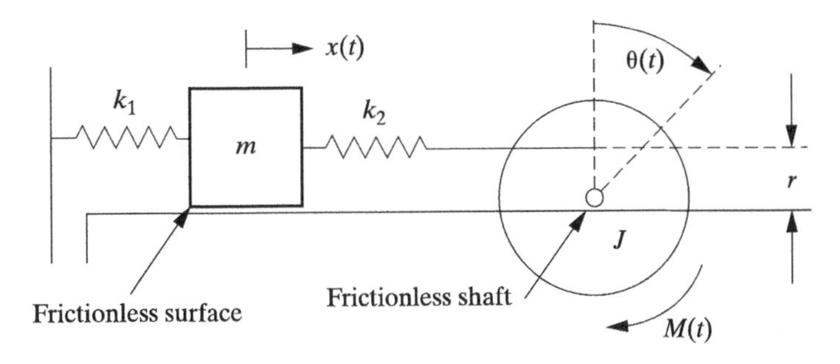

Figure 2.9 Vibration model of a simple machine part. The quantity $M(t)$ denotes an applied moment. The disk rotates without translation.

2.7 Consider the wing vibration model of Figure 2.10. Using the vertical motion of the point of attachment of the springs, $x(t)$, and the rotation of this point, $\theta(t)$, determine the equations of motion using Lagrange's method. Use the small-angle approximation and write the equations in matrix form. Note that G denotes the center of mass and e denotes the distance between the point of rotation and the center of mass. Ignore the gravitational force.

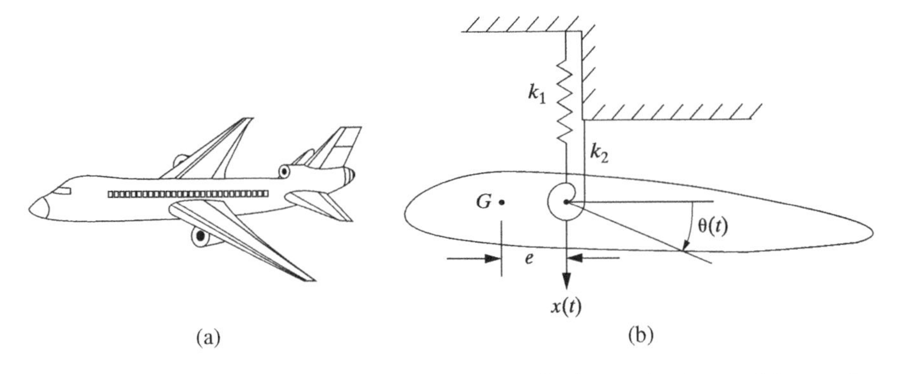

Figure 2.10 An airplane in flight (a) presents a number of different vibration models, one of which is given in part (b). In (b) a vibration model of a wing in flight is sketched, which accounts for bending and torsional motion by modeling the wing as attached to ground (the aircraft body in this case) through a linear spring k_1 and a torsional spring k_2.

2.8 Discuss the definiteness of the matrix K in Example 2.5.3.

2.9 A and B are two real square matrices. Show by example that there exists a matrix A and B, such that $AB \neq BA$. State some conditions on A and B for which $AB = BA$.

2.10 Show that the ij^{th} element of the matrix C, where $C = AB$, the product of the matrix A with the matrix B, is the inner product of the vector consisting of the ith row of the matrix A and the vector consisting of the jth column of the matrix B.

2.11 Calculate the solution of Equation (2.37) to the initial conditions given by $\mathbf{q}^T(0) = \mathbf{0}$ and $\dot{\mathbf{q}}(0) = [0 \;\; 1 \;\; 0]^T$.

2.12 **a)** Calculate the equation of motion in matrix form for the system of Figure 2.6 if the force applied at $f_1 = -g_1x_2 - g_2\dot{x}_2$ and $f_2 = -g_3x_1 - g_4\dot{x}_1$.
b) Calculate f_1 and f_2 so that the resulting closed loop system is diagonal (decoupled).

2.13 Show that if A and B are any two real square matrices, then $(A + B)^T = A^T + B^T$.

2.14 Show, by using the definition in Equation (2.4), that if $\mathbf{x}$ is a real vector and a is any real scalar, then $(a\mathbf{x})^T = a\mathbf{x}^T$.

2.15 Using the definition of the matrix product, show that $(AB)^T = B^TA^T$.

2.16 Show that Equation (2.19) can be written in symmetric first order from $A\dot{\mathbf{x}} + B\mathbf{x} = \mathbf{F}$, where $\mathbf{x} = [\mathbf{q}^T \;\; \dot{\mathbf{q}}^T]^T$, $\mathbf{F} = [\mathbf{f}^T \;\; \mathbf{0}]^T$ and A and B are symmetric.

2.17 Compute the equilibrium positions for the following system of Example 2.7.2 for the case that the masses m_1 and m_2 are arbitrary and not equal.

2.18 Calculate the equilibrium position for the nonlinear system defined by

$$\ddot{x} + x - \beta^2x^3 = 0$$

3

Matrices and the Free Response

3.1 Introduction

As illustrated in Chapter 1, the nature of the free response of a single degree of freedom (SDOF) system is determined by the roots of the characteristic equation (Equation (1.8)). In addition, the exact solution is calculated using these roots. A similar situation exists for the multiple degree of freedom (MDOF) systems described in the previous chapter. Motivated by the SDOF system, this chapter examines the problem of characteristic roots for systems in matrix notation and extends many of the ideas discussed in Chapter 1 to the MDOF systems described in Chapter 2. The mathematical tools needed to extend the ideas of Chapter 1 are those of linear algebra (and in particular matrix theory), which are introduced here in an informal way, as needed.

Chapter 2 illustrated that many types of mechanical systems can be characterized by vector differential equations with matrix coefficients. Just as the nature of the scalar coefficients in the SDOF case determines the form of the response, the nature of the matrix coefficients determines the form of the response of MDOF systems.

In fact, if we attempt to follow the method of solving SDOF vibration problems in solving MDOF systems, we are led immediately to a standard matrix problem called the algebraic eigenvalue problem. This chapter introduces the matrix eigenvalue problem and applies it to the MDOF vibration problems introduced in Chapter 2. The eigenvalues and eigenvectors can be used to determine the time response to initial conditions by the process called modal analysis, which is introduced here. The use of high-level codes such as MATLAB is introduced to compute mode shapes and natural frequencies. The chapter concludes with simulation of the time response to initial condition disturbances using numerical integration as an alternative to modal analysis.

3.2 Eigenvalues and Eigenvectors

This section introduces topics from linear algebra and the matrix eigenvalue problem needed to study the vibrations of MDOF systems. Consider first the simple

Vibration with Control, Second Edition. Daniel John Inman.

Companion Website: www.wiley.com/go/inmanvibrationcontrol2e

conservative vibration problem of Equation (2.17) repeated here

$$M\ddot{\mathbf{q}} + K\mathbf{q} = \mathbf{0}$$

for the free response case that $\mathbf{f}(t) = \mathbf{0}$. Since M is assumed to be positive definite, it has an inverse. Premultiplying the equation of motion by the matrix M^{-1} yields the following equation for the free response

$$\ddot{\mathbf{q}} + M^{-1}K\mathbf{q} = \mathbf{0}$$

Following the mathematical approach of Section 1.2 and the physical notion that the solution should oscillate suggests that a solution may exist of the form of nonzero constant $\mathbf{u}$, in this case a vector, times the exponential $e^{\mu jt}$, i.e. $\mathbf{q}(t) = \mathbf{u}e^{\mu jt}$. Substitution of this expression into the preceding equation yields

$$-\mu^2\mathbf{u} + A\mathbf{u} = \mathbf{0}, \mathbf{u} \neq \mathbf{0}$$

where $A = M^{-1}K$. Rearrangement of this expression yields the equation

$$A\mathbf{u} = \lambda\mathbf{u}, \ \mathbf{u} \neq \mathbf{0}$$

where $\lambda = \mu^2$ and $\mathbf{u}$ cannot be zero. This expression is exactly a statement of the matrix eigenvalue problem. As in the case of the SDOF system, the constants $\lambda = \mu^2$ characterize the natural frequencies of the system. With this as a motivation, the matrix eigenvalue problem is described in detail in this section and applied to the linear vibration problem in Section 3.3. Computational considerations are discussed in Section 3.7.

Square matrices can be characterized by their eigenvalues and eigenvectors, defined in this section. Let A denote an $n \times n$ square matrix. The scalar λ is defined to be an *eigenvalue* of the matrix A with corresponding *eigenvector* $\mathbf{u}$, which must be nonzero, if λ and $\mathbf{u}$ satisfy the equation

$$A\mathbf{u} = \lambda\mathbf{u}, \quad \mathbf{u} \neq \mathbf{0} \tag{3.1}$$

Geometrically, this means that the action of the matrix A on the vector $\mathbf{u}$ only changes the length of the vector $\mathbf{u}$ and does not change its direction or orientation in space. Physically, the eigenvalue λ will yield information about the natural frequencies of the system described by the matrix A. It should be noted that if the vector $\mathbf{u}$ is an eigenvector of A, then so is the vector $\alpha\mathbf{u}$, where α is any scalar. Thus the magnitude of an eigenvector is arbitrary.

A rearrangement of Equation (3.1) yields

$$(\mathrm{A} - \lambda I)\mathbf{u} = \mathbf{0} \tag{3.2}$$

where I is the $n \times n$ identity matrix. Since $\mathbf{u}$ cannot be zero, by the definition of an eigenvector, the inverse of the matrix $(A - \lambda I)$ must not exist. That is, there cannot exist a matrix $(A - \lambda I)^{-1}$ such that $(A - \lambda I)^{-1}(A - \lambda I) = I$. Otherwise, premultiplying Equation (3.2) by this inverse would yield that the only solution to Equation (3.2) is $\mathbf{u} = \mathbf{0}$, violating the definition of an eigenvector. Matrices that do not have inverses are said to be *singular*, and those that do have an inverse are called *nonsingular*.

Whether or not a matrix is singular can also be determined by examining the *determinant* of the matrix. The determinant of an $n \times n$ matrix A is defined and denoted by

$$\det A = |A| = \sum_{s=1}^{n} a_{rs}|A_{rs}| \tag{3.3}$$

for any fixed r, where a_{rs} is the element of A at the intersection of the rth row and s^{th} column of A and $|A_{rs}|$ is the determinant of the matrix formed from A by striking out the rth row and s^{th} column multiplied by $(-1)^{r+s}$. An illustration of this for $n = 2$ is given in Section 2.4. The value of the determinant of a matrix is a unique scalar. In addition, it is a simple matter to show that

$$|A| = |A^T| \tag{3.4}$$

$$|AB| = |A||B| \tag{3.5}$$

Whether or not the determinant of a matrix is zero is significant and useful. The following five statements are entirely equivalent:

1) A is nonsingular
2) A^{-1} exists
3) $\det A \neq 0$
4) The only solution of the equation $A\mathbf{u} = \mathbf{0}$ is $\mathbf{u} = \mathbf{0}$
5) Zero is not an eigenvalue of A.

Note that if $\det(A) = 0$, then A^{-1} does not exist, A is singular and $A\mathbf{u} = \mathbf{0}$ has a nontrivial solution, i.e. zero is an eigenvalue of A.

Example 3.2.1

Calculate The determinant of the matrix A given by

$$A = \begin{bmatrix} a & b & c \\ d & e & f \\ g & h & l \end{bmatrix}$$

from Equation (3.3).

Solution:

$$\det A = \begin{bmatrix} a & b & c \\ d & e & f \\ g & h & l \end{bmatrix} = a\begin{vmatrix} e & f \\ h & l \end{vmatrix} - b + \begin{vmatrix} d & f \\ g & l \end{vmatrix} + c\begin{vmatrix} d & e \\ g & n \end{vmatrix}$$

$$= ael + bfg + cdh - ceg - bdl - afh$$

Applying the concept of the determinant of a matrix to the eigenvalue problem stated in Equation (3.2) indicates that if λ is to be an eigenvalue of the matrix A, then λ must satisfy the equation

$$\det(A - \lambda I) = 0 \tag{3.6}$$

This expression results in a polynomial in λ, which is called the *characteristic equation* of the matrix A.

Since A is an $n \times n$ matrix, Equation (3.6) will have n roots (or A will have n eigenvalues), which are denoted by λ_i. Then Equation (3.6) can be rewritten as

$$\det(A - \lambda I) = \prod_{i=1}^{n} (\lambda - \lambda_i) = 0 \tag{3.7}$$

If λ_i happens to be a root that is repeated m_i times, then this becomes

$$\det(A - \lambda I) = \prod_{i=1}^{k} (\lambda - \lambda_i)^{m_i}, \text{ where } \sum_{i=1}^{k} m_i = n \tag{3.8}$$

Also, note from examination of Equation (3.2) that any given eigenvalue may have many eigenvectors associated with it. As already mentioned, if $\mathbf{u}$ is an eigenvector of A with corresponding eigenvalue λ and α is any scalar, $\alpha\mathbf{u}$ is also an eigenvector of A with corresponding eigenvalue λ. Eigenvectors have several other interesting properties, many of which are useful in calculating the free response of a vibrating system.

The first property has to do with the concept of linear independence. A set of vectors, denoted by

$$\{\mathbf{e}_i\}_{i=1}^{n} = \{\mathbf{e}_1, \mathbf{e}_2, \cdots, \mathbf{e}_n\}$$

is said to be linearly independent, or just independent, if

$$\alpha_1\mathbf{e}_1 + \alpha_2\mathbf{e}_2 + \cdots + \alpha_n\mathbf{e}_n = \mathbf{0} \tag{3.9}$$

implies that each of the scalars α_i is zero. If this is not the case, i.e. if there exists one or more nonzero scalars α_i satisfying Equation (3.9), then the set of vectors $\{\mathbf{u}_i\}$ is said to be *linearly dependent.* The set of all linear combinations of all n-dimensional real vectors is called the *span* of the set of all n-dimensional real vectors. A set of n linearly independent vectors, $\{\mathbf{e}_1, \mathbf{e}_2, \ldots, \mathbf{e}_n\}$, is said to form a *basis* for the span of vectors of dimension n. This means that if $\mathbf{x}$ is any vector of dimension n, then there exists a unique representation of the vector $\mathbf{x}$ in terms of the basis vectors $\mathbf{e}_i$, given by

$$\mathbf{x} = a_1\mathbf{e}_1 + a_2\mathbf{e}_2 + \cdots + a_n\mathbf{e}_n \tag{3.10}$$

The coefficients, a_i, are sometimes called the *coordinates* of the vector $\mathbf{x}$ in the basis $\{\mathbf{e}_i\}_{i=1}^{n}$. One familiar basis is that consisting of unit vectors $(\hat{\mathbf{i}}, \hat{\mathbf{j}}, \hat{\mathbf{k}})$ of a rectangular coordinate system, which forms a basis for the set of all three-dimensional real vectors.

Another important use of the idea of linear independence is contained in the concept of the *rank* of a matrix. The rank of a matrix is defined to be the number of independent rows (or columns) of the matrix when the rows (columns) are treated like vectors. This property is used in stability analysis in Chapter 4, and in control in Chapter 6. Note that a square $n \times n$ matrix is nonsingular if and only if its rank is n (i.e. if and only if it has *full rank*).

If the scalar product, or dot product, of two vectors is zero, i.e. if $\mathbf{x}_i^T\mathbf{x}_j = 0$, then the two vectors are said to be *orthogonal*. If $\mathbf{x}_i^T\mathbf{x}_i = 1$, the vector $\mathbf{x}_i$ is called a *unit vector*. If a set of unit vectors is also orthogonal, i.e. if

$$\mathbf{x}_i^T\mathbf{x}_j = \delta_{ik=j} = \begin{cases} 0 & i \neq j \\ 1 & i = j \end{cases}$$

they are said to be an *orthonormal* set. Here, δ_{ij} is the Kronecker delta. Again, the familiar unit vectors of rectangular coordinate systems used in introductory mechanics are an orthonormal set of vectors. Also, as will be discussed later, the eigenvectors of a symmetric matrix can be used to form an orthonormal set. This property is used in this chapter and again in Chapter 5 to solve various vibration problems.

Another important property of eigenvectors is as follows. If A is a square matrix and if the eigenvalues of A are distinct, then the eigenvectors associated with those eigenvalues are independent. If A is also symmetric, then an independent set of eigenvectors exist even if the eigenvalues are repeated. Furthermore, if zero is not an eigenvalue of A and A has eigenvalues λ_i with corresponding eigenvectors $\mathbf{u}_i$, then the eigenvectors of A^{-1} are also $\mathbf{u}_i$ and the eigenvalues are λ_i^{-1}. Thus, A and A^{-1} have related eigenvalue problems. Yet another useful result for the eigenvalues, λ_i, of a matrix A is that the eigenvalues of $(A \pm \beta I)$ are just $\lambda_i \pm \beta$ where β is any scalar (called a shift).

The matrix A is *similar* to the matrix B if there exists a nonsingular matrix P such that

$$A = PBP^{-1} \tag{3.11}$$

In this case, P is referred to as a *similarity transformation* (matrix) and may be used to change vibration problems from one coordinate system, which may be complicated, to another coordinate system that has a simple (or canonical) form.

The reason that similarity transformations are of interest is that if two matrices are similar, they will have the same eigenvalues. Another way to state this is that similarity transformations preserve eigenvalues, or that eigenvalues are *invariant* under similarity transformations. Some square matrices are similar to diagonal matrices. Diagonal matrices consist of all zero elements except for those on the diagonal, making them easy to manipulate. The algebra of diagonal matrices is much like that of scalar algebra. This class of matrices is examined in detail next and forms the essence of the approach to solving vibration problems called *modal analysis* (Section 5.3).

If the matrix A is similar to a diagonal matrix, denoted by Λ, then A can be written as

$$A = P\Lambda P^{-1} \tag{3.12}$$

Post-multiplying this expression by P yields

$$AP = P\Lambda \tag{3.13}$$

Now, let the vectors $\mathbf{p}_i$, $i = 1, 2, \ldots, n$, be the columns of the matrix P, i.e.

$$P = [\mathbf{p}_1 \quad \mathbf{p}_2 \quad \mathbf{p}_3 \quad \cdots \quad \mathbf{p}_n] \tag{3.14}$$

Note that no $\mathbf{p}_i$ can be a zero vector, since P is nonsingular. If λ_{ii} denotes the ith diagonal element of the diagonal matrix, Λ, Equation (3.13) can be rewritten as the n separate equations

$$A\mathbf{p}_i = \lambda_{ii}\mathbf{p}_i,\ i = 1, 2, \ldots, n. \tag{3.15}$$

Equations (3.15) state that $\mathbf{p}_i$ is the ith eigenvector of the matrix A and that λ_{ii} is the associated eigenvalue, λ_i. The preceding observation can be summarized as:

1) If A is similar to a diagonal matrix, the diagonal elements of that matrix are the eigenvalues of A (i.e. $\lambda_i = \lambda_{ii}$).
2) A is similar to a diagonal matrix if and only if A has a set of n linearly independent eigenvectors.
3) If A has distinct eigenvalues, then it is similar to a diagonal matrix.

As an important note for vibration analysis: if A is a real symmetric matrix, then there exists a matrix P such that Equation (3.12) holds.

If the eigenvectors of A are linearly independent, they can be used to form an orthonormal set. Let $\mathbf{s}_i$ denote the orthonormal eigenvectors of A so that $\mathbf{s}_i^T\mathbf{s}_j = \delta_{ij}$, the Kronecker delta. Forming a matrix out of this set of normalized eigenvectors yields

$$S = [\mathbf{s}_1 \quad \mathbf{s}_2 \quad \mathbf{s}_3 \quad \cdots \quad \mathbf{s}_n] \tag{3.16}$$

Here, note that expanding the matrix product S^TS, yields

$$S^TS = I, \tag{3.17}$$

where I is the $n \times n$ identity matrix, because of the orthonormality of the rows and columns of S. Equation (3.17) implies immediately that $S^T = S^{-1}$. Such real-valued matrices are called *orthogonal matrices* and Equation (3.12) can be written as

$$A = S\Lambda S^T \tag{3.18}$$

In this case, A is said to be *orthogonally similar* to Λ. If S is complex valued, then $S^*S = I$, where the asterisk indicates the complex conjugate transpose of S, and S is called a Hermitian matrix. Orthonormal sets are used to compute the time response of vibrating systems from the eigenvalues and eigenvectors.

Often it is convenient in vibration analysis to modify the concept of orthogonally similar matrices by introducing the concept of a weighting matrix. To this end, the eigenvectors of a matrix K can be normalized with respect to a second positive definite matrix, which in this case is chosen to be the matrix M. That is, the magnitude of the eigenvectors of K, $\mathbf{x}_i$, are chosen such that

$$\mathbf{x}_i^TM\mathbf{x}_j = \delta_{ij} \tag{3.19}$$

In this case the weighted transformation, denoted by S_m, has the following properties

$$S_m^TMS_m = I \tag{3.20}$$

$$S_m^TKS_m = \text{diag}[\omega_i^2] \tag{3.21}$$

where ω_i^2 denote the eigenvalues of the matrix K. This is not to be confused with the diagonal matrix S^TKS, where S is made up of the (not weighted) eigenvectors of the matrix K.

3.3 Natural Frequencies and Mode Shapes

As mentioned previously, the concept of the eigenvalue of a matrix is closely related to the concept of natural frequency of vibration in mechanical structures, just as the roots of the characteristic equation and natural frequency of an SDOF system are related. To make the connection formally, consider again the undamped nongyroscopic conservative system described by

$$M\ddot{\mathbf{q}}(t) + K\mathbf{q}(t) = \mathbf{0} \tag{3.22}$$

subject to initial conditions $\mathbf{q}_0$ and $\dot{\mathbf{q}}_0$. Here the matrices M and K are assumed to be symmetric and positive definite.

In an attempt to solve Equation (3.22), a procedure similar to the method used to solve an SDOF system is employed by assuming a solution of the form

$$\mathbf{q}(t) = \mathbf{u}e^{\mu jt} \tag{3.23}$$

Here, $\mathbf{u}$ is a nonzero, unknown vector of constants, μ is a scalar value to be determined, $j = \sqrt{-1}$, and t is, of course, the time. Substitution of Equation (3.23) into Equation (3.22) yields

$$(-M\mu^2 + K)\mathbf{u}e^{\mu jt} = \mathbf{0} \tag{3.24}$$

This is identical to the procedure used in Section 1.2 for SDOF systems. Since $e^{\mu jt}$ is never zero for any value of μ or t, Equation (3.24) holds if and only if

$$(-M\mu^2 + K)\mathbf{u} = \mathbf{0} \tag{3.25}$$

This is starting to look very much like the eigenvalue problem posed in Equation (3.2). To make the analogy more complete, let $\mu^2 = \lambda$, so that Equation (3.25) becomes

$$(K - \lambda M)\mathbf{u} = \mathbf{0} \tag{3.26}$$

Since it is desired to calculate nonzero solutions of Equation (3.22), the vector $\mathbf{u}$ should be nonzero. This corresponds very well to the definition of an eigenvector, i.e. that it be nonzero. Eigenvalue problems stated in terms of two matrices of the form $A\mathbf{x} = \lambda B\mathbf{x}$, $\mathbf{x} \neq \mathbf{0}$, are called *generalized eigenvalue problems*. Now recall that a nonzero solution $\mathbf{u}$ of Equation (3.26) exists, if and only if the matrix $(K - \lambda M)$ is singular or if and only if

$$\det(K - \lambda \mathrm{M}) = 0 \tag{3.27}$$

Next, note that since M is positive definite, it must have an inverse. To see this, note that if M^{-1} does not exist, then there is a nonzero vector $\mathbf{u}$ such that

$$M\mathbf{u} = \mathbf{0},\ \mathbf{u} \neq \mathbf{0} \tag{3.28}$$

Premultiplying by $\mathbf{u}^T$ results in

$$\mathbf{u}^T M\mathbf{u} = 0, \ \mathbf{u} \neq \mathbf{0} \tag{3.29}$$

which clearly contradicts the fact that M is positive definite (recall the end of Section 2.1).

Since M^{-1} exists, $\det(M^{-1}) \neq 0$ and we can multiply Equation (3.27) by $\det(M^{-1})$ to get (invoking Equation (3.5))

$$\det(M^{-1}K - \lambda I) = 0 \tag{3.30}$$

which is of the same form as Equation (3.6) used to define eigenvalues and yields a polynomial in λ of order n. As will be illustrated, each root of Equation (3.30), or eigenvalue of the matrix $M^{-1}K$, is the square of one of the natural frequencies of Equation (3.22).

There are several alternative ways to relate the eigenvalue Equation (3.1) to the natural frequency problem of Equation (3.25). For instance, since M is positive definite, it has a positive definite square root. That is, there exists a unique positive definite matrix $M^{1/2}$ such that $M^{1/2}M^{1/2} = M$. The eigenvalues of $M^{1/2}$ are $\beta_i^{1/2}$, where β_i are the eigenvalues of M. Both M and its matrix square root have the same eigenvectors. Furthermore, if P is the matrix of eigenvectors of M, then

$$M^{1/2} = P\Lambda_M^{1/2}P^{-1} \tag{3.31}$$

where $\Lambda_M^{1/2}$ is a diagonal matrix with diagonal elements $\beta_i^{1/2}$. Many times in modeling systems, M is already diagonal, in which case the matrix square root is calculated by taking the square root of each of the diagonal elements. Systems with non-diagonal mass matrix are called *dynamically coupled* systems. The existence of this matrix square root provides an important alternative relationship between matrix eigenvalues and vibrational natural frequencies and allows a direct analogy to the SDOF case. Matrix factorizations, such as the square root, lead to more computationally efficient algorithms (Section 3.8).

Since $M^{1/2}$ is positive definite, it has an inverse $M^{-1/2}$, pre- and post-multiplying Equation (3.27) by $\det(M^{-1/2})$ and factoring out -1, yields

$$\det(\lambda I - M^{-1/2}KM^{-1/2}) = 0 \tag{3.32}$$

Equation (3.32) is an alternative way of expressing the eigenvalue problem. The difference between Equations (3.32) and (3.30) is that the matrix $\tilde{K} = M^{-1/2}KM^{-1/2}$ is symmetric and positive definite, whereas $M^{-1}K$ is not necessarily symmetric. Matrix symmetry provides both a theoretical and computational advantage. Specifically, a symmetric matrix is similar to a diagonal matrix consisting of its eigenvalues along the diagonal, and the eigenvectors of a symmetric matrix are linearly independent and orthogonal. The corresponding differential equation then becomes

$$I\ddot{\mathbf{r}}(t) + M^{-1/2}KM^{-1/2}\mathbf{r}(t) = \mathbf{0} \tag{3.33}$$

where $\mathbf{q}(t) = M^{-1/2}\mathbf{r}(t)$ has been substituted into Equation (3.22) and the result premultiplied by $M^{-1/2}$.

As expected and shown later, the numbers λ_i are directly related to the natural frequencies of vibration of the system described by Equation (3.22): $\omega_i^2 = \mu_i^2 = \lambda_i$. It is expected, as in the case of SDOF systems with no damping, that the natural frequencies will be such that the motion oscillates without decay. Mathematically, this result follows from realizing that the matrix $\tilde{K} = M^{-1/2}KM^{-1/2}$ is symmetric and positive definite, ensuring the nature of the natural frequencies and eigenvectors.

To see that a real, symmetric, positive definite matrix such as $\tilde{K} = M^{-1/2}KM^{-1/2}$ has positive real eigenvalues (and, hence, real eigenvectors) requires some simple manipulation of the definitions of these properties. First, note that if $\mathbf{u}$ is an eigenvector of A with corresponding eigenvalue λ, then

$$A\mathbf{u} = \lambda\mathbf{u} \tag{3.34}$$

Assuming that λ and $\mathbf{u}$ are complex and taking the conjugate transpose of this expression yields (because A is symmetric)

$$\mathbf{u}^*A = \mathbf{u}^*\lambda^* \tag{3.35}$$

Premultiplying Equation (3.34) by $\mathbf{u}^*$, postmultiplying Equation (3.35) by $\mathbf{x}$, and subtracting the two yields

$$0 = \mathbf{u}^*A\mathbf{u} - \mathbf{u}^*A\mathbf{u} = (\lambda - \lambda^*)\mathbf{u}^*\mathbf{u}$$

or since $\mathbf{u} \neq \mathbf{0}$, that $\lambda = \lambda^*$. Hence, λ must be real valued.

Recall from Equation (3.18) that A can be written as $A = S\Lambda S^T$. Therefore, for any and all arbitrary vectors $\mathbf{x}$

$$\mathbf{x}^TA\mathbf{x} = \mathbf{x}^TS\Lambda S^T\mathbf{x} = \mathbf{y}^T\Lambda\mathbf{y}$$

where $\mathbf{y} = S^T\mathbf{x}$ is also free to take on any real value. This can be expressed as

$$\mathbf{y}^T\Lambda\mathbf{y} = \sum_{i=1}^{n} \lambda_i y_i^2 > 0$$

since A is positive definite. This inequality holds for any and all vectors $\mathbf{y}$. In particular, if the vectors $\mathbf{y}_1 = [1\ 0\ 0\ \dots\ 0]^T$, $\mathbf{y}_2 = [0\ 1\ 0\ \dots\ 0]^T$, ... $\mathbf{y}_n = [0\ 0\ 0\ \dots\ 1]^T$ are, in turn, substituted into this last inequality, the result is $\lambda_i > 0$, for each of the n values of the index, i. Hence, a positive definite symmetric matrix has positive real eigenvalues (the converse is also true).

Applying this fact to Equation (3.32) indicates that each eigenvalue of the mass normalized stiffness matrix $\tilde{K} = M^{-1/2}KM^{-1/2}$ is a positive real number. From Equation (3.25) we see that the natural frequencies of Equation (3.22) are $\mu = \omega$, where $\omega^2 = \lambda$, a positive real number. Hence, the coefficient of t in Equation (3.23) has the form $\omega = \pm\sqrt{\lambda}j$, just as in the SDOF case. The square roots of the λ_i are the natural frequencies of the system, i.e. $\omega_i = \sqrt{\lambda_i}$, where i ranges from 1 to n, n being the number of degrees of freedom. There is one natural frequency for each degree of freedom.

The concept of a positive definite matrix can also be related to conditions on the elements of the matrix in a useful manner. Namely, it can be shown that a symmetric matrix A is positive definite if and only if the leading principal minors of A are positive. That is, if

$$A = \begin{bmatrix} a_{11} & a_{12} & \cdots & a_{1n} \\ a_{21} & a_{22} & \cdots & a_{2n} \\ \vdots & \vdots & \vdots & \vdots \\ a_{n1} & a_{n2} & \cdots & a_{nn} \end{bmatrix}$$

then A is positive definite if and only if

$$a_{11} > 0$$

$$\det \begin{bmatrix} a_{11} & a_{12} \\ a_{21} & a_{22} \end{bmatrix} > 0$$

$$\det \begin{bmatrix} a_{11} & a_{12} & a_{13} \\ a_{21} & a_{22} & a_{23} \\ a_{31} & a_{32} & a_{33} \end{bmatrix} > 0$$

$$\vdots$$

$$\det A > 0$$

This condition provides a connection between the condition that a matrix be positive definite and the physical parameters of the system. For example, the stiffness matrix of Example 2.5.1 will be positive definite if and only if $k_1 + k_2 > m\Omega^2$ and $2k_2 > m\Omega^2$, by the preceding principle minor condition. That is, for $A = K$ in Example 2.5.1, the first two conditions yield the two inequalities in k_i, m and Ω. This provides physical insight, as it indicates that stability may be lost if the system spins faster (Ω) than the stiffness k_i can handle. These inequalities are useful in vibration design and in stability analysis.

Another interesting fact about symmetric matrices is that their eigenvectors form a complete set, or a basis. Recall that a set of real vectors $\{\mathbf{u}_i\}$ of dimension n is a *basis* for the set of all real n-dimensional vectors, if and only if they are linearly independent and every other real vector of dimension n can be written as a linear combination of the $\mathbf{u}_i$. Thus, the solution $\mathbf{q}(t)$ can be expanded in terms of these eigenvectors. The set of eigenvectors of the matrix $\tilde{K} = M^{-1/2}KM^{-1/2}$ forms a linearly independent set, such that any vector of dimension n can be written as a linear combination of these vectors. In particular, the solution of the vibration problem can be expanded in terms of this basis.

Combining the preceding matrix results leads to the following solution for the response $\mathbf{r}(t)$. There are n solutions of Equation (3.33) of the form

$$\mathbf{r}_k(t) = \mathbf{u}_k e^{\mu_k jt} \tag{3.36}$$

As just shown, under the assumption that $M^{-1/2}KM^{-1/2}$ is positive definite, the numbers μ_k must all be of the form

$$\mu_k = \pm\sqrt{\lambda_k} \tag{3.37}$$

where the λ_k are the positive eigenvalues of the matrix $M^{-1/2}KM^{-1/2}$. Combining Equations (3.36) and (3.37), it is seen that each $\mathbf{r}_k(t)$ must have the form

$$\mathbf{r}_k(t) = (a_k e^{-\sqrt{\lambda_k} jt} + b_k e^{+\sqrt{\lambda_k} jt})\mathbf{u}_k \tag{3.38}$$

where a_k and b_k are arbitrary constants. Since the $\mathbf{u}_k$ are the eigenvectors of a symmetric matrix, they form a basis, so the n-dimensional vector $\mathbf{r}(t)$ can be expressed as a linear combination of these. That is

$$\mathbf{r}(t) = \sum_{k=1}^{n} (a_k e^{-\sqrt{\lambda_k} jt} + b_k e^{\sqrt{\lambda_k} jt})\mathbf{u}_k \tag{3.39}$$

where the arbitrary constants a_k and b_k can be determined from the initial conditions $\mathbf{r}(0)$ and $\dot{\mathbf{r}}(0)$. This amounts to solving the $2n$ algebraic equations given by

$$\begin{aligned} \mathbf{r}(0) &= \sum_{k=1}^{n} (a_k + b_k)\mathbf{u}_k \\ \dot{\mathbf{r}}(0) &= j\sum_{k=1}^{n} \sqrt{\lambda_k}(b_k - a_k)\mathbf{u}_k \end{aligned} \tag{3.40}$$

for the $2n$ constants a_k and b_k, $k = 1,\ldots, n$.

Since the symmetric properties of the matrix $M^{-1/2}KM^{-1/2}$ were used to develop the solution given by Equation (3.39), note that the solution expressed in Equation (3.39) is the solution of a slightly different problem than the solution $\mathbf{q}(t)$ of Equation (3.22). The two are related by the transformation

$$\mathbf{q}(t) = M^{-1/2}\mathbf{r}(t) \tag{3.41}$$

which also specifies how the initial conditions in the original coordinates are to be transformed.

Equation (3.39) can be manipulated, using Euler's formulas for trigonometric functions, to become

$$\mathbf{r}(t) = \sum_{k=1}^{n} c_k \sin(\omega_k t + \phi_k)\mathbf{u}_k \tag{3.42}$$

where c_k and ϕ_k are constants determined by initial conditions. This form clearly indicates the oscillatory nature of the system and defines the concept of natural frequency. Here, $\omega_k = +\sqrt{\lambda_k}$ denotes the undamped natural frequencies. Note that the frequencies are always positive because the Euler formula transformation from $e^{\pm\sqrt{\lambda_k}t}$ to $\sin \omega_k t$ effectively uses the $\pm$ sign in defining oscillation at the (positive) frequency, ω_k. This expression extends the undamped SDOF result to undamped MDOF systems.

To evaluate the constants c_k and ϕ_k, the orthonormality of the vectors $\mathbf{u}_k$ is again used. Applying the initial conditions to Equation (3.42) yields

$$\mathbf{r}(0) = \sum_{k=1}^{n} c_k \sin(\phi_k)\mathbf{u}_k \tag{3.43}$$

and

$$\dot{\mathbf{r}}(0) = \sum_{k=1}^{n} c_k \omega_k \cos(\phi_k)\mathbf{u}_k \tag{3.44}$$

Equation (3.41) is used to yield $\mathbf{r}(0) = M^{1/2}\mathbf{q}(0)$ and $\dot{\mathbf{r}}(0) = M^{1/2}\dot{\mathbf{q}}(0)$ from the given initial conditions of $\mathbf{q}(0)$ and $\dot{\mathbf{q}}(0)$. Premultiplying Equation (3.43) by $\mathbf{u}_i^T$ yields

$$\mathbf{u}_i^T \mathbf{r}(0) = \sum_{k=1}^{n} c_k \sin(\phi_k)\mathbf{u}_i^T \mathbf{u}_k$$

Invoking the orthonormality for the vectors $\mathbf{u}_i$ yields

$$c_i \sin \phi_i = \mathbf{u}_i^T \mathbf{r}(0) \tag{3.45}$$

Likewise, Equation (3.44) yields

$$c_i \cos \phi_i = \frac{\mathbf{u}_i^T \dot{\mathbf{r}}(0)}{\omega_i} \tag{3.46}$$

Combining Equations (3.45) and (3.46) and renaming the index yields

$$\phi_i = \tan^{-1}\left\{\frac{\omega_i \mathbf{u}_i^T \mathbf{r}(0)}{\mathbf{u}_i^T \dot{\mathbf{r}}(0)}\right\}$$

and

$$c_i = \frac{\mathbf{u}_k^T \mathbf{r}(0)}{\sin \phi_i}$$

Note that if the initial position, $\mathbf{r}(0)$, is zero, then Equation (3.45) would imply that $\phi_i = 0$ for each i, and then Equation (3.46) is used to compute the coefficients, c_i. Once the constants c_i and ϕ_i are determined, then the index is changed to k to fit into the sum of Equation (3.42), which is written in terms of c_k and ϕ_k. Next consider the eigenvectors $\mathbf{u}_k$ to see how they represent the physical motion of the system. Suppose the initial conditions $\mathbf{r}(0)$ and $\dot{\mathbf{r}}(0)$ are chosen in such a way that $c_k = 0$, for $k = 2, 3, \ldots, n$, $c_1 = 1$, and $\phi_k = 0$ for all k. Then, the expansion in Equation (3.42) reduces to one simple term, namely

$$\mathbf{r}(t) = \sin(\omega_1 t)\mathbf{u}_1 \tag{3.47}$$

This implies that every mass is vibrating with frequency ω_1 or is stationary and that the relative amplitude of vibration of each of the masses is the value of the corresponding element of $\mathbf{u}_1$. Thus, the size and sign of each element of the eigenvector indicates the positions of each mass from its equilibrium position, i.e. the "shape" of the vibration at any instant of time. Transforming this vector back into the physical coordinate system via $\mathbf{v}_1 = M^{-1/2}\mathbf{u}_1$ allows the interpretation that the vector $\mathbf{v}_1$ is the first *mode shape* of the system, or the mode shape corresponding to the first natural frequency. This can clearly be repeated for each of the subscripts k, so that $\mathbf{v}_k$ is the kth mode shape. Hence, the transformed eigenvectors are referred to as the *modes* of vibration of the system. Since eigenvectors are arbitrary to within a multiplicative constant, so are the mode shapes. If the arbitrary constant is chosen so that $\mathbf{v}_k$ is normalized, i.e. so that $\mathbf{v}_k{}^T\mathbf{v}_k = 1$, and the vector $\mathbf{v}_k$ is real, the $\mathbf{v}_k$ is called a *normal mode* of the system. The constants c_k in Equation (3.42) are called *modal participation factors*, because their relative magnitudes indicate how much the indexed mode influences the response of the system.

Note that if the system has three or more degrees of freedom, one of the elements of a particular mode shape has the potential to be zero. This defines what is called a *node* of a mode and implies that no motion occurs at that location at that frequency corresponding to the mode shape.

The just-described procedure constitutes a theoretical *modal analysis* of the system of Equation (3.22). Some researchers refer to Equations (3.39) and (3.42) as the *expansion theorem*. They depend on the completeness of the eigenvectors associated with the system, i.e. of the matrix, $M^{-1/2}KM^{-1/2}$.

Example 3.3.1

It should be obvious from Equations (3.27) and (3.32) how to calculate the eigenvalues and hence the natural frequencies of the system, as they are the roots of the characteristic polynomial following from $\det(\tilde{K} - \lambda I) = 0$. To calculate the eigenvectors, however, may not be as obvious and is illustrated here. Let λ_1 be an eigenvalue of A; then λ_1 and $\mathbf{u}_1 = [x_1 \; x_2]^T$ satisfy the vector equation

$$\begin{bmatrix} a_{11} - \lambda_1 & a_{12} \\ a_{21} & a_{22} - \lambda_1 \end{bmatrix} \begin{bmatrix} x_1 \\ x_2 \end{bmatrix} = \begin{bmatrix} 0 \\ 0 \end{bmatrix} \tag{3.48}$$

This represents two dependent equations in x_1 and x_2, the two components of the eigenvector $\mathbf{u}_1$. Hence only their ratio can be determined. Proceeding with the first equation in Equation (3.48) yields

$$(a_{11} - \lambda_1)x_1 + a_{12}x_2 = 0$$

which is solved for the ratio x_1/x_2. Then, the vector $\mathbf{u}_1$ is "normalized" so that $\mathbf{u}_1^T\mathbf{u}_1 = x_1^2 + x_2^2 = 1$. The normalization yields specific values for x_1 and x_2. As a consequence of the singularity of $(A - \lambda I)$, the second equation in Equation (3.48), $a_{21} + x_1 + (a_{22} - \lambda)x_2 = 0$, is dependent on the first and does not yield new information.

Example 3.3.2

This example illustrates the procedure for calculating the free vibrational response of an MDOF system by using a *modal expansion*. The procedure is illustrated by a two degrees of freedom system, since the procedure for a larger number of degrees of freedom is the same. The purpose of this example is to develop an understanding of the eigenvector problem and is not intended to imply that this is the most efficient way to calculate the time response of a system (it is not).

Consider the system described in Figure 2.1 with $m_1 = 9$, $m_2 = 1$, $k_1 = 24$ and $k_2 = 3$. Then the equation of motion becomes

$$\begin{bmatrix} 9 & 0 \\ 0 & 1 \end{bmatrix} \ddot{\mathbf{q}} + \begin{bmatrix} 27 & -3 \\ -3 & 3 \end{bmatrix} \mathbf{q} = \mathbf{0}$$

subject to the initial condition $\mathbf{q}(0) = [1\ 0]^T$ and $\dot{\mathbf{q}}(0) = [0\ \ 0]^T$ in some set of consistent units. The matrix $\tilde{K} = M^{-1/2}KM^{-1/2}$ becomes

$$\begin{bmatrix} 1/3 & 0 \\ 0 & 1 \end{bmatrix} \begin{bmatrix} 27 & -3 \\ -3 & 3 \end{bmatrix} \begin{bmatrix} 1/3 & 0 \\ 0 & 1 \end{bmatrix} = \begin{bmatrix} 3 & -1 \\ -1 & 3 \end{bmatrix}$$

The characteristic Equation (3.32) becomes

$$\lambda^2 - 6\lambda + 8 = 0$$

which has roots $\lambda_1 = 2$, $\lambda_2 = 4$.

To compute the eigenvectors, follow Equation (3.48) and calculate

$$\left(\tilde{K} - \lambda_1 I\right)\mathbf{u}_1 = 0 \Rightarrow \begin{bmatrix} 3-2 & -1 \\ -1 & 3-2 \end{bmatrix} \begin{bmatrix} x_1 \\ x_2 \end{bmatrix} = 0$$

This represents two equations. Taking the “top” equation yields

$$x_1 - x_2 = 0 \Rightarrow x_1 = x_2 \text{ so that } \mathbf{u}_1 = \begin{bmatrix} 1 \\ 1 \end{bmatrix}$$

where 1 is arbitrarily chosen to satisfy the first equation. To normalize $\mathbf{u}_1$, divide $\mathbf{u}_1$ by its magnitude

$$\hat{\mathbf{u}}_1 = \frac{1}{\sqrt{\mathbf{u}_1^T\mathbf{u}_1}} \begin{bmatrix} 1 \\ 1 \end{bmatrix} = \frac{1}{\sqrt{2}} \begin{bmatrix} 1 \\ 1 \end{bmatrix}$$

Likewise for λ_2, the normalized eigenvector is computed using Equation (3.48) with the value of $\lambda_2 = 4$ to become

$$\hat{\mathbf{u}}_2 == \frac{1}{\sqrt{2}} \begin{bmatrix} -1 \\ 1 \end{bmatrix}$$

Note that

$$\hat{\mathbf{u}}_2 == \frac{1}{\sqrt{2}} \begin{bmatrix} 1 \\ -1 \end{bmatrix}$$

also satisfies Equation (3.48) and this choice is arbitrary. The arbitrary nature of the mode shapes’ numerical values does not affect the final result (see Problem 3.9). It is only the ratio of the elements in the mode shapes that matter.

Thus the orthogonal matrix of eigenvectors is

$$S = [\,\hat{\mathbf{u}}_1 \quad \hat{\mathbf{u}}_2\,] = \frac{1}{\sqrt{2}} \begin{bmatrix} 1 & -1 \\ 1 & 1 \end{bmatrix}$$

Also note that $S^T(M^{-1/2}KM^{-1/2})S = \text{diag}[2\ 4]$, as it should. The transformed initial conditions become

$$\mathbf{r}(0) = \begin{bmatrix} 3 & 0 \\ 0 & 1 \end{bmatrix} \begin{bmatrix} 1 \\ 0 \end{bmatrix} = \begin{bmatrix} 3 \\ 0 \end{bmatrix}$$

and of course $\dot{\mathbf{r}}(0) = [0 \ \ 0]^T$. The values of the constants in Equation (3.42) are found from

$$\phi_1 = \tan^{-1}\left[\frac{\omega_1 \mathbf{u}_1^T \mathbf{r}(0)}{\mathbf{u}_1^T \dot{\mathbf{r}}(0)}\right] = \tan^{-1} \infty = \frac{\pi}{2}$$

$$c_1 = \frac{\mathbf{u}_1^T \mathbf{r}(0)}{\sin \phi_1} = \frac{3}{\sqrt{2}}$$

$$\phi_2 = \tan^{-1}\left[\frac{\omega_2 \mathbf{u}_2^T \mathbf{r}(0)}{\mathbf{u}_2^T \dot{\mathbf{r}}(0)}\right] = \tan^{-1} \infty = \frac{\pi}{2}$$

$$c_2 = \frac{\mathbf{u}_2^T \mathbf{r}(0)}{\sin \phi_2} = \frac{-3}{\sqrt{2}}$$

Hence the solution $\mathbf{r}(t)$ is given by

$$\mathbf{r}(t) = \frac{3}{\sqrt{2}} \cos \sqrt{2}t \begin{bmatrix} 1 \\ 1 \end{bmatrix} - \frac{3}{\sqrt{2}} \cos 2t \begin{bmatrix} -1 \\ 1 \end{bmatrix}$$

In the original coordinates this becomes

$$\mathbf{q}(t) = M^{-1/2}\mathbf{r}(t)$$

$$\mathbf{q}(t) = \frac{3}{\sqrt{2}} \cos \sqrt{2}t \begin{bmatrix} 1/3 \\ 1 \end{bmatrix} - \frac{3}{\sqrt{2}} \cos 2t \begin{bmatrix} -1/3 \\ 1 \end{bmatrix}$$

Multiplying this out yields the motion of the individual masses

$$q_1(t) = \frac{1}{\sqrt{2}} \cos \sqrt{2}t + \frac{1}{\sqrt{2}} \cos 2t$$

$$q_2(t) = \frac{3}{\sqrt{2}} \cos \sqrt{2}t - \frac{3}{\sqrt{2}} \cos 2t$$

The two mode shapes are ($\mathbf{v}_i = M^{1/2}\mathbf{u}_i$)

$$\mathbf{v}_1 = \frac{1}{\sqrt{2}} \begin{bmatrix} 1/3 \\ 1 \end{bmatrix} \quad \text{and} \quad \mathbf{v}_2 = \frac{1}{\sqrt{2}} \begin{bmatrix} -1/3 \\ 1 \end{bmatrix}$$

If K is not symmetric, i.e. if we have a system of the form

$$M\ddot{\mathbf{q}}(t) + (K + H)\mathbf{q}(t) = \mathbf{0}$$

then proceed by solving an eigenvalue problem of the form

$$M^{-1}(K + H)\mathbf{x} = \lambda \mathbf{x}$$

or

$$A\mathbf{u} = \lambda \mathbf{u}$$

where A is not a symmetric matrix. In this case, the eigenvalues λ_i and the eigenvectors $\mathbf{u}_i$ are in general complex numbers. Also, because of the asymmetry, the matrix A has a *left eigenvector*, $\mathbf{w}_k$, which satisfies

$$\mathbf{w}_k^T A = \lambda_k \mathbf{w}_k^T$$

and, in general, may not equal the *right eigenvector*, $\mathbf{u}_k$. Now let $\mathbf{w}_k$ be a left eigenvector and $\mathbf{u}_i$ be a right eigenvector, then

$$A\mathbf{u}_i = \lambda_i \mathbf{u}_i \quad \text{or} \quad \mathbf{w}_k^T A \mathbf{u}_i = \lambda_i \mathbf{w}_k^T \mathbf{u}_i \tag{3.50}$$

and

$$\mathbf{w}_k^T A = \lambda_k \mathbf{w}_k^T \quad \text{or} \quad \mathbf{w}_k^T A \mathbf{u}_i = \lambda_k \mathbf{w}_k^T \mathbf{u}_i \tag{3.51}$$

where the ith eigenvalues of $\mathbf{x}_i$ and $\mathbf{y}_i$ are the same. Subtracting Equation (3.51) from Equation (3.50) yields $(\lambda_i - \lambda_k)\mathbf{w}_k^T \mathbf{u}_i = 0$, so that if $\lambda_i \neq \lambda_k$, then $\mathbf{w}_k^T \mathbf{u}_i = 0$. This is called *biorthogonality.*

For distinct eigenvalues, the right and left eigenvectors of A each form a linearly independent set and can then be used to express any $n \times 1$ vector, i.e. an expansion theorem still exists. These relations are useful for treating gyroscopic systems, systems with constraint damping, systems with follower forces and feedback control systems.

3.4 Canonical Forms

The diagonal matrix of eigenvalues of Section 3.3 is considered a canonical, or simple, form of a symmetric matrix. This is so because of the ease of manipulation of a diagonal matrix. For instance, the square root of a diagonal matrix is just the diagonal matrix with nonzero elements equal to the square root of the diagonal elements of the original matrix.

From the point of view of vibration analysis, the diagonal form provides an immediate record of the systems' natural frequencies of vibration. In addition, the similarity transformation Equation (3.12) can be used to solve the undamped vibration problem of Equation (3.33). To see this, let S be the orthogonal similarity transformation associated with the symmetric matrix $\tilde{K} = M^{-1/2} K M^{-1/2}$. Substitution of $\mathbf{r}(t) = S\mathbf{y}(t)$ into Equation (3.33) and premultiplying by S^T yields

$$\ddot{\mathbf{y}}(t) + \Lambda \mathbf{y}(t) = 0 \tag{3.52}$$

where Λ is diagonal. Thus, Equation (3.52) represents n scalar equations (modal equations), each of the form

$$\ddot{y}_i(t) + \omega_i^2 y_i(t) = 0 \quad i = 1, 2, \ldots, n \tag{3.53}$$

These expressions can be integrated separately using the initial conditions $\mathbf{y}(0) = S^T \mathbf{r}(0)$ and $\dot{\mathbf{y}}(0) = S^T \dot{\mathbf{r}}(0)$ to yield a solution equivalent to Equation (3.42). This argument forms the crux of what is called *modal analysis* and is repeated many times in the following chapters.

Unfortunately, not every square matrix is similar to a diagonal matrix. However, every square matrix is similar to an upper triangular matrix. That is, let the matrix A have eigenvalues $\lambda_1, \lambda_2, \ldots, \lambda_n$; then there exists a nonsingular matrix P such that

$$P^{-1}AP = \begin{bmatrix} \lambda_1 & t_{12} & 0 & \cdots & 0 & 0 \\ 0 & \lambda_2 & t_{23} & \cdots & 0 & 0 \\ \vdots & \vdots & \vdots & & \vdots & \vdots \\ 0 & 0 & 0 & \cdots & \lambda_{n-1} & t_{n-1,n} \\ 0 & 0 & 0 & \cdots & 0 & \lambda_n \end{bmatrix} \tag{3.54}$$

The matrix $P^{-1}AP$ is said to be *upper triangular*. If the matrix is symmetric, then the t_{ij}'s in Equation (3.54) are all zero, and the upper triangular matrix becomes a diagonal matrix. Upper or lower triangular matrices have distinct numerical advantages in solving vibration problems for the case where the mass matrix is not diagonal.

If the system is simple, the mass matrix is typically diagonal. In this case, computing its inverse is just a matter of taking the reciprocal of the diagonal elements. However, for more complicated problems, the mass matrix is not diagonal. For instance, in finite element modeling the mass is often not diagonal. Systems with non-diagonal mass matrices are called *dynamically coupled*, and computing the inverse is numerically more complicated. A numerically inexpensive way to compute the inverse square root of the mass matrix in a dynamically coupled system is to use the Cholesky decomposition. This factorization of the mass matrix is based on the matrix theory result that any symmetric, positive definite matrix M can be written as $M = LL^T$, where the matrix L is lower triangular. A lower triangular matrix is a matrix that has all the elements above the diagonal equal to zero. Alternately, any symmetric, positive definite matrix M can be written as $M = U^TU$, where the matrix U is upper triangular.

To use the Cholesky factorization to compute the modal equations for an undamped system, consider writing the equations of motion in the form

$$LL^T\ddot{\mathbf{q}}(t) + K\mathbf{q}(t) = 0$$

Multiplying this expression by L^{-1} and setting $\mathbf{q} = L^{-T}\mathbf{r}$, where the superscript "$-T$" refers to taking the inverse of the transpose, yields

$$LL^{-T}\ddot{\mathbf{r}}(t) + L^{-1}KL^{-T}\mathbf{r}(t) = 0 \Rightarrow \ddot{\mathbf{r}}(t) + L^{-1}KL^{-T}\mathbf{r}(t) = 0$$

Since the matrix $L^{-1}\,KL^{-T}$ is symmetric, it can be further transformed into modal coordinates by an additional orthogonal transformation rendering the equations of motion decoupled.

Example 3.4.1

Fully populated mass matrices are often produced in finite element modeling (Chapter 11). A classic example is proportional to

$$M = \begin{bmatrix} 2 & 1 \\ 1 & 2 \end{bmatrix}$$

Find the Cholesky factors of this matrix.

Solution: Let L have the form

$$L = \begin{bmatrix} a & 0 \\ b & c \end{bmatrix} \Rightarrow L^T = \begin{bmatrix} a & 0 \\ b & c \end{bmatrix} \Rightarrow LL^T = \begin{bmatrix} a^2 & ab \\ ab & b^2 + c^2 \end{bmatrix}$$

Thus for this L to be a Cholesky factor of M, solve $M = LL^T$ or

$$\begin{bmatrix} 2 & 1 \\ 1 & 2 \end{bmatrix} = \begin{bmatrix} a^2 & ab \\ ab & b^2 + c^2 \end{bmatrix}$$

$$\Rightarrow a^2 = 2, ab = 1, \text{ and } b^2 + c^2 = 2$$

Solving for a, b and c yields: $a = 1.414$, $b = 0.707$, $c = 1.225$. Thus

$$L = \begin{bmatrix} 1.414 & 0 \\ 0.707 & 1.225 \end{bmatrix}$$

In MATLAB, the command is Chol(M), or Chol(M,'lower'), as discussed in Section 3.7.

Example 3.4.2
Show that the matrix $L^{-1}KL^{-T}$ is in fact symmetric.

Solution: To see that $L^{-1}KL^{-T}$ is symmetric, it suffices to show that it is equal to its transpose. Thus, taking the transpose of $L^{-1}KL^{-T}$ yields

$$(L^{-1}KL^{-T})^T = (L^{-T})^T(L^{-1}K)^T$$

using the rule that $(AB)^T = B^TA^T$ (Appendix B). Noting that the transpose of a transpose is the original matrix and applying again $(AB)^T = B^TA^T$ yields

$$(L^{-1}KL^{-T})^T = (L^{-T})^T(L^{-1}K)^T = L^{-1}KL^{-T}$$

since $(K^T)^T = K$. Thus, the transpose of $L^{-1}KL^{-T}$ is equal to itself and it is therefore symmetric.

A classic result in the theory of matrices is known as *Jordan's Theorem* and states the following: Let A be $n \times n$ with eigenvalues λ_i of multiplicities m_i, so that

$$\det(A - \lambda I) = \prod_{i=1}^{k} (\lambda_i - \lambda)^{m_i}, \text{ where } \sum_{i=1}^{k} m_i = n$$

Then, every matrix A is similar to a block-diagonal matrix of the form

$$J = \begin{bmatrix} \Lambda_1 & 0 & 0 & \cdots & 0 \\ 0 & \Lambda_2 & 0 & \cdots & 0 \\ 0 & 0 & \Lambda_3 & \cdots & \vdots \\ \vdots & \vdots & \cdots & \ddots & 0 \\ 0 & 0 & \cdots & 0 & \Lambda_i \end{bmatrix} \tag{3.55}$$

where each block Λ_i is of the form

$$\Lambda_i = \begin{bmatrix} \lambda_i & a & 0 & \cdots & 0 \\ 0 & \lambda_i & a & \ddots & \vdots \\ \vdots & 0 & \ddots & \ddots & 0 \\ 0 & \cdots & \ddots & \lambda_i & a \\ 0 & 0 & \cdots & 0 & \lambda_i \end{bmatrix}$$

Here $a = 0$ or 1, depending on whether or not the associated eigenvectors are dependent. The value of a is determined as follows: if λ_i are distinct, then $a = 0$, always. If λ_i is repeated m_i times but has m_i linearly independent eigenvectors, then $a = 0$. If $\mathbf{u}_i$ is *dependent* (degenerate), then $a = 1$. If the preceding matrix describes a vibration problem, the value of a determines whether or not a given system can be diagonalized. Note, then, that in general it is eigenvector "degeneracy" that causes problems in vibration analysis – not just repeated eigenvalues.

Next, recall again that the determinant of a matrix is invariant under a similarity transformation. Expanding the determinant yields

$$\det(A - \lambda I) = (-1)^n(\lambda - \lambda_1)(\lambda - \lambda_2)\ldots(\lambda - \lambda_n)$$

which is the characteristic polynomial and hence is equal to

$$\det(A - \lambda I) = (-1)^n(\lambda^n + c_1\lambda^{n-1} + \ldots + c_{n-1}\lambda + c_n) \tag{3.56}$$

Thus, the coefficients c_i of the characteristic polynomial must also be invariant under similarity transformations. This fact is used to some advantage.

The *trace* of a matrix A is defined as

$$\mathrm{tr}(A) = \sum_{i=1}^{n} a_{ii} \tag{3.57}$$

That is, the trace is the sum of the diagonal entries of the matrix. Some manipulation yields that

$$c_1 = -\mathrm{tr}(A) \tag{3.58}$$

and

$$\mathrm{tr}(A) = \sum_{i=1}^{n} \lambda_i \tag{3.59}$$

Thus, the trace of a matrix is invariant under similarity transformations. Some additional properties of the trace are

$$\mathrm{tr}(AB) = \mathrm{tr}(BA) \tag{3.60}$$

For P nonsingular

$$\mathrm{tr}(A) = \mathrm{tr}(P^{-1}AP) \tag{3.61}$$

For α and β scalars

$$\mathrm{tr}(\alpha A + \beta B) = \alpha\mathrm{tr}(A) + \beta\mathrm{tr}(\mathrm{B}) \tag{3.62}$$

and

$$\mathrm{tr}(A) = \mathrm{tr}(A^T) \tag{3.63}$$

It is interesting to note that the tr(A) and det(A) can be used to check computational accuracy, because they are invariant under similarity transformations.

Example 3.4.3

Consider the matrix A and the orthogonal matrix P given by

$$A = \begin{bmatrix} 3 & -1 \\ -1 & 2 \end{bmatrix},\ P = \frac{1}{\sqrt{2}}\begin{bmatrix} 1 & -1 \\ 1 & 1 \end{bmatrix}$$

And show by direct calculation that $\mathrm{tr}(A) = \mathrm{tr}(P^{-1}AP)$.

Solution: First compute tr(A) = 3 + 2 = 5. Next note that since P is orthogonal it is equal to its transpose. To check this, note that

$$PP^T = \frac{1}{\sqrt{2}}\begin{bmatrix} 1 & -1 \\ 1 & 1 \end{bmatrix}\frac{1}{\sqrt{2}}\begin{bmatrix} 1 & 1 \\ -1 & 1 \end{bmatrix} = \frac{1}{2}\begin{bmatrix} 2 & 0 \\ 0 & 2 \end{bmatrix} = \begin{bmatrix} 1 & 0 \\ 0 & 1 \end{bmatrix}$$

Compute $P^{-1}A = P^TA$

$$P^TA = \frac{1}{\sqrt{2}}\begin{bmatrix} 1 & 1 \\ -1 & 1 \end{bmatrix}\begin{bmatrix} 3 & -1 \\ -1 & 2 \end{bmatrix} = \frac{1}{2}\begin{bmatrix} 2 & 1 \\ -4 & 3 \end{bmatrix}$$

Next compute P^TAP

$$(P^TA)P = \frac{1}{\sqrt{2}}\begin{bmatrix} 2 & 1 \\ -4 & 3 \end{bmatrix}\frac{1}{\sqrt{2}}\begin{bmatrix} 1 & -1 \\ 1 & 1 \end{bmatrix} = \frac{1}{2}\begin{bmatrix} 3 & -1 \\ -1 & 7 \end{bmatrix}$$

Thus the $\mathrm{tr}(P^{-1}AP) = \mathrm{tr}(P^TAP) = (1/2)(3+7) = 5 = \mathrm{tr}(A)$, as it should.

Example 3.4.4

Prove by direct calculation that the $\mathrm{tr}(AB) = \mathrm{tr}(BA)$ for all 2×2 matrices.

Solution: Define general 2×2 matrices A and B by an array of symbols a_i and b_i representing arbitrary real numbers by

$$A = \begin{bmatrix} a_1 & a_2 \\ a_3 & a_4 \end{bmatrix} \text{ and } B = \begin{bmatrix} b_1 & b_2 \\ b_3 & b_4 \end{bmatrix}$$

Then just compute the two products AB and BA

$$AB = \begin{bmatrix} a_1 & a_2 \\ a_3 & a_4 \end{bmatrix}\begin{bmatrix} b_1 & b_2 \\ b_3 & b_4 \end{bmatrix} = \begin{bmatrix} a_1b_1 + a_2b_3 & a_1b_2 + a_1b_4 \\ a_3b_1 + a_4b_3 & a_3b_2 + a_4b_4 \end{bmatrix}$$

$$\text{so that: } \mathrm{tr}\,(AB) = a_1b_1 + a_2b_3 + a_3b_2 + a_4b_4$$

$$BA = \begin{bmatrix} b_1 & b_2 \\ b_3 & b_4 \end{bmatrix}\begin{bmatrix} a_1 & a_2 \\ a_3 & a_4 \end{bmatrix} = \begin{bmatrix} a_1b_1 + a_3b_2 & a_2b_1 + a_4b_2 \\ a_1b_3 + a_3b_4 & a_2b_3 + a_4b_4 \end{bmatrix}$$

$$\text{so that: } \mathrm{tr}\,(BA) = a_1b_1 + a_3b_2 + a_2b_3 + a_4b_4 = \mathrm{tr}\,(AB)$$

3.5 Lambda Matrices

Since many structures exhibit velocity-dependent forces, the ideas in Section 3.4 need to be extended to equations of the form

$$A_1\ddot{\mathbf{q}} + A_2\dot{\mathbf{q}} + A_3\mathbf{q} = \mathbf{0} \tag{3.64}$$

Of course, this expression could be placed in the state space form of Equation (2.25), and the methods of Section 3.4 can be applied. In fact, many numerical algorithms do exactly that. However, the second-order form does retain more of the physical identity of the problem and hence is worth developing.

Again, assume solutions of Equation (3.64) of the form $\mathbf{q}(t) = \mathbf{u}e^{\lambda t}$, where $\mathbf{u}$ is a nonzero vector of constants. Then Equation (3.64) becomes

$$(\lambda^2 A_1 + \lambda A_2 + A_3)\mathbf{u}e^{\lambda t} = \mathbf{0}$$

or, since $e^{\lambda t}$ is never zero

$$(\lambda^2 A_1 + \lambda A_2 + A_3)\mathbf{u} = \mathbf{0}$$

This last expression can be written as

$$D_2(\lambda)\mathbf{u} = \mathbf{0} \tag{3.65}$$

where $D_2(\lambda)$ is referred to as a *lambda matrix* and $\mathbf{u}$ is referred to as a *latent vector*. In fact, in this case, $\mathbf{u}$ is called the right latent vector (Lancaster, 1966).

Here it is important to distinguish between the concept of eigenvalues and eigenvectors of a *matrix* Equation (3.1), and eigenvalues and eigenvectors of a *system* Equation (3.65). Lancaster (1966) has suggested referring to λ and $\mathbf{u}$ of the system as latent roots and latent vectors, respectively, in order to make this distinction clear. Unfortunately, this did not catch on in the engineering literature. In order to be compatible with the literature, the distinction between eigenvectors (of a single matrix) and latent vectors (of the system) must be made from context. Equation (3.65) expresses the system eigenvectors and occasionally is referred to as a nonlinear eigenvalue problem, a matrix polynomial problem, or a lambda matrix problem.

For the existence of nonzero solutions of Equation (3.65), the matrix $D_2(\lambda)$ must be singular, so that

$$\det(D_2(\lambda)) = 0 \tag{3.66}$$

The solutions to this $2n$-degree polynomial in λ are called *latent roots*, eigenvalues, or characteristic values and contain information about the system's natural frequencies. Note that the solution of Equation (3.66) and the solution of $\det(A - \lambda I) = 0$, where A is the state matrix (Equation (2.25), given by

$$A = \begin{bmatrix} 0 & I \\ -A_1^{-1}A_3 & -A_1^{-1}A_2 \end{bmatrix} \tag{3.67}$$

are the same. Also, the eigenvectors of A are just $[\mathbf{u}_i \quad \lambda_i\mathbf{u}_i]^T$, where $\mathbf{u}_i$ are the latent vectors of Equation (3.65) and λ_i are the solutions of Equation (3.66).

An $n \times n$ lambda matrix, $D_2(\lambda)$, is said to be *simple* if A_1^{-1} exists and if for each eigenvalue (latent root) λ_i satisfying Equation (3.65), the rank of $D_2(\lambda_i)$ is $n - \alpha_i$; where α_i is the multiplicity of the eigenvalue λ_i. If this is not true, then $D_2(\lambda)$ is said to be *degenerate*. If each of the coefficient matrices are real and symmetric and if $D_2(\lambda)$ is simple, the solution of Equation (3.64) is given by

$$\mathbf{q}(t) = \sum_{i=1}^{2n} c_i \mathbf{u}_i e^{\lambda_i t} \tag{3.68}$$

Here the c_i are $2n$ constants to be determined from the initial conditions and the $\mathbf{u}_i$ are the right eigenvectors (latent vectors) of $D_2(\lambda)$. Note that if $A_2 = 0$, Equation (3.65) collapses to the eigenvalue problem of a matrix. The definitions of degenerate and simple still hold in this case.

Since, in general, $\mathbf{u}_i$ and λ_i are complex, the solution $\mathbf{q}(t)$ will be complex. The physical interpretation is as follows: The displacement is the real part of $\mathbf{q}(t)$, and the velocity is the real part of $\dot{\mathbf{q}}(t)$. The terms *modes* and *natural frequencies* can again be used if care is taken to interpret their meaning properly. The damped natural frequencies of the system are again related to the λ_i in the sense that if the initial conditions $\mathbf{q}(0)$ and $\dot{\mathbf{q}}(0)$ are chosen such that $c_i = 0$ for all values of i except $i = 1$, each coordinate $q_i(t)$ will oscillate (if underdamped) at a frequency determined by λ_i. Furthermore, if the $\mathbf{u}_i$ are normalized, i.e. if $\mathbf{u}_i^* \mathbf{u}_i = 1$, then the elements of $\mathbf{u}_i$ indicate the relative displacement and phase of each mass when the system vibrates at that frequency. Here $\mathbf{u}^*$ denotes the complex conjugate of the transpose of the vector $\mathbf{u}$.

In many situations, the coefficient matrices are symmetric and the damping matrix C is chosen to be of a form that allows the solution of Equation (2.19), repeated here

$$M\ddot{\mathbf{q}}(t) + C\dot{\mathbf{q}}(t) + K\mathbf{q}(t) = \mathbf{0} \tag{3.69}$$

to be expressed as a linear combination of the normal modes, or eigenvectors of the matrix $\tilde{K}$, which, of course, are real. In this case, the matrix of eigenvectors decouples the equations of motion. In fact, the main reason for this assumption is the convenience offered by the analysis of systems that decouple. The advantage in the normal mode case is that the eigenvectors are all real valued. To this end, consider the symmetric damped system of Equation (3.69) and note the following:

1) If $C = \alpha M + \beta K$, where α and β are any real scalars, then the eigenvectors (latent vectors) of Equation (3.65) are the same as the eigenvectors of the same eigenvalue problem with $C = 0$.
2) If $C = \sum_{i=1}^{n} \beta_{i-1} K^{i-1}$, where β_i are real scalars, then the eigenvectors of Equation (2.13) are the same as the eigenvectors of the undamped system ($C = 0$).
3) The eigenvectors of Equation (2.13) are the same as those of the undamped system (with $C = 0$), if and only if $CM^{-1}K = KM^{-1}C$ (Caughey and O'Kelly, 1965).

Systems satisfying any of the above rules are said to be *proportionally damped*, to have *Rayleigh damping*, or to be *normal mode systems*. Such systems can be decoupled by the modal matrix associated with the matrix $\tilde{K}$.

Of the cases just mentioned, (3) is the most general and includes the other two as special cases. It is interesting to note that case 3 follows from a linear algebra theorem that states that two symmetric matrices have the same eigenvectors, if and only if they commute (Bellman, 1970), i.e. if and only if there exists a similarity transformation simultaneously diagonalizing both matrices. It is also worth noting that in the normal mode case the eigenvectors are real, but the reverse is not true (see the discussion of overdamping Barkwell and Lancaster, 1992). That is, some structures with real-valued eigenvectors are not normal mode systems, because the matrix of modal vectors does not decouple the equations of motion (i.e. diagonalize the coefficient matrices). The significance of complex eigenvectors is that the elements are not in phase with each other as they are in the normal mode case. Some researchers have incorrectly stated that if the damping is small in value, normal modes can be assumed. However, even small amounts of damping can cause condition 3 above to be violated, resulting in complex mode shapes (Lallement and Inman, 1995).

As a generic illustration of a normal mode system, let S_m be the matrix of eigenvectors of K normalized with respect to the mass matrix M (i.e. $S_m = M^{-1/2}S$) so that

$$\begin{aligned} S_m^T M S_m &= I \\ S_m^T K S_m &= \Lambda_K = \text{diag}\left[\omega_i^2\right] \end{aligned} \tag{3.70}$$

where ω_i^2 are the eigenvalues of the matrix K and correspond to the square of the natural frequencies of the undamped system. If case 3 holds, then the damping is also diagonalized by the transformation S_m, so that

$$S_m^T C S_m = \text{diag}[2\zeta_i\omega_i] \tag{3.71}$$

where ζ_i are called the modal damping ratios. Then Equation (3.64) can be transformed into a diagonal system via the following: Let $\mathbf{q}(t) = S_m\mathbf{y}(t)$ in Equation (2.13) and premultiply by S_m^T to get

$$\ddot{y}_i(t) + 2\zeta_i\omega_i\dot{y}_i(t) + \omega_i^2 y_i(t) = 0,\ i = 1, 2, \ldots, n \tag{3.72}$$

where $y_i(t)$ denotes the ith component of the vector $\mathbf{y}(t)$. Each of the n Equations (3.72) is a scalar, which can be analyzed by the methods of Chapter 1 for SDOF systems. In this case, the ζ_i are called *modal damping ratios* and the ω_i are the undamped *natural frequencies*, or modal frequencies.

Alternately, the modal decoupling described in the above paragraph can be obtained by using the mass normalized stiffness matrix. To see this, substitute $\mathbf{q} = M^{-1/2}\mathbf{r}$ into Equation (3.69), multiply by $M^{-1/2}$ to form $\tilde{K} = M^{-1/2}KM^{-1/2}$, compute the normalized eigenvectors of $\tilde{K}$ and use these to form the columns of the orthogonal matrix S. Next, use the substitution $\mathbf{r} = S\mathbf{y}$ in the equation of motion, premultiply by S^T and Equation (3.72) results. This procedure is illustrated in the following example.

Example 3.5.1

Let the coefficient matrices of Equation (3.69) have the values

$$M = \begin{bmatrix} 9 & 0 \\ 0 & 1 \end{bmatrix}, \; C = \begin{bmatrix} 9 & -1 \\ -1 & 1 \end{bmatrix}, \; K = \begin{bmatrix} 27 & -3 \\ -3 & 3 \end{bmatrix}$$

Calculating $CM^{-1}K$ yields

$$CM^{-1}K = \begin{bmatrix} 30 & -6 \\ -6 & \dfrac{10}{3} \end{bmatrix}$$

which is symmetric and hence equal to $KM^{-1}C$, so that condition 3 is satisfied. From Example 3.3.2, the eigenvectors and eigenvalues of the matrix $\tilde{K}$ are

$$\mathbf{u}_1^T = [1 \quad 1]/\sqrt{2}, \lambda_1 = 2, \mathbf{u}_2^T = [-1 \quad 1]/\sqrt{2}, \text{ and } \lambda_2 = 4$$

Then $S^T M^{-1/2} C M^{-1/2} S = \text{diag}\left[2/3 \quad 4/3\right]$, and $S^T M^{-1/2} K M^{-1/2} S = \text{diag}\ [2 \quad 4]$. Hence, the solution to Equation (3.69) is captured in the two scalar equations given by

$$\ddot{y}_1(t) + (2/3)\dot{y}_1(t) + 2y_1(t) = 0$$

and

$$\ddot{y}_2(t) + (4/3)\dot{y}_2(t) + 4y_2(t) = 0$$

each of which can easily be solved by the methods described in Chapter 1. From the displacement coefficient, the frequencies are

$$\omega_1 = \sqrt{2} \text{ rad/s, and } \omega_2 = \sqrt{4} = 2 \text{ rad/s}$$

and from the velocity coefficients. the damping ratios are

$$\zeta_1 = \frac{2}{3}\frac{1}{2\omega_1} = \frac{1}{3\sqrt{2}}, \quad \text{and} \quad \zeta_2 = \frac{4}{3}\frac{1}{2\omega_2} = \frac{1}{3}.$$

To regain the solution in the physical coordinate system, use the transformation

$$\mathbf{q}(t) = M^{-1/2} S\mathbf{y}(t)$$

3.6 Eigenvalue Estimates

In many instances, it is enough to know an approximate value, or estimate, of a particular eigenvalue or how changes in certain parameters affect the natural frequencies. Methods that require less computation than solving the characteristic equation of a given system, but yet yield some information about the eigenvalues of the system, may be useful. As an example, consider the SDOF spring-mass system driven by $F_0 \sin \omega t$. If, in a given design situation, one wanted to avoid resonance, it would be enough to know that the natural frequency is less than the driving frequency, ω. Also, since the free response of the system is a function of the eigenvalues, estimates of eigenvalues yield some estimates of the nature of the free response of the structure and may lead to design inequalities.

One of the most basic estimates of the eigenvalues of a symmetric matrix is given by Rayleigh's principle. This principle states that if λ_{min} is the smallest eigenvalue of the symmetric matrix A and λ_{max} is its largest, then for any nonzero vector $\mathbf{x}$

$$\lambda_{min} \leq \frac{\mathbf{x}^T A\mathbf{x}}{\mathbf{x}^T\mathbf{x}} \leq \lambda_{max} \tag{3.73}$$

This quotient defines what is called the *Rayleigh quotient* for the matrix A, i.e. the Rayleigh quotient is defined as the scalar ratio $R(\mathbf{x}) = \mathbf{x}^T A\mathbf{x}/\mathbf{x}^T\mathbf{x}$ (see Huseyin (1978) for a proof).

The variational characterization of Rayleigh's quotient can also be used to characterize the other eigenvalues of A. If the minimization of the Rayleigh quotient is carried out over all vectors orthogonal to the first eigenvector, the second eigenvalue results. The ith eigenvalue is calculated by

$$\lambda_i = \min_{\substack{\mathbf{x}^T\mathbf{x}=1 \\ \mathbf{x}^T\mathbf{x}_k=0}} (\mathbf{x}^T A\mathbf{x}),\ k = 1,\ 2,\ \ldots,\ i-1 \tag{3.74}$$

which states that the ith eigenvalue is obtained by taking the minimum value of $\mathbf{x}^T A\mathbf{x}$ over all vectors $\mathbf{x}$ that satisfy $\mathbf{x}^T\mathbf{x} = 1$ and that are orthogonal to the first $(i - 1)$ eigenvectors.

To apply Rayleigh's quotient to the vibration problem of a conservative system

$$M\ddot{\mathbf{q}} + K\mathbf{q} = \mathbf{0} \tag{3.75}$$

requires little manipulation. Recall that the eigenvalue problem for Equation (3.75) can be written as

$$\lambda M\mathbf{u} = \mathrm{K}\mathbf{u}$$

or

$$R(\lambda, \mathbf{u}) = \frac{\mathbf{u}^T K\mathbf{u}}{\mathbf{u}^T M\mathbf{u}} \tag{3.76}$$

where the notation $R(\lambda,\mathbf{u})$ denotes the Rayleigh quotient. Equation (3.76) can be examined for all vectors, such that $\mathbf{u}^T M\mathbf{u} = 1$. Alternatively, $R(\lambda,\mathbf{u})$ can be formed for the system of equations in Equation (3.33) to yield

$$R(\lambda,\mathbf{q}) = \mathbf{q}^T M^{-1/2} K M^{-1/2}\mathbf{q} \tag{3.77}$$

which can be examined for all vectors $\mathbf{q}$ with $\|\mathbf{q}\| = \sqrt{\mathbf{q}^T\mathbf{q}} = 1$, called the *norm* of $\mathbf{q}$.

Example 3.6.1

Consider the system of Figure 2.1 with $m_1 = 1$, $m_2 = 4$, $k_1 = 2$ and $k_2 = 1$. The nondimensional equation of motion is then given by

$$\begin{bmatrix} 1 & 0 \\ 0 & 4 \end{bmatrix}\ddot{\mathbf{q}} + \begin{bmatrix} 3 & -1 \\ -1 & 1 \end{bmatrix}\mathbf{q} = 0$$

where $\mathbf{q} = [x_1\ x_2]^T$. Since M is diagonal

$$M^{-1/2} = \begin{bmatrix} 1 & 0 \\ 0 & 0.5 \end{bmatrix}$$

and $R(\lambda, \mathbf{q})$ from Equation (3.77) becomes

$$R(\lambda, \mathbf{q}) = \mathbf{q}^T \begin{bmatrix} 3 & -1/2 \\ -1/2 & 1/4 \end{bmatrix} \mathbf{q}$$

If a trial vector is chosen (out of thin air and then normalized) of $\mathbf{q} = [0.243\ 0.970]^T$, then $R(\lambda, \mathbf{q}) = 0.176$. Since the actual value is $\lambda_1 = 0.1619$, the Rayleigh quotient appears to be a reasonable estimate.

Again, note that the Rayleigh method provides an estimate of λ_1 without having to solve for the roots of the characteristic equation. It should also be noted that the method is not as accurate as it may sometimes appear from the usual textbook examples. If the trial vector $\mathbf{q}$ is "near" the first eigenvector, the estimate will be fairly close. If not, the estimate will not be as good. For instance, if $\mathbf{q} = [1\ 0]^T$ is chosen in the preceding example, then $R(\lambda, \mathbf{q}) = 3$, which is not a very good estimate of λ_1. The Rayleigh Quotient however is of use in analysis and design, while of little computational value for computing frequencies.

Several other results of interest involving eigenvalue inequalities are useful in vibration analysis. One is a method of determining the effect that truncating degrees of freedom of a system on the eigenvalues of the system. Let the symmetric matrix A be $n \times n$ with eigenvalues, λ_i, ordered as

$$\lambda_1 \leq \lambda_2 \leq \ldots \leq \lambda_n$$

and let the matrix B be formed from the matrix A by deleting a row and column. Hence, B is $(n - 1)$ by $(n - 1)$, so it will have $n - 1$ eigenvalues, which are denoted by

$$\gamma_1 < \gamma_2 < \ldots < \gamma_{n-1}$$

It can be shown that these two sets of eigenvalues are *interlaced*, i.e. that

$$\lambda_1 \leq \gamma_1 \leq \lambda_2 \leq \gamma_2 \leq \lambda_3 \leq \ldots \leq \gamma_{n-1} \leq \lambda_\mathrm{n} \tag{3.78}$$

This last statement shows that the natural frequencies of a system decrease as the number of degrees of freedom increase. In fact, if A_r denotes a symmetric $r \times r$ matrix and $\lambda_i(A_r)$ denotes the ith eigenvalue of the matrix A_r, then

$$\lambda_i(A_{r+1}) \leq \lambda_\mathrm{i}(A_\mathrm{r}) \leq \lambda_{i+1}(\mathrm{A}_{\mathrm{r}+1}) \tag{3.79}$$

This is referred to as a *Sturmian Separation Theorem* (Bellman, 1970) and is useful in illustrating how the order of a vibration model affects the natural frequencies, such as when model reduction is used (defined in Section 6.12).

Another useful result reported by Bellman (1970) is that if A and B are $n \times n$ symmetric matrices, then

$$\lambda_k(A + B) \geq \lambda_k(A),\ k = 1, 2, \ldots, n \tag{3.80}$$

if B is positive semidefinite, and

$$\lambda_k(A+B) > \lambda_k(A), k = 1, 2, \ldots, n \tag{3.81}$$

if B is positive definite. Here $\lambda_k(A + B)$ refers to the kth eigenvalue of the matrix $A + B$, etc.

Many times the physical parameters of a system are known only to a certain precision. For instance, mass and stiffness coefficients may be measured accurately for most systems, but viscous damping coefficients are very hard to measure and are not always known to a high degree of accuracy.

A symmetric matrix with error in its elements can be written as the sum

$$B = A + E_e \tag{3.82}$$

where B is a known symmetric matrix with known eigenvalues

$$\beta_1 \leq \beta_2 \leq \ldots \leq \beta_n$$

and A is a symmetric matrix with unknown eigenvalues

$$\lambda_1 \leq \lambda_2 \leq \ldots \leq \lambda_n$$

and E_e is a symmetric matrix representing the errors in the matrix B. The object is to estimate λ_i given the numbers β_i, without knowing too much about the matrix E_e. It can be shown that

$$|\beta_i - \lambda_i| \leq \|E_e\| \tag{3.83}$$

where $\|E_e\|$ denotes the Euclidian norm of the matrix E_e, defined as the square root of the sum of the squares of each elements of E_e. It is easy to see that $\|E_e\| < n\varepsilon$, where n is the dimension of E_e and ε is the absolute value of the largest element in the matrix E_e. Combining these two inequalities yields

$$|\beta_i - \lambda_i| \leq n\varepsilon \tag{3.84}$$

Inequality Equation (3.95) can be used to measure the effects of errors in the parameters of a physical system on the system's eigenvalues. For instance, let $\tilde{K}$ be the mass normalized stiffness matrix of the actual system associated with Equation (3.33), which is measured by some experiment. Let B denote the matrix consisting of all measured values and let E_e be the matrix consisting of all the measured errors. Then from Equation (3.84), with $A = \tilde{K}$ and, with eigenvalues ω_i^2, the inequality becomes $|\beta_i - \omega_i^2| \leq n\varepsilon$, or $-n\varepsilon \leq \omega_i^2 - \beta_i \leq n\varepsilon$, which in turn can be written as

$$\beta_i - n\varepsilon \leq \omega_i^2 \leq \beta_i + n\varepsilon \tag{3.85}$$

This last expression indicates how the *actual* natural frequencies, ω_i, are related to the calculated natural frequencies, $\beta_i^{1/2}$, and the measurement error, ε. Note that the assumption of symmetry will be satisfied for the matrix E_i, since each element is the sum of the errors of the stiffness elements in that position, so that the ij^{th} element of E_i will contain the same measurement error as the ji^{th} element of E_i.

A fundamental theorem from linear algebra that yields simple estimates of the eigenvalues of a matrix from knowledge only of its elements is attributed to Gerschgorin (Todd, 1962). Simply stated, let a_{ij} denote the ij^{th} element of a matrix A. Then every eigenvalue of A lies inside of at least one of the circles in the complex plane centered at a_{ii} of radius

$$r_i = \sum_{\substack{j=1 \\ j \neq i}}^{n} |a_{ij}| \tag{3.86}$$

If a disc has no point in common with any other disc, it contains only one eigenvalue. The following example serves to illustrate the statement of Gerschgorin's theory for a symmetric matrix.

Example 3.6.2

Let the matrix A be

$$A = \begin{bmatrix} 2.5 & -1 & 0 \\ -1 & 5 & -\sqrt{2} \\ 0 & -\sqrt{2} & 10 \end{bmatrix}$$

Then using the formula in Equation (3.86), define three circles in the plane. The first one has center 2.5 and radius $r_1 = |a_{12}| + |a_{13}| = 1$, the second has its center at 5 with radius $r_2 = |a_{21}| + |a_{23}| = (1 + \sqrt{2})$, and the third is centered at 10 with radius $\sqrt{2}$. The circles are illustrated in Figure 3.1. The actual eigenvalues of the system are

$$\lambda_1 = 2.1193322$$
$$\lambda_2 = 5.00$$
$$\lambda_3 = 10.380678$$

which lie inside the Gerschgorin circles, as illustrated in Figure 3.1.

In the course of the development of a prototype, a system is built, analyzed and finally tested. At that point, small adjustments are made in the design to fine-tune the system so that the prototype satisfies all the response specifications. Once these design changes are made, it may not be desirable or efficient to recalculate

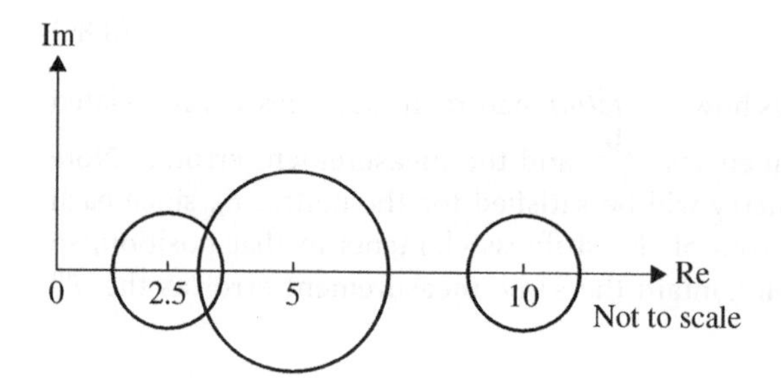

Figure 3.1 The relationship between Gerschgorin circles and eigenvalues.

the eigensolution. Instead, a perturbation technique may be used to show how small changes in the elements of a matrix affect its eigensolution.

Perturbation methods are based on approximations of a function obtained by writing down a Taylor series expansion (see any introductory calculus text) for a function about some point. The equivalent statement for matrix and vector functions is more difficult to derive. However, with proper assumptions, a similar expansion can be written down for the eigenvalue problem.

In the following, let A denote an $n \times n$ symmetric matrix with distinct eigenvalues, denoted by μ_i, and refer to A as the *unperturbed matrix*. Define the matrix $A(\varepsilon)$ by $A(\varepsilon) = A + \varepsilon B$. The matrix $A(\varepsilon)$ is called the *perturbed matrix*. Note that $A(0) = A$. Furthermore, denote the eigenvalues of $A(\varepsilon)$ by $\lambda_i(\varepsilon)$ and the corresponding eigenvectors by $\mathbf{x}_i(\varepsilon)$. It is clear that as ε approaches zero, $\lambda_i(\varepsilon)$ approaches μ_i and $\mathbf{x}_i(\varepsilon)$ approaches $\mathbf{x}_i$ for each value of the index i. Here μ_i and $\mathbf{x}_i$ are the eigenvalues and eigenvectors of A, respectively (Lancaster, 1969). For sufficiently small ε and symmetric A and B, the expansions for $\lambda_i(\varepsilon)$ and $\mathbf{x}_i(\varepsilon)$ are

$$\lambda_i(\varepsilon) = \lambda_i + \varepsilon\lambda_i^{(1)} + \varepsilon^2\lambda_i^{(2)} + \cdots \tag{3.87}$$

and

$$\mathbf{x}_i(\varepsilon) = \mathbf{x}_i + \varepsilon\mathbf{x}_i^{(1)} + \varepsilon^2\mathbf{x}_i^{(2)} + \cdots \tag{3.88}$$

where $\mathbf{x}_i^T\mathbf{x}_i = 1$. Here, the parenthetical superscript (k) denotes the kth derivative, with respect to the parameter ε, evaluated at $\varepsilon = 0$ and multiplied by $(1/k!)$. That is

$$\lambda_i^{(k)} = \left(\frac{1}{k!}\right)\left[\frac{d^k\lambda_i}{d\varepsilon^k}\right]_{\varepsilon=0}$$

Here, differentiation of the vector $\mathbf{x}$ is defined by differentiating each element of $\mathbf{x}$.

Next, consider the ith eigenvalue problem for the perturbed matrix

$$A(\varepsilon)\mathbf{x}_i(\varepsilon) = \lambda_i(\varepsilon)\mathbf{x}_i(\varepsilon) \tag{3.89}$$

Substitution of Equations (3.87) and (3.88) into Equation (3.89) yields

$$(A + \varepsilon B)(\mathbf{x}_i + \varepsilon\mathbf{x}_i^{(1)} + \varepsilon^2\mathbf{x}_i^{(2)} + \cdots) = (\lambda_i + \varepsilon\lambda_i^{(1)} + \varepsilon^2\lambda_i^{(2)} + \cdots)(\mathbf{x}_i + \varepsilon\mathbf{x}_i^{(1)} + \cdots) \tag{3.90}$$

Multiplying this last expression out, and comparing coefficients of the powers of ε yields several useful relationships. The result of comparing the coefficients of ε^0 is just the eigenvalue problem for the unperturbed system. The coefficient of ε^1, however, yields the expression

$$(\lambda_i I - A)\mathbf{x}_i^{(1)} = (B - \lambda_i^{(1)}I)\mathbf{x}_i \tag{3.91}$$

Premultiplying this by $\mathbf{x}_i^T$ results in (suppressing the index)

$$\mathbf{x}^T(B - \lambda^{(1)}I)\mathbf{x} = \mathbf{x}^T(\lambda I - A)\mathbf{x}^{(1)} = 0 \tag{3.92}$$

The last term in Equation (3.92) is zero, since $\mathbf{x}^T$ is the left eigenvector of A, i.e. $\mu\mathbf{x}_i^T = \mathbf{x}_i^T A$. Hence, the first term in the perturbation of the eigenvalue becomes (recall that $\mathbf{x}^T\mathbf{x} = 1$)

$$\lambda_i^{(1)} = \mathbf{x}_i^T B\mathbf{x}_i \tag{3.93}$$

Equation (3.93) indicates how the eigenvalues of a matrix, and hence the natural frequencies of an undamped system, change as the result of a small change, εB, in the matrix values. This is illustrated in Example 3.6.3. The preceding formulas can be used to calculate the eigenvalues of the perturbation matrix in terms of the perturbation matrix itself and the known eigensolution of the unperturbed system defined by A. Equations (3.87) can be used to yield the eigenvalues of the "new", or perturbed, system by making the approximations $\lambda_i(\varepsilon) = \mu_i + \varepsilon\lambda_i^{(1)}$ and using Equation (3.93). This method is good for small values of ε.

Perturbation schemes can also be used to calculate the effect of the perturbation on the eigenvectors as well. In addition, the method can be easily used for nongyroscopic conservative systems of the forms given in Equation (3.32). It has also been used for damped systems and for systems with gyroscopic forces. Example 3.6.3 illustrates its use for systems of the form of Equation (3.33).

Example 3.6.3

This example illustrates the use of perturbation calculations to find the result of making a small perturbation to a given system (here A is perturbed to $A(\varepsilon)$)

$$M^{-1/2}KM^{-1/2} = A = \begin{bmatrix} 3 & -1 & 0 \\ -1 & 1 & -1 \\ 0 & -1 & 5 \end{bmatrix}, \text{ and } A(\varepsilon) = \begin{bmatrix} 3.1 & -1.1 & 0 \\ -1.1 & 1.1 & -1 \\ 0 & -1 & 5 \end{bmatrix},$$

Suppose the eigensolution of A is known, i.e.

$$\begin{aligned} \lambda_1 &= 0.3983 \\ \lambda_2 &= 3.3399 \\ \lambda_3 &= 5.2618 \\ \mathbf{x}_1 &= [0.3516 \quad 0.9148 \quad 0.1988]^T \\ \mathbf{x}_2 &= [-0.9295 \quad 0.3159 \quad 0.1903]^T \\ \mathbf{x}_3 &= [0.1113 \quad -0.2517 \quad 0.9614]^T \end{aligned}$$

Given this information, the eigensolution of the new system $A(\varepsilon)$ is desired, where

$$\varepsilon B = A(\varepsilon) - A = (0.1)\begin{bmatrix} 1 & -1 & 0 \\ -1 & 1 & 0 \\ 0 & 0 & 0 \end{bmatrix}$$

Here $\varepsilon = 0.1$ is small, so that the series in Equation (3.87) converges and can be truncated. Equation (3.93) yields

$$\begin{aligned} \varepsilon\lambda_1^{(1)} &= \mathbf{x}_1^T \varepsilon B \mathbf{x}_1 = 0.03172 \\ \varepsilon\lambda_2^{(1)} &= \mathbf{x}_2^T \varepsilon B \mathbf{x}_2 = 0.15511 \\ \varepsilon\lambda_3^{(1)} &= \mathbf{x}_3^T \varepsilon B \mathbf{x}_3 = 0.01317 \end{aligned}$$

Then the new (perturbed) eigenvalues are

$$\lambda_1(\varepsilon) = 0.43002\ (0.4284)$$
$$\lambda_2(\varepsilon) = 3.55410\ (3.4954)$$
$$\lambda_3(\varepsilon) = 5.27497\ (5.2762)$$

Here the actual values are given in parentheses for comparison.

The methods presented in this section are not really needed to compute eigenvalues. Rather the methods of the following section should be used for computing accurate eigenvalues and modal data. The eigenvalue approximations and bounds presented in this section are significant analytical tools that can be used in design and redesign to understand how changes in the system or the system's model effect modal data.

3.7 Computation Eigenvalue Problems in MATLAB

The availability of cheap, high-speed computing and the subsequent development of high-level mathematically-oriented computer codes (MATLAB, Mathcad and Mathematic, in particular) almost negates the need for eigenvalue approximation methods and schemes presented in the previous section. The very nature of many computational schemes demands that the analytical formulation change. The following presents some alternative formulations to matrix related computations based on the available codes. The details of the various algorithms used in these codes are left to the references (Datta, 1995; Golub and Van Loan, 1996; Meirovitch, 1980). Table 3.1 lists various MATLAB commands useful in computing natural frequencies, damping ratios and mode shapes.

Table 3.1 Sample MATLAB matrix commands for solving the eigenvalue problem.

Command	Description
`M= [1 0; 0 4]`	creates the mass matrix of Example 3.6.1
`Chol(M)`	computes the Cholesky factor of the matrix M
`Sqrtm(M)`	computes the matrix square root of M
`inv(M)`	computes the inverse of the matrix M
`M\I`	computes the inverse of the matrix M using Gaussian elimination
`d=eig(A)`	returns a vector d containing the eigenvalues of A
`[V D]=eig(A)`	returns a matrix V of eigenvectors and matrix D of eigenvalues
`[V D]=eig(A'nobalance')`	returns a matrix V of eigenvectors and matrix D of eigenvalues without balancing
`d=eig(A B)`	returns a vector d of eigenvalues using the generalized problem $A\mathbf{u} = \lambda B\mathbf{u}$ (works for B singular)
`[V D]=eig(A B)`	returns a matrix D of eigenvalues and matrix V of eigenvectors solving the generalized problem $A\mathbf{u} = \lambda B\mathbf{u}$

The best way to compute a matrix inverse is not to. Rather Gaussian elimination can be used to effectively solve for the inverse of a matrix. The matrix inverse can be thought of as the solution to a system of n linear equations in n variables written in the matrix form: $A\mathbf{x} = \mathbf{b}$. Solving this by Gaussian elimination yields the effective inverse:

$$\mathbf{x} = A^{-1}\mathbf{b}$$

The best way to compute the eigenvalues and eigenvectors of a matrix is to use one of the many eigenvalue routines developed by the numerical linear algebra community and packaged nicely in a variety of commercial codes. These are both numerically superior to computing the roots of the polynomial derived from $\det(\lambda I - A)$, and applicable to much larger order systems.

The matrix square root can be computed by using the function of a matrix approach, which is trivial for diagonal matrices (as often, but not always, is the case for the mass matrix). However, for nondiagonal matrices, the square root involves solving the eigenvalue problem for the matrix. This is given in Equation (3.31) and repeated here. If M is a positive definite matrix, then its eigenvalues μ_i are all positive numbers, its eigenvectors, $\mathbf{u}_i$ form an orthonormal set and can be used to form an orthogonal matrix $S = [\mathbf{u}_1\ \mathbf{u}_2\ \dots\ \mathbf{u}_n]$, such that $S^TMS = \text{diag}(\mu_i)$. Then any scalar function f of the matrix M can be computed by

$$f(M) = S\text{diag}\left[f(\mu_1)\quad f(\mu_2)\quad \cdots \quad f(\mu_n)\right] S^T \tag{3.94}$$

In particular, the inverse and matrix square root of any positive definite matrix can be computed with Equation (3.94).

An alternative to the eigenvalue decomposition of Equation (3.94) is to use the Cholesky decomposition, or Cholesky factors of a positive definite matrix. As introduced earlier, recall that any positive definite matrix can be factored into the product of an upper triangular matrix U and its transpose, $M = U^TU$. In this case, it follows that

$$(U^T)^{-1}MU^{-1} = I$$

Hence, the Cholesky factor U behaves like a square root. In fact, if M is diagonal, $U = U^T$ is the square root of M.

The most efficient way to compute the undamped eigenvalues is to use the Cholesky factors. In this case, the transformation of Equations (3.33) and (3.69) become

$$\tilde{K} = (U^T)^{-1}KU^{-1}, \quad \text{and } \tilde{C} = (U^T)^{-1}CU^{-1}$$

So far, several different approaches to computing the natural frequencies and mode shapes of a conservative system have been presented. These are summarized in Table 3.2, along with a computational "time" measured by listing the floating-point operations per second (flops) for a given example in MATLAB. The command "flops" is no longer available in MATLAB.

Note from Table 3.2 that using the Cholesky factor U requires the least flops to produce the eigenvalues and eigenvectors. The next "fastest calculation" is using

Table 3.2 Comparison of the computing "time" required to calculate eigenvalues and eigenvectors for the various methods for a conservative system.

Method	Flops
`inv(U')*K*inv(U)`	118
`M\K`	146
`inv(M)*K`	191
`inv(sqrtm(M))*K* inv(sqrtm(M))`	228
`[V,D]=eig(K,M)`	417

Gaussian elimination to compute $M^{-1}K$, but this becomes an asymmetric matrix so that the eigenvectors are not orthogonal and hence an additional computational step is required.

The eigenvalue problem can also be placed into a number of state matrix forms and these are now presented. The first and most common case is given by Equation (2.28). The associated eigenvalue problem for the state matrix is asymmetric and in general gives complex eigenvalues and eigenvectors. In addition, the eigenvectors of the state matrix are twice as long and related to the eigenvectors $\mathbf{u}_i$ in the second-order form by

$$A\mathbf{z} = \lambda\mathbf{z},\ A = \begin{bmatrix} 0 & I \\ -M^{-1}K & -M^{-1}C \end{bmatrix} \Rightarrow \mathbf{z}_i = \begin{bmatrix} \mathbf{u}_i \\ \lambda_i\mathbf{u}_i \end{bmatrix} \tag{3.95}$$

The eigenvalues however are exactly the same.

Other state space approaches can be formulated by rearranging the equations of motion in state space form. For instance, in Equation (3.69), let

$$\mathbf{y}_1 = \mathbf{q} \quad \text{and} \quad \mathbf{y}_2 = \dot{\mathbf{q}}$$

This then implies that

$$\dot{\mathbf{y}}_1 = \mathbf{y}_2 \text{ and hence : } -K\dot{\mathbf{y}}_1 = -K\mathbf{y}_2$$

Then the equation of motion can be written as

$$M\dot{\mathbf{y}}_2 = -C\mathbf{y}_2 - K\mathbf{y}_1$$

Combining the last two expressions yields the state space system and symmetric generalized eigenvalue problem

$$\underbrace{\begin{bmatrix} -K & 0 \\ 0 & M \end{bmatrix}}_{A} \begin{bmatrix} \dot{\mathbf{y}}_1 \\ \dot{\mathbf{y}}_2 \end{bmatrix} = \underbrace{\begin{bmatrix} 0 & -K \\ -K & -C \end{bmatrix}}_{B} \begin{bmatrix} \mathbf{y}_1 \\ \mathbf{y}_2 \end{bmatrix} \Rightarrow \lambda A\mathbf{y} = B\mathbf{y}$$

which does not require a matrix inverse.

Alternate forms of solving the eigenvalue problem can be useful for special cases, such as a nearly singular mass matrix. Such formulas can also be useful for analysis. Once the state space eigenvalue problem is solved, the data needs to be related to natural frequencies, damping ratios and mode shapes of the physical system. This can be done in the case of an underdamped system by representing all of the eigenvalues as the complex pairs

$$\lambda_i = -\zeta_i\omega_i - \omega_i\sqrt{1-\zeta_i^2}j \text{ and } \lambda_{i+1} = -\zeta_i\omega_i + \omega_i\sqrt{1-\zeta_i^2}j$$

Comparing this form to the complex form $\lambda_i = \alpha_i + \beta_i j = \text{Re}(\lambda_i) + \text{Im}(\lambda_i)j$ yields that the modal frequencies and damping ratios can be determined by

$$\begin{aligned} \omega_i &= \sqrt{\alpha_i^2 + \beta_i^2} = \sqrt{\text{Re}(\lambda_i)^2 + \text{Im}(\lambda_i)^2} \\ \zeta_i &= \frac{-\alpha_i}{\sqrt{\alpha_i^2 + \beta_i^2}} = \frac{-\text{Re}(\lambda_i)}{\sqrt{\text{Re}(\lambda_i)^2 + \text{Im}(\lambda_i)^2}} \end{aligned} \tag{3.96}$$

The mode shapes are taken as the first n values of the $2n$ state vector by the relationship given in Equation (3.95). The mode shapes in this case are likely to be complex valued, even if the condition for normal modes to exist is satisfied ($CM^{-1}K = KM^{-1}C$). In this case, there will be a normalizing condition on $\mathbf{u}$ in Equation (3.95) that will normalize the modes to be real valued. If, however, $CM^{-1}K \neq KM^{-1}C$, then the vector $\mathbf{u}$ will be complex meaning that the masses pass through their equilibrium out of phase with each other.

3.8 Numerical Simulation of the Time Response in MATLAB™

The time response can be computed by calculating the eigenvalues and eigenvectors of the system and then forming the summation of modes, as outlined in Example 3.3.2. This same procedure also works for the damped case, as long as the damping is proportional. However, for systems that do not have proportional damping ($KM^{-1}C$ not symmetric), then the modal summations are over-complex values, which can occasionally lead to confusion. In these cases, numerical simulation can be performed to compute the time response directly without computing the eigenvalues and eigenvectors. The method follows directly from the material in Section 1.11 with the state spaced model of Equations (2.28) and (3.97). For any class of second-order systems, the equations of motion can be written in the state space form, as given in Equation (2.28) and repeated here (for the free response case, $\mathbf{f}(t) = \mathbf{0}$): $\dot{\mathbf{x}} = A\mathbf{x}$, $\mathbf{x}(0) = \mathbf{x}_0$, where

$$\mathbf{x} = \begin{bmatrix} \mathbf{q} \\ \dot{\mathbf{q}} \end{bmatrix}, \text{ and } A = \begin{bmatrix} 0 & I \\ -M^{-1}(K+H) & -M^{-1}(C+G) \end{bmatrix}$$

To solve this using numerical integration, the Runge-Kutta ode command in MATLAB is used. The ode command uses a fifth-order Runge-Kutta, automated time step method for numerically integrating the equation of motion (Inman, 2014). The following example illustrates the procedure.

Example 3.8.1

Compute the response of the following system

$$M = \begin{bmatrix} 4 & 0 \\ 0 & 3 \end{bmatrix}, \; C = \begin{bmatrix} 2 & -1 \\ -1 & 1 \end{bmatrix}, \; G = \begin{bmatrix} 0 & 1 \\ -1 & 0 \end{bmatrix}, \; K = \begin{bmatrix} 10 & -4 \\ -4 & 4 \end{bmatrix}$$

to the initial conditions of

$$\mathbf{x}(0) = \begin{bmatrix} 0.1 \\ 0 \end{bmatrix} \text{m}, \quad \dot{\mathbf{x}}(0) = \begin{bmatrix} 0 \\ 0 \end{bmatrix} \text{m/s},$$

using numerical integration by MATLAB.

In order to numerically integrate the equations of motion in MATLAB using Runge-Kutta, an M-file must first be created containing the system dynamics and stored (Example 1.11.2). The following file sets up the equations of motion in state space form:

```
---------------
function v=f391(t,x)
M=[4 0; 0 3];C=[2 -1;-1 1];G=[0 1; -1 0];K=[10 -4;-4 4];
A=[zeros(2) eye(2);-inv(M)*K -inv(M)*(C+G)];
v=A*x;
---------------
```

This function must be saved under the name f391.m. Note that the command `zeros(n)` produces and nxn matrix of zeros and that the matrix `eye(n)` creates an nxn identity matrix. Once this is saved, the following is typed in the command window:

```
>>clear all
>>xo=[0.1;0;0;0];
>>ts=[0 40];
>>[t,x]=ode45('f391',ts,xo);
>>xlabel('Time (s)'); ylabel('Displacement (m)')
>>plot(t,x(:,1),t,x(:,2),'-')
```

This returns the plot shown in Figure 3.2. Note the command `x(:,1)` pulls off the record for $x_1(t)$ and the command ode45 calls a fifth-order Runge-Kutta program. The command "`ts=[0 40];`" tells the code to integrate from 0 to 40 time units (seconds in this case).

The plot illustrated in Figure 3.2 can also be labeled and titled using additional plotting commands in MATLAB. For instance, typing "`,title('displacement versus time')`" after the plot command in the code in Example 3.8.1 would add a title to the plot.

This numerical solution technique also still applies if the system is nonlinear. In this case, the state space formulation becomes a nonlinear vector rather then a matrix. This form was illustrated in Equations (1.81) and (1.82), and again in

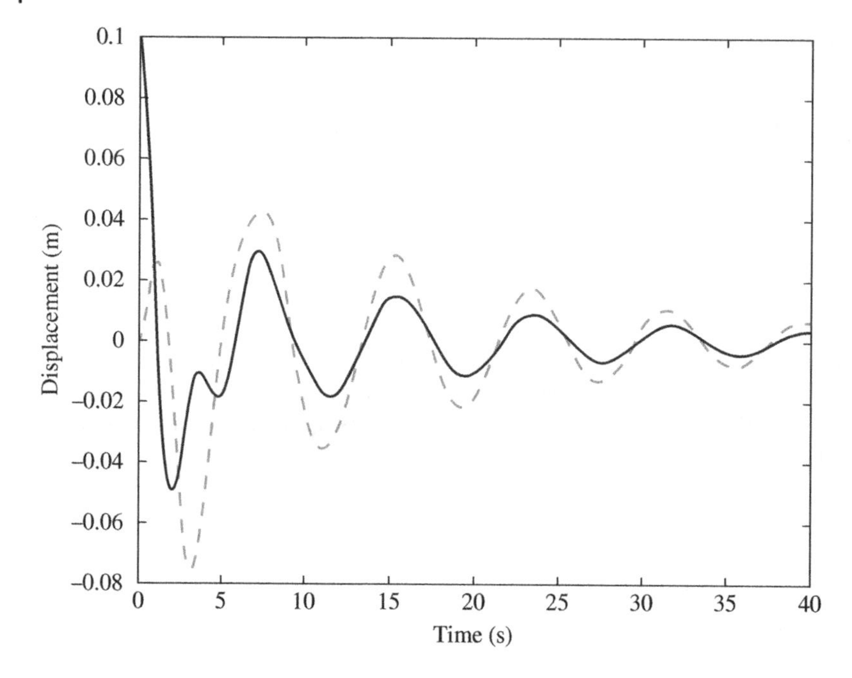

Figure 3.2 The response $q_1(t)$ versus time (solid line) and $q_2(t)$ versus time (dashed line), as computed in MATLAB using numerical integration.

Section 2.7. An example of the state space form of a nonlinear system is given in Example 2.7.2.

Chapter Notes

The material of Section 3.2 can be found in any text concerning linear algebra or matrices, such as Lancaster (1969). An excellent quick summary of relevant matrix results is available in the first chapter of Huseyin (1978). A very good historical account and development can be found in Bellman (1960, 1970). An explanation of mode shapes and undamped natural frequencies in Section 3.3 can be found in any modern vibration text. Most linear algebra and matrix texts devote several chapters to canonical forms (Section 3.4); for instance both Lancaster (1966) and Bellman (1970) do. The development of lambda matrices of Section 3.5 stems mostly from the book and work of Lancaster (1966), who has published extensively in that area. The idea of decoupling the equations of motion is based on the result of commuting matrices discussed in Bellman (1960) and was set straight in the engineering literature by Caughey and O'Kelly (1965). The material of Section 3.6 follows the pattern presented in Meirovitch (1980); however, Rayleigh quotients are discussed in every vibration text and most texts on matrices – in particular, Bellman (1970) and Lancaster (1969). Bellman (1970) also treats the lacing of eigenvalues in a rigorous fashion. Gerschgorin's result is also to be found in many texts on matrices. An excellent treatment of perturbation methods can be found in Kato (1966). The results presented in Section 3.6 on perturbation of eigenvalues are due to Lancaster (1969). Other applications of perturbation results to vibration problems are presented in Hagedorn (1983) and Meirovitch and Ryland (1979). Key papers in the development of linear systems and control using linear algebra can be found in Patel et al. (1994). Information and sample codes for solving dynamics problems in MATLAB can be found in Soutas-Little and Inman (1999), or by simply typing "matlab" into Google.

References

Barkwell, L. and Lancaster P. (1992) Overdamped and gyroscopic vibrating systems. *ASME Journal of Applied Mechanics*, 59, 176–181.

Bellman, R. (1960) *Introduction to Matrix Analysis*, 1st Edition. McGraw-Hill, New York.

Bellman, R. (1968) Some inequalities for the square root of a positive definite matrix. *Linear Algebra and its Applications*, 1, 321–324.

Bellman, R. (1970) *Introduction to Matrix Analysis*, 2nd Edition. McGraw-Hill, New York.

Caughey, T. K. and O'Kelley, M. E. J. (1965) Classical normal modes in damped linear dynamic systems. *ASME Journal of Applied Mechanics*, 32, 583–588.

Datta, B. N. (1995) *Numerical Linear Algebra and Applications*. Brooks/Cole, Pacific Grove, CA.

Golub, G. H. and Van Loan, C. F. (1996) *Matrix Computations*, 3rd Edition. Johns Hopkins University Press, Baltimore.

Hagedorn, P. (1983) The eigenvalue problem for a certain class of discrete linear systems: a perturbation approach. *Proceedings of the 4th Symposium on Dynamics and Control of Large Structures*, pp. 355–372.

Huseyin, K. (1978) *Vibrations and Stability of Multiple Parameter Systems*. Sijthoff Noordhoff International Publishers, Alphen aan den Rijn.

Inman, D. J. (2014) *Engineering Vibrations*, 4th Edition. Pearson, New Jersey.

Kato, T. (1966) *Perturbation Theory for Linear Operators*. Springer-Verlag, New York.

Lancaster, P. (1966) *Lambda Matrices and Vibrating Systems*. Pergamon Press, Elmsford, NY.

Lancaster, P. (1969) *Theory of Matrices*. Academic Press, New York.

Lallament, G. and Inman, D. J. (1995) A tutorial on complex modes. *Proceedings of the 13th International Modal Analysis Conference*, Society of Experimental Mechanics, pp. 490–495.

Meirovitch, L. and Ryland, G. (1979) Response of slightly damped gyroscopic system. *Journal of Sound and Vibration*, 46(1), 149.

Meirovitch, L. (1980) *Computational Methods in Structural Dynamics*. Sijthoff & Noordhoff International Publishers, Alphen aan den Rijn.

Patel, R. V., Laub, A. J. and Van Dooren, P. M. (eds) (1994) *Numerical Linear Algebra Techniques for Systems and Control*. Institute of Electrical and Electronic Engineers, Inc., New York.

Soutas-Little, R. W. and, D. J. (1999) *MATLAB Supplement to Engineering Mechanics, Dynamics*. Prentice Hall, New Jersey.

Todd, J. (1962) *Survey of Numerical Analysis*. McGraw-Hill, New York.

Problems

3.1 The determinant of the generic matrix A

$$A = \begin{bmatrix} a & b \\ c & d \end{bmatrix}$$

is $ad - cd$. Use this simple formula to verify Equations (3.4) and (3.5) for

$$A = \begin{bmatrix} 1 & 2 \\ 3 & 4 \end{bmatrix} \text{ and } B = \begin{bmatrix} 2 & -1 \\ -1 & 1 \end{bmatrix}$$

3.2 Compute the eigenvalues of the matrix B in Problem 3.1.

3.3 Compute the determinate of

$$A = \begin{bmatrix} 1 & 3 & -2 \\ 0 & 1 & 1 \\ 2 & 5 & 3 \end{bmatrix}$$

3.4 Determine whether the matrix

$$A = \begin{bmatrix} 1 & 1 & 1 & 1 \\ 1 & -1 & 0 & 1 \\ 1 & 1 & 2 & -1 \\ 1 & 1 & 1 & 2 \end{bmatrix}$$

is singular or not by calculating the value of its determinant.

3.5 Check if the four vectors given by

$$\mathbf{x}_1 = [1 \quad 1 \quad 1 \quad 1]^T,\ \mathbf{x}_2 = [1 - 1 \quad 1 \quad 1]^T,\ \mathbf{x}_3 = [1 \quad 0 \quad 2 \quad 1]^T$$

and

$$\mathbf{x}_4 = [1 \quad 1 - 1 \quad 2]^T$$

are independent.

3.6 Determine the rank of the matrix

$$A = \begin{bmatrix} 1 & 2 & 0 & 1 \\ 1 & -1 & 3 & 1 \\ 1 & 1 & 2 & -1 \end{bmatrix}$$

3.7 Find the value of α such that the two vectors $\mathbf{x}_1 = [1 \quad 2 \quad 3]^T$ and $\mathbf{x}_2 = [3 \ -1 \ \alpha]^T$ are orthogonal.

3.8 Consider the three two matrices A and P

$$A = \begin{bmatrix} 1 & 0 \\ 0 & 2 \end{bmatrix} \text{ and } P = \begin{bmatrix} 2 & -1 \\ -1 & 1 \end{bmatrix}$$

and show that A and $B{=}P^{-1}AP$ have the same eigenvalues.

3.9 Consider Example 3.3.2 and show that if the other eigenvector

$$\hat{\mathbf{u}}_2 == \frac{1}{\sqrt{2}}\begin{bmatrix} 1 \\ -1 \end{bmatrix}$$

is chosen, that the solution does not change.

3.10 Consider the following system

$$\begin{bmatrix} 1 & 1 \\ 1 & 4 \end{bmatrix}\ddot{\mathbf{x}} + \begin{bmatrix} 3 & -1 \\ -1 & 1 \end{bmatrix}\mathbf{x} = \mathbf{0}$$

with initial conditions

$$\mathbf{x}(0) = [0 \quad 1]^T \text{ and } \dot{\mathbf{x}}(0) = [0 \quad 0]^T$$

a) Calculate the eigenvalues of the system.
b) Calculate the eigenvectors and normalize them.
c) Use (a) and (b) to write the solution $\mathbf{x}(t)$ for the preceding initial conditions.
d) Sketch $x_1(t)$ versus t and $x_2(t)$ versus t.
e) What is the solution if $\mathbf{x}(0) = [0 \quad 0]^T$ and $\dot{\mathbf{x}}(0) = [0 \quad 0]^T$?

3.11 Calculate the natural frequencies of the following system

$$\begin{bmatrix} 4 & 0 & 0 \\ 0 & 2 & 0 \\ 0 & 0 & 1 \end{bmatrix}\ddot{\mathbf{x}} + \begin{bmatrix} 4 & -1 & 0 \\ -1 & 2 & -1 \\ 0 & -1 & 1 \end{bmatrix}\mathbf{x} = \mathbf{0}$$

3.12 Consider the matrix

$$A = \begin{bmatrix} 1 & 1 \\ 0 & 2 \end{bmatrix}$$

and calculate its eigenvalues and eigenvectors. Are the left and right eigenvectors the same? Are they orthogonal? Are they biorthogonal?

3.13 Does the following system have normal modes (i.e. does it decouple)?

$$\begin{bmatrix} 1 & 0 \\ 0 & 1 \end{bmatrix}\ddot{\mathbf{x}} + \begin{bmatrix} 15 & -3 \\ -3 & 3 \end{bmatrix}\dot{\mathbf{x}} + \begin{bmatrix} 5 & -1 \\ -1 & 1 \end{bmatrix}\mathbf{x} = \mathbf{0}$$

3.14 Does the system in Problem 3.13 oscillate? Why or why not?

3.15 Consider the following system

$$\begin{bmatrix} 1 & 0 \\ 0 & 1 \end{bmatrix}\ddot{\mathbf{x}} + \begin{bmatrix} 3 & -1 \\ -1 & 3 \end{bmatrix}\dot{\mathbf{x}} + \begin{bmatrix} 4 & -2 \\ -2 & 4 \end{bmatrix}\mathbf{x} = \mathbf{0}$$

a) Calculate the eigenvalues of the system.
b) Calculate the system eigenvectors and normalize them.
c) Show that the eigenvectors can be used to diagonalize the system.
d) Calculate the modal damping ratios, damped and undamped natural frequencies.

e) Calculate the free response for $\mathbf{x}^T(0) = [1 \quad 0], \dot{\mathbf{x}}^T(0) = [0 \quad 0]$.
f) Plot the responses $x_1(t)$ and $x_2(t)$ as well as $\|\mathbf{x}(t)\|$.

3.16

a) Calculate the eigenvalues of the matrix

$$A = \begin{bmatrix} 3 & -1 \\ -1 & 2 \end{bmatrix},$$

b) What are the eigenvalues of the matrix

$$A = \begin{bmatrix} 5 & -1 \\ -1 & 4 \end{bmatrix}?$$

Think before you calculate anything.

3.17 For the matrix in Problem 3.16, calculate $\mathbf{x}_1^T A\mathbf{x}_1/\mathbf{x}_1^T\mathbf{x}_1$ and $\mathbf{x}_2^T A\mathbf{x}_2/\mathbf{x}_2^T\mathbf{x}_2$, where $\mathbf{x}_1$ and $\mathbf{x}_2$ are the eigenvectors of A. Next, choose *five* different values of a vector $\mathbf{x}$ and calculate the five scalars $\mathbf{x}^T A\mathbf{x}/\mathbf{x}^T\mathbf{x}$ for your five choices. Compare all of these numbers with the values of the eigenvalues computed in Problem 3.16. Can you draw any conclusions?

3.18 Consider the following model of a machine part which has equations of motion given by

$$\begin{bmatrix} 1 & 0 \\ 0 & 4 \end{bmatrix}\ddot{\mathbf{x}} + \begin{bmatrix} k_1 + k_2 & -k_2 \\ -k_2 & k_2 \end{bmatrix}\mathbf{x} = \mathbf{0}.$$

Let $k_1 = 2$ and $k_2 = 1$. The elements of M are known precisely, whereas the elements of K are known only to within 0.01 at worst. (Everything here is dimensionless.) Note that the machine will fail if it is disturbed by a driving frequency equal to one of the natural frequencies of the system. If there is a disturbance to this system of frequency $\sqrt{0.15}j \quad (\lambda = 0.15)$, will this system fail? Why or why not? Try to work this out with a minimum of calculation.

3.19 Referring to Problem 3.18, suppose that in order to satisfy a given manufacturing change, the spring coefficient k_1 is required to change from 2 to 2.1 units. How will this affect the natural frequencies of the system? Give a quantitative answer without recalculating the eigenvalues, that is, use perturbation results.

3.20 If m_1 is neglected in Problem 3.18, i.e. if the order of the model is reduced by one, by what would you expect the natural frequency of the new system to be bounded? Check your result by calculation.

3.21 Show that Gerschgorin's theory works for the matrices

$$R = \begin{bmatrix} 3 & -1 \\ -1 & 2 \end{bmatrix}, \quad A_1 = \begin{bmatrix} 1 & 1 \\ 0 & 2 \end{bmatrix}, \quad A_2 = \begin{bmatrix} 1 & 1 \\ -1 & 2 \end{bmatrix}$$

3.22 Prove the if A is similar to a diagonal matrix, then the eigenvectors of A form a linearly independent set.

3.23 Derive the relationship between S_m of Equation (3.20) and the matrix S of Equation (3.17).

3.24 Show that the matrices M, $M^{1/2}$, M^{-1} and M^2 all have the same eigenvectors. How are the eigenvalues related?

3.25 Prove that if a real symmetric matrix has positive eigenvalues, then it must be positive definite.

3.26 Derive Equation (3.40). Let $n = 3$, and solve symbolically for the constants of integration.

3.27 Derive Equation (3.42) from Equation (3.39).

3.28 Let S be the matrix of eigenvectors of the symmetric matrix A. Show that S^TAS is diagonal and compare it to SAS^T.

3.29 Derive the relationship between the modal matrix S in Example 3.3.2 and the matrix S_m in Equation (3.21).

3.30 Use perturbation to calculate the effect on the eigenvalues of the matrix A given in Example 3.6.2 by making the following changes in A: Change a_{11} by 0.1, a_{12} and a_{21} by 0.1, and a_{22} by 0.2.

3.31 A geometric interpretation of the eigenvector problem for a 2×2 matrix is that the eigenvectors determine the principal axis of an ellipse. Calculate the matrix A for the quadratic form

$$2x_1^2 + 2x_1x_2 + 2x_2^2 = 3 = \mathbf{x}^T A\mathbf{x}.$$

Then use the eigenvector of A to determine the principal axis for the ellipse.

3.32 Show that the eigenvalues for the first-order form, Equation (2.27), are equivalent to the latent roots of Equation (3.65) by noting that

$$\det\begin{bmatrix} A & D \\ C & B \end{bmatrix} = \det A \det[B - CA^{-1}D]$$

as long as A^{-1} exists, for the case that $G = H = 0$.

3.33 Show that the generic system of the form

$$\begin{bmatrix} m_1 & 0 \\ 0 & m_2 \end{bmatrix}\ddot{\mathbf{x}}(t) + \begin{bmatrix} c_1 + c_2 & -c_2 \\ -c_2 & c_2 \end{bmatrix}\dot{\mathbf{x}}(t) + \begin{bmatrix} k_1 + k_2 & -k_2 \\ -k_2 & k_2 \end{bmatrix}\mathbf{x}(t) = 0$$

has normal modes, if and only if

$$\frac{c_1}{c_2} = \frac{k_1}{k_2}.$$

3.34 Consider the system defined by following coefficient matrices

$$M = \begin{bmatrix} 100 & 0 & 0 \\ 0 & 200 & 0 \\ 0 & 0 & 200 \end{bmatrix}, \; K = \begin{bmatrix} 2000 & -1000 & 0 \\ -1000 & 2000 & -1000 \\ 0 & -1000 & 1000 \end{bmatrix}$$

Compute the eigenvalues, eigenvectors and natural frequencies.

3.35 Consider again the system of Problem 3.34 with a damping matrix of the form

$$C = \begin{bmatrix} 10 & -10 & 0 \\ -10 & 30 & -20 \\ 0 & -20 & 20 \end{bmatrix}$$

added to the system. Does the system have normal modes?

3.36 Compute the eigenvalues and eigenvectors for the damped system of Problem 3.35. Also compute the natural frequencies and mode shapes using a code.

3.37 Compute the response of the system defined in Problem 3.34 to the initial displacement of $\mathbf{x}(0) = [\,0.01 \quad 0 \quad 0 \quad -0.01\,]^T$ and zero initial velocity using numerical simulation.

3.38 Consider the system of Problem 3.34 with a gyroscopic term added of the form

$$G = \begin{bmatrix} 0 & 1 & 0 \\ -1 & 0 & 1 \\ 0 & -1 & 0 \end{bmatrix}$$

Compute the eigenvalues and eigenvectors. What are the natural frequencies?

3.39 Compute the time response of the system of Problem 3.35 to the initial displacement of $\mathbf{x}(0) = [\,0.01 \quad 0 \quad 0 \quad -0.01\,]^T$ and zero initial velocity using numerical simulation.

3.40 Consider the following system and compute the natural frequencies and damping ratios. Note that the system will not uncouple into modal equations and has complex modes. The system is given by

$$\begin{bmatrix} 2 & 0 \\ 0 & 1 \end{bmatrix} \ddot{\mathbf{x}}(t) + \begin{bmatrix} 1 & -0.5 \\ -0.5 & 0.5 \end{bmatrix} \dot{\mathbf{x}}(t) + \begin{bmatrix} 3 & -1 \\ -1 & 1 \end{bmatrix} \mathbf{x}(t) = 0$$

4

Stability

4.1 Introduction

A rough idea concerning the concept of stability was introduced for single degree of freedom (SDOF) systems in Section 1.8. It was pointed out that the sign of the coefficients of the acceleration, velocity and displacement terms determined the stability behavior of a given SDOF, linear system. That is, if the coefficients have the proper sign, the motion will always remain within a given bound. This idea is extended in this chapter to the multiple degree of freedom (MDOF) systems described in the previous two chapters. For the linear, MDOF case, the criteria based on the sign of the coefficients is translated into a criterion based on the definiteness of certain coefficient matrices.

It should be noted that variations of the definition of stability are adopted, depending on the nature of the particular problem under consideration. Specifically, definitions used in vibration are slightly different than those used in control and system theory treatments. However, all definitions of stability are concerned with the response of a system to certain disturbances and whether or not the response stays within certain bounds.

4.2 Lyapunov Stability

The majority of the work done on the stability behavior of dynamical systems is based on a formal definition of stability given by Lyapunov (Hahn, 1963). This definition is stated with reference to the equilibrium point, $\mathbf{x}_0$, of a given system. In the case of the linear systems, the equilibrium point can always be taken to be the origin in state space, i.e. zero initial position and velocity. The definition of Lyapunov is stated in terms of the state vector of a given system rather than in physical coordinates directly, so that the equilibrium point refers to both the position and velocity. An equilibrium position is also referred to as a critical point in mathematical discussions of stability in nonlinear systems.

Let $\mathbf{x}(0)$ represent the vector of initial conditions for a given system (both position and velocity). The system is said to have a *stable equilibrium* if for any arbitrary positive number, ε, there exists some positive number $\delta(\varepsilon)$, such that whenever

Vibration with Control, Second Edition. Daniel John Inman.

Companion Website: www.wiley.com/go/inmanvibrationcontrol2e

$\|\mathbf{x}(0)\| < \delta$, then $\|\mathbf{x}(t)\| < \varepsilon$ for all values of $t > 0$. A physical interpretation of this mathematical definition is that if the initial state is within a certain value, i.e. $\|\mathbf{x}(0)\| < \delta(\varepsilon)$, then the motion stays within another bound for all time, i.e. $\|\mathbf{x}(t)\| < \varepsilon$. Here $\|\mathbf{x}(t)\|$, called the norm of $\mathbf{x}$, is defined by $\|\mathbf{x}(t)\| = (\mathbf{x}^T\mathbf{x})^{1/2}$.

To apply this definition to the SDOF system of Equation (1.1), note that $\mathbf{x}(t) = [x(t)\ \ \dot{x}(t)]^T$. Hence

$$\|\mathbf{x}(t)\| = (\mathbf{x}^T\mathbf{x})^{1/2} = \sqrt{x^2(t) + \dot{x}^2(t)}$$

For the sake of illustration, let the initial conditions be given by $x(0) = 0$ and $\dot{x}(0) = \omega_n = \sqrt{k/m}$. Then the solution is given by $x(t) = \sin\omega_n t$. Intuitively, this system has a stable response as the displacement response is bounded by 1, and the velocity response is bounded by ω_n. The following simple calculation illustrates how this solution satisfies the Lyapunov definition of stability.

First, note that

$$\|\mathbf{x}(0)\| = (x^2(0) + \dot{x}^2(0))^{1/2} = (0 + \omega_n^2)^{1/2} = \omega_n \tag{4.1}$$

and that

$$\|\mathbf{x}(t)\| = [\sin^2\omega_n t + \omega_n^2\cos^2\omega_n t]^{1/2} < (1 + \omega_n^2)^{1/2} \tag{4.2}$$

These expressions show exactly how to choose δ as a function of ε for this system. From Equation (4.2), note that if $(1 + \omega_n^2)^{1/2} < \varepsilon$, then $\|x(t)\| < \varepsilon$. From Equation (4.1), note that if $\delta(\varepsilon)$ is chosen to be $\delta(\varepsilon) = \varepsilon\omega_n(1 + \omega_n^2)^{-1/2}$, then the definition can be followed directly to show that if

$$\|\mathbf{x}(0)\| = \omega_n \le \delta(\varepsilon) = \frac{\varepsilon\omega_n}{\sqrt{1 + \omega_n^2}}$$

is true, then $\omega_n < \varepsilon\omega_n/(\sqrt{1 + \omega_n^2})$. This last expression yields

$$\sqrt{1 + \omega_n^2} < \varepsilon$$

That is, if $\|\mathbf{x}(0)\| < \delta(\varepsilon)$, then $\sqrt{1 + \omega_n^2} < \varepsilon$ must be true, and Equation (4.2) yields that

$$\|\mathbf{x}(t)\| \le \sqrt{1 + \omega_n^2} < \varepsilon$$

Hence, by a judicious choice of the function $\delta(\varepsilon)$, it has been shown that if $\|\mathbf{x}(0)\| < \delta(\varepsilon)$, then $\|\mathbf{x}(t)\| < \varepsilon$ for all $t > 0$. This is true for any arbitrary choice of the positive number, ε.

The preceding argument demonstrates that the undamped harmonic oscillator has solutions that satisfy the formal definition of Lyapunov stability. If dissipation, such as viscous damping, is included in the formulation, then not only is this definition of stability satisfied, but also

$$\lim_{t\to\infty} \|\mathbf{x}(t)\| = 0. \tag{4.3}$$

Such systems are said to be *asymptotically stable*. As in the SDOF case, if a system is asymptotically stable it is also stable. In fact, by definition, a system is asymptotically stable if it is stable and the norm of its response goes to zero as t gets large. This can be seen by examining the definition of a limit (Hahn, 1963).

The procedure for calculating $\delta(\varepsilon)$ is similar to that of calculating ε and δ for limits and continuity in beginning calculus. As in the case of limits in calculus, this definition of stability does not provide the most efficient means of checking the stability of a given system. Hence, the remainder of this chapter develops methods to check a given system's stability properties that require less effort than applying the definition directly.

There are many theories that apply to the stability of MDOF systems, some of which are discussed here. The most common method of analyzing the stability of such systems is to show the existence of a Lyapunov function for the system. A *Lyapunov function*, denoted by $V(\mathbf{x})$, is a real scalar function of the vector $\mathbf{x}(t)$, which has continuous first partial derivatives and satisfies the two following conditions:

1) $V(\mathbf{x}) > 0$ for all values of $\mathbf{x}(t) \neq 0$.
2) $\dot{V}(\mathbf{x}) \leq 0$ for all values of $\mathbf{x}(t) \neq 0$.

Here $\dot{V}(\mathbf{x})$ denotes the time derivative of the function $V(\mathbf{x})$. Based on this definition of a Lyapunov function, several extremely useful stability results can be stated. The first result states that if there exists a Lyapunov function for a given system, then that system is *stable*. If, in addition, the function $\dot{V}(\mathbf{x})$ is strictly less than zero, the system is *asymptotically stable*. This is called the ***direct***, or ***second method***, of Lyapunov. It should be noted that if a Lyapunov function cannot be found, nothing can be concluded about the stability of the system, as the Lyapunov theorems are the only sufficient conditions for stability.

The stability of a system can also be characterized by the eigenvalues of that system. In fact, it can easily be shown that a given linear system is stable if and only if it has no eigenvalue with positive real part. Furthermore, the system will be asymptotically stable if and only if all of its eigenvalues have negative real parts (no zero real parts allowed). These statements are certainly consistent with the discussion in Section 4.1. The correctness of the statements can be seen by examining the solution using the expansion theorem (modal analysis) of the previous chapter (Equation 3.68). The eigenvalue approach to stability has the attraction of being both necessary and sufficient. However, calculating all the eigenvalues of the state matrix of a system is not always desirable and is difficult to extend to design. In fact, use of the eigenvalue criteria requires almost as much calculation as computing the solution of the system.

The interest in developing various different stability criteria is to find conditions that:

1) are easier to check than calculating the solution;
2) are stated in terms of the physical parameters of the system; and
3) can be used to help design and/or control systems to be stable.

Again, these goals can be exemplified by recalling the SDOF case, where it was shown that the sign of the coefficients m, c and k determine the stability behavior of the system. To this end, more convenient stability criteria are examined based on the classifications of a given physical system stated in Chapter 2.

4.3 Conservative Systems

For conservative systems of the form

$$M\ddot{\mathbf{q}} + K\mathbf{q} = \mathbf{0} \tag{4.4}$$

where M and K are symmetric, a simple stability condition results – namely, if M and K are positive definite, the eigenvalues of K are all positive, and hence the eigenvalues of the system are all purely imaginary. The solutions are then all linear combinations of terms of the form $e^{\pm\omega_n jt}$, or – by invoking Euler's formulas – all terms are of the form $A\sin(\omega_n t + \phi)$. Thus, from the preceding statement, the system of Equation (4.4) is stable, since both the displacement and velocity response of the system are always less than some constant (A and $\omega_n A$, respectively) for all time and for any initial conditions.

Also note that if K has a negative eigenvalue, then the system has a positive real exponent. In this case, one mode has a temporal coefficient of the form e^{at}, $a > 0$, which grows without bound, causing the system to become unstable (note that in this case, $\delta(\varepsilon)$ cannot be found).

The condition that K be positive definite can be coupled with the determinant condition, discussed in Section 3.3, to yield inequalities in the system's parameters. In turn, these inequalities can be used as design criteria. It should be pointed out that in most mechanical systems, K will be positive definite or positive semidefinite, unless some applied or external force proportional to displacement is present. In control theory, the applied control force is often proportional to position, as indicated in Equation (2.24) and Example 2.5.4.

It is instructive to note that the function $V(\mathbf{q})$ defined by (the energy in the system)

$$V(\mathbf{q}) = \frac{1}{2}[\dot{\mathbf{q}}^T(t)M\dot{\mathbf{q}}(t) + \mathbf{q}^T(t)K\mathbf{q}(t)] \tag{4.5}$$

serves as a Lyapunov function for the system in Equation (4.4). To see this, note first that $V(\mathbf{q}) > 0$, since M and K are positive definite, and that

$$\frac{d}{dt}[V(\mathbf{q})] = \dot{\mathbf{q}}^T M\ddot{\mathbf{q}} + \dot{\mathbf{q}}^T K\mathbf{q} \tag{4.6}$$

Now, if $\mathbf{q}(t)$ is a solution of Equation (4.4), it must certainly satisfy Equation (4.4). Thus, premultiplying Equation (4.4) by $\dot{\mathbf{q}}^T$ yields

$$\dot{\mathbf{q}}^T M\ddot{\mathbf{q}} + \dot{\mathbf{q}}^T K\mathbf{q} = 0 \tag{4.7}$$

This, of course, shows that $\dot{V}(\mathbf{q}) = 0$. Hence, $V(\mathbf{q})$ is a Lyapunov function and, by the second method of Lyapunov, the equilibrium of the system described by Equation (4.4) is stable. The condition that M and K are symmetric and positive definite is referred to as a stable system in vibration analysis, but is referred to as marginally stable in the control literature.

In cases where K may be positive semidefinite, the motion corresponding to the zero eigenvalue of K is called a *rigid body mode* and corresponds to a translational motion. Note that in this case Equation (4.5) is not a Lyapunov function because $V(\mathbf{q}) = 0$ for $\mathbf{q} \neq \mathbf{0}$, corresponding to the singularity of the matrix K. Since the other modes are purely imaginary, such systems may still be considered well-behaved because they consist of stable oscillations superimposed on the translational motion. This is common with moving mechanical parts. This explains why the concept of stability is defined differently in different situations. For instance, in aircraft stability, some rigid body motion is desirable.

4.4 Systems with Damping

As in the SDOF system case, if damping is added to the stable system in Equation (4.4), the resulting system can become asymptotically stable. In particular, if M, C and K are all symmetric and positive definite, then the system

$$M\ddot{\mathbf{q}} + C\dot{\mathbf{q}} + K\mathbf{q} = \mathbf{0} \tag{4.8}$$

is asymptotically stable. Each of the eigenvalues of Equation (4.8) can be shown to have a negative real part. Again, since the conditions of stability are stated in terms of the definiteness of the coefficient matrices, the stability condition can be directly stated in terms of inequalities involving the physical constants of the system.

To see that this system has a stable equilibrium by using the Lyapunov direct method, note that $V(\mathbf{q})$ as defined by Equation (4.5) is still a Lyapunov function for the damped system of Equation (4.8). In this case, the solution $\mathbf{q}(t)$ must satisfy

$$\dot{\mathbf{q}}^T M\ddot{\mathbf{q}} + \dot{\mathbf{q}}^T K\mathbf{q} = -(\dot{\mathbf{q}}^T C\dot{\mathbf{q}}) \tag{4.9}$$

which comes directly from Equation (4.8) by premultiplying by $\dot{\mathbf{q}}^T(t)$. This means that the time derivative of the proposed Lyapunov function, $\dot{V}(\mathbf{q})$, is given by Equation (4.9) to be

$$\frac{d}{dt}V(\mathbf{q}(t)) = -\dot{\mathbf{q}}^T C\dot{\mathbf{q}} < 0 \tag{4.10}$$

This is negative for all nonzero values of $\mathbf{q}(t)$, because the matrix C is positive definite. Hence, $V(\mathbf{q})$ defined by Equation (4.5) is in fact a Lyapunov function for the system described by Equation (4.8), and the system equilibrium is stable. Furthermore, since the inequality in Equation (4.10) is strict, the system's equilibrium is asymptotically stable.

An illustration of an asymptotically stable system is given in Example 2.5.4. The matrices M, C and K are all positive definite. In addition, the solution of Problem 3.15 shows that both elements of the vector $\mathbf{q}(t)$ are combinations of the functions $e^{-at}\sin\omega_n t$, $a > 0$. Hence, each element goes to zero as t increases to infinity, as the definition in Equation (4.3) indicates it should.

4.5 Semidefinite Damping

An interesting situation occurs when the damping matrix in Equation (4.8) is only positive semidefinite. Then the above argument for the existence of a Lyapunov function is still valid, so that the system is stable. However, it is not clear whether or not the system is asymptotically stable. There are two equivalent answers to this question of asymptotic stability for systems with a semidefinite damping matrix.

The first approach is based on the *null space* of the matrix C. The null space of a matrix C is the set of all nonzero vectors $\mathbf{x}$ such that $C\mathbf{x} = \mathbf{0}$, i.e. the set of those vectors corresponding to the zero eigenvalues of the matrix C. Since C is semidefinite in this situation, there exists at least one nonzero vector $\mathbf{x}$ in the null space of C. Moran (1970) showed that if C is semidefinite in Equation (4.8), then the equilibrium of Equation (4.8) is asymptotically stable if and only if none of the eigenvectors of the matrix K lie in the null space of C. This provides a convenient necessary and sufficient condition for asymptotic stability of semidefinite systems, but it requires the computation of the eigenvectors for K or at least the null space of C.

Physically, this result makes sense because if there is an eigenvector of the matrix K in the null space of C, the vector also becomes an eigenvector of the system. Furthermore, this eigenvector results in a zero damping mode for the system, and hence a set of initial conditions exists that excites the system into an undecaying harmonic motion.

The second approach avoids having to solve an eigenvector problem to check for asymptotic stability. Walker and Schmitendorf (1973) showed that the system of Equation (4.8) with semidefinite damping will be asymptotically stable if and only if

$$\text{Rank}\begin{bmatrix} C \\ CK \\ CK^2 \\ \vdots \\ CK^{n-1} \end{bmatrix} = n \tag{4.11}$$

where n is the number of degrees of freedom of the system. The rank of a matrix is the number of linearly independent rows (or columns) the matrix has (Appendix B). This type of rank condition comes from control theory considerations and is used and explained again in Chapter 6.

These two approaches are equivalent. They essentially comment on whether or not the system can be transformed into a coordinate system in which one or more modes are undamped. It is interesting to note that if C is semidefinite and $KM^{-1}C$ is symmetric, then the system is not asymptotically stable. This results since, as was pointed out in Section 3.5, if $KM^{-1}C = CM^{-1}K$, then $\tilde{K}$, and $\tilde{C}$ have the same eigenvectors and the system can be decoupled. In this decoupled form, there will be at least one equation with no velocity term corresponding to the zero eigenvalue of C. The solution of this equation will not go to zero with time, and hence the system cannot be asymptotically stable.

4.6 Gyroscopic Systems

The stability properties of gyroscopic systems provide some very interesting and unexpected results. First, consider an undamped gyroscopic system of the form

$$M\ddot{\mathbf{q}} + G\dot{\mathbf{q}} + K\mathbf{q} = \mathbf{0} \tag{4.12}$$

where M and K are both positive definite and symmetric and where G is skew-symmetric. Since the quadratic form $\dot{\mathbf{q}}^T G\dot{\mathbf{q}}$ is zero for any choice of $\mathbf{q}$, the Lyapunov function for the previous system, Equation (4.5), still works for Equation (4.12), and the equilibrium of Equation (4.12) is stable.

If the matrix K in Equation (4.12) is indefinite, semidefinite or negative definite, the system may still be stable. This is a reflection of the fact that gyroscopic forces can sometimes be used to stabilize an unstable system. A child's spinning top provides an example of such a situation. The vertical position of the top is unstable until the top is spun, providing a stabilizing gyroscopic force.

Easy-to-use conditions are not available to check if Equation (4.12) is stable when K is not positive definite. Hagedorn (1975) has been able to show that if K is negative definite and if the matrix $4K - GM^{-1}G$ is negative definite, then the system is definitely unstable. Several authors have examined the stability of Equation (4.12) when the dimension of the system is $n = 2$. Teschner (1977) showed that if $n = 2$, K is negative definite, and $4K - GM^{-1}G$ is positive definite, then the system is stable. Inman and Saggio (1985) showed that if $n = 2$, K is negative definite, det $K > 0$, and the trace of $4K - GM^{-1}G$ is positive, then the system is stable.

Huseyin et al. (1983) showed that for any degree of freedom system, if $4K - GM^{-1}G$ is positive definite and if the matrix $(GM^{-1}K - KM^{-1}G)$ is positive semidefinite, then the system is stable. In addition, they showed that if $GM^{-1}K = KM^{-1}G$, then the system is stable if and only if the matrix $4K - GM^{-1}G$ is positive definite. These represent precise conditions for the stability of undamped gyroscopic systems. Most of these ideas result from Lyapunov's direct method. The various results on gyroscopic systems are illustrated in Example 4.6.1. Bernstein and Bhat (1995) and Adegas and Stoustrup (2015) give additional examples and summarize known stability conditions up to 1994 and 2014, respectively.

Example 4.6.1

Consider a simplified model of a mass mounted on a circular, weightless rotating shaft that is also subjected to an axial compression force. This system is described by Equation (4.12) with

$$M = I, G = 2\gamma \begin{bmatrix} 0 & -1 \\ 1 & 0 \end{bmatrix}, K = \begin{bmatrix} k_1 - \gamma^2 - \eta & 0 \\ 0 & k_2 - \gamma^2 - \eta \end{bmatrix}$$

where γ represents the angular velocity of the shaft and η the axial force. The parameters k_1 and k_2 represent the flexural stiffness in two principal directions, noted in Figure 4.1.

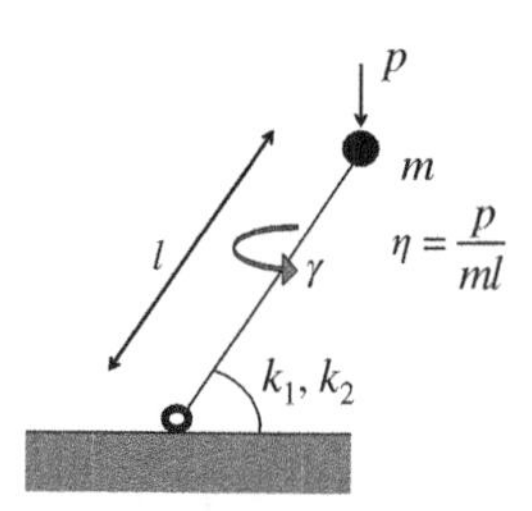

Figure 4.1 Schematic of a rotating shaft subject to an axial compression force.

It is instructive to consider this problem first for fixed rotational speed ($\gamma = 2$) and for $\eta = 3$. Then the relevant matrices become ($M = I$)

$$K = \begin{bmatrix} k_1 - 7 & 0 \\ 0 & k_2 - 7 \end{bmatrix}$$

$$4K - GM^{-1}G = 4\begin{bmatrix} k_1 - 3 & 0 \\ 0 & k_2 - 3 \end{bmatrix}$$

Figure 4.2 shows plots of stable and unstable choices of k_1 and k_2 using the previously mentioned theories. To get the various regions of stability illustrated in Figure 4.2, consider the following calculations:

1) K positive definite implies $k_1 - 7 > 0$ and $k_2 - 7 > 0$, a region of stable operation.
2) $\det K > 0$ implies that $(k_1 - 7)(k_2 - 7) > 0$, or that $k_1 < 7$, $k_2 < 7$. The $\mathrm{tr}(4K - GM^{-1}G) > 0$ implies that $4[(k_1 - 3) + (k_2 - 3)] > 0$, or that $k_1 + k_2 > 6$, which again yields a region of stable operation.
3) $4K - GM^{-1}G$ negative definite implies $k_1 < 3$ and $k_2 < 3$, a region of unstable operation.

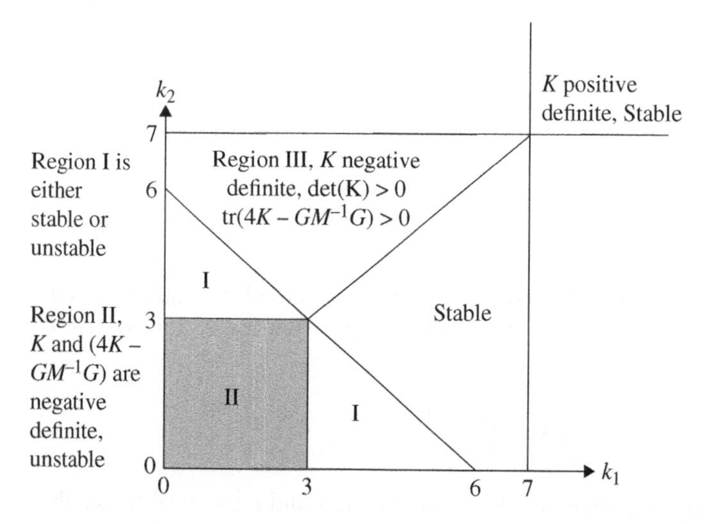

Figure 4.2 Stability regions for the system of the conservative gyroscopic system of Figure 4.1, as a function of the stiffness.

4) $4K - GM^{-1}G$ positive definite implies that $k_1 > 3$ and $k_2 > 3$, a region of either stable or unstable operation depending on other considerations. If, in addition, the matrix

$$GM^{-1}K - KM^{-1}G = 4\begin{bmatrix} 0 & k_1 - k_2 \\ k_1 - k_2 & 0 \end{bmatrix}$$

is zero, i.e. if $k_1 = k_2$, then the system is stable. Thus, the line $k_1 = k_2$ represents a region of stable operation for $k_1 = k_2 \geq 3$ and unstable operation for $k_1 = k_2 < 3$.

4.7 Damped Gyroscopic Systems

As the previous section illustrated, gyroscopic forces can be used to stabilize an unstable system. The next logical step is to consider adding damping to the system. Since added positive definite damping has caused stable symmetric systems to become asymptotically stable, the same effect is expected here. However, this turns out *not* to be the case in all circumstances.

Consider a damped gyroscopic system of the form

$$M\ddot{\mathbf{q}} + (C + G)\dot{\mathbf{q}} + K\mathbf{q} = \mathbf{0} \tag{4.13}$$

where $M = M^T > 0$, $C = C^T$, $G = -G^T$ and $K = K^T$. The following results are due to Kelvin, Tait and Chetaev, and are referred to as the KTC theorem by Zajac (1964, 1965).

1) If K and C are both positive definite, the system is asymptotically stable.
2) If K is not positive definite and C is positive definite, the system is unstable.
3) If K is positive definite and C is positive semidefinite, the system may be stable or asymptotically stable. The system is asymptotically stable if and only if none of the eigenvectors of the undamped gyroscopic system are in the null space of C. Also, proportionally damped systems will be stable.

Hughes and Gardner (1975) showed that the Walker and Schmitendorf rank condition, Equation (4.11), also applies to gyroscopic systems with semidefinite damping and positive definite stiffness. In particular, let the state matrix A and the "observer" matrix C_O be defined and denoted by

$$A = \begin{bmatrix} 0 & I \\ -M^{-1}K & -M^{-1}G \end{bmatrix}, \quad C_O = [0 \quad C]$$

Then the equilibrium of Equation (4.13) is asymptotically stable if the rank of the $2n \times 2n^2$ matrix $R^T = [C_O^T \quad A^T C_O^T \quad \cdots \quad (A^T)^{n-1} C_O^T]$ is full, i.e. rank $R^T = 2n$. Systems that satisfy either this rank condition or Equation (4.11) are said to be *pervasively damped,* meaning that the influence of the damping matrix C pervades each of the system's coordinates. Each mode of a pervasively damped system is damped, and such systems are asymptotically stable.

Note that condition (2) points out that if one attempts to stabilize an unstable system by adding gyroscopic forces to the system and at the same time introduces viscous damping, the system will remain unstable. A physical example of this is again given by the spinning top if the friction in the system is modeled as viscous damping. With dissipation considered, the top is in fact unstable and eventually falls over after precessing because of the effects of friction.

4.8 Circulatory Systems

Next consider those systems that have asymmetries in the coefficient of the displacement term. Such systems are called *circulatory*. A physical example is given in Example 2.5.3. Other examples can be found in the fields of aeroelasticity, thermoelastic stability and in control (Example 2.5.4). The equation of motion of such systems takes the form

$$M\ddot{\mathbf{q}} + (K + H)\mathbf{q} = \mathbf{0} \tag{4.14}$$

where $M = M^T$, $K = K^T$ and $H = -H^T$. Here K is the symmetric part of the position coefficient and H is the skew-symmetric part. Results and stability conditions for circulatory systems are not as well developed as for symmetric conservative systems.

Since damping is not present, the stability of Equation (4.14) will be entirely determined by the matrix $A_3 = K + H$, as long as M is nonsingular. In fact, it can be shown (Huseyin, 1978: p. 174) that Equation (4.14) is stable if and only if there exists a symmetric and positive definite matrix P such that $PM^{-1}A_3$ is symmetric and positive definite. Furthermore, if the matrix $PM^{-1}A_3$ is symmetric, there is no flutter instability. On the other hand, if such a matrix P does not exist, Equation (4.14) can be unstable, both by flutter and by divergence. In this case, the system will have some complex eigenvalues with positive real parts.

The preceding results are obtained by considering an interesting subclass of circulatory systems that results from a factorization of the matrix, $M^{-1}A_3$. Taussky (1972) showed that any real square matrix can be written as the product of two symmetric matrices. That is, there exist two real symmetric matrices S_1 and S_2, such that $M^{-1}A_3 = S_1S_2$. With this factorization in mind, all asymmetric matrices $M^{-1}A_3$ can be classified into two groups; those for which at least one of the matrix factors, such as S_1, is positive definite and those for which neither of the factors is positive definite. Matrices for which S_1 (or S_2) is positive definite are called *symmetrizable matrices*, or *pseudosymmetric matrices*. The corresponding systems are referred to as *pseudoconservative systems*, *pseudosymmetric systems* or *symmetrizable systems*, and behave essentially like symmetric systems. One can think of this transformation as a change of coordinate systems to one in which the physical properties are easily recognized.

In fact, for $M^{-1}A_3 = S_1S_2$, S_1 positive definite, the system described by Equation (4.14) is stable if and only if S_2 is positive definite. Furthermore, if S_2 is not positive definite, instability can only occur through divergence, and no flutter instability is possible. Complete proofs of these statements can be found in Huseyin (1978) along with a detailed discussion. The proof follows from the simple idea that if $M^{-1}A_3$ is symmetrizable, then the system is mathematically similar to a symmetric system. Thus, the stability problem is reduced to considering that of the symmetric matrix S_2.

The similarity transformation is given by the matrix $S_2^{1/2}$, the positive definite square root of the matrix S_1. To see this, premultiply Equation (4.14) by $S_1^{-1/2}$, which is nonsingular. This yields

$$S_1^{-1/2}\ddot{\mathbf{q}} + S_1^{-1/2}(M^{-1}A_3)\mathbf{q} = \mathbf{0} \tag{4.15}$$

which becomes

$$S_1^{-1/2}\ddot{\mathbf{q}} + S_1^{-1/2}S_1S_2\mathbf{q} = \mathbf{0}$$

or

$$S_1^{-1/2}\ddot{\mathbf{q}} + S_1^{1/2}S_2\mathbf{q} = \mathbf{0} \tag{4.16}$$

Substitution of $\mathbf{q} = S_1^{1/2}\mathbf{y}$ into this last expression yields the equivalent symmetric system

$$\ddot{\mathbf{y}} + S_1^{1/2}S_2S_1^{1/2}\mathbf{y} = \mathbf{0} \tag{4.17}$$

Thus, there is a nonsingular transformation $S_1^{1/2}$ relating the solution of symmetric problems given by Equation (4.17) to the asymmetric problem of Equation (4.14). Because the transformation is nonsingular, the eigenvalues of Equations (4.14) and (4.17) are the same. Thus, the two representations have the same stability properties. Here the matrix $S_1^{1/2}S_2S_1^{1/2}$ is seen to be symmetric by taking its transpose, i.e. $(S_1^{1/2}S_2S_1^{1/2})^T = S_1^{1/2}S_2S_1^{1/2}$. Thus, if S_2 is positive definite then $S_1^{1/2}S_2S_1^{1/2}$ is positive definite (and symmetric), so that Equation (4.17) is stable. Methods for calculating the matrices S_1 and S_2 are discussed in the next section.

Note that if the system is not symmetrizable, i.e. if S_1 is not positive definite, then $S_1^{1/2}$ does not exist and the preceding development fails. In this case, instability of Equation (4.14) can be caused by either flutter or divergence.

4.9 Asymmetric Systems

For systems that have both asymmetric velocity and stiffness coefficients not falling into any of the previously mentioned classifications, several different approaches are available. The first approach discussed here follows the idea of a pseudosymmetric system introduced in the previous section, and the second approach follows methods of constructing Lyapunov functions. The systems considered in this section are of the most general form (Equation 2.7 with $\mathbf{f} = \mathbf{0}$)

$$A_1\ddot{\mathbf{q}} + A_2\dot{\mathbf{q}} + A_3\mathbf{q} = \mathbf{0} \tag{4.18}$$

where A_1 is assumed to be nonsingular, $A_2 = C + G$, and $A_3 = K + H$. Since A_1 is nonsingular and since A_2 and A_3 are symmetric, it is sufficient to consider the equivalent system

$$\ddot{\mathbf{q}} + A_1^{-1}A_2\dot{\mathbf{q}} + A_1^{-1}A_3\mathbf{q} = \mathbf{0} \tag{4.19}$$

The system described by Equation (4.19) can again be split into two classes, by examining the factorization of the matrices $A_1^{-1}A_2$ and $A_1^{-1}A_3$ in a fashion similar to the previous section. First note that there exists a factorization of these matrices of the form $A_1^{-1}A_2 = T_1T_2$ and $A_1^{-1}A_3 = S_1S_2$, where the matrices S_1, S_2 T_1 and T_2 are all symmetric. This is always possible because of the result of Taussky just mentioned, i.e. any real square matrix can always be written as the product of two symmetric matrices. Then the system in Equation (4.19) is similar to a symmetric system if and only if there exists at least one factorization of $A_1^{-1}A_2$ and $A_1^{-1}A_3$, such that $S_1 = T_1$, which is positive definite. Such systems are called *symmetrizable*. Under this assumption, it can be shown that the equilibrium position of Equation (4.18) is asymptotically stable if the eigenvalues of the matrix $A_1^{-1}A_2$ and the matrix $A_1^{-1}A_3$ are all positive real numbers. This corresponds to requiring the matrices S_2 and T_2 to be positive definite.

Deciding if the matrices $A_1^{-1}A_2$ and $A_1^{-1}A_3$ are symmetrizable is, in general, not an easy task. However, if the matrix A_2 is proportional, i.e. if $A_2 = \alpha A_1 + \beta A_3$, where α and β are scalars, and if $A_1^{-1}A_3$ is symmetrizable, then $A_1^{-1}A_2$ is also symmetrizable, and there exists a common factor S_1T_1. It can also be shown that if two real matrices commute and one of them is symmetrizable, then the other matrix is also symmetrizable, and they can be reduced to a symmetric form simultaneously.

Several of the usual stability conditions stated for symmetric systems can now be stated for symmetrizable systems. If $A_1^{-1}A_2$ has nonnegative eigenvalues (i.e. zero is allowed) and if $A_1^{-1}A_3$ has positive eigenvalues, Equation (4.18) is asymptotically stable if and only if the $n^2 \times n$ matrix

$$R = \begin{bmatrix} A_1^{-1}A_2 \\ A_1^{-1}A_2\left(A_1^{-1}A_3\right) \\ A_1^{-1}A_2\left(A_1^{-1}A_3\right)^2 \\ \vdots \\ A_1^{-1}A_2\left(A_1^{-1}A_3\right)^{n-1} \end{bmatrix} \tag{4.20}$$

has rank n. This, of course, is equivalent to the statement made by Moran (1970) for symmetric systems, mentioned in Section 4.5, that the system is asymptotically stable if and only if none of the eigenvectors of $A_1^{-1}A_3$ lie in the null space of $A_1^{-1}A_2$.

The KTC theorem can also be extended to systems with asymmetric but symmetrizable coefficients. However, the extension is somewhat more complicated. Consider the matrix $S = (A_1^{-1}A_2)T_1 + T_1(A_1^{-1}A_2)^T$, and note that S is symmetric. If S is positive definite and if A_3 is nonsingular, then Equation (4.18) is stable if and only if all of the eigenvalues of $A_1^{-1}A_3$ are positive numbers. If $S_1 \neq T_1$, the matrix $A_1^{-1}A_2$ contains a gyroscopic term, and this result states the equivalent problem faced in using gyroscopic forces to stabilize an unstable system, that it cannot be done in the presence of damping. Hence, in the case of $S_1 \neq T_1$, the stability of the system is determined by the eigenvalues of T_2 (which are the eigenvalues of A_3) for systems with symmetrizable stiffness coefficient matrix.

The following two examples serve to illustrate the above discussion, as well as indicate the level of computation required.

Example 4.9.1

The preceding results are best understood by considering some examples. First, consider a system described by

$$\begin{bmatrix} 1 & 0 \\ 0 & 1 \end{bmatrix} \ddot{\mathbf{q}} + \begin{bmatrix} 2 & 4 \\ 4 & 2 \end{bmatrix} \dot{\mathbf{q}} + \begin{bmatrix} 10 & 8 \\ 0 & 1 \end{bmatrix} \mathbf{q} = \mathbf{0}$$

Here note that

$$A_1^{-1}A_2 = \begin{bmatrix} 2 & 4 \\ 1 & 2 \end{bmatrix} = \begin{bmatrix} 1.2461 & -0.2769 \\ -0.2769 & 0.3115 \end{bmatrix} \begin{bmatrix} 2.8889 & 5.7778 \\ 5.7778 & 11.5556 \end{bmatrix}$$

$$A_1^{-1}A_3 = \begin{bmatrix} 10 & 8 \\ 0 & 1 \end{bmatrix} = \begin{bmatrix} 1.2461 & -0.2769 \\ -0.2769 & 0.3115 \end{bmatrix} \begin{bmatrix} 10 & 8.8889 \\ 8.8889 & 11.1111 \end{bmatrix}$$

so that $T_1 = S_1$, and the coefficient matrices have a common factor. Then the eigenvalue problem associated with this system is similar to a symmetric eigenvalue problem. An illustration on how to calculate the symmetric factors of a matrix is given in Example 4.9.3.

According to the previous theorems, the stability of this equation may be indicated by calculating the eigenvalues of $A_1^{-1}A_2$ and of $A_1^{-1}A_3$. The eigenvalues of $A_1^{-1}A_2$ in this example are $\lambda_{1,2} = 0, 4$, and those of $A_1^{-1}A_3$ and $\lambda_{1,2} = 1, 10$. Hence, $A_1^{-1}A_3$ has positive real eigenvalues and $A_1^{-1}A_2$ has nonnegative real eigenvalues. Because of the singularity of the matrix $A_1^{-1}A_2$, knowledge of the rank of the matrix Equation (4.20) is required in order to determine if the system is asymptotically stable or just stable. The matrix of Equation (4.20) is

$$\begin{bmatrix} 2 & 4 \\ 1 & 2 \\ 20 & 20 \\ 10 & 10 \end{bmatrix} \sim \begin{bmatrix} 0 & 0 \\ 1 & 2 \\ 0 & 0 \\ 1 & 1 \end{bmatrix} \sim \begin{bmatrix} 0 & 0 \\ 1 & 0 \\ 0 & 0 \\ 0 & 1 \end{bmatrix}$$

which obviously has rank $= 2$, the value of n. Here, the symbol $\sim$ denotes column (or row) equivalence, as discussed in Appendix B. Thus, the previous result states that the equilibrium of this example is asymptotically stable. This is in agreement with the eigenvalue calculation for the system, which yields

$$\lambda_{1,2} = -1 \pm 2j$$

$$\lambda_{3,4} = -1 \pm j$$

showing clearly that the equilibrium is in fact asymptotically stable, as predicted by the theory.

Example 4.9.2

As a second example, consider the asymmetric problem given by

$$\begin{bmatrix} 1 & 1 \\ 0 & 1 \end{bmatrix} \ddot{\mathbf{q}} + \begin{bmatrix} 9 & 20 \\ 3 & 8 \end{bmatrix} \dot{\mathbf{q}} + \begin{bmatrix} -5 & -6 \\ -4 & 0 \end{bmatrix} \mathbf{q} = \mathbf{0}$$

Premultiplying this by A_1^{-1} yields

$$I\ddot{\mathbf{q}} + \begin{bmatrix} 6 & 12 \\ 3 & 8 \end{bmatrix}\dot{\mathbf{q}} + \begin{bmatrix} -1 & -6 \\ -4 & 0 \end{bmatrix}\mathbf{q} = \mathbf{0}$$

The eigenvalues of $A_1^{-1}A_2$ are $\lambda_{1,2} = 7 \pm (1/2)\sqrt{148}$ and those of $A_1^{-1}A_3$ are $\lambda_{1,2} = -1/2 \pm (1/2)\sqrt{97}$. Thus, both coefficient matrices have real distinct eigenvalues and are therefore symmetrizable. However, a simple computation shows that there does not exist a factorization of $A_1^{-1}A_2$ and $A_1^{-1}A_3$ such that $T_1 = S_1$.

Thus, the generalized KTC theorem must be applied. Accordingly, if the matrix $(A_1^{-1}A_2)T_1 + T_1(A_1^{-1}A_2)^T$ is positive definite, then the equilibrium of this system is unstable, since $A_1^{-1}A_3$ has a negative eigenvalue. To calculate T_2, note that $(A_1^{-1}A_3) = T_1T_2$, where T_1 is positive definite and hence nonsingular. Thus, multiplying by T_1^{-1} from the right yields that the matrix T_2 is given by $T_2 = T_1^{-1}(A_1^{-1}A_3)$.

Let T_1^{-1} be a general generic symmetric matrix denoted by

$$T_1^{-1} = \begin{bmatrix} a & b \\ b & c \end{bmatrix}$$

where it is desired to calculate a, b and c so that T_1 is positive definite. Thus

$$T_2 = \begin{bmatrix} a & b \\ b & c \end{bmatrix}\begin{bmatrix} -1 & -6 \\ -4 & 0 \end{bmatrix} = \begin{bmatrix} -a-4b & -6a \\ -b-4c & -6b \end{bmatrix}$$

Requiring T_2 to be symmetric and T_1 to be positive definite yields the following relationships for a, b and c

$$ac > b^2$$
$$6a = b + 4c$$

This set of equations has multiple solutions; one convenient solution is $a = 2$, $b = 0$ and $c = 3$. Then T_1 becomes

$$T_1 = \begin{bmatrix} \frac{1}{2} & 0 \\ 0 & \frac{1}{3} \end{bmatrix}$$

Thus $(A_1^{-1}A_2)T_1 + T_1(A_1^{-1}A_2)^T$ becomes

$$(A_1^{-1}A_2)T_1 + T_1(A_1^{-1}A_2)^T = \begin{bmatrix} 6 & \frac{11}{12} \\ \frac{11}{2} & \frac{16}{3} \end{bmatrix}$$

which is positive definite. Thus, the equilibrium must be unstable.

This analysis again agrees with a calculation of the eigenvalues, which are $\lambda_1 = 0.3742$, $\lambda_2 = -13.5133$ and $\lambda_{3,4} = -0.4305 \pm 0.2136j$, indicating an unstable equilibrium, as predicted.

Example 4.9.3

The question of how to calculate the factors of a symmetrizable matrix is discussed by Huseyin (1978) and Ahmadian and Chou (1987). Here it is shown that the matrices $A_1^{-1}A_3 = S_1S_2$ and $A_1^{-1}A_2 = T_1T_2$ of Example 4.9.2 do not have any common factorization such that $T_1 = S_1$. Hence $A_1^{-1}A_2$ and $A_1^{-1}A_3$ cannot be simultaneously symmetrized by the same transformation.

It is desired to find a symmetric positive definite matrix P, such that $PA_1^{-1}A_2$ and $PA_1^{-1}A_3$ are both symmetric. To that end, let

$$P = \begin{bmatrix} a & b \\ b & c \end{bmatrix}$$

Then

$$PA_1^{-1}A_2 = \begin{bmatrix} 6a + 3b & 12a + 8b \\ 6b + 3c & 12b + 8c \end{bmatrix}$$

and

$$PA_1^{-1}A_3 = \begin{bmatrix} -a - 4b & -6a \\ -b - 4c & -6b \end{bmatrix}$$

Symmetry of both matrices then requires that

$$\begin{aligned} 6b + 3c &= 12a + 8b \\ b + 4c &= 6a \end{aligned} \tag{4.21}$$

Positive definiteness of P requires

$$\begin{aligned} a &> 0 \\ ac &> b^2 \end{aligned} \tag{4.22}$$

It will be shown that the problem posed by Equations (4.21) and (4.22) does not have a solution. Equations (4.21) may be written in matrix form as

$$\begin{bmatrix} -2 & 3 \\ 1 & 4 \end{bmatrix} \begin{bmatrix} b \\ c \end{bmatrix} = 6a \begin{bmatrix} 2 \\ 1 \end{bmatrix}$$

which has the unique **solution**

$$\begin{bmatrix} b \\ c \end{bmatrix} = a \begin{bmatrix} 2.73 \\ 2.18 \end{bmatrix}$$

for all values of a. Thus $b = 2.73a$ and $c = 2.18a$, so that $b^2 = 7.45a^2$ and $ac = 2.18a^2$. Then

$$ac = 2.18a^2 < 7.45a^2 = b^2$$

and the condition in Equation (4.22) cannot be satisfied.

Other alternatives exist for analyzing the stability of asymmetric systems. Walker (1970), approaching the problem by looking for Lyapunov functions, was able to state several results for the stability of Equation (4.18) in terms of a fourth matrix R. If there exists a symmetric positive definite matrix R, such that $RA_1^{-1}A_3$ is symmetric and positive definite (this is the same as requiring $A_1^{-1}A_3$ to be symmetrizable), then the system is stable if the symmetric part of $RA_1^{-1}A_2$ is positive semidefinite and asymptotically stable if the symmetric part of $RA_1^{-1}A_2$ is strictly positive definite. This result is slightly more general than the symmetrizable results just stated, in that it allows the equivalent symmetric systems to have gyroscopic forces.

In addition to these results, Walker (1970) showed that if there exists a symmetric matrix R such that $RA_1^{-1}A_2$ is skew-symmetric and $RA_1^{-1}A_3$ is symmetric and such that R and $RA_1^{-1}A_3$ have the same definiteness, then the system is stable but not asymptotically stable.

Another approach to the stability of Equation (4.18), not depending on symmetrizable coefficients, has been given by Mingori (1970). He showed that if the coefficient matrices M, C, G, H and K satisfy the commutivity conditions

$$HC^{-1}M = MC^{-1}H$$
$$HC^{-1}G = GC^{-1}H$$
$$HC^{-1}K = KC^{-1}H$$

then the stability of the system is determined by the matrix

$$Q = HC^{-1}MC^{-1}H - GC^{-1}H + K$$

This theory states the system is stable, asymptotically stable or unstable, if the matrix Q possesses nonnegative, positive or at least one negative eigenvalue, respectively. Although the problem addressed is general, the restrictions are severe. For instance, this method cannot be used for systems with semidefinite damping (C^{-1} does not exist).

Other more complicated and more general stability conditions are due to Walker (1974) and an extension of his work by Ahmadian and Inman (1986). The methods are developed by using Lyapunov functions to derive stability and instability conditions based on the direct method. These are stated in terms of the symmetry and definiteness of certain matrices consisting of various combinations of the matrices A_1, A_2 and A_3. These conditions offer a variety of relationships among the physical parameters of the system, which can aid in designing a stable or asymptotically stable system.

4.10 Feedback Systems

One of the major reasons for using feedback control is to stabilize the system response. However, most structures are inherently stable to begin with and control is applied to improve performance. Unfortunately, the introduction of active control can effectively destroy the symmetry and definiteness of system, introducing the possibility of instability. Thus, checking the stability of a system after a control is designed is an important step. A majority of the work in control takes place in the state space (first-order form). However, it is interesting to treat the control problem specifically in "mechanical" or physical coordinates, in order to take advantage of the natural symmetries and definiteness in the system. Lin (1981) developed a theory for closed-loop asymptotic stability for mechanical structures being controlled by velocity and position feedback. The systems considered here have the form (Section 2.3)

$$M\ddot{\mathbf{q}} + A_2\dot{\mathbf{q}} + K\mathbf{q} = \mathbf{f}_f + \mathbf{f} \tag{4.23}$$

where $M = M^T$ is positive definite, $A_2 = C + G$ is asymmetric and K is symmetric. Here the vector $\mathbf{f}$ represents external disturbance forces (taken to be zero in this section) and the vector $\mathbf{f}_f$ represents the control force derived from the action of r force actuators represented by

$$\mathbf{f}_f = B_f\mathbf{u} \tag{4.24}$$

where the $r \times 1$ vector $\mathbf{u}$ denotes the r inputs, one for each control device (actuator), and B_f denotes the $n \times r$ matrix of weighting factors (influence coefficients or actuator gains) with structure determined by the actuator locations. In order to be able to feedback the position and velocity, let $\mathbf{y}$ be a $s \times 1$ vector of sensor outputs denoted and defined by

$$\mathbf{y} = C_p\mathbf{q} + C_v\dot{\mathbf{q}} \tag{4.25}$$

Here C_p and C_v are $s \times n$ matrices of displacement and velocity influenced coefficients, respectively, with structure determined by the sensor locations and where s is the number of sensors. Equation (4.25) represents those coordinates that are measured as part of the control system and is a mathematical model of the transducer and signal processing used to measure the system's response. The input vector $\mathbf{u}$ is chosen to be of the special form

$$\mathbf{u}(t) = -G_f\mathbf{y} = -G_fC_p\mathbf{q} - G_fC_v\dot{\mathbf{q}} \tag{4.26}$$

where the $r \times s$ matrix G_f consists of constant feedback gains. This form of control law is called *output feedback*, because the input is proportional to the measured output or response, $\mathbf{y}$.

In Equation (4.24), the matrix B_f reflects the location on the structure of any actuator or device being used to supply the forces $\mathbf{u}$. For instance, if an electromechanical or piezoelectric actuator is attached to mass m_1 of Figure 2.6 and if it supplies a force of the form $F_0 \sin \omega t$, the vector $\mathbf{u}$ reduces to the scalar $u = F_0\sin(\omega t)$ and the matrix B_f reduces to the vector $B_f^T = [1 \;\; 0]$. Alternately, the control force can be written as a column vector $\mathbf{u}$, and B_f written as a matrix

$$\mathbf{u} = \begin{bmatrix} F_0 \sin \omega t \\ 0 \end{bmatrix}, \quad \text{and} \quad B_f = \begin{bmatrix} 1 & 0 \\ 0 & 0 \end{bmatrix}.$$

If, on the other hand, there are two actuators, one attached to m_1 supplying a force $F_1 \sin(\omega_1 t)$ and one at m_2 supplying a force $F_2 \sin(\omega_2 t)$, the vector $\mathbf{u}$ becomes $\mathbf{u}^T = [F_1 \sin(\omega_1 t) \;\; F_2 \sin(\omega_2 t)]$ and the matrix B_f becomes $B_f = I$, the 2×2 identity matrix. Likewise, if the positions x_1 and x_2 are measured, the matrices in

Equation (4.25) become $C_p = I$ and $C_v = 0$, the 2×2 matrix of zeros. If only the position x_1 is measured and the control force is applied to x_2 then

$$B_f = \begin{bmatrix} 0 & 0 \\ 0 & 1 \end{bmatrix}, \quad \text{and} \quad C_p = \begin{bmatrix} 1 & 0 \\ 0 & 0 \end{bmatrix}$$

Making the appropriate substitutions in the preceding equations and assuming no external disturbance (i.e. $\mathbf{f} = \mathbf{0}$) yields an equivalent homogeneous system, which includes the effect of the controls. It has the form

$$M\ddot{\mathbf{q}} + (C + G + B_f G_f C_v)\dot{\mathbf{q}} + (K + B_f G_f C_p)\mathbf{q} = \mathbf{0} \tag{4.27}$$

For the sake of notation, define the matrices $C^* = B_f G_f C_v$ and $K^* = B_f G_f Cp$. Note that since the number of actuators r is usually much smaller than the number of modeled degrees of freedom n (the dimension of the system), the matrices K^* and C^* are usually singular. Since, in general, $C + C^*$ and $K + K^*$ may not be symmetric or positive definite, it is desired to establish constraints on any proposed control law to ensure the symmetry and definiteness of the coefficient matrices and hence the stability of the system (Problem 4.11). These constraints stem from requiring $C + C^*$ and $K + K^*$ to be symmetric positive definite. The problem of interest in control theory is how to choose the matrix G_f so that the response $\mathbf{q}$ has some desired property (performance and stability). Interest in this section focuses on finding constraints on the elements of G_f so that the response $\mathbf{q}$ is asymptotically stable or at least stable. The stability methods of this chapter can be applied to Equation (4.27) to develop these constraints. Note that the matrices $B_f G_f C_p$ and $B_f G_f C_v$ are represented as the matrices K_p and K_v, respectively, in Equation (2.24).

Collocated control refers to the case that the sensors are located at the same physical location as the actuators. If the sensors or the actuators add no additional dynamics, then collocated controllers provide improved stability of the closed loop system. As seen above, the closed loop system coefficients C^* and K^* generally lose their symmetry for many choices of B_f, C_f, B_v and C_v. If, however, the gain matrices G_f and G_v are symmetric and if $B_f^T = C_f$ and $B_v^T = C_v$ then the matrices C^* and K^* remain symmetric. The symmetry then results in the possibility of choosing the gain matrices so that C^* and K^* remain positive definite, insuring closed loop stability for stable open loop systems (C and K positive definite). Placing sensors and actuators at the same location causes $B_f^T = C_f$ and $B_v^T = C_v$ so that collocated control enhances closed loop stability. The controller design consists of choosing gains G_f and G_v that are symmetric and positive definite (or at least semidefinite), with collocated sensors and actuators to insure a stable closed loop response.

Example 4.10.1

Consider the two degree-of-freedom system of Figure 2.6, with a control force applied to m_1 and a measurement made of x_2 so that the control system is not collocated. Then the input matrix, output matrix and symmetric control gain matrix are

$$B_f = \begin{bmatrix} 1 & 0 \\ 0 & 0 \end{bmatrix}, \quad C_p = \begin{bmatrix} 0 & 1 \\ 0 & 0 \end{bmatrix}, \quad \text{and} \quad G = \begin{bmatrix} g_1 & 0 \\ 0 & g_2 \end{bmatrix}$$

Note that this is not collocated because $B_f^T \neq C_f$. The closed loop system of Equation (4.27) becomes

$$\begin{bmatrix} m_1 & 0 \\ 0 & m_2 \end{bmatrix}\begin{bmatrix} \ddot{x}_1 \\ \ddot{x}_2 \end{bmatrix} + \begin{bmatrix} c_1+c_2 & -c_2 \\ -c_2 & c_2 \end{bmatrix}\begin{bmatrix} \dot{x}_1 \\ \dot{x}_2 \end{bmatrix} + \begin{bmatrix} k_1+k_2 & -k_2+g_1 \\ -k_2 & k_2 \end{bmatrix}\begin{bmatrix} x_1 \\ x_2 \end{bmatrix} = \begin{bmatrix} 0 \\ 0 \end{bmatrix}$$

This has an asymmetric displacement coefficient implying the potential loss of stability. If, on the other hand, a control force is applied to m_1 and a measurement made of x_1, then the control system is collocated and the input matrix, output matrix and control gain matrix are

$$B_f = \begin{bmatrix} 1 & 0 \\ 0 & 0 \end{bmatrix}, \; C_p = \begin{bmatrix} 1 & 0 \\ 0 & 0 \end{bmatrix}, \quad \text{and} \quad G = \begin{bmatrix} g_1 & 0 \\ 0 & g_2 \end{bmatrix}$$

so that $B_f^T = C_f$ and the closed loop system of Equation (4.27) becomes

$$\begin{bmatrix} m_1 & 0 \\ 0 & m_2 \end{bmatrix}\begin{bmatrix} \ddot{x}_1 \\ \ddot{x}_2 \end{bmatrix} + \begin{bmatrix} c_1+c_2 & -c_2 \\ -c_2 & c_2 \end{bmatrix}\begin{bmatrix} \dot{x}_1 \\ \dot{x}_2 \end{bmatrix} + \begin{bmatrix} k_1+k_2+g_1 & -k_2 \\ -k_2 & k_2 \end{bmatrix}\begin{bmatrix} x_1 \\ x_2 \end{bmatrix} = \begin{bmatrix} 0 \\ 0 \end{bmatrix}$$

which is symmetric and stable for any choice of g_1, such that $k_1 + k_2 + g_1 > 0$.

The topic of control and Equation (4.27) is discussed in more detail in Chapter 6. Historically, most of the theory developed in the literature for the control of systems has been done using a state space model of the structure. The next section considers the stability of systems in the state variable coordinate system.

4.11 Stability in the State Space

In general, if none of the just-mentioned stability results are applicable, the problem can be cast in first-order form as given in Section 2.4, Equation (2.27) with $\mathbf{f}(t) = 0$. The system of Equation (4.18) then has the form

$$\dot{\mathbf{x}} = A\mathbf{x} \tag{4.28}$$

where A is a $2n \times 2n$ state matrix and $\mathbf{x}$ is a $2n$ state vector. In this setting, it can easily be shown that the system is asymptotically stable if all the eigenvalues of A have negative real parts and unstable if A has one or more eigenvalues with positive real parts.

The search for stability by finding a Lyapunov function in first-order form leads to the *Lyapunov equation*

$$A^T B + BA = -Q \tag{4.29}$$

where Q is positive semidefinite and B is the symmetric positive definite, unknown matrix of the desired (scalar) Lyapunov function

$$V(\mathbf{x}) = \mathbf{x}^T B\mathbf{x} \tag{4.30}$$

Do not confuse the arbitrary matrices B used here with the B used for input and output matrices. To see that $V(\mathbf{x})$ is, in fact, the desired Lyapunov function, note that differentiation of Equation (4.30) yields

$$\frac{d}{dt}[V(\mathbf{x})] = \dot{\mathbf{x}}^T B\mathbf{x} + \mathbf{x}^T B\dot{\mathbf{x}} \tag{4.31}$$

Substitution of the state Equation (4.28) into Equation (4.31) yields

$$\begin{aligned}\frac{d}{dt}[V(\mathbf{x})] &= \mathbf{x}^T A^T B\mathbf{x} + \mathbf{x}^T BA\mathbf{x} \\ &= \mathbf{x}^T (A^T B + BA)\mathbf{x} \\ &= -\mathbf{x}^T Q\mathbf{x}\end{aligned} \tag{4.32}$$

Here, taking the transpose of Equation (4.28) yields $\dot{\mathbf{x}}^T = \mathbf{x}^T A^T$, which is used to remove the time derivative in the second term. Hence, if $V(\mathbf{x})$ is to be a Lyapunov function, the matrix Q must be positive semidefinite. The problem of showing stability by this method for a system represented by the matrix A then becomes, given the symmetric positive semidefinite matrix Q, to find a positive definite matrix B such that Equation (4.29) is satisfied. This approach involves solving a system of linear equations for the $n(n+1)/2$ elements b_{ik} of the matrix B.

As explained by Walker (1974), Hahn (1963) has shown that for a given choice of a symmetric positive definite matrix Q, there exists a unique solution, i.e. there exists a symmetric matrix B satisfying Equation (4.29) if the eigenvalues of A, λ_i satisfy $\lambda_i + \lambda_k \neq 0$ for all $i, k = 1, 2 \ldots, 2n$. Furthermore, the matrix B is positive definite if and only if each eigenvalue of A has negative real part, in which case the system is asymptotically stable. B is indefinite if and only if at least one eigenvalue of A has a positive real part, in which case the equilibrium of the system is unstable. Many theoretical and numerical calculations in stability theory are based on the solution of Equation (4.29). Walker (1974) has shown that this system of linear equations has a unique solution.

Solving for the eigenvalues of Equation (4.28) can involve writing out the system's characteristic equation. In such cases, where this can be done analytically and the coefficients of the characteristic equation are available, a simple stability condition exists, namely, if the characteristic equation is written in the form

$$\lambda^n + a_1\lambda^{n-1} + a_2\lambda^{n-2} + \cdots + a_n = 0 \tag{4.33}$$

then the system is asymptotically stable if and only if the principal minors of the $n \times n$ *Hurwitz matrix* defined by

$$\begin{bmatrix} a_1 & 1 & 0 & 0 & \cdots & & & 0 \\ a_3 & a_2 & a_1 & 1 & \cdots & & & 0 \\ a_5 & a_4 & a_3 & a_2 & a_1 & 1 & \cdots & 0 \\ \cdot & & & & & & & \cdot \\ \cdot & & & & & & & \cdot \\ \cdot & & & & & & & \cdot \\ \cdot & \cdot & \cdot & \cdot & \cdot & \cdot & \cdot & a_n \end{bmatrix}$$

are all positive. In addition, if any of the coefficients a_i are nonpositive (i.e. negative or zero), then the system may be unstable. This is called the *Hurwitz test.*

Writing out the (determinant) principal minors of the Hurwitz matrix yields nonlinear inequalities in the coefficients that provide relationships in the physical parameters of the system. If these inequalities are satisfied, asymptotic stability is insured.

Example 4.11.1

As an illustration of the Hurwitz method, consider determining the asymptotic stability of the system with the characteristic equation

$$\lambda^3 + a_1\lambda^2 + a_2\lambda + a_3 = 0$$

The Hurwitz matrix is

$$\begin{bmatrix} a_1 & 1 & 0 \\ a_3 & a_2 & a_1 \\ 0 & 0 & a_3 \end{bmatrix}$$

From the Hurwitz test, $a_1 > 0$, $a_2 > 0$ and $a_3 > 0$ must be satisfied. From the principal minors of the Hurwitz matrix, the inequalities

$$\begin{aligned} a_1 &> 0 \\ a_1a_2 - a_3 &> 0 \\ a_1(a_2a_3) - a_3^2 = a_1a_2a_3 - a_3^2 &> 0 \end{aligned}$$

must be satisfied. The above set reduces to the conditions that $a_1 > 0$, $a_2 > 0$, $a_3 > 0$, and $a_1a_2 - a_3 > 0$ be satisfied for the system to be asymptotically stable.

4.12 Stability of Nonlinear Systems

A major and unique feature of nonlinear systems versus linear systems is the existence of multiple equilibrium points, as discussed in Sections 1.10 and 2.7. This greatly complicates the discussion of the stability of nonlinear systems, because each equilibrium position may have a different stability property. These solutions emanating from initial conditions near one equilibrium point may have a very different behavior than solutions starting near a different equilibrium point, even though the systems physical parameters are the same. The discussion of stability for nonlinear systems also differs from that of linear systems, because stability for linear systems refers to stability of the response, whereas in nonlinear systems stability properties are assigned to an equilibrium position.

To build an intuitive notion of stability in nonlinear systems, consider the rotating pendulum equation that models the system sketched in Figure 4.3. Referring to the free-body diagram of the pendulum and moments around the origin per Equation (2.11) yields

$$J\ddot{\theta}(t) = \sum_{i=1}^{2} m_{0i}(t) \Rightarrow J\ddot{\theta}(t) = -mgl\sin\theta \tag{4.34}$$

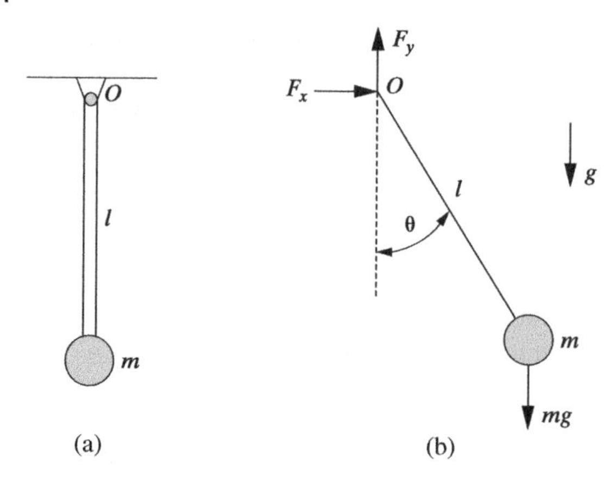

Figure 4.3 (a) A pendulum made of a massless rod and tip mass-free to rotate 360° around the point O. (b) The corresponding free body diagram.

In this case, $J = ml^2$, so the equation of motion becomes

$$\ddot{\theta} = -\frac{g}{l}\sin\theta$$

If friction is modeling in the joint O as a viscous damping term of constant value c, the equation of motion becomes

$$\ddot{\theta} = -\frac{g}{l}\sin\theta - \frac{c}{ml^2}\dot{\theta}$$

Letting $\theta = x_1$ and $\dot{\theta} = x_2$ allows the undamped equation of motion to be put into first-order form

$$\begin{aligned}\dot{x}_1 &= x_2\\ \dot{x}_2 &= -\frac{g}{l}\sin x_1 - \frac{c}{ml^2}x_2\end{aligned} \Rightarrow \dot{\mathbf{x}} = \mathbf{F}(\mathbf{x}) = \begin{bmatrix} x_2 \\ -\frac{g}{l}\sin x_1 - \frac{c}{ml^2}x_2 \end{bmatrix}$$

Setting this last expression to zero yields that the equilibrium positions are

$$\mathbf{x}_e = \begin{bmatrix} x_{1e} \\ x_{2e} \end{bmatrix} = \begin{bmatrix} n\pi \\ 0 \end{bmatrix} = \begin{bmatrix} 0 \\ 0 \end{bmatrix}, \begin{bmatrix} \pi \\ 0 \end{bmatrix}, \begin{bmatrix} 2\pi \\ 0 \end{bmatrix}, \begin{bmatrix} 3\pi \\ 0 \end{bmatrix} \cdots$$

Intuitively the equilibrium points corresponding to $x_1 = \theta =$ odd multiples of π, which corresponds to the pendulum in the straight upright position, will be unstable in the sense that any motion started with an initial condition near the top position will move away from the $[\pi \;\; 0\,]^T$ position and fall away from it. That is if $\|\mathbf{x}(0)\| < \delta(\varepsilon)$, then the motion does not stay within another bound for all time, i.e. $\|\mathbf{x}(t)\| < \varepsilon$. Also any motion started near the zero or even multiples of π with zero initial velocity will likely oscillate near that equilibrium. That is $\|\mathbf{x}(0)\| < \delta(\varepsilon)$, then the motion stays within another bound for all time, i.e. $\|\mathbf{x}(t)\| < \varepsilon$. Thus common sense tells us that $[0 \;\; 0\,]^T$ is likely a stable equilibrium, while $[\pi \;\; 0]^T$ is likely an unstable equilibrium.

The pendulum example can be more formally analyzed by looking at a Taylor series expansion of the equation of motion in state space form. Consider the system of the form $\dot{\mathbf{x}} = \mathbf{F}(\mathbf{x})$, where $\mathbf{F}$ is differentiable and $\mathbf{x}_0$ denotes and equilibrium position. The behavior of the system near $\mathbf{x}_0$ is related to the eigenvalues of the Jacobian of $\mathbf{F}$ evaluated at $\mathbf{x}_0$. To see this, note that a multi-variable function can be expanded in a Taylor series about $\mathbf{x}_0$ of the form

$$\mathbf{F}(\mathbf{x}) = \mathbf{F}(\mathbf{x}_0) + J(\mathbf{x}_0)(\mathbf{x} - \mathbf{x}_0) + o(\|\mathbf{x} - \mathbf{x}_0\|) \tag{4.35}$$

where J denotes the Jacobian of $\mathbf{F}$ and $o(\ldots)$ refers to additional terms of negligible magnitude. The Jacobian matrix J is formed by the various partial derivatives of the function $\mathbf{F}$. For example, if $\mathbf{F}$ is a function of 4 variables

$$J(u, x, y, z) = \begin{bmatrix} \frac{\partial f_1}{\partial u} & \frac{\partial f_1}{\partial x} & \frac{\partial f_1}{\partial y} & \frac{\partial f_1}{\partial z} \\ \frac{\partial f_2}{\partial u} & \frac{\partial f_2}{\partial x} & \frac{\partial f_2}{\partial y} & \frac{\partial f_2}{\partial z} \\ \frac{\partial f_3}{\partial u} & \frac{\partial f_3}{\partial x} & \frac{\partial f_3}{\partial y} & \frac{\partial f_3}{\partial z} \\ \frac{\partial f_4}{\partial u} & \frac{\partial f_4}{\partial x} & \frac{\partial f_4}{\partial y} & \frac{\partial f_4}{\partial z} \end{bmatrix} \tag{4.36}$$

Here f_1 refers to the first entry in $\mathbf{F}$, f_2 the second, etc., each a function of u, x, y and z.

The behavior of the system near an equilibrium point is related to the Jacobian of $\mathbf{F}$ evaluated at the equilibrium point through its eigenvalues. If the eigenvalues all have negative real parts, then the system is stable near that equilibrium point, if any eigenvalue has a positive real part, then the equilibrium point is unstable. If the largest real part of the eigenvalues is zero, the Jacobian matrix test fails to determine the stability. In particular:

1) If the eigenvalues are negative or complex with negative real part, then the equilibrium point is a sink meaning that is all the solutions will die out and return to that equilibrium point. This motion is similar to asymptotic stability defined in Figure 1.20 for a linear system. This equilibrium is referred to as a stable solution.
2) If the eigenvalues are imaginary (zero real parts), the solution will oscillate around the equilibrium similar to the stable response of Figure 1.19 for a linear system.
3) If the eigenvalues are positive or complex with positive real part, then the equilibrium point is a source (i.e. all the solutions will move away from the equilibrium point). Note that if the eigenvalues are complex, then the solutions will move away from the equilibrium point similar to Figure 1.22 for a linear system and if positive real the response will be similar to Figure 1.21.
4) If the eigenvalues are real numbers with different signs (one positive and one negative), then the equilibrium point is a "saddle" point.

Note that these conditions mimic the stability conditions for the eigenvalues of the equivalent linear system; in fact, that is how the theory was developed.

Example 4.12.1
Consider the pendulum of Figure 4.3 and discuss the stability of the first two equilibrium points using the Jacobian.

Solution: From the equations of motion in state space form, $f_1 = x_2$ and $f_2 = -(g/l)\sin(x_1)$, so that

$$J(x_1, x_2) = \begin{bmatrix} \dfrac{\partial f_1}{\partial x_1} & \dfrac{\partial f_1}{\partial x_2} \\ \dfrac{\partial f_2}{\partial x_1} & \dfrac{\partial f_2}{\partial x_2} \end{bmatrix} = \begin{bmatrix} 0 & 1 \\ -\dfrac{g}{l}\cos x_1 & -\dfrac{c}{ml^2} \end{bmatrix}$$

Next, evaluate the Jacobian at the first equilibrium position

$$J(\mathbf{x}_0) = \begin{bmatrix} 0 & 1 \\ -\dfrac{g}{l} & -\dfrac{c}{ml^2} \end{bmatrix}$$

and compute its eigenvalues

$$\det\begin{bmatrix} 0-\lambda & 1 \\ -\dfrac{g}{l} & -\dfrac{c}{ml^2}x_2 - \lambda \end{bmatrix} = 0 \Rightarrow \lambda^2 + \frac{c}{ml^2}\lambda + \frac{g}{l} = 0$$

$$\Rightarrow \lambda = \frac{-c}{2ml^2} \pm \frac{1}{2}\sqrt{\frac{c^2}{ml^2} - 4\frac{g}{l}}$$

Depending on the value of c, these eigenvalues are either negative real, or complex with negative real parts. Thus the equilibrium is asymptotically stable and solutions with initial conditions near $[0 \;\; 0]^T$ will return to that equilibrium position in time. Note that if the damping is set to zero, then the roots are a complex conjugate pair with zero real part corresponding to an oscillation around the equilibrium point.

Next consider the equilibrium at the top, $[\pi \;\; 0]^T$. The Jacobian becomes

$$J(\pi, 0) = \begin{bmatrix} 0 & 1 \\ -\dfrac{g}{l}\cos\pi & -\dfrac{c}{ml^2} \end{bmatrix} = \begin{bmatrix} 0 & 1 \\ \dfrac{g}{l} & -\dfrac{c}{ml^2} \end{bmatrix}$$

The eigenvalues of $J(\pi, 0)$ are

$$\det\begin{bmatrix} -\lambda & 1 \\ \dfrac{g}{l} & -\dfrac{c}{ml^2} - \lambda \end{bmatrix} = \lambda^2 + \frac{c}{ml^2}\lambda - \frac{g}{l} = 0$$

$$\Rightarrow \lambda = -\frac{c}{2ml^2} \pm \frac{1}{2}\sqrt{\frac{c^2}{m^2l^4} + 4\frac{g}{l}}$$

The values of λ are all positive numbers, indicating that equilibrium corresponding to $[n\pi \;\; 0]^T$ are unstable and solutions starting near $[n\pi \;\; 0]^T$ will fall away from that equilibrium and not return.

Another and more popular approach to determining the stability of nonlinear systems is to use the Lyapunov theory discussed in Section 4.2. Recall that the main idea is to find a Lyapunov function for the system, which is positive definite with a negative semidefinite derivative. The following example illustrates the process.

Example 4.12.2

Consider again the pendulum of Example 4.12.2 and discuss the systems stability by finding the Lyapunov function for the system.

Solution: The difficulty in using the Lyapunov approach is in finding a Lyapunov function for the system, as they are not unique and there is no certain way of deriving them. However, the energy in a system is often a good choice. For the pendulum of Figure 4.3, the energy is the sum of the kinetic plus potential energy

$$V(\mathbf{x}) = E = \frac{1}{2}m(\dot{\theta}l)^2 + mgl(1 - \cos\theta) = \frac{1}{2}m(x_2 l)^2 + mgl(1 - \cos x_1)$$

Next, compute the time derivative of V

$$\dot{V}(\mathbf{x}) = m x_2 l^2 \dot{x}_2 + mgl(\sin x_1)\dot{x}_1 = -mgx_2 l \sin x_1 + mgl(\sin x_1)x_2 = 0$$

Thus, the derivative is semidefinite for all values of x_1 and x_2. However, $V(\mathbf{x})$ is not positive definite for all values of $\mathbf{x}$. If the values of x_1 are restricted to be in the interval

$$-\frac{\pi}{2} \le x_1 = \theta \le \frac{\pi}{2}$$

then the $(1 - \cos\theta) > 0$ and $V(\mathbf{x})$ is positive definite and the system is stable.

Chapter Notes

The classification of systems in this chapter is motivated by the text of Huseyin (1978). This text provides a complete list of references for each type of system mentioned here, with the exception of the material on control systems. In addition, Huseyin's text provides an in-depth discussion of each topic. The material of Section 4.2 is standard Lyapunov (also spelled Liapunov in older literature) stability theory and the definitions are available in most texts. The reader who understands limits and continuity from elementary calculus should be able to make the connection to the definition of stability. The material in Sections 4.3 and 4.4 is also standard fare and can be found in most texts considering stability of mechanical systems. The material of Section 4.5 on semidefinite damping results from several papers (as referenced) and is not usually found in text form. The material on gyroscopic systems presented in Section 4.6 is also from several papers. The material on damped gyroscopic systems is interesting, because it violates instinct by illustrating that adding damping to a structure may not always make it "more stable".

The next section deals with asymmetric but symmeterizable systems. The material of Section 4.9 is taken from Inman (1983). The major contributors to the theories (Walker and Huseyin) developed separate methods, which turned out to be quite similar and, in fact, related. The paper by Ahmadian and Chou (1987) should be consulted for methods of calculating symmetric factors of a matrix. Feedback-control systems are presented in this chapter (Section 4.10) before they are formally introduced in Chapter 6, to drive home the fact that the introduction of a control to a system adds energy to it and can make it unstable. The topics of Section 4.11 also present material from control texts. It is important to note that the controls community generally thinks of a stable system as one that is defined as asymptotically stable in this text, i.e. one with eigenvalues with negative real parts. Systems are said to be *marginally stable* by the controls community if the eigenvalues are all purely imaginary. This is called stable in this and other vibration texts. Survey articles on stability of second-order systems are provided by Bernstein and Bhat (1995), Nicholson and Lin (1996) and Adegas and Stoustrup (2015). Vidyasagar (2002) includes detailed stability analysis for both linear and non linear systems as well as definitions of other types of stability.

References

Adegas, F. D. and Stoustrup, J. (2015) Linear matrix inequalities for analysis and control of linear vector second-order systems. *International Journal of Robust and Nonlinear Control*, doi 10.1002/rnc.3242

Ahmadian, M. and Chou, S. -H. (1987) A new method for finding symmetric forms of asymmetric finite dimensional dynamic systems. *ASME Journal of Applied Mechanics*, 54(3), 700–705.

Ahmadian, M. and Inman, D. J. (1986) Some stability results for general linear lumped parameter dynamic systems. *ASME Journal of Applied Mechanics*, 53(1), 10–15.

Bernstien, D. and Bhat (1995) Lyapunov stability, semistability and asymptotic stability of second-order matrix systems. *ASME Journal of Mechanical Design and Vibration and Acoustics Special 50th Anniversary Edition*, 117, 145–153.

Hagedorn, P. (1975) Uber die Instabilitate konservativer Systemenmit Gyroskopischen Kraften. *Archives of Rational Mechanics and Analysis*, 58(1), 1–7.

Hahn, W. (1963) *Theory and Application of Lyapunov's Direct Method*. Prentice Hall, Englewood Cliffs, NJ.

Hughes, P. C. and Gardner, L. T. (1975) Asymptotic stability of linear stationary mechanical systems. *ASME Journal of Applied Mechanics*, 42, 28–29.

Huseyin, K. (1978.) *Vibrations and Stability of Multiple Parameter Systems*. Sigthoff and Noordhoff International Publishers, Alphen aan den Rijn.

Huseyin, K., Hagedorn, P. and Teschner, W. (1983) On the stability of linear conservative gyroscopic systems. *Zietschoift fur augewand Mathematik und Mechanick*, 34(6), 807–815.

Inman, D. J. (1983) Dynamics of asymmetric non-conservative systems. *ASME Journal of Applied Mechanics*, 50, 199–203.

Inman, D. J. and Saggio, F. III (1985) Stability analysis of gyroscopic systems by matrix methods. *AIAA Journal of Guidance, Control and Dynamics*, 8(1), 150–151.

Junkins, J. (1986) Private communication.

Lin, J. (1981) *Active Control of Space Structures*. C.S. Draper Lab Final Report, No. R-1454.

Mingori, D. L. (1970) A stability theorem for mechanical systems with constraint damping. *ASME Journal of Applied Mechanics*, 37, 253–258.

Moran, T. J. (1970) A simple alternative to the Routh-Hurwitz criterion for symmetric systems. *ASME Journal of Applied Mechanics*, 37, 1168–1170.

Nicholson, D. W. and Lin, B. (1996) Stable response of non-classically damped mechanical systems, II: ASME. *Applied Mechanics Reviews*, 49(10(Part 2)), S49–S54.

Taussky, O. (1972) The role of symmetric matrices in the study of general matrices. *Linear Algebra and Its Applications*, 2, 147–153.

Teschner, W. (1977) Instabilitat bei Nichtlinearen Konservativer Systemenmit Gyroskopischen kraften. PhD thesis, Technical University Darmstadt.

Vidyasagar, M. (2002) *Nonlinear Systems Analysis*. Philadelphia, Society of Industrial and Applied Mathematics.

Walker, J. A. (1970) On the dynamic stability of linear discrete dynamic systems. *ASME Journal of Applied Mechanics*, 37, 271–275.

Walker, J. A. (1974) On the application of Lyapunov's direct method to linear lumped parameter elastic systems. *ASME Journal of Applied Mechanics*, 41, 278–284.

Walker, J. A. and Schmitendorf, W. E. (1973) A simple test for asymptotic stability in partially dissipative symmetric systems. *ASME Journal of Applied Mechanics*, 40, 1120–1121.

Zajac, E. E. (1964) The Kelvin-Taft-Chetaev theorem and extensions. *Journal of the Astronautical Sciences*, 11(2), 46–49.

Zajac, E. E. (1965) Comments on stability of damped mechanical systems and a further extension. *AIAA Journal*, 3, 1749–1750.

Problems

4.1 Consider the system of Figure 2.6 with $c_1 = c$, $c_2 = 0, f = 0$, $k_1 = k_2 = k$ and $m_1 = m_2 = m$. The equation of motion is

$$ml\ddot{\mathbf{x}} + \begin{bmatrix} c & 0 \\ 0 & 0 \end{bmatrix}\dot{\mathbf{x}} + \begin{bmatrix} 2k & -k \\ -k & k \end{bmatrix}\mathbf{x} = \mathbf{0}$$

Use Moran's theorem to see if this system is asymptotically stable.

4.2 Repeat problem number 4.1 by using Walker and Schmitendorf's theorem.

4.3 Again consider the system of Figure 2.6, this time with

$$c_1 = 0, c_2 = c, f = 0, k_1 = k_2 = k \text{ and } m_1 = m_2 = m$$

The equation of motion is

$$ml\ddot{\mathbf{x}} + \begin{bmatrix} c & -c \\ -c & c \end{bmatrix}\dot{\mathbf{x}} + \begin{bmatrix} 2k & -k \\ -k & k \end{bmatrix}\mathbf{x} = \mathbf{0}$$

Is this system asymptotically stable? Use any method.

4.4 Discuss the stability of the system of Equation (2.32) of Example 2.5.3 using any method. Note that your answer should depend on the relative values of η, m, E, I and l.

4.5 Calculate a Lyapunov function for the system of Example 4.9.1.

4.6 Show that for a system with symmetric coefficients, that if C is positive semidefinite and $CM^{-1}K = KM^{-1}C$, then the system is *not* asymptotically stable.

4.7 Calculate the matrices and vectors B_f, $\mathbf{u}$, C_p and C_v and R defined in Section 4.10 for the system Figure 2.6, for the case that the velocities of m_1 and m_2 are measured and the actuator (f_2) at m_2 supplies a force of $-g_1\dot{x}_1$. Discuss the stability of this closed-loop system as the gain g_1 is changed.

4.8 The characteristic equation of a given system is

$$\lambda^4 + 10\lambda^3 + \lambda^2 + 15\lambda + 3 = 0$$

Is this system asymptotically stable or unstable? Use a root solver to check your answer.

4.9 Consider the system defined by

$$\begin{bmatrix} m_1 & 0 \\ 0 & m_2 \end{bmatrix}\ddot{\mathbf{q}} + \begin{bmatrix} 0 & 0 \\ 0 & 0 \end{bmatrix}\dot{\mathbf{q}} + \begin{bmatrix} k_1 & -k_3 \\ k_3 & k_2 \end{bmatrix}\mathbf{q} = \mathbf{0}$$

Assume a value of the matrix R from the theory of Walker (1970) of the form

$$R = \begin{bmatrix} m_1 & g \\ g & m_2 \end{bmatrix}$$

and calculate relationships between the parameters m_i, k and g that guarantee the stability of the system. Can the system be asymptotically stable?

4.10 Let

$$A_2 = \begin{bmatrix} c_1 & c_4 \\ c_3 & c_2 \end{bmatrix}$$

in Problem 4.9 and repeat the analysis.

4.11 Let $\mathbf{y} = C_pB_f^T\mathbf{q} + C_vB_f^T\dot{\mathbf{q}}$, where C_p and C_v are restricted to be symmetric and show that the resulting closed loop system with $G_f = I$ and $G = 0$ in Equation (4.27) has symmetric coefficients (Junkins, 1986)

$$M\ddot{\mathbf{q}} + C\dot{\mathbf{q}} + K\mathbf{q} = -B_fG_f(C_vB_f^T\dot{\mathbf{q}}C_pB_f^T\mathbf{q})$$

4.12 For the system of Problem 4.10, choose feedback matrices B_f, G_f, C_p and C_v that make the system symmetric and stable (see Problem 4.11 for a hint).

4.13 Consider the system defined by

$$I\ddot{\mathbf{x}} + \begin{bmatrix} 1 & -1 \\ -1 & 1 \end{bmatrix}\dot{\mathbf{x}} + \begin{bmatrix} 1 & -1 \\ -1 & 1 \end{bmatrix}\mathbf{x} = \mathbf{0}$$

Show that it is not stable. Then use collocated control to force it to be asymptotically stable.

4.14 The characteristic equation of a two-link structure with stiffness at each joint and loaded at the end by p_2 and the joint by p_1 is

$$2\lambda^4 + p_2^2 + \lambda^2 p_1 + 4\lambda^2 p_2 + p_1 p_2 - 8\lambda^2 - p_1 - 4p_2 + 2 = 0$$

where the structure's parameters are all taken to be unity (Huseyin (1978: p. 84). Calculate and sketch the divergence boundary in the $p_1 - p_2$ space. Discuss the flutter condition.

4.15 Use the Hurwitz test to discuss the stability of the system in Problem 4.14.

4.16 Consider the system of Problem 4.1 and compute the damping ratios.

4.17 Consider the system of Problem 4.1 with $ml = 1, c = 2$ and $k = 2$ and compute a collocated control law acting on mass m_1 such that the damping ratios of the closed loop system are $\zeta_1 = 0.345$ and $\zeta_2 = 0.707$.

4.18 Consider the system of Problem 4.1 with $ml = 1, c = 2$ and $k = 2$ and compute a collocated control law acting on mass m_2, such that one of the damping ratios of the closed loop ratios is are $\zeta = 0.707$.

4.19 Consider the system of Problem 4.5 and calculate the eigenvalues in MATLAB of the state matrix to determine the stability.

4.20 Consider the nonlinear system defined by the equation of motion given by

$$\ddot{x}(t) + k(x - x^3) = 0$$

and determine its equilibrium points and use the Jacobian to determine the stability of each.

4.21 Show that the equilibrium point [0 0] of the damped pendulum equation is asymptotically stable by using a Lyapunov function.

5

Forced Response of Lumped Parameter Systems

5.1 Introduction

Up to this point, with the exception of Section 1.4 and some brief comments about feedback control, only the free response of a system has been discussed. In this chapter, the forced response of a system is considered in detail. Such systems are called *nonhomogeneous*. Here an attempt is made to extend the concepts used for the forced response of a single degree of freedom (SDOF) system to the forced response of a general lumped parameter system. In addition, the concept of stability of the forced response, as well as bounds on the forced response, are discussed. The beginning sections of this chapter are devoted to the solution for the forced response of a system by modal analysis, and the later sections are devoted to introducing the use of a forced modal response in measurement and testing. The topic of experimental modal testing is considered in detail in Chapter 12. This chapter ends with an introduction to numerical simulation of the response to initial conditions and an applied force.

Since only linear systems are considered, the superposition principle can be employed. This principle states that the total response of the system is the sum of the free response (homogeneous solution) plus the forced response (the nonhomogeneous solution). Hence, the form of the transient responses calculated in Chapter 3 are used again as part of the solution of a system subject to external forces and nonzero initial conditions. The numerical integration technique presented at the end of the chapter may also be used to simulate nonlinear system response, although that is not presented.

5.2 Response via State Space Methods

This section considers the state space representation of a structure given by

$$\dot{\mathbf{x}}(t) = A\mathbf{x}(t) + \mathbf{f}(t) \tag{5.1}$$

where A is a $2n \times 2n$ matrix containing the generalized mass, damping and stiffness matrices, as defined in Equation (2.27). The $2n \times 1$ state vector $\mathbf{x}(t)$ contains

Vibration with Control, Second Edition. Daniel John Inman.

Companion Website: www.wiley.com/go/inmanvibrationcontrol2e

both the velocity and position vectors and will be referred to as the response in this case. Equation (5.1) reflects the fact that any set of n differential equations of second order can be written as a set of $2n$ first-order differential equations. In some sense, Equation (5.1) represents the most convenient form for solving for the forced response, since a great deal of attention has been focused on solving state space descriptions numerically (i.e. the Runge-Kutta method) as discussed in Sections 1.11, 3.8 and 5.9, as well as analytically. In fact, several software packages are available for solving Equation (5.1) numerically on virtually every computing platform. The state space form is also the form of choice for solving controls problems (Chapter 6).

Only a few of the many approaches to solving this system are presented here; the reader is referred to texts on numerical integration and systems theory for other methods. More attention is paid in this chapter to developing methods that cater to the special form of mechanical systems, i.e. systems written in terms of position, velocity and acceleration rather than in state space.

The first method presented here is simply that of solving Equation (5.1) by using the Laplace transform. Let $\mathbf{X}(0)$ denote the Laplace transform of the initial conditions. Taking the Laplace transform of Equation (5.1) yields

$$s\mathbf{X}(s) = A\mathbf{X}(s) + \mathbf{F}(s) + \mathbf{X}(0) \tag{5.2}$$

where $\mathbf{X}(s)$ denotes the Laplace transform of $\mathbf{x}(t)$ and is defined by

$$\mathbf{X}(s) = \int_0^\infty \mathbf{x}(t)e^{-st}dt \tag{5.3}$$

Here s is a complex scalar. Algebraically solving Equation (5.2) for $\mathbf{X}(s)$ yields

$$\mathbf{X}(s) = (sI - A)^{-1}\mathbf{X}(0) + (sI - A)^{-1}\mathbf{F}(s) \tag{5.4}$$

The matrix $(sI - A)^{-1}$ is referred to as the *resolvent matrix*. The inverse Laplace transform of Equation (5.4) then yields the solution, $\mathbf{x}(t)$. The form of Equation (5.4) clearly indicates the superposition of the transient solution, which is the first term on the right-hand side of Equation (5.4), and the forced response, which is the second term on the right-hand side of Equation (5.4). The inverse Laplace transform is defined by

$$\mathbf{x}(t) = \pounds^{-1}[X(s)] = \lim_{a\to\infty} \frac{1}{2\pi j} \int_{c-ja}^{c+ja} \mathbf{X}(s)e^{st}ds \tag{5.5}$$

where $j = \sqrt{-1}$. In many cases, Equation (5.5) can be evaluated by use of a table such as the one found in Thomson (1960) or a symbolic code. If the integral in Equation (5.5) cannot be found in a table or calculated, then numerical integration can be used to solve Equation (5.1), as presented in Section 5.9.

Example 5.2.1

Consider a simple SDOF system, $\ddot{x} + 3\dot{x} + 2x = \mu(t)$, where $\mu(t)$ is the unit step function, written in the state space form defined by Equation (5.1) with

$$A = \begin{bmatrix} 0 & 1 \\ -2 & -3 \end{bmatrix}, \quad \mathbf{f}(t) = \begin{bmatrix} f_1(t) \\ f_2(t) \end{bmatrix}, \quad \mathbf{x}(t) = \begin{bmatrix} x(t) \\ \dot{x}(t) \end{bmatrix}$$

subject to a force given by $f_1 = 0$ and

$$f_2(t) = \begin{cases} 0, & t < 0 \\ 1, & t \geq 0 \end{cases}$$

the unit step function, and initial condition given by $\mathbf{x}(0) = [0 \ \ 1]^T$. To solve this, first calculate

$$(sI - A) = \begin{bmatrix} s & -1 \\ 2 & s+3 \end{bmatrix}$$

and then determine the resolvent matrix

$$(sI - A)^{-1} = \frac{1}{s^2 + 3s + 2} \begin{bmatrix} s+3 & 1 \\ -2 & s \end{bmatrix}$$

Equation (5.4) becomes

$$\mathbf{X}(s) = \frac{1}{s^2 + 3s + 2} \begin{bmatrix} s+3 & 1 \\ -2 & s \end{bmatrix} \begin{bmatrix} 0 \\ 1 \end{bmatrix} + \frac{1}{s^2 + 3s + 2} \begin{bmatrix} s+3 & 1 \\ -2 & s \end{bmatrix} \begin{bmatrix} 0 \\ \frac{1}{s} \end{bmatrix}$$
$$= \begin{bmatrix} \dfrac{s+1}{s^3 + 3s^2 + 2s} \\ \dfrac{s+1}{s^2 + 3s + 2} \end{bmatrix}$$

Taking the inverse Laplace transform by using a table yields

$$\mathbf{x}(t) = \begin{bmatrix} \frac{1}{2} - \frac{1}{2} e^{-2t} \\ e^{-2t} \end{bmatrix}$$

This solution is in agreement with the fact that the system is overdamped. Also note that $\dot{x}_1 = x_2$, as it should in this case, and that setting $t = 0$ satisfies the initial conditions.

A second method for solving Equation (5.1) imitates the solution of a first-order scalar equation, following what is referred to as the method of "variation of parameters" (Boyce and DiPrima, 2012). For the matrix case, the solution depends on defining the exponential of a matrix. The *matrix exponential* of a matrix A is defined by the infinite series

$$e^A = \sum_{k=0}^{\infty} \frac{A^k}{k!} \tag{5.6}$$

where $k!$ denotes k factorial with $0! = 1$ and $A^0 = I$, the $n \times n$ identity matrix. This series converges for all square matrices A. By using the definition of a scalar multiplied times a matrix of Section 2.1, the time-dependent matrix e^{At} is similarly defined to be

$$e^{At} = \sum_{k=0}^{\infty} \frac{A^k t^k}{k!} \tag{5.7}$$

which also converges. The time derivative of Equation (5.7) yields

$$\frac{d}{dt}(e^{At}) = Ae^{At} = e^{At}A \tag{5.8}$$

Note that the matrix A and the matrix e^{At} commute, because a matrix commutes with its powers.

Following the method of variation of parameters, assume that the solution of Equation (5.1) is of the form

$$\mathbf{x}(t) = e^{At}\mathbf{c}(t) \tag{5.9}$$

where $\mathbf{c}(t)$ is an unknown vector function of time. The time derivative of Equation (5.9) yields

$$\dot{\mathbf{x}}(t) = Ae^{At}\mathbf{c}(t) + e^{At}\dot{\mathbf{c}}(t) \tag{5.10}$$

This results from the product rule and Equation (5.8). Substitution of Equation (5.9) into Equation (5.1) yields

$$\dot{\mathbf{x}}(t) = Ae^{At}\mathbf{c}(t) + \mathbf{f}(t) \tag{5.11}$$

Subtracting Equation (5.11) from Equation (5.10) yields

$$e^{At}\dot{\mathbf{c}}(t) = \mathbf{f}(t) \tag{5.12}$$

Premultiplying Equation (5.12) by e^{-At} (which always exists) yields

$$\dot{\mathbf{c}}(t) = e^{-At}\mathbf{f}(t) \tag{5.13}$$

Simple integration of the differential Equation (5.13) yields the solution for $\mathbf{c}(t)$

$$\mathbf{c}(t) = \int_0^t e^{-A\tau}\mathbf{f}(\tau)d\tau + \mathbf{c}(0) \tag{5.14}$$

Here, the integration of a vector is defined as integration of each element of the vector, just as differentiation is defined on a per-element basis. Substitution of Equation (5.14) into the assumed solution of Equation (5.9) produces the solution of Equation (5.1) as

$$\mathbf{x}(t) = e^{At}\int_0^t e^{-A\tau}\mathbf{f}(\tau)d\tau + e^{At}\mathbf{c}(0) \tag{5.15}$$

Here, $\mathbf{c}(0)$ is the initial condition on $\mathbf{x}(t)$. That is, substitution of $t = 0$ into Equation (5.9) yields

$$\mathbf{x}(0) = e^0\mathbf{c}(0) = I\mathbf{c}(0) = \mathbf{c}(0)$$

so that $\mathbf{c}(0) = \mathbf{x}(0)$, the initial conditions on the state vector. The complete solution of Equation (5.1) can then be written as

$$\mathbf{x}(t) = e^{At}\mathbf{x}(0) + \int_0^t e^{A(t-\tau)}\mathbf{f}(\tau)d\tau \tag{5.16}$$

The first term represents the response due to the initial conditions, i.e. the free response of the system. The second term represents the response due to the applied force, i.e. the steady-state response. Note that the solution given in Equation (5.16) is independent of the nature of the viscous damping in the system (i.e. proportional or not) and gives both the displacement and velocity time response.

The matrix e^{At} is often called the *state transition matrix* of the system defined by Equation (5.1). The matrix e^{At} "maps" the initial condition $\mathbf{x}(0)$ into the new or next position $\mathbf{x}(t)$. While Equation (5.16) represents a closed-form solution of Equation (5.1) for any state matrix A, use of this form centers on the calculation of the matrix e^{At}. Many papers have been written on different methods of calculating e^{At} (Moler and Van Loan, 1978).

One method of calculating e^{At} is to realize that e^{At} is equal to the inverse Laplace transform of the resolvent matrix for A. In fact, a comparison of Equations (5.16) and (5.4) yields

$$e^{At} = \pounds^{-1}\{(sI - A)^{-1}\} \tag{5.17}$$

Another interesting method of calculating e^{At} is restricted to those matrices A with diagonal Jordan form. Then it can be shown that (recall Equation (3.31))

$$e^{At} = Ue^{\Lambda t}U^{-1} \tag{5.18}$$

where U is the matrix of eigenvectors of A and Λ is the diagonal matrix of eigenvalues of A. Here $e^{\Lambda t} = \text{diag}[e^{\lambda_1 t} \; e^{\lambda_2 t} \; \cdots \; e^{\lambda_n t}]$, where the λ_i denote the eigenvalues of the matrix A.

Example 5.2.2

Compute the matrix exponential, e^{At}, for the state matrix

$$A = \begin{bmatrix} -2 & 3 \\ -3 & -2 \end{bmatrix}$$

Using the Laplace transform approach of Equation (5.17) requires forming the matrix $(SI - A)$

$$sI - A = \begin{bmatrix} s+2 & -3 \\ 3 & s+2 \end{bmatrix}$$

Calculating the inverse of this matrix yields

$$(sI - A)^{-1} = \begin{bmatrix} \dfrac{s+2}{(s+2)^2+9} & \dfrac{3}{(s+2)^2+9} \\ \dfrac{-3}{(s+2)^2+9} & \dfrac{s+2}{(s+2)^2+9} \end{bmatrix}$$

The matrix exponential is now computed by taking the inverse Laplace transform of each element. This results in

$$e^{At} = \pounds^{-1}\left\{(sI - A)^{-1}\right\} = \begin{bmatrix} e^{-2t}\cos 3t & e^{-2t}\sin 3t \\ -e^{-2t}\sin 3t & e^{-2t}\cos 3t \end{bmatrix}$$

$$= e^{-2t}\begin{bmatrix} \cos 3t & \sin 3t \\ -\sin 3t & \cos 3t \end{bmatrix}$$

5.3 Decoupling Conditions and Modal Analysis

An alternative approach to solving for the response of a system by transform or matrix exponent methods is to use the eigenvalue and eigenvector information from the free response as a tool for solving for the forced response. This provides a useful theoretical tool as well as a computationally different approach. Approaches based on the eigenvectors of the system are referred to as *modal analysis* and also form the basis for understanding modal test methods (Chapter 12). Modal analysis can be carried out in either the state vector coordinates of the first-order form or the physical coordinates defining the second-order form.

First consider the system described by Equation (5.1). If the matrix A has a diagonal Jordan form (Section 3.4), which happens when it has distinct eigenvalues for example, then the matrix A can be diagonalized by its modal matrix. In this circumstance, Equation (5.1) can be reduced to $2n$ independent first-order equations. To see this, let $\mathbf{u}_i$ be the eigenvectors of the state matrix A with eigenvalues λ_i. Let $U = [\mathbf{u}_1\ \mathbf{u}_2\ \ldots\ \mathbf{u}_{2n}]$ be the modal matrix of the matrix of A. Then substitute $\mathbf{x} = U\mathbf{z}$ into Equation (5.1) to obtain

$$U\dot{\mathbf{z}} = AU\mathbf{z} + \mathbf{f} \tag{5.19}$$

Premultiplying Equation (5.19) by U^{-1} yields the decoupled system

$$\dot{\mathbf{z}} = U^{-1}AU\mathbf{z} + U^{-1}\mathbf{f} \tag{5.20}$$

each element of which is of the form

$$\dot{z}_i = \lambda_i z_i + F_i \tag{5.21}$$

where z_i is the ith element of the vector $\mathbf{z}$ and F_i is the ith element of the vector $\mathbf{F} = U^{-1}\mathbf{f}$. Equation (5.21) can now be solved using scalar integration of each of the equations subject to the initial condition $z_i(0) = [U^{-1}\mathbf{x}(0)]_i$ In this way, the vector $\mathbf{z}(t)$ can be calculated, and the solution $\mathbf{x}(t)$ in the original coordinates becomes

$$\mathbf{x}(t) = U\mathbf{z}(t) \tag{5.22}$$

The amount of effort required to calculate the solution via this method is comparable to that required to calculate the solution via Equation (5.16). The modal form offered by Equations (5.20) and (5.21) provides a tremendous analytical advantage.

The above process can also be used on systems described in the second-order form. The differences are that the eigenvector-eigenvalue problem is solved in n dimensions instead of $2n$ dimensions and the solution is the position vector instead of the state vector of position and velocity. In addition, the modal vectors used to decouple the equations of motion have important physical significance when viewed in second-order form, which is not as apparent in the state space form.

Consider examining the forced response in the physical or spatial coordinates defined by Equation (2.7). First consider the simplest problem, that of calculating the forced response of an undamped nongyroscopic system of the form

$$M\ddot{\mathbf{q}}(t) + K\mathbf{q}(t) = \mathbf{f}(t) \tag{5.23}$$

where M and K are $n \times n$ real positive definite matrices, $\mathbf{q}(t)$ is the vector of displacements, and $\mathbf{f}(t)$ is a vector of applied forces. The system is also subject to an initial position given by $\mathbf{q}(0)$ and an initial velocity given by $\dot{\mathbf{q}}(0)$.

To solve Equation (5.23) by eigenvector expansions, one must first solve the eigenvalue-eigenvector problem for the corresponding homogeneous system. That is, one must calculate λ_i and $\mathbf{u}_i$ such that ($\lambda_i = \omega_i^2$)

$$\lambda_i \mathbf{u}_i = \tilde{K}\mathbf{u}_i \tag{5.24}$$

Note $\mathbf{u}_i$ now denotes an $n \times 1$ eigenvector of the mass normalized stiffness matrix. From this, the modal matrix S_m is calculated and normalized such that

$$\begin{aligned} S_m^T M S_m &= I \\ S_m^T K S_m &= \Lambda = \text{diag}\left(\omega_i^2\right) \end{aligned} \tag{5.25}$$

This procedure is the same as that of Section 3.3, except that in the case of the forced response, the form that the temporal part of the solution will take is not known. Hence, rather than assuming that the dependence is of the form $\sin(\omega t)$, the temporal form is computed from a generic temporal function designated as $y_i(t)$.

Since the eigenvectors $\mathbf{u}_i$ form a basis in an n-dimensional space, any vector $\mathbf{q}(t_1)$, t_1 some fixed but arbitrary time, can be written as a linear combination of the vectors $\mathbf{u}_i$, thus

$$\mathbf{q}(t_1) = \sum_{i=1}^{n} y_i(t_1)\mathbf{u}_i = S_m \mathbf{y}(t_1) \tag{5.26}$$

where $\mathbf{y}(t_1)$ is an n-vector with components $y_i(t_1)$ to be determined. Since t_1 is arbitrary, it is reasoned that Equation (5.26) must hold for any t. Therefore

$$\mathbf{q}(t) = \sum_{i=1}^{n} y_i(t)\mathbf{u}_i = S_m \mathbf{y}(t) \quad t \geq 0 \tag{5.27}$$

This must be true for any n-dimensional vector $\mathbf{q}$. In particular, this must hold for solutions of Equation (5.23). Substitution of Equation (5.27) into Equation (5.23) shows that the vector $\mathbf{y}(t)$ must satisfy

$$M S_m \ddot{\mathbf{y}}(t) + K S_m \mathbf{y}(t) = \mathbf{f}(t) \tag{5.28}$$

Premultiplying by S_m^T yields

$$\ddot{\mathbf{y}}(t) + \Lambda \mathbf{y}(t) = S_m^T \mathbf{f}(t) \tag{5.29}$$

Equation (5.29) represents n decoupled equations, each of the form

$$\ddot{y}_1(t) + \omega_i^2 y_i(t) = f_i(t) \tag{5.30}$$

where $f_i(t)$ denotes the ith element of the vector $S_m^T \mathbf{f}(t)$.

If K is assumed to be positive definite, each ω_i^2 is a positive real number. Denoting the "modal" initial conditions by $y_i(0)$ and $\dot{y}_i(0)$, the solutions of Equation (5.30) are calculated by the method of variation of parameters to be

$$y_i(t) = \frac{1}{\omega_i}\int_0^t f_i(t-\tau)\sin(\omega_i\tau)d\tau + y_i(0)\cos(\omega_i t) + \frac{\dot{y}_i(0)}{\omega_i}\sin(\omega_i t) \tag{5.31}$$
$$i = 1, 2, 3, \ldots n$$

(See Boyce and DePrima, 2012, for a derivation.)

If K is semidefinite, one or more values of ω_i^2 might be zero. Then Equation (5.30) would become

$$\ddot{y}_i(t) = f_i(t) \tag{5.32}$$

Integrating Equation (5.32) then yields

$$y_i(t) = \int_0^t \left[\int_0^\tau f_i(s)ds\right] d\tau + y_i(0) + \dot{y}_i(0)t \tag{5.33}$$

which represents a rigid body motion.

The initial conditions for the new coordinates are determined from the initial conditions for the original coordinates by the transformation

$$\mathbf{y}(0) = S_m^{-1}\mathbf{q}(0) \tag{5.34}$$

and

$$\dot{\mathbf{y}}(0) = S_m^{-1}\dot{\mathbf{q}}(0) \tag{5.35}$$

This method is often referred to as *modal analysis* and differs from the state space modal approach, in that the computations involve matrices and vectors of size n rather than $2n$. They result in a solution for the position vector rather than the $2n$-dimensional state vector. The coordinates defined by the vector $\mathbf{y}$ are called *modal coordinates, normal coordinates, decoupled coordinates,* and (sometimes) *natural coordinates.* Note that in the case of a free response, i.e. $f_i = 0$, then $y_i(t)$ is just $e^{\pm\omega_i jt}$, where ω_i is the ith natural frequency of the system, as discussed in Section 3.3.

Alternately, the modal decoupling described in the above paragraphs can be obtained by using the mass normalized stiffness matrix. To see this, substitute $\mathbf{q} = M^{-1/2}\mathbf{r}$ into Equation (2.17), multiply by $M^{-1/2}$ to form $\tilde{K} = M^{-1/2}KM^{-1/2}$, compute the normalized eigenvectors of $\tilde{K}$ and use these to form the columns of the orthogonal matrix S. Next use the substitution $\mathbf{r} = S\mathbf{y}$ in the equation of motion, premultiply by S^T and Equation (5.30) results. This procedure is illustrated in the following example.

Example 5.3.1

Consider the undamped system of Example 3.3.2 and Figure 2.6 subject to a harmonic force applied to m_2, given by

$$\mathbf{f}(t) = \begin{bmatrix} 0 \\ 1 \end{bmatrix} \sin 3t$$

Following the alternate approach, **compute** the modal force by multiplying the physical force by $S^T M^{-1/2}$

$$S^T M^{-1/2}\mathbf{f}(t) = \frac{1}{\sqrt{2}} \begin{bmatrix} 1 & 1 \\ -1 & 1 \end{bmatrix} \begin{bmatrix} 1/3 & 0 \\ 0 & 1 \end{bmatrix} \begin{bmatrix} 0 \\ 1 \end{bmatrix} \sin 3t = \frac{1}{\sqrt{2}} \begin{bmatrix} 1 \\ 1 \end{bmatrix} \sin 3t$$

Combining this with the results of Example 3.3.2 yields that the modal Equations (5.30) are

$$\ddot{y}_1(t) + 2y_1(t) = \frac{1}{\sqrt{2}} \sin 3t,$$

$$\ddot{y}_2(t) + 4y_2(t) = \frac{1}{\sqrt{2}} \sin 3t.$$

This of course is subject to the transformed initial conditions and each equation can be solved by the methods of Chapter 1. For instance, if the initial conditions in the physical coordinates are

$$\mathbf{q}(0) = \begin{bmatrix} 0.1 \\ 0 \end{bmatrix}, \quad \text{and} \quad \dot{\mathbf{q}}(0) = \begin{bmatrix} 0 \\ 0 \end{bmatrix}$$

then in the modal coordinates the initial velocity remains at zero, but the initial displacement is transformed to become (solving $\mathbf{q} = M^{-1/2}\mathbf{r}$ and $\mathbf{r} = S\mathbf{y}$, for $\mathbf{y}$)

$$\mathbf{y}(0) = S^T M^{1/2}\mathbf{q}(0) = \frac{1}{\sqrt{2}} \begin{bmatrix} 1 & 1 \\ -1 & 1 \end{bmatrix} \begin{bmatrix} 3 & 0 \\ 0 & 1 \end{bmatrix} \begin{bmatrix} 0.1 \\ 0 \end{bmatrix} = \frac{0.3}{\sqrt{2}} \begin{bmatrix} 1 \\ -1 \end{bmatrix}$$

Solving for y_1 proceeds by first computing the particular solution (see Section 1.4 for zero damping) by assuming $y(t) = X\sin 3t$ in the modal equation for y_1 to get

$$-9X \sin 3t + 2X \sin 3t = \frac{1}{\sqrt{2}} \sin 3t$$

Solving for X yields that the particular solution is $y_{1p}(t) = \left(-1/(7\sqrt{2})\right) \sin 3t$. The total solution is then (from Equation (1.35) with zero damping)

$$y_1(t) = A \sin \sqrt{2}t + B \cos \sqrt{2}t - \frac{1}{7\sqrt{2}} \sin 3t$$

Applying the modal initial conditions yields

$$y_1(0) = B = \frac{0.3}{\sqrt{2}}, \quad \dot{y}_1(0) = \sqrt{2}A - \frac{3}{7\sqrt{2}} = 0 \Rightarrow A = \frac{3}{14}$$

Thus the solution of the first modal equation is

$$y_1(t) = \frac{3}{14}\sin\sqrt{2}t + \frac{0.3}{\sqrt{2}}\cos\sqrt{2}t - \frac{1}{7\sqrt{2}}\sin 3t$$

Likewise the solution to the second modal equation is

$$y_2(t) = \frac{0.3}{\sqrt{2}}\sin 2t - \frac{0.3}{\sqrt{2}}\cos 2t - \frac{0.2}{\sqrt{2}}\sin 3t$$

The solution in physical coordinates is then found from $\mathbf{q}(t) = M^{-1/2}S\mathbf{y}(t)$.

5.4 Response of Systems with Damping

The key to using modal analysis to solve for the forced response of systems with velocity-dependent terms is whether or not the system can be decoupled. As in the case of the free response, discussed in Section 3.5, this will happen for symmetric systems if and only if the coefficient matrices commute, i.e. if $KM^{-1}C = CM^{-1}K$. Ahmadian and Inman (1984a) reviewed previous work on decoupling and extended the commutivity condition to systems with asymmetric coefficients. Inman (1982) and Ahmadian and Inman (1984b) used the decoupling condition to carry out modal analysis for general asymmetric systems with commuting coefficients. In each of these cases, the process is the same, with an additional transformation into symmetric coordinates as introduced in Section 4.9. Hence, only the symmetric case is illustrated here.

To this end, consider the problem of calculating the forced response of the nongyroscopic damped linear system given by

$$M\ddot{\mathbf{q}} + C\dot{\mathbf{q}} + K\mathbf{q} = \mathbf{f}(t) \tag{5.36}$$

where M and K are symmetric and positive definite and C is symmetric and positive semidefinite. In addition, it is assumed that $KM^{-1}C = CM^{-1}K$. Let S_m be the modal matrix of K normalized with respect to M, as defined by Equations (3.70) and (3.71). Then the commutivity of the coefficient matrices yields

$$\begin{aligned} S_m^T M S_m &= I \\ S_m^T C S_m &= \Lambda_C = \text{diag}(2\zeta_i\omega_i) \\ S_m^T K S_m &= \Lambda_K = \text{diag}(\omega_i^2) \end{aligned} \tag{5.37}$$

where Λ_C and Λ_K are diagonal matrices as indicated. Making the substitution $\mathbf{q} = \mathbf{q}(t) = S_m\mathbf{y}(t)$ into Equation (5.36) and premultiplying by S_m^T as before yields

$$I\ddot{\mathbf{y}} + \Lambda_C\dot{\mathbf{y}} + \Lambda_K\mathbf{y} = S_m^T\mathbf{f}(t) \tag{5.38}$$

Equation (5.39) is diagonal and can be written as n decoupled equations of the form

$$\ddot{y}_i(t) + 2\zeta_i\omega_i\dot{y}_i(t) + \omega_i^2 y_i(t) = f_i(t) \tag{5.39}$$

Here $\zeta_i = \lambda_i(C)/2\omega_i$, where $\lambda_i(C)$ denotes the eigenvalues of the matrix C. In this case, these are the nonzero elements of Λ_C. This expression is the nonhomogeneous counterpart of Equation (3.71).

If it is assumed that $4\tilde{K} - \tilde{C}^2$ is positive definite, then $0 < \zeta_i < 1$ (Inman and Andry, 1980) and the solution of Equation (5.40) is (assuming all initial conditions are zero)

$$y_i(t) = \frac{1}{\omega_{di}} \int_0^t e^{-\zeta_i\omega_i\tau} f_i(t-\tau)\sin(\omega_{di}\tau)d\tau,\ i = 1, 2, 3, \ldots, n \tag{5.40}$$

where $\omega_{di} = \omega_i\sqrt{1-\zeta_i^{(2)}}$. If $4\tilde{K} - \tilde{C}^2$ is not positive definite, other forms of $y_i(t)$ result, depending on the eigenvalues of the matrix $4\tilde{K} - \tilde{C}^2$.

In addition to the forced response given by Equation (5.40), there will be a transient response, or homogeneous solution, due to any nonzero initial conditions. If this response is denoted by y_i^H, then the total response of the system in the decoupled coordinate system is the sum $y_i(t) + y_i^H(t)$. This solution is related to the solution in the original coordinates by the modal matrix S_m and is given by

$$\mathbf{q}(t) = S_m[\mathbf{y}(t) + \mathbf{y}^H(t)] \tag{5.41}$$

For asymmetric systems, the procedure is similar, with the exception of computing a second transformation; this transforms the asymmetric system into an equivalent symmetric system, as done in Section 4.9.

For systems in which the coefficient matrices do not commute, i.e. for which $KM^{-1}C \neq CM^{-1}K$ in Equation (5.36), modal analysis of a sort is still possible without resorting to state space. To this end consider the symmetric case given by the system

$$M\ddot{\mathbf{q}} + C\dot{\mathbf{q}} + K\mathbf{q} = \mathbf{f}(t) \tag{5.42}$$

where M, C and K are symmetric.

Let $\mathbf{u}_i$ be the eigenvectors of the lambda matrix

$$(M\lambda_i^2 + C\lambda_i + K)\mathbf{u}_i = \mathbf{0} \tag{5.43}$$

with associated eigenvalue λ_i. Let n be the number of degrees of freedom (so there are $2n$ eigenvalues), let $2s$ be the number of real eigenvalues, and let $2(n-s)$ be the number of complex eigenvalues. Assuming that $D_2(\lambda)$ is simple and that the $\mathbf{u}_i$ are normalized so that

$$\mathbf{u}_i^T(2M\lambda_i + C)\mathbf{u}_i = 1 \tag{5.44}$$

a particular solution of Equation (5.42) is given by Lancaster (1966) in terms of the generalized modes $\mathbf{u}_i$ to be

$$\mathbf{q}(t) = \sum_{k=1}^{2s} \mathbf{u}_k\mathbf{u}_k^T \int_0^t e^{-\lambda_k(t+\tau)}\mathbf{f}(\tau)d\tau + \sum_{k=2s+1}^{2n} \int_0^t Re\left\{e^{\lambda_k(t-\tau)}\mathbf{u}_k\mathbf{u}_k^T\right\}\mathbf{f}(\tau)d\tau \tag{5.45}$$

This expression is more difficult to compute but does offer some insight into the form of the solution that is useful in modal testing, as will be illustrated in Chapter 12. Note that the eigenvalues indexed λ_1 through λ_{2s} are real, whereas those labeled λ_{2s+1} to λ_{2n} are complex. The complex eigenvalues λ_{2s+1} and $\lambda_{2(s+1)}$ are conjugates of each other. Also, note that the nature of the matrices C and K completely determine the value of s. In fact, if $4\tilde{K} - \tilde{C}^2$ is positive definite, $s = 0$ in Equation (5.44). Also note that if λ_k is real, so is the corresponding $\mathbf{u}_k$. On the other hand, if λ_k is complex, the corresponding eigenvector $\mathbf{u}_k$ is real if and only if $KM^{-1}C = CM^{-1}K$; otherwise the eigenvectors are complex valued.

The particular solution Equation (5.45) has the advantage of being stated in the original, or physical, coordinate system. To get the total solution, the transient response developed in Section (3.4) must be added to Equation (5.45). This should be done, unless steady-state conditions prevail.

Example 5.4.1

Consider Example 5.3.1 with a damping force applied of the form $C = 0.1\ K$ (proportional damping). In this case, $4\tilde{K} - \tilde{C}^2$ is positive definite so that each mode will be underdamped (Inman and Andry, 1980). The alternate transformation used in Example 5.3.1 is employed here to find the modal equations given in Equation (5.40). The equations of motion and initial conditions (assuming compatible units) are

$$\begin{bmatrix} 9 & 0 \\ 0 & 1 \end{bmatrix} \ddot{\mathbf{q}}(t) + \begin{bmatrix} 2.7 & -0.3 \\ -0.3 & 0.3 \end{bmatrix} \dot{\mathbf{q}}(t) + \begin{bmatrix} 27 & -3 \\ -3 & 3 \end{bmatrix} \mathbf{q}(t) = \begin{bmatrix} 0 \\ 1 \end{bmatrix} \sin 3t,$$

$$\mathbf{q}(0) = \begin{bmatrix} 1 \\ 0 \end{bmatrix}, \quad \dot{\mathbf{q}}(0) = \begin{bmatrix} 0 \\ 0 \end{bmatrix}$$

Since the damping is proportional, the undamped transformation computed in Examples 3.3.2 and 5.3.1 can be used to decouple the equations of motion. Using the transformation $\mathbf{y}(t) = S^T M^{1/2} \mathbf{q}(t)$ and premultiplying the equations of motion by $S^T M^{-1/2}$ yields the uncoupled modal equations

$$\ddot{y}_1(t) + 0.2\dot{y}_1(t) + 2y_1(t) = \frac{1}{\sqrt{2}} \sin 3t, \ y_1(0) = \frac{3}{\sqrt{2}}, \ \dot{y}_1(0) = 0$$

$$\ddot{y}_2(t) + 0.4\dot{y}_2(t) + 4y_2(t) = \frac{1}{\sqrt{2}} \sin 3t, \ y_2(0) = \frac{-3}{\sqrt{2}}, \ \dot{y}_2(0) = 0$$

From the modal equations, the frequencies and damping rations are evident

$$\omega_1 = \sqrt{2} = 1.414 \text{ rad/s}, \ \zeta_1 = \frac{0.2}{2\sqrt{2}} = 0.071 < 1,$$

$$\omega_{1d} = \omega_1 \sqrt{1 - \zeta_1^2} = 1.41 \text{ rad/s}$$

$$\omega_2 = \sqrt{4} = 2 \text{ rad/s}, \ \zeta_2 = \frac{0.4}{(2)(2)} = 0.1 < 1,$$

$$\omega_{2d} = \omega_2 \sqrt{1 - \zeta_2^2} = 1.99 \text{ rad/s}$$

Solving the two modal equations using the approach of Example 1.4.1 (y_2 is solved there) yields

$$y_1(t) = e^{-0.1t}(0.3651 \sin 1.41t + 2.1299 \cos 1.41t) - 0.1015 \sin 3t - 0.0087 \cos 3t$$

$$y_2(t) = -e^{-0.2t}(0.0084 \sin 1.99t + 2.0892 \cos 1.99t) - 0.1325 \sin 3t - 0.032 \cos 3t$$

This forms the solution in the modal coordinates, to regain the solution in physical coordinates using the transformation $\mathbf{q}(t) = M^{-1/2}S\mathbf{y}(t)$. Note that the transient term is multiplied by a decaying exponential in time and will decay off leaving the steady state to persist.

5.5 Stability of the Forced Response

In the previous chapter several types of stability for the free response of a system were defined and discussed in great detail. In this section the concept of stability as it applies to the forced response of a system is discussed. In particular, systems are examined in the state space form given by Equation (5.1).

The stability of the forced response of a system is defined in terms of bounds of the response vector $\mathbf{x}(t)$. Hence, it is important to recall that a vector $\mathbf{x}(t)$ is bounded if

$$\|\mathbf{x}(t)\| = \sqrt{\mathbf{x}^T\mathbf{x}} \leq M \tag{5.46}$$

where M is some finite positive real number. The quantity $\|\mathbf{x}\|$ just defined is called the *norm* of $\mathbf{x}(t)$. The response $\mathbf{x}(t)$ is also referred to as the *output* of the system, hence the phrase bounded-output stability.

A fundamental classification of stability of forced systems is called *bounded-input, bounded-output* (BIBO) *stability*. The system described by Equation (5.1) is called BIBO stable if any bounded forcing function $\mathbf{f}(t)$, called the input, produces a bounded response $\mathbf{x}(t)$, i.e. a bounded output, regardless of the bounded initial condition $\mathbf{x}(0)$. An example of a system that is not BIBO stable is given by the SDOF oscillator

$$\ddot{y} + \omega_n^2 y = \sin \omega t \tag{5.47}$$

In state space form this becomes

$$\dot{\mathbf{x}} = \begin{bmatrix} 0 & 1 \\ -\omega_n^2 & 0 \end{bmatrix} \mathbf{x} + \mathbf{f}(t) \tag{5.48}$$

where $\mathbf{x} = [y \ \ \dot{y}]^T$ and $\mathbf{f}(t) = [0 \ \ \sin \omega t]^T$. This system is not BIBO stable, since for any bounded initial condition $y(t)$, and hence $\mathbf{x}(t)$, blows up when $\omega = \omega_n$ (i.e. at resonance).

A second classification of stability is called *bounded stability*, or *Lagrange stability*, and is a little weaker than BIBO stability. The system described in

Equation (5.1) is said to be *Lagrange stable* with respect to a *given* input $\mathbf{f}(t)$, if the response $\mathbf{x}(t)$ is bounded for any bounded initial condition $\mathbf{x}(0)$. Referring to the example of the previous paragraph, if $\omega \neq \omega_n$, then the system described by Equation (5.48) is bounded with respect to $\mathbf{f}(t) = [0 \ \ \sin \omega t]^T$, because when $\omega \neq \omega_n$, $\mathbf{x}(t)$ does not blow up. Note that if a given system is BIBO stable, it will also be Lagrange stable. However, a system that is Lagrange stable may not be BIBO stable.

As an example of a system that is BIBO stable, consider adding damping to the preceding system. The result is an SDOF damped oscillator that has state matrix

$$A = \begin{bmatrix} 0 & 1 \\ -k/m & -c/m \end{bmatrix} \tag{5.49}$$

Recall that the damping term prevents the solution $\mathbf{x}(t)$ from becoming unbounded at resonance. Hence, $y(t)$ and $\dot{y}(t)$ are bounded for any bounded input $\mathbf{f}(t)$, and the system is BIBO stable as well as Lagrange stable.

The difference in the two examples is due to the stability of the free response of each system. The undamped oscillator is stable but not asymptotically stable, and the forced response *is not* BIBO stable. On the other hand, the damped oscillator is asymptotically stable and *is* BIBO stable. To some extent this is true in general. Namely, it is shown by Müller and Schlehlen (1977) that if the forcing function $\mathbf{f}(t)$ can be written as a constant matrix B times a vector $\mathbf{u}$, i.e. $\mathbf{f}(t) = B\mathbf{u}$, then if

$$\text{rank}[B \quad AB \quad A^2B \quad A^3B \quad \cdots \quad A^{2n-1}B] = 2n \tag{5.50}$$

where $2n$ is the dimension of the matrix A, the system in Equation (5.1) is BIBO stable if and only if the free response is asymptotically stable. If $\mathbf{f}(t)$ does not have this form or does not satisfy the rank condition Equation (5.50), then asymptotically stable systems are BIBO stable, and BIBO stable systems have a stable-free response.

Another way to look at the difference between the above two examples is to consider the phenomenon of resonance. The undamped SDOF oscillator of Equation (5.48) experiences an infinite amplitude at $\omega = \omega_n$, the resonance condition, which is certainly unstable. However, the underdamped SDOF oscillator of Equation (5.49) is bounded at the resonance condition, as discussed in Section 1.4. Hence, the damping "lowers" the peak response at resonance from infinity to some finite, or bounded, value, resulting in a system that is BIBO stable.

Patel and Toda (1980) proposed the following concept for the forced response of systems with asymptotically stable autonomous dynamics, such as Equation (4.28). If $\dot{\mathbf{x}}(t) = A\mathbf{x}(t)$ is asymptotically stable, then $\dot{\mathbf{x}}(t) = A\mathbf{x}(t) + \mathbf{f}(t, \mathbf{x}(t))$ remains asymptotically stable for all small perturbations

$$\frac{\|\mathbf{f}\|}{\|\mathbf{x}\|} \leq \frac{\min \lambda(Q)}{\max \lambda(B)}$$

Here B and Q satisfy the Lyapunov Equation (4.29) repeated here

$$A^T B + BA = -Q$$

provided $\mathbf{f}$ is zero at the state space origin. The notation min $\lambda(Q)$ refers to the smallest eigenvalue of Q. The obvious use of the preceding conditions is to use the stability results of Chapter 4 for the free response to guarantee BIBO stability or boundedness of the forced response, $\mathbf{x}(t)$. To this extent, other concepts of stability of systems subject to external forces are not developed.

5.6 Response Bounds

Given that a system is either BIBO stable or at least bounded, it is sometimes of interest to calculate bounds for the forced response of the system without actually calculating the response itself. A summary of early work on the calculation of bounds is given in review papers by Nicholson and Inman (1983) and Nicholson and Lin (1996). More recent work is given in Hu and Eberhard (1999). A majority of the work reported there examines bounds for systems in the physical coordinates $\mathbf{q}(t)$ in the form of Equation (5.36). In particular, if $CM^{-1}K = KM^{-1}C$ and if the forcing function or input is of the form (periodic)

$$\mathbf{f}(t) = \mathbf{f}_0 e^{j\omega t} \tag{5.51}$$

where $\mathbf{f}_0$ is an $n \times 1$ vector of constants, $j^2 = -1$, and ω is the driving frequency, then

$$\frac{||\mathbf{q}(t)||}{||\mathbf{f}_0||} \leq \max_j \begin{cases} \dfrac{1}{\lambda_j(K)} & \text{if } \dfrac{\lambda_j(K)}{\lambda_j(M)} \leq \dfrac{\lambda_j^2(C)}{2\lambda_j^2(M)} \\ \sqrt{\dfrac{\lambda_i(M)}{\lambda_i^2(C)\lambda_i(K)}} & \text{otherwise} \end{cases} \tag{5.52}$$

Here, $\lambda_i(M)$, $\lambda_i(C)$ and $\lambda_i(K)$ are used to denote the ordered eigenvalues of the matrices M, C and K, respectively. The first inequality in Equation (5.52) is satisfied if the free system is overdamped, and the bound $[\lambda_i^2(C)\lambda_i(K)/\lambda_i(M)]^{-1/2}$ is applied for underdamped systems.

Bounds on the forced response are also available for systems that do not decouple, i.e. for systems with coefficient matrices such that $CM^{-1}K \neq KM^{-1}C$. One way to approach such systems is to write the system in the normal coordinates of the undamped system. Then the resulting damping matrix can be written as the sum of a diagonal matrix and an off-diagonal matrix, which clearly indicates the degree of decoupling.

Substituting $\mathbf{q}(t) = S_m\mathbf{x}$ into Equation (5.36) and premultiplying by S_m^T, where S_m is the modal matrix for K, yields

$$I\ddot{\mathbf{x}} + (\Lambda_C + C_1)\dot{\mathbf{x}} + \Lambda_K\mathbf{x} = \tilde{\mathbf{f}}(t) \tag{5.53}$$

The matrix Λ_C is the diagonal part of $S_m^T CS_m$, and C_1 is the matrix of off-diagonal elements of $S_m^T CS_m$, Λ_K is the diagonal matrix of squared undamped natural frequencies, and $\tilde{\mathbf{f}} = S_m^T\mathbf{f}$.

The steady-state response of Equation (5.53) with a sinusoidal input, i.e. $\tilde{\mathbf{f}} = \mathbf{f}_0 e^{j\omega t}$ and Equation (5.53) underdamped, is given by

$$\frac{\|\mathbf{x}(t)\|}{\|\mathbf{f}_0(t)\|} < \frac{2}{\lambda_{\min}(\Lambda_C\Lambda_{CK})} e^{\beta\|C_1\|/\lambda_{\min}} \tag{5.54}$$

Here, $\lambda_{\min}(\Lambda_C \Lambda_{CK})$ denotes the smallest eigenvalue of the matrix $\Lambda_C \Lambda_{CK}$, where $\Lambda_{CK} = (4\Lambda_K - \Lambda_C^2)^{1/2}$. Also, $\lambda_{\min}$ is the smallest eigenvalue of the matrix Λ_C, $\|C_1\|$ is the matrix norm defined by the maximum value of the square root of the largest eigenvalue of $C_1^T C_1 = C_1^2$ and β is defined by

$$\beta = \sqrt{\left\| I + (\Lambda_{CK}^{-1} \Lambda_C)^2 \right\|} \tag{5.55}$$

Examination of the bound in Equation (5.54) shows that the greater the coupling in the system characterized by $\|C_1\|$, the larger the bound. Thus, for small values of $\|C_1\|$, i.e. small coupling, the bound is good, whereas for large values of $\|C_1\|$ or very highly coupled systems, the bound will be very large and too conservative to be of practical use. This is illustrated in Example 5.6.1.

Example 5.6.1

Consider a system defined by Equation (5.37) with $M = I$

$$K = \begin{bmatrix} 5 & -1 \\ -1 & 1 \end{bmatrix}, \quad C = 0.5K + 0.5I + \xi \begin{bmatrix} 0 & 1 \\ 1 & 0 \end{bmatrix}$$

subject to a sinusoidal driving force applied to the first mass so that $\mathbf{f}_0 = [1\ 0]^T$. The parameter ξ clearly determines the degree of proportionality or coupling in the system. The bounds are tabulated in Table 5.1 for various values of ξ, along with a comparison to the exact solution.

Examination of Table 5.1 clearly illustrates that as the degree of coupling increases (larger ξ), the bound gets farther away from the actual response. Note that the value given in the "exact" column is the largest value obtained by the exact response.

Table 5.1 Forced response bounds.

ξ	Exact Solution	Bound
0	1.30	1.50
−0.1	1.56	2.09
−0.2	1.62	3.05
−0.3	1.70	4.69
−0.4	1.78	8.60

5.7 Frequency Response Methods

This section attempts to extend the concept of frequency response introduced in Sections 1.5 and 1.6 to multiple degree of freedom (MDOF) systems. In so doing, the material in this section makes the connection between analytical modal analysis and experimental modal analysis discussed in Chapter 12. The development starts by considering the response of a structure due to a harmonic or sinusoidal input, denoted by $\mathbf{f}(t) = \mathbf{f}_0 e^{j\omega t}$. The equations of motion in spatial or physical coordinates given by Equation (5.36) with no damping ($C = 0$) or Equation (5.23) are considered first. In this case, an oscillatory solution of Equation (5.23) of the form

$$\mathbf{q}(t) = \mathbf{u}e^{j\omega t} \tag{5.56}$$

is assumed. This is equivalent to the frequency response theorem stated in Section 1.5. That is, if a system is harmonically excited, the response will consist of a steady-state term that oscillates at the driving frequency with different amplitude and phase.

Substitution of the assumed oscillatory solution into Equation (5.36) with $C = 0$ yields

$$(K - \omega^2 M)\mathbf{u}e^{j\omega t} = \mathbf{f}_0 e^{j\omega t} \tag{5.57}$$

Dividing through by the nonzero scalar $e^{j\omega t}$ and solving for $\mathbf{u}$ yields

$$\mathbf{u} = (K - \omega^2 M)^{-1}\mathbf{f}_0 \tag{5.58}$$

Note that the matrix inverse of $(K - \omega^2 M)$ exists as long as ω is not one of the natural frequencies of the structure. This is consistent with the fact that without damping, the system is Lagrange stable and not BIBO stable. The matrix coefficient of Equation (5.58) is defined to be the *receptance matrix*, denoted by $\alpha(\omega)$, i.e.

$$\alpha(\omega) = (K - \omega^2 M)^{-1} \tag{5.59}$$

Equation (5.58) can be thought of as the *response model* of the structure. Solution of Equation (5.58) yields the vector, $\mathbf{u}$, which, coupled with Equation (5.56), yields the steady-state response of the system to the input force, $\mathbf{f}(t)$.

Each element of the response matrix can be related to a single-frequency response function by examining the definition of matrix multiplication. In particular, if all the elements of the vector $\mathbf{f}_0$, denoted by f_i, except the jth element are set equal to zero, then the ij^{th} element of $\alpha(\omega)$ is just the receptance transfer function between u_i, the ith element of the response vector $\mathbf{u}$ and f_j. That is

$$\alpha_{ij}(\omega) = \frac{u_i}{f_j},\ f_i = 0,\ i = 0, \dots, n,\ i \neq j \tag{5.60}$$

Note that since $\alpha(\omega)$ is symmetric, this interpretation implies that $u_i/f_j = u_j/f_i$. Hence, a force applied at position j yields the same response at point i as a force applied at i does at point k. This is called *reciprocity.*

An alternative to computing the inverse of the matrix $(K - \omega^2 M)$ is to use the modal decomposition of $\alpha(\omega)$. Recalling Equations (3.20) and (3.21) from Section 3.2, the matrices M and K can be rewritten as

$$M = S_m^{-T} S_m^{-1} \tag{5.61}$$

$$K = S_m^{-T}\operatorname{diag}(\omega_r^2) S_m^{-1} \tag{5.62}$$

where ω_i are the natural frequencies of the system and S_m is the matrix of modal vectors normalized with respect to the mass matrix. Substitution of these "modal" expressions into Equation (5.59) yields

$$\begin{aligned}\alpha(\omega) &= \left\{S_m^{-T}[\text{diag}(\omega_r^2) - \omega^2 I]S_m^{-1}\right\}^{-1} \\ &= S_m\left\{\text{diag}[\omega_r^2 - \omega^2]^{-1}\right\}S_m^T\end{aligned} \tag{5.63}$$

Expression (5.63) can also be written in summation notation by considering the ik^{th} element of $\alpha(\omega)$, recalling formula (2.6), and partitioning the matrix S_m into columns, denoted by $\mathbf{s}_r$. The vectors $\mathbf{s}_r$ are, of course, the eigenvectors of the matrix K normalized with respect to the mass matrix M. This yields

$$\alpha(\omega) = \sum_{r=1}^{n}(\omega_r^2 - \omega^2)^{-1}\mathbf{s}_r\mathbf{s}_r^T \tag{5.64}$$

The ij^{th} element of the receptance matrix becomes

$$\alpha_{ij}(\omega) = \sum_{r=1}^{n}(\omega_r^2 - \omega^2)^{-1}[\mathbf{s}_r\mathbf{s}_r^T]_{ij} \tag{5.65}$$

where the matrix element $[\mathbf{s}_r\mathbf{s}_r^T]_{ij}$ is identified as the *modal constant* or *residue* for the rth mode and the matrix $\mathbf{s}_r\mathbf{s}_r^T$ is called the *residue matrix*. Note that the right-hand side of Equation (5.65) can also be rationalized to form a single fraction consisting of the ratio of two polynomials in ω^2. Hence, $[\mathbf{s}_r\mathbf{s}_r^T]_{ij}$ can also be viewed as the matrix of constants in the partial fraction expansion of Equation (5.60).

Next, consider the same procedure applied to Equation (5.36) with nonzero damping. As always, consideration of damped systems results in two cases: those systems that decouple and those that do not.

First consider Equation (5.36) with damping such that $CM^{-1}K = KM^{-1}C$, so that the system decouples and the system eigenvectors are real. In this case, the eigenvectors of the undamped system are also eigenvectors for the damped system, as was established in Section 3.5. The definition of the receptance matrix takes on a slightly different form to reflect the damping in the system. Under the additional assumption that the system is underdamped, i.e. that the modal damping ratios ζ_r are all between 0 and 1, Equation (5.58) becomes

$$\mathbf{u} = (K + j\omega C - \omega^2 M)^{-1}\mathbf{f}_0 \tag{5.66}$$

Because the system decouples, the matrix D can be written as

$$C = S_m^{-T}\text{diag}(2\zeta_r\omega_r)S_m^{-1} \tag{5.67}$$

Substitution of Equations (5.61), (5.62) and (5.67) into Equation (5.66) yields

$$\mathbf{u} = S_m[\text{diag}(\omega_r^2 + 2j\zeta_r\omega_r\omega - \omega^2)^{-1}]S_m^T\mathbf{f}_0 \tag{5.68}$$

This expression defines the complex receptance matrix given by

$$\alpha(\omega) = \sum_{r=1}^{n}(\omega_r^2 + 2j\zeta_r\omega_r\omega - \omega^2)^{-1}\mathbf{s}_r\mathbf{s}_r^T \tag{5.69}$$

Next consider the general viscously damped case. In this case the eigenvectors $\mathbf{s}_r$ are complex and the receptance matrix is given (Lancaster 1966) as

$$\alpha(\omega) = \sum_{r=1}^{n} \left\{ \frac{\mathbf{s}_r \mathbf{s}_r^T}{j\omega - \lambda_r} + \frac{\mathbf{s}_r^* \mathbf{s}_r^{*T}}{j\omega - \lambda_r^*} \right\} \tag{5.70}$$

Here the asterisk denotes the conjugate; the λ_r are the complex system eigenvalues and the $\mathbf{s}_r$ are the system eigenvectors.

The expressions for the receptance matrix and the interpretation of an element of the receptance matrix given by Equation (5.60) form the background for modal testing. In addition, the receptance matrix forms a response model for the system. Considering the most general case, Equation (5.70), the phenomenon of resonance is evident. In fact, if the real part of λ_r is small, $j\omega - \lambda_r$ is potentially small, and the response will be dominated by the associated mode $\mathbf{s}_r$. The receptance matrix is a generalization of the frequency response function of Section 1.5. In addition, like the transition matrix of Section 5.2, the receptance matrix maps the input of the system into the output of the system.

5.8 Stability of Feedback Control

This section examines the stability behavior of systems subject to a forced response and being controlled by an active feedback system. In most control literature, feedback is added to either enhance performance or stabilize the system. However, most structures are inherently stable (either BIBO or Lagrange stable) unless undamped and excited at resonance. However, if feedback control is added in order to improve the performance of the structure, unstable response is possible. The following example illustrates the potential for a feedback control system intended to add stiffness to a structure that may cause an unstable response.

Example 5.8.1
Discuss the stability properties of the feedback control system constructed in Example 2.5.4 repeated here with numerical values except for the control parameters

$$\begin{bmatrix} 1 & 0 \\ 0 & 1 \end{bmatrix} \ddot{\mathbf{q}} + \begin{bmatrix} 0.02 & -0.01 \\ 0.01 & 0.01 \end{bmatrix} \dot{\mathbf{q}} + \begin{bmatrix} 2 & -1 \\ -1 + k_{p1} & -1 + k_{p2} \end{bmatrix} \mathbf{q} = \begin{bmatrix} 0 \\ 0 \end{bmatrix}$$

where k_{p1} and k_{p2} are the effective closed loop gains.

Solution: Here the open loop system is asymptotically stable because all the coefficient matrices are symmetric and positive definite so that the energy is a Lyapunov function. Note that the closed loop stiffness matrix becomes asymmetric, so that the energy is no longer a Lyapunov function and stability is in question. Recall that finding a Lyapunov function only insures stability and that not finding one yields no information. Note that if k_{p1} is zero, the stiffness matrix is still symmetric and positive definite ($k_{p2} > 0$), the energy is still a Lypunov function and the system remains asymptotically stable. Since k_{p2} alone does not obviously affect the stability, assume it to be zero. Then it remains to find values of k_{p1} for which the stiffness matrix $K(k_{p1})$ remains positive definite since the energy, i.e. the Lypunov function, is

$$V(\mathbf{x}) = \frac{1}{2}\left(\dot{\mathbf{x}}^T I \mathbf{x} + \mathbf{x}^T \begin{bmatrix} 2 & -1 \\ -1 + k_{p1} & 1 \end{bmatrix} \mathbf{x}\right)$$

Recall that a matrix will be positive definite if all its eigenvalues are positive real numbers. Computing the eigenvalues of K with k_{p2} set to zero yields

$$\det \begin{bmatrix} 2 - \lambda & -1 \\ -1 + k_{p1} & 1 - \lambda \end{bmatrix} = 0 \Rightarrow \lambda^2 - 3\lambda + 1 + k_{p1} = 0$$

$$\Rightarrow \lambda_{1,2} = \frac{3}{2} \pm \frac{1}{2}\sqrt{5 - 4k_{p1}}$$

These eigenvalues will be positive real for all values of k_{p1} satisfying

$$-1 < k_{p1} < \frac{5}{4}$$

If k_{p1} falls within this range, the closed loop system will be asymptotically stable. If k_{p1} falls outside this region, the system could be stable or not. Recall from the discussion in Section 4.9 that a system is asymptotically stable if the velocity and displacement coefficient matrices have positive real eigenvalues.

There are a few structural situations where the applied force can cause the system to be unstable. The follower force of Example 2.5.3 and Equation (2.32) illustrates this situation. In this case, feedback control can be used to stabilize the forced response. The following example illustrates this.

Example 5.8.2

Consider the following forced system and a) determine if its forced response is stable or not, and b) if not stable, add a feedback control law to make it stable

$$\begin{bmatrix} 9 & 0 \\ 0 & 1 \end{bmatrix} \ddot{\mathbf{x}}(t) + \begin{bmatrix} 27 & -3 \\ -3 & 3 \end{bmatrix} \mathbf{x}(t) = \begin{bmatrix} 0 \\ 3\cos 2t \end{bmatrix}$$

Solution: With no damping in the system and with the mass and stiffness matrices being symmetric and positive definite, the system could at best be Lagrange stable. If the driving frequency, 2 rad/s, is equal to one of the natural frequencies, then the system is unstable. Calculating the systems eigenvalues yields

$$M^{-1/2} = \begin{bmatrix} 1/3 & 0 \\ 0 & 1 \end{bmatrix} \Rightarrow \tilde{K} = M^{-1/2} K M^{-1/2} = \begin{bmatrix} 3 & -1 \\ -1 & 3 \end{bmatrix} \Rightarrow \lambda_{1,2} = 2, 4$$

Thus, the natural frequencies are $\sqrt{2}$ and 2 rad/s. Hence the forced system is unstable because of resonance. To render the system Lagrange stable, a control system that modifies the natural frequencies, hence the stiffness, would remove the unbounded response by removing resonance. A controller that added damping could further improve the situation by making it BIBO stable. Consider then the controller with velocity and position feedback defined by

$$\begin{bmatrix} 9 & 0 \\ 0 & 1 \end{bmatrix} \ddot{\mathbf{x}}(t) + \begin{bmatrix} 27 & -3 \\ -3 & 3 \end{bmatrix} \mathbf{x}(t) = \begin{bmatrix} 0 \\ 3\cos 2t \end{bmatrix} + B_1\mathbf{x}(t) + B_2\dot{\mathbf{x}}(t)$$

Here a reasonable choice of B_1 and B_2 might be

$$B_1 = \begin{bmatrix} -g_1 & 0 \\ 0 & 0 \end{bmatrix} \text{ and } B_2 = \begin{bmatrix} -g_2 & 0 \\ 0 & -g_3 \end{bmatrix}$$

where the gains g_1, g_2 and g_3 are chosen to remove the resonance and to add damping while keeping the closed loop matrices symmetric.

First consider just removing the resonance condition by setting $B_2 = 0$ and solving for g_1 that renders the system Lagrange stable. The equivalent open loop system becomes

$$\begin{bmatrix} 9 & 0 \\ 0 & 1 \end{bmatrix} \ddot{\mathbf{x}}(t) + \begin{bmatrix} 27 + g_1 & -3 \\ -3 & 3 \end{bmatrix} \mathbf{x}(t) = \begin{bmatrix} 0 \\ 3\cos 2t \end{bmatrix}$$

To determine the definiteness of the closed loop stiffness matrix in terms of the feedback gain g_1 given by

$$K(g_1) = \begin{bmatrix} 27 + g_1 & -3 \\ -3 & 3 \end{bmatrix}$$

use the principle minor condition of Section 3.3. Since the $\det[K(g_1)] > 0$ and $27 + g_1 > 0$ for all $g_1 > 0$, the system is Lagrange stable as long as either of the eigenvalues of $M^{-1/2}K(g_1)M^{-1/2}$ are not equal to 4, the driving frequency. Intuition says that any positive nonzero choice of g_1 will cause the eigenvalues to move away from 2 and 4. For instance, consider the gain as a perturbation in Equation (3.87) with the notation

$$A = \begin{bmatrix} 3 & -3 \\ -3 & 3 \end{bmatrix}, \ B = \begin{bmatrix} 1 & 0 \\ 0 & 0 \end{bmatrix} \text{ and } \varepsilon = g_1$$

Then Equation (3.87) suggests that both eigenvaules shift away from 2 and 4 and hence the natural frequencies shift away from $\sqrt{2}$ and 2 rad/s so that no resonance occurs. However, an exact calculation should be made to make sure that a chosen value of g_1 does not result in resonance.

Next consider the feedback control defined by B_2 which effectively adds damping. This means the open loop system is asymptotically stable and hence the forced response will be BIBO.

5.9 Numerical Simulations in MATLAB

This section extends Section 3.8 to include simulation of systems subject to an applied force. The method is the same as described in Section 3.8, with the exception that the forcing function is now included in the equations of motion. All the codes mentioned in Section 3.8 have the ability to numerically integrate the equations of motion, including both the effects of the initial conditions and the effects of any applied forces. Numerical simulation provides an alternative to computing the time response by modal methods as done in Equation (5.45). The approach is to perform numerical integration following the material in Sections 1.11 and 3.8 with the state spaced model. For any class of second-order systems, the equations of motion can be written in the state space form as by Equation (5.1) subject to appropriate initial conditions on the position and velocity. While numerical solutions are a discrete time approximation, they are systematic and relatively easy to compute with modern high-level codes. The following example illustrates the procedure in MATLAB.

Example 5.9.1

Consider the system in physical coordinates defined by

$$\begin{bmatrix} 5 & 0 \\ 0 & 1 \end{bmatrix}\begin{bmatrix} \ddot{q}_1 \\ \ddot{q}_2 \end{bmatrix} + \begin{bmatrix} 3 & -0.5 \\ -0.5 & 0.5 \end{bmatrix}\begin{bmatrix} \dot{q}_1 \\ \dot{q}_2 \end{bmatrix} + \begin{bmatrix} 3 & -1 \\ -1 & 1 \end{bmatrix}\begin{bmatrix} q_1 \\ q_2 \end{bmatrix} = \begin{bmatrix} 1 \\ 1 \end{bmatrix}\sin(4t),$$

$$\mathbf{q}(0) = \begin{bmatrix} 0 \\ 0.1 \end{bmatrix}, \ \dot{\mathbf{q}}(0) = \begin{bmatrix} 1 \\ 0 \end{bmatrix}.$$

In order to use the Runge-Kutta numerical integration, first put the system into the state space form. Computing the inverse of the mass matrix and defining the state vector $\mathbf{x}$ by

$$\mathbf{x} = \begin{bmatrix} \mathbf{q} \\ \dot{\mathbf{q}} \end{bmatrix} = \begin{bmatrix} q_1 & q_2 & \dot{q}_1 & \dot{q}_2 \end{bmatrix}^T,$$

the state equations become

$$\dot{\mathbf{x}} = \begin{bmatrix} 0 & 0 & 1 & 0 \\ 0 & 0 & 0 & 1 \\ -3/5 & 1/5 & -3/5 & 1/2 \\ 1 & -1 & 1/2 & -1/2 \end{bmatrix}\mathbf{x} + \begin{bmatrix} 0 \\ 0 \\ 1/5 \\ 1 \end{bmatrix}\sin 4t, \ \mathbf{x}(0) = \begin{bmatrix} 0 \\ 0.1 \\ 1 \\ 0 \end{bmatrix}$$

The steps to solve this numerically in MATLAB follow those of Example 3.8.1 with the additional term for the forcing function. The corresponding M-file is

```
---------------
function v=f591(t,x)
M=[5 0; 0 5];C=[3 -0.5;-0.5 0.5]; K=[3 -1;-1 1];
A=[zeros(2) eye(2);-inv(M)*K -inv(M)*C];b=[0;0;0.2;1];
v=A*x+b*sin(4*t);
---------------
```

This function must be saved under the name `f591.m`. Once this is saved, the following is typed into the command window

```
EDU>clear all
EDU>xo=[0;0.1;1;0];
EDU>ts=[0 50];
EDU>[t,x]=ode45('f591',ts,xo);
EDU>plot(t,x(:,1),t,x(:,2),'--'),title('x1,x2 versus time')
```

This returns the plot shown in Figure 5.1. Note the command x(:,1) pulls off the record for $x_1(t)$ and the command ode45 calls a fifth-order Runge-Kutta program. The command "`ts=[0 50];`" tells the code to integrate from 0 to 50 time units (seconds in this case).

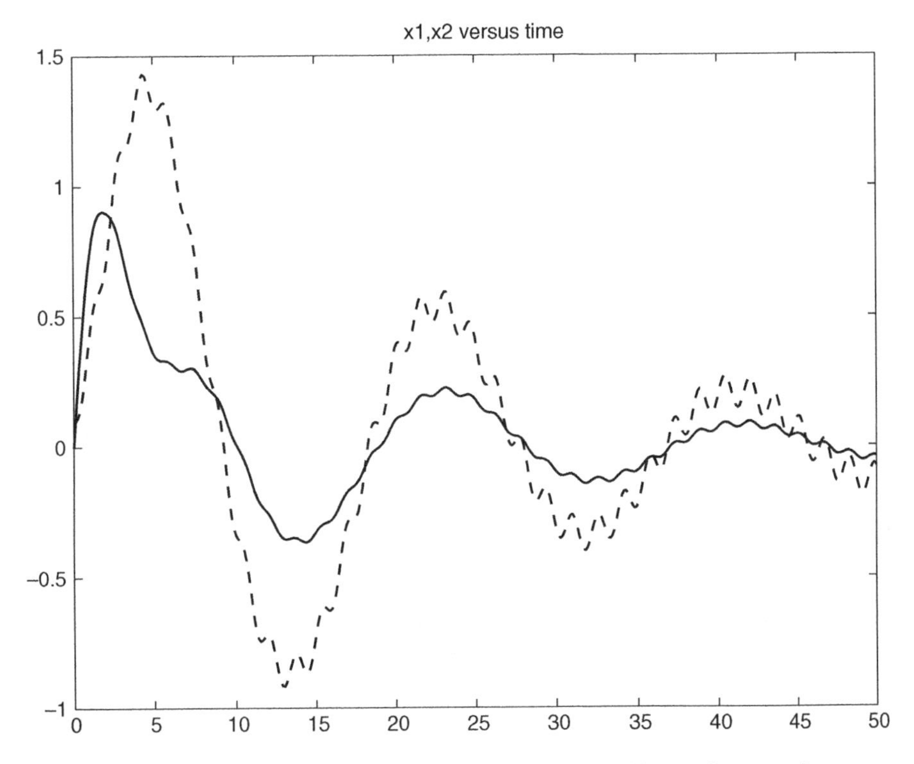

Figure 5.1 The displacement response to the initial conditions and forcing function of Example 5.9.1.

Chapter Notes

The field of systems theory and control has advanced the idea of using matrix methods for solving large systems of differential equations (Patel et al., 1994). Thus, the material in Section 5.2 can be found in most introductory systems theory texts such as Chen (1998) or Kailath (1980). In addition, those texts contain material on modal decoupling of the state matrix, as covered in Section 5.3. Theoretical modal analysis (Section 5.3) is just a method of decoupling the equations of motion of a system into a set of simple-to-solve SDOF equations. This method is extended in Section 5.4 and generalized to equations that cannot be decoupled. For such systems, modal analysis of the solution is simply an expansion of the solution in terms of its eigenvectors. This material parallels the development of the free response in Section 3.5. The material of Section 5.6 on bounds is not widely used. However, it does provide some methodology for design work. The results presented in Section 5.6 are from Yae and Inman (1987). The material on frequency response methods presented in Section 5.7 is essential in understanding experimental modal analysis and testing and is detailed in Ewins (2000). Section 5.9 is a brief introduction to the important concept of numerical simulation of dynamic systems.

References

Ahmadian, M. and Inman, D. J. 1984(a) Classical normal modes in asymmetric nonconservative dynamic systems. *AIAA Journal,* 22(7), 1012–1015.

Ahmadian, N. and Inman, D. J. 1984(b) Modal analysis in non-conservative dynamic systems. *Proceedings of the 2nd International Conference on Modal Analysis,* 1: 340–344.

Boyce, E. D. and DiPrima, R. C. 2012. *Elementary Differential Equation and Boundary Value Problem,* 10th Edition. John Wiley & Sons, Hoboken, NJ.

Chen, C. T. (1998) *Linear System Theory and Design,* 3rd Edition. Oxford University Press, Oxford, UK.

Ewins. D. J. (2000) *Modal Testing, Theory, Practice and Application,* 2nd Edition. Research Studies Press, Hertfordshire, UK.

Hu, B and Eberhard, P. (1999) Response bounds for linear damped systems. ASME *Journal of Applied Mechanic,* 66, 997–1003.

Inman, D. J. and Andry, Jr., A. N. (1980) Some results on the nature of eigenvalues of discrete damped linear systems, *ASME Journal of Applied Mechanics,* 47(4), 927–930.

Inman, D. J. (1982) Modal analysis for asymmetric systems. *Proceedings of the 1st International Conference on Modal Analysis,* pp. 705–708.

Kailath, T. 1980. *Linear Systems.* Prentice Hall, Englewood Cliffs, NJ.

Lancaster, P. (1966) *Lambda Matrices and Vibrating Systems.* Pergamon Press, Elmsford, NY.

Müller, D. C. and Schiehlen, W. D. (1977) *Forced Linear Vibrations.* Springer-Verlag, New York.

Moler, C. B. and Van Loan, C. F. (1978) Nineteen dubious ways to compute the exponential of a matrix. *SIAM Review,* 20, 801–836.

Nicholson, D. W. and Inman, D. J. (1983) Stable response of damped linear systems. *Shock and Vibration Digest,* 15(11), 19–25.

Nicholson, D. W. and Lin, B. (1996) Stable response of non-classically damped systems, Part II. *Applied Mechanics Reviews,* 49(10), S41–S48.

Patel, R. V. and Toda, M. (1980) Quantitative measures of robustness for multivariable systems. *Joint Automatic Control Conference*, 17, 35.

Patel, R. V., Laub, A. J. and Van Dooren, P. M. (eds) (1994) *Numerical Linear Algebra Techniques for Systems and Control*. Institute of Electrical and Electronic Engineers, Inc., New York.

Thomson, W. T. (1960) *Laplace Transforms*, 2nd Edition. Prentice Hall, Englewood Cliffs, NJ.

Yae, K. H. and Inman, D. J. (1987) Response bounds for linear underdamped systems. *ASME Journal of Applied Mechanics*, 54(2), 419–423.

Problems

5.1 Use the resolvent matrix to calculate the solution of

$$\ddot{x} + \dot{x} + 2x = \sin 3t$$

with zero initial conditions.

5.2 Calculate the transition matrix e^{At} for the system of Problem 5.1.

5.3 Prove that $e^{-A} = (e^{A})^{-1}$ and show that $e^{A}e^{-A} = I$.

5.4 Compute e^{At} where $A \begin{bmatrix} 1 & 1 \\ 0 & 0 \end{bmatrix}$

5.5 Show that if $\mathbf{z}(t) = ae^{-j\phi}e^{(\mu+j\omega)t}\mathbf{f}(t)$, where a and $\mathbf{f}(t)$ are real and $j^2 = -1$, then $\text{Re}(\mathbf{z}) = a\mathbf{f}(t)e^{\mu t}\cos(\omega t - \phi)$.

5.6 Consider a spring-mass system of the form

$$\begin{pmatrix} 4 & 0 & 0 \\ 0 & 2 & 0 \\ 0 & 0 & 1 \end{pmatrix}\ddot{\mathbf{x}} + \begin{pmatrix} 4 & -1 & 0 \\ -1 & 2 & -1 \\ 0 & -1 & 1 \end{pmatrix}\mathbf{x} = \begin{bmatrix} \mu(t) \\ 0 \\ 0 \end{bmatrix}$$

Where $\mu(t)$ is the unit step function, calculate the response of that system with the given applied force and zero initial conditions.

5.7 Solve for $\mathbf{x}(t)$ for the following spring-mass-damper system

$$\begin{bmatrix} 1 & 0 \\ 0 & 1 \end{bmatrix}\ddot{\mathbf{x}} + \begin{bmatrix} 3 & -1 \\ -1 & 3 \end{bmatrix}\dot{\mathbf{x}} + \begin{bmatrix} 4 & -2 \\ -2 & 4 \end{bmatrix}\mathbf{x} = \begin{bmatrix} \sin t \\ 0 \end{bmatrix},$$

$$\mathbf{x}(0) = \begin{bmatrix} 1 \\ 0 \end{bmatrix} \text{ and } \dot{\mathbf{x}} = \begin{bmatrix} 0 \\ 0 \end{bmatrix}$$

5.8 Calculate a bound on the forced response of system given in Problem 5.7. Which was easier to calculate, the bound or the actual response?

5.9 Calculate the receptance matrix for the system of Example 5.6.1 with $\xi = -0.1$.

5.10 Discuss the similarities between the receptance matrix, the transition matrix and the resolvent matrix.

5.11 Using the definition of the matrix exponential, prove each of the following:
a) $(e^{At})^{-1} = e^{-At}$
b) $e^0 = I$
c) $e^A e^B = e^{(A+B)}$, if $AB = BA$.

5.12 Develop the formulation for modal analysis of symmetrizable systems by applying the transformations of Section 4.9 to the procedure following Equation (5.24).

5.13 Using standard methods of differential equations, solve Equation (5.40) to get Equation (5.41).

5.14 Plot the response of the system in Example 5.6.1 along with the bound indicated in Equation (5.55) for the case $\xi = -0.1$.

5.15 Derive the solution of Equation (5.40) for the case $\zeta_i > 1$.

5.16 Show that $\mathbf{f}_0 e^{j\omega t}$ is in fact periodic.

5.17 Compute the modal equations for the system described by

$$\begin{bmatrix} 1 & 0 \\ 0 & 4 \end{bmatrix} \ddot{\mathbf{q}}(t) + \begin{bmatrix} 5 & -1 \\ -1 & 1 \end{bmatrix} \mathbf{q}(t) = \begin{bmatrix} 0 \\ 1 \end{bmatrix} \sin 2t$$

subject to the initial conditions of zero initial velocity and an initial displacement of $\mathbf{x}(0) = [0 \quad 1]^T$ mm.

5.18 Repeat Problem 5.17 for the same system with damping matrix defined by $C = 0.1K$.

5.19 Derive the relationship between the transformations S and S_m.

5.20 Consider Example 5.4.1 and compute the total response in physical coordinates.

5.21 Consider Example 5.4.1 and plot the response in physical coordinates.

5.22 Consider the problem of Example 5.4.1 and use the method of numerical integration discussed in Section 5.9 to solve and plot the solution. Compare your results to the analytical solution found in Problem 5.21.

5.23 Consider the following undamped system

$$\begin{bmatrix} 4 & 0 \\ 0 & 9 \end{bmatrix} \begin{bmatrix} \ddot{x}_1 \\ \ddot{x}_2 \end{bmatrix} + \begin{bmatrix} 30 & -5 \\ -5 & 5 \end{bmatrix} \begin{bmatrix} x_1 \\ x_2 \end{bmatrix} = \begin{bmatrix} 0.23500 \\ 2.97922 \end{bmatrix} \sin(2.756556t)$$

a) Compute the natural frequencies and mode shapes and discuss whether or not the system experiences resonance.
b) Compute the modal equations
c) Simulate the response numerically.

5.24 For the system of Example 5.4.1, plot the frequency response function over a range of frequencies from zero to 8 rad/s.

5.25 Compute and plot the frequency response function for the system of Example 5.4.1 for damping matrix having the value $C = \alpha K$ for several different values of α ranging from 0.1 to 1. Discuss your results. What happens to the peaks?

5.26 Prove that there is no value of $g_1 > 0$ that would make the problem of Example 5.8.2 unstable, by using the characteristic equations for both the closed loop system and the open loop system.

6

Vibration Suppression

6.1 Introduction

Vibrations are a major cause of fatigue and subsequent failure and have accounted for numerous disasters. Unwanted vibrations can also ruin the performance of a structure. Hence vibrations are undesirable in the majority of structures and machines. This motivates the need for methods to suppress, mitigate or redirect unwanted vibrations. When a structure or machine is initially designed, it is often done so using only static models with dynamic response accounted for by factors of safety because of the complexity of dynamic modeling. As a result, many of the methods of suppressing vibrations are added on after the initial design, because the negative effect of vibrations is often not evident until the structure is in service. This chapter focuses on both passive means of reducing vibrations and active means. The best way to suppress vibrations is to build a good design considering the dynamic response. Here, design is used to denote an educated method of choosing and adjusting the physical parameters of a vibrating system in order to obtain a more favorable response.

This chapter discusses passive means first by introducing vibration isolation and absorber concepts. Often the process of designing a part or structure requires a trade off between various parameters such as mass and stiffness, hence the discipline of optimization is introduced as an aide in calculating the best choice of physical parameters. The emerging technology of "metastructures" is introduced based on absorber principles. This section is followed by a discussion of sensitivity which helps determine how small changes in a given parameter, such as introduced in manufacturing, affect the original design.

The remainder of the chapter formalizes the aspects of control theory introduced in previous chapters and applies the theory to vibration suppression of structures. This topic is usually called *structural control* and has become increasingly important, as the design of mechanisms and structures has become more precise, lighter weight and less tolerant of transient vibrations. Many structures, such as tall buildings, robotic manipulator arms and flexible spacecraft, have been designed using active vibration suppression as part of the total design. Active control provides an important tool for the vibration engineer.

Vibration with Control, Second Edition. Daniel John Inman.

Companion Website: www.wiley.com/go/inmanvibrationcontrol2e

6.2 Isolators and Absorbers

Isolation of a vibrating mass refers to designing the connection of a mass (machine part or structure) to ground in such a way as to reduce unwanted effects or disturbances through that connection. Vibration absorption, on the other hand, refers to adding an additional degree of freedom (spring and mass) to the structure to *absorb* the unwanted disturbance. The typical model used in vibration isolation design is the simple single degree of freedom (SDOF) system of Figure 1.1 (a), without damping, or Figure 1.4 (a) with damping. The idea here is two-fold. First, if a harmonic force is applied to the mass through movement of the ground (i.e. as the result of nearby rotating machine), the values of c and k should be chosen to minimize the resulting response of the mass. The design *isolates* the mass from the effects of ground motion. The springs on an automobile serve this purpose.

A second use of the concept of isolation is that the mass represents the mass of a machine, causing an unwanted harmonic disturbance. In this case the values of m, c and k are chosen so that the disturbance force passing through the spring and dashpot to the ground is minimized. This isolates the ground from the effects of the machine. The motor mounts in an automobile are examples of this type of isolation.

In either case, the details of the governing equations for the isolation problem consist of analyzing the steady-state forced harmonic response of equations of the form (1.21). For instance, if it is desired to isolate the mass of Figure 1.8 from the effects of a disturbance, $F_0 \sin(\omega t)$, then the magnification curves of Figure 1.9 indicate how to choose the damping ζ and the isolator's frequency ω_n so that the amplitude of the resulting vibration is as small as possible. Curves similar to the magnification curves, called transmissibility curves, are commonly used in isolation problems.

The ratio of the amplitude of the force transmitted through the connection between the ground and the mass to the amplitude of the driving force is called the *transmissibility*. For the system of Figure 1.8, the force transmitted to ground is transmitted through the spring, k, and the damper, c. From Equation (1.22), these forces at steady state are

$$F_k = kx_{ss}(t) = kX \sin(\omega t - \phi) \tag{6.1}$$

and

$$F_c = c\dot{x}_{ss}(t) = c\omega X \cos(\omega t - \phi) \tag{6.2}$$

Here F_k and F_c denote the force in the spring and the force in the damper, respectively, and X is the magnitude of the steady-state response, as given in Section 1.4. The magnitude of the transmitted force is the magnitude of the vector sum of these two forces, denoted by F_T, and is given by

$$F_T^2 = \left|kx_{ss} + c\dot{x}_{ss}\right|^2 = [(kX)^2 + (c\omega X)^2] \tag{6.3}$$

Thus the magnitude of transmitted force becomes

$$F_T = kX \left[1 + \left(\frac{c\omega}{k}\right)^2\right]^{1/2} \tag{6.4}$$

The amplitude of the applied force is just F_0, so that the transmissibility ratio, denoted by TR, becomes

$$TR = \frac{F_T}{F_0} = \frac{\sqrt{1 + (2\zeta\omega/\omega_n)^2}}{\sqrt{[1 - (\omega/\omega_n)^2]^2 + [2\zeta\omega/\omega_n]^2}} \tag{6.5}$$

Plots of expression (6.5) versus the frequency ratio ω/ω_n for various values of ζ are called *transmissibility curves*. One such curve is illustrated in Figure 6.1. This curve indicates that for values of $\omega/\omega_n > \sqrt{2}$(i.e. TR <1), vibration isolation occurs, whereas for values of $\omega/\omega_n < \sqrt{2}$ ($TR \geq 1$) an amplification of vibration occurs. Of course, the largest increase in amplitude occurs at resonance.

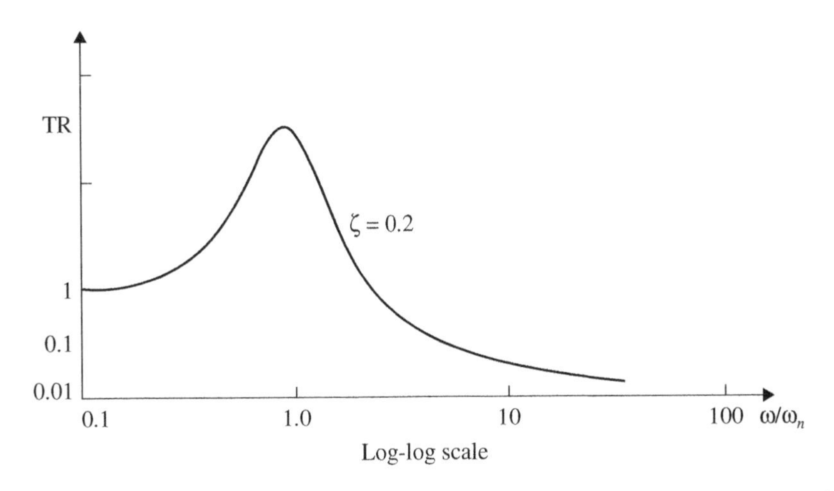

Figure 6.1 Transmissibility curve used in determining frequency values for vibration isolation.

If the physical parameters of a system are constrained such that isolation is not feasible, a vibration absorber may be included in the design. A vibration absorber consists of an attached second mass, spring and damper, forming a two degree of freedom system. The second spring-mass system is then "tuned" to resonate and hence absorb all the vibrational energy of the system.

The basic method of designing a vibration absorber is illustrated here by examining the simple case with no damping. To this end consider the two degree of freedom system of Figure 2.6 with $c_1 = c_2 = f_2 = 0$, $m_2 = m_a$, the absorber mass, $m_1 = m$, the primary mass, $k_1 = k$, the primary stiffness, and $k_2 = k_a$, the absorber spring constant. In addition, let $x_1 = x$, the displacement of the primary mass, and $x_2 = x_a$, the displacement of the absorber. Also, let the driving force $F_0 \sin(\omega t)$ be applied to the primary mass, m. The absorber is designed to the steady state response of this mass by choosing the values of m_a and k_a. Recall that the steady-state response of a harmonically excited system is found by assuming a solution that is proportional to a harmonic term of the same frequency as the driving frequency.

From Equation (2.35) the equations of motion of the two-mass absorber system are

$$\begin{bmatrix} m & 0 \\ 0 & m_a \end{bmatrix}\begin{bmatrix} \ddot{x} \\ \ddot{x}_a \end{bmatrix} + \begin{bmatrix} k+k_a & -k_a \\ -k_a & k_a \end{bmatrix}\begin{bmatrix} x \\ x_a \end{bmatrix} = \begin{bmatrix} F_0 \\ 0 \end{bmatrix}\sin\omega t \tag{6.6}$$

Assuming that in the steady state the solution of Equation (6.6) will be of the form

$$\begin{bmatrix} x(t) \\ x_a(t) \end{bmatrix} = \begin{bmatrix} X \\ X_a \end{bmatrix}\sin\omega t \tag{6.7}$$

and substituting into Equation (6.6) yields

$$\begin{bmatrix} k+k_a-m\omega^2 & -k_a \\ -k_a & k_a-m_a\omega^2 \end{bmatrix}\begin{bmatrix} X \\ X_a \end{bmatrix}\sin\omega t = \begin{bmatrix} F_0 \\ 0 \end{bmatrix}\sin\omega t \tag{6.8}$$

Solving for the magnitudes X and X_a yields

$$\begin{bmatrix} X \\ X_a \end{bmatrix} = \frac{1}{(k+k_a-m\omega^2)(k_a-m_a\omega^2)-k_a^2}\begin{bmatrix} (k_a-m_a\omega^2) & k_a \\ k_a & (k+k_a-m\omega^2) \end{bmatrix}\begin{bmatrix} F_0 \\ 0 \end{bmatrix} \tag{6.9}$$

or

$$X = \frac{(k_a-m_a\omega^2)F_0}{(k+k_a-m\omega^2)(k_a-m_a\omega^2)-k_a^2} \tag{6.10}$$

and

$$X_a = \frac{k_aF_0}{(k+k_a-m\omega^2)(k_a-m_a\omega^2)-k_a^2} \tag{6.11}$$

As can be seen by examining Equation (6.10), if k_a and m_a are chosen such that $k_a - m_a\omega^2 = 0$, i.e. such that $\sqrt{k_a/m_a} = \omega$, then the magnitude of the steady-state response of the primary m is zero, i.e. $X = 0$. Hence, if the added absorber mass, m_a, is "tuned" to the driving frequency, ω, then the amplitude of the steady-state vibration of the primary mass, X, is zero and the absorber mass effectively absorbs the energy in the system.

The addition of damping into the absorber-mass system provides two more parameters to be adjusted to improve the response of the mass m. However, with damping, the magnitude X cannot be made exactly zero. The equations of motion for the damped absorber (Inman, 2014) yield a damped version of Equation (6.10), which can be written in terms of dimensionless ratios. The amplitude X of the primary system is written in terms of the static deflection $\Delta = F_0/k$ of the primary system. In addition, consider the mixed "damping ratio" defined by

$$\zeta = \frac{c_a}{2m_a\omega_p} \tag{6.12}$$

where $\omega_p = \sqrt{k/m}$ is the original natural frequency of the primary system without the absorber attached. Using the standard frequency ratio $r = \omega/\omega_p$, the ratio of natural frequencies $\beta = \omega_a/\omega_p$, where $\omega_a = \sqrt{k_a/m_a}$ (the absorber natural frequency) and the mass ratio $\mu = m_a/m$ the normalized amplitude of the primary system becomes

$$\frac{X}{\Delta} = \frac{Xk}{F_0} = \sqrt{\frac{(2\zeta r)^2 + \left(r^2 - \beta^2\right)^2}{(2\zeta r)^2\left(r^2 - 1 + \mu r^2\right)^2 + \left[\mu r^2\beta^2 - \left(r^2 - 1\right)\left(r^2 - \beta^2\right)\right]^2}} \tag{6.13}$$

which expresses the dimensionless amplitude of the primary system. Figure 6.2 illustrates the effect of an absorber on the response of the system in the frequency domain.

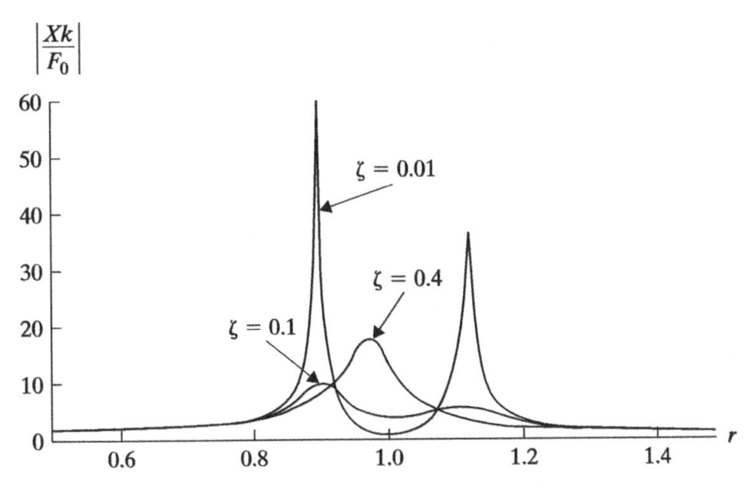

Figure 6.2 Normalized amplitude of vibration of the primary mass as a function of the frequency ratio for several values of the damping in the absorber system for the case of negligible damping in the primary system (i.e. a plot of Equation (6.13) for $\mu = 0.25$ and $\beta = 1$ as r varies).

Note that for small damping ($\zeta = 0.01$), the absorption is completely destroyed and two substantial peaks appear (recall with the absorber the system is now two degrees of freedom). As the damping increases, the peak amplitude decreases until at relatively high damping ($\zeta = 0.4$) the two peaks converge to one. The value of $\zeta = 0.1$ seems to give the lowest response magnitude and hence the best absorption. This implies there are some optimal parameters to be determined. The next section illustrates methods of choosing the design parameters to make X as small as possible.

6.3 Optimization Methods

Optimization methods (Gill et al., 1981) can be used to obtain the "best" choice of the physical parameters m_a, c_a and k_a in the design of a vibration absorber or, for that matter, any degree of freedom vibration problem (Vakakis and Paipetis, 1986). The basic optimization problem is described in the following and then applied to the damped vibration absorber problem mentioned in the preceding section.

The general form for standard nonlinear programming problems is to minimize some scalar function of the vector of design variables $\mathbf{y}$, denoted by $J(\mathbf{y})$, subject to p inequality constraints and q equality constraints, denoted by

$$g_s(\mathbf{y}) < 0 \quad s = 1, 2, \ldots, p \tag{6.14}$$

$$h_r(\mathbf{y}) = 0 \quad r = 1, 2, \ldots, q \tag{6.15}$$

respectively. The function $J(\mathbf{y})$ is referred to as the *objective function*, or *cost function*. The process is an extension of the constrained minimization problems studied in beginning calculus.

There are many methods available to solve such optimization problems. The intention of this section is not to present these various methods of design optimization but rather to introduce the use of optimization techniques as a vibration design method. The reader should consult one or more of the many texts on optimization for details of the various methods.

A common method of solving optimization problems with equality constraints is to use the method of *Lagrange multipliers.* This method defines a new vector $\theta = [\theta_1 \quad \theta_2 \quad \cdots \quad \theta_q]^T$ called the vector of Lagrange multipliers (in optimization literature they are sometimes denoted by λ_i), and the constraints are added directly to the objective function by using the scalar term $\theta^T\mathbf{h}$. The new cost function becomes $J' = J(\mathbf{y}) + \theta^T\mathbf{h}(\mathbf{y})$, which is then minimized as a function of y_i and θ_i. This is illustrated in the following example.

Example 6.3.1

Suppose it is desired to find the smallest value of the damping ratio ζ and the frequency ratio $r = \omega/\omega_n$, such that the transmissibility ratio is 0.1. The problem can be formulated as follows. Since $TR = 0.1$, then $(TR)^2 = 0.01$ or $(TR)^2 - 0.01 = 0$, which is the constraint $h_1(\mathbf{y})$ in this example. The vector $\mathbf{y}$ becomes $\mathbf{y} = [\zeta \; r]^T$, and the vector θ is reduced to the scalar θ. The cost function J' then becomes

$$\begin{aligned} J' &= \zeta^2 + r^2 + \theta(TR^2 - 0.01) \\ &= \zeta^2 + r^2 + \theta[0.99 + .02r^2 - 0.01r^4 + 3.96\zeta^2 r^2] \end{aligned}$$

The necessary (but not sufficient) conditions for a minimum are that the first derivatives of the cost function with respect to the design variables must vanish. This yields

$$\begin{aligned} J'_\zeta &= 2\zeta + \theta(7.92\zeta r^2) = 0 \\ J'_r &= 2r + \theta[0.04r - 0.04r^3 + 7.92\zeta^2 r] = 0 \\ J'_\theta &= 0.99 + 0.02r^2 - 0.01r^4 + 3.96\zeta^2 r^2 = 0 \end{aligned}$$

where the subscripts on J' denote partial differentiation with respect to the given variable. These three nonlinear algebraic equations in the three unknowns ζ, r and θ can be solved numerically to yield

$$\zeta = 0.037, r = 3.956 \text{ and } \theta = -0.016$$

The question arises as to how to pick the objective function $J(\mathbf{y})$. The choice is arbitrary, but the function should be chosen to have a single global minimum, hence the choice of the quadratic form ($\zeta^2 + r^2$) in the previous example. The following discussion on absorbers indicates how the choice of the cost function affects the result of the design optimization. The actual minimization process can follow several formulations; the results presented next follow Fox (1971).

Soom and Lee (1983) examined several possible choices of the cost function $J(\mathbf{y})$ for the absorber problem and provided a complete analysis of the absorber problem (for a two degree of freedom system) given by

$$\begin{bmatrix} m_1 & 0 \\ 0 & m_a \end{bmatrix}\begin{bmatrix} \ddot{x}_1 \\ \ddot{x}_a \end{bmatrix} + \begin{bmatrix} c_1 + c_a & -c_a \\ -c_a & c_a \end{bmatrix}\begin{bmatrix} \dot{x}_1 \\ \dot{x}_a \end{bmatrix} + \begin{bmatrix} k_1 + k_a & -k_a \\ -k_a & k_a \end{bmatrix}\begin{bmatrix} x_1 \\ x_a \end{bmatrix} = \begin{bmatrix} f \\ 0 \end{bmatrix}\cos \omega t \tag{6.16}$$

These equations are first non-dimensionalized by defining the following new variables and constants

$$\begin{aligned}
\omega_1 &= \sqrt{\frac{k_1}{m_1}} & \zeta_1 &= \frac{c_1}{2\sqrt{m_1 k_1}} \\
\omega &= \frac{\omega}{\omega_1} & \zeta_2 &= \frac{c_a}{2\sqrt{m_1 k_1}} \\
\tau &= \omega_1 & P &= \frac{f}{k_1 L}, \ L \text{ the static deflection of } x_1 \\
\mu &= \frac{m_a}{m_1} & z_1 &= \frac{x_1}{L} \\
k &= \frac{k_a}{k_1} & z_2 &= \frac{x_a}{L}
\end{aligned}$$

Substitution of these into Equation (6.16) and dividing by $k_1 L$ yields the dimensionless equations

$$\begin{bmatrix} 1 & 0 \\ 0 & \mu \end{bmatrix}\begin{bmatrix} \ddot{z}_1 \\ \ddot{z}_2 \end{bmatrix} + \begin{bmatrix} 2(\zeta_1 + \zeta_2) & -2\zeta_2 \\ -2\zeta_2 & 2\zeta_2 \end{bmatrix}\begin{bmatrix} \dot{z}_1 \\ \dot{z}_2 \end{bmatrix} + \begin{bmatrix} 1 + k & -k \\ -k & k \end{bmatrix}\begin{bmatrix} z_1 \\ z_2 \end{bmatrix} = \begin{bmatrix} P \\ 0 \end{bmatrix}\cos \omega\tau \tag{6.17}$$

where the overdots now indicate differentiation with respect to τ. As before, the steady-state responses of the two masses are assumed to be of the form

$$\begin{aligned}
z_1 &= |A_1|\cos(\omega\tau + \phi_1) \\
z_2 &= |A_2|\cos(\omega\tau + \phi_2)
\end{aligned} \tag{6.18}$$

Substitution of Equation (6.18) into Equation (6.17) and solving for the amplitudes $|A_1|$ and $|A_2|$ yields

$$|A_2| = \sqrt{(a/q)^2 + (b/q)^2}|A_1| \tag{6.19}$$

$$|A_1| = \frac{P}{\sqrt{(1 - \omega^2 - r/q)^2 + (2\zeta_1\omega + s/q)^2}} \tag{6.20}$$

where the constants, a, b, q, r and s are defined by

$$a = k^2 + 4\zeta_2^2\omega^2 - \mu k\omega^2$$
$$b = -2\zeta_2\mu\omega^3$$
$$q = (k - \mu\omega^2)^2 + 4\zeta_2^2\omega^2$$
$$r = \mu k^2\omega^2 - \mu^2 k\omega^4 + 4\zeta_2^2\omega^4$$
$$s = 2\zeta_2\mu^2\omega^5$$

Note that Equations (6.19) and (6.20) are similar in form to Equations (6.10) and (6.11) for the undamped case. However, the tuning condition is no longer obvious, and there are many possible design choices to make. This marks the difference between an absorber with damping and one without. The damped case is, of course, a much more realistic model of the absorber dynamics.

Optimization methods are used to make the best design choice among all the physical parameters. The optimization is carried out using the design variable defined by the design vector. Here, α is the tuning condition defined by

$$\alpha = \sqrt{\frac{k}{\mu}}$$

and ζ_2' is a damping ratio defined by

$$\zeta_2' = \frac{\zeta_2}{\sqrt{\mu k}}$$

The quantity ζ_2' is the damping ratio of the "added-on" absorber system of mass m_a. The tuning condition α is the ratio of the two undamped natural frequencies of the two masses.

The designer has the choice of making up objective functions. So, in this sense, the optimization produces an arbitrary best design. Choosing the objective function is the art of optimal design. However, several cost or objective functions can be used and the results of each optimization compared. Soom and Lee (1983) considered several different objective functions:

- J_1 = the maximum value of $|A_1|$, the magnitude of the displacement response in the frequency domain;
- $J_2 = \sum(|A_1| - 1)^2$ for frequencies where $|A_1| > 1$ and where the sum runs over a number of discrete points on the displacement response curves of mass m_1;
- J_3 = maximum $(\omega|A_1|)$, the maximum velocity of m_1;
- $J_4 = \sum|A_1|^2$, the mean squared displacement response;
- $J_5 = \sum(\omega|A_1|)^2$, the mean squared velocity response.

These objective function were all formed by taking 100 equally spaced points in the frequency range from $\omega = 0$ to 2. The only constraints imposed are that the stiffness and damping coefficients be positive.

Solutions to the various optimizations yield the following interesting design conclusions:

1) From minimizing J_1, the plot of J_1 versus ζ_1 for various mass ratios is given in Figure 6.3 and shows that one would not consider using a dynamic absorber

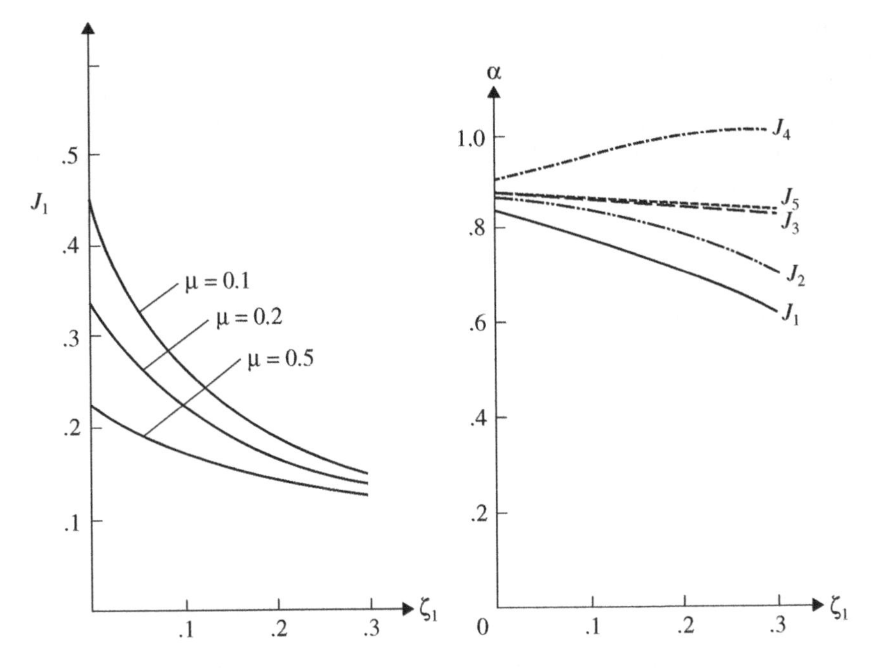

Figure 6.3 A plot of the cost function J_1, versus the damping ratio ζ_1 for various mass ratios.

Figure 6.4 The tuning parameter α versus the first mode damping ration ζ_1 for each cost function J indicating a broad range of optimal values of α.

for a system with a damping ratio much greater than $\zeta_1 = 0.2$. The plots clearly show that not much reduction in magnitude can be expected for systems with large damping in the main system.

2) For large values of damping, $\zeta_1 = 0.3$, the different objective functions lead to different amounts of damping in the absorber mass, ζ_2, and the tuning ratio, α. Thus, the choice of the objective function changes the optimum point. This is illustrated in Figure 6.4.
3) The peak, or maximum value, of $z_1(t)$ at resonance also varies somewhat, depending on the choice of the cost function. The lowest reduction in amplitude occurs with objective function J_1, as expected, which is 30% lower than the value for J_4.

It was concluded from this study that while optimal design can certainly improve the performance of a device, a certain amount of ambiguity exists in an optimal design based on the choice of the cost function. Thus, the cost function must be chosen with some understanding of the design objectives as well as physical insight.

6.4 Metastructures

A recently emerging topic in vibration suppression is the use of metastructures. Originally, metamaterials were defined simply as engineered materials created to have properties not found in naturally occurring materials. An implication encountered by those who first investigated metamaterials was that such materials would be assembled from microscopic elements arranged in a periodic fashion to produce unique wave propagation behavior leading to concepts of negative index of refraction, negative permeability, electromagnetic band gaps and other unique behaviors. Acoustic metamaterials have been created to control noise and sound. Here we broadly interpret a *metastructure* as a structure with inserts designed to produce excellent vibration suppression in certain frequency ranges to avoid resonance across a range of excitation frequencies (Sun et al., 2010). The basic idea is that of creating a structure with multiple vibration absorbers that are embedded into the structure, rather then attached to it as an added-on device.

The basic idea is explained by examining the longitudinal vibrations of a solid bar compared to the same bar with many vibration absorbers embedded into it, as suggested in Figure 6.5. A simple lumped parameter model of such a device is given in Figure 6.6.

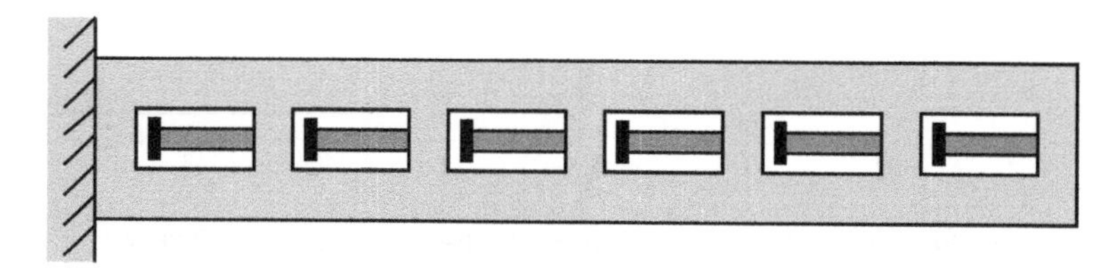

Figure 6.5 An illustration of bar with absorbers embedded into it, suggesting the concept of a metastructure.

Using the methods of Chapter 5, the response of the system of Figure 6.6 can be computed and studied for various values of the absorber parameters. An example of the effect that the inserted absorbers have on the response of the bar is illustrated in Figure 6.7 in both the frequency domain (frequency response) and in the time domain (impulse response).

The basic concept of a metastructure is to reduce mass by not having to add on parts, as in typical vibration suppression approaches. For instance, the two systems featured in Figures 6.6 and 6.7 both have the same mass. However, the concept of inserts may introduce other issues such as reduces stiffness, strength and fatigue resilience. These are all current research topics.

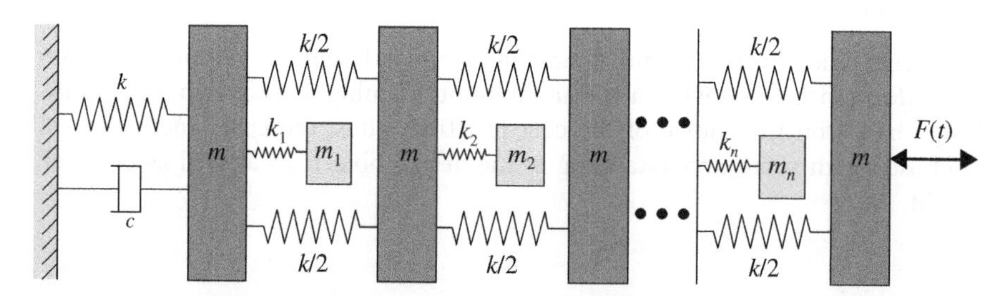

Figure 6.6 A lumped parameter model of a bar with embedded absorbers of mass m_n and stiffness k_n. The mass, damping and stiffness properties of the host structure are indicated as m, c and k respectively and distributed throughout the model.

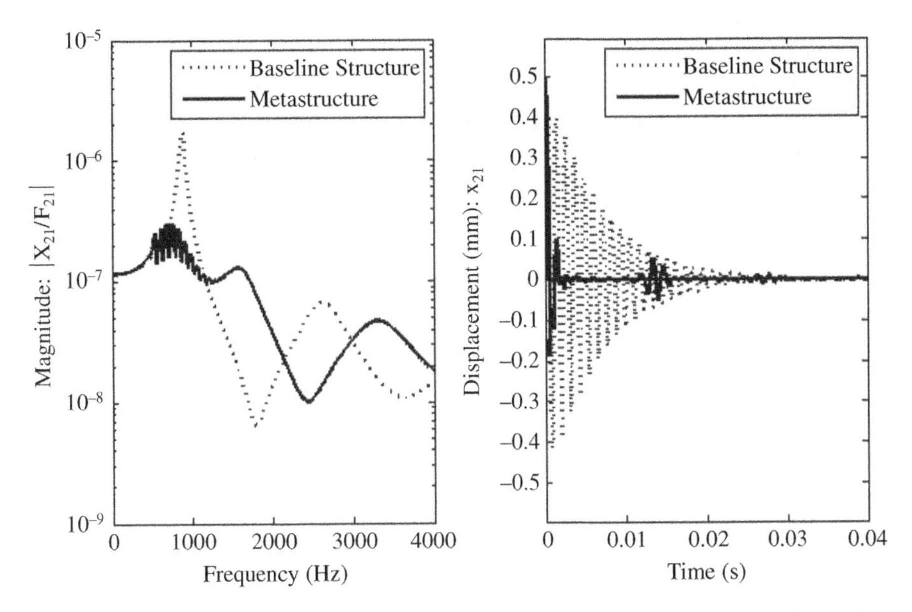

Figure 6.7 The dashed line is the response of the plain beam (no VAs) and the solid line is the response with 20 distributed and embedded absorbers (with VAs). The plot on the left is the frequency response and on the right is the impulse response (Reichl and Inman, 2016).

6.5 Design Sensitivity and Redesign

Design sensitivity analysis usually refers to the study of the effect of parameter changes on the result of an optimization procedure or an eigenvalue-eigenvector computation. For instance, in the optimization procedure presented in Section 6.3, the nonabsorber damping ratio ζ_1 was not included as a parameter in the optimization. How the resulting optimum changes as ζ_1 changes is the topic of sensitivity analysis for the absorber problem. The eigenvalue and eigenvector perturbation analysis of Section 3.6 is an example of design sensitivity for the eigenvalue problem. This can, on the other hand, be interpreted as the *redesign* problem, which poses the question of how much does the eigenvalue and eigenvector solution change as a specified physical parameter changes because of some other design process. In particular, if a design change causes a system parameter to change, the eigensolution can be computed without having to recalculate the entire eigenvalue/eigenvector set. This is also referred to as a *reanalysis* procedure and sometimes falls under the heading of *structural modification.* These methods are all fundamentally similar to the perturbation methods introduced in Section 3.6. This section develops the equations for discussing the sensitivity of natural frequencies and mode shapes for conservative systems.

The motivation for studying such methods comes from examining the large-order dynamical systems often used in current vibration technology (Antoulas, 2005). Making changes in large systems is part of the design process. However, large amounts of computer time are required to find the solution of the redesigned system. It makes sense, then, to develop efficient methods to update existing solutions when small design changes are made in order to avoid a complete reanalysis. In addition, this approach can provide insight into the design process.

Several approaches are available for performing a sensitivity analysis. The one presented here is based on parameterizing the eigenvalue problem. Consider a conservative n degree of freedom system defined by

$$M(\alpha)\ddot{\mathbf{q}}(t) + K(\alpha)\mathbf{q}(t) = \mathbf{0} \tag{6.21}$$

where the dependence of the coefficient matrices on the design parameter α is indicated. The parameter α is considered to represent a change in the matrix M and/or the matrix K. The related eigenvalue problem is

$$M^{-1}(\alpha)K(\alpha)\mathbf{u}_i(\alpha) = \lambda_i(\alpha)\mathbf{u}_i(\alpha) \tag{6.22}$$

Here, the eigenvalue $\lambda_i(\alpha)$ and the eigenvector $\mathbf{u}_i(\alpha)$ will also depend on the parameter α. The mathematical dependence is discussed in detail by Whitesell (1980). It is assumed that the dependence is such that M, K, λ_i and $\mathbf{u}_i$ are all twice differentiable with respect to the parameter α.

Proceeding, if $\mathbf{u}_i$ is normalized with respect to the mass matrix, differentiation of Equation (6.22) with respect to the parameter α yields

$$\frac{d}{d\alpha}(\lambda_i) = \mathbf{u}_i^T \left[\frac{d}{d\alpha}(K) - \lambda_i \frac{d}{d\alpha}(M)\right] \mathbf{u}_i \tag{6.23}$$

Here the dependence of α has been suppressed for notational convenience. The second derivative of λ_i can also be calculated as

$$\begin{aligned} \frac{d^2}{d\alpha^2}\lambda_i = {} & 2\mathbf{u}_i'^T \left[\frac{d}{d\alpha}(K) - \lambda_i \frac{d}{d\alpha}(M)\right] \mathbf{u}_i' \\ & + \mathbf{u}_i^T \left[\frac{d^2}{d\alpha^2}(K) - \frac{d}{d\alpha}(\lambda_i)\frac{d}{d\alpha}(M) - \lambda_i \frac{d^2}{d\alpha^2}(M)\right] \mathbf{u}_i \end{aligned} \tag{6.24}$$

The notation $\mathbf{u}'$ denotes the derivative of the eigenvector with respect to α. The expression for the second derivative of λ_i requires the existence and computation of the derivative of the corresponding eigenvector. For the special case that M is a constant and with some manipulation (Whitesell, 1980), the eigenvector derivative can be calculated from the related problem for the eigenvector $\mathbf{v}_i$ from the formula

$$\frac{d}{d\alpha}(\mathbf{v}_i) = \sum_{k=1}^{n} c_k(i,\alpha)\mathbf{v}_k \tag{6.25}$$

where the vectors $\mathbf{v}_k$ are related to the $\mathbf{u}_k$ by the mass transformation $\mathbf{v}_k = M^{1/2}\mathbf{u}_k$. The coefficients $c_k(i,\alpha)$ in this expansion are given by

$$c_k(i,\alpha) = \begin{cases} 0 & i = k \\ \dfrac{1}{\lambda_i - \lambda_k}\mathbf{u}_k^T \dfrac{dA}{d\alpha}\mathbf{u}_i & i \neq k \end{cases} \tag{6.26}$$

where the matrix A is the symmetric matrix $M^{-1/2}KM^{-1/2}$ depending on α.

Equations (6.23) and (6.25) yield the sensitivity of the eigenvalues and eigenvectors of a conservative system to changes in the stiffness matrix. More general and computationally efficient methods for computing these sensitivities are available in the literature. Adhikari and Friswell (2001) give formulas for damped systems and reference to additional methods.

Example 6.5.1

Consider the system discussed previously in Example 3.3.2. Here, take $M = I$ and $\tilde{K}$ becomes

$$\tilde{K} = \begin{bmatrix} 3 & -1 \\ -1 & 3 \end{bmatrix} = K$$

The eigenvalues of the matrix are $\lambda_{1,2} = 2,4$, and the normalized eigenvectors are

$$\mathbf{u}_1 = \mathbf{v}_1 = (1/\sqrt{2})\begin{bmatrix} 1 & 1 \end{bmatrix}^T \text{ and } \mathbf{u}_2 = \mathbf{v}_2 = (1/\sqrt{2})\begin{bmatrix} -1 & 1 \end{bmatrix}^T$$

It is desired to compute the sensitivity of the natural frequencies and mode shapes of this system as the result of a parameter change in the stiffness of the spring attached to ground. To this end, suppose the new design results in a new stiffness matrix of the form

$$K(\alpha) = \begin{bmatrix} 3+\alpha & -1 \\ -1 & 3 \end{bmatrix}$$

then

$$\frac{d}{d\alpha}(M) = \begin{bmatrix} 0 & 0 \\ 0 & 0 \end{bmatrix} \text{ and } \frac{d}{d\alpha}(K) = \begin{bmatrix} 1 & 0 \\ 0 & 0 \end{bmatrix}$$

Following Equations (6.23) and (6.25), the derivatives of the eigenvalues and eigenvectors become

$$\frac{d(\lambda_1)}{d\alpha} = 0.5, \frac{d(\lambda_2)}{d\alpha} = 0.5, \frac{d(\mathbf{u}_1)}{d\alpha} = \frac{1}{4\sqrt{2}}\begin{bmatrix} -1 \\ 1 \end{bmatrix}, \text{ and } \frac{d(\mathbf{u}_2)}{d\alpha} = \frac{-1}{4\sqrt{2}}\begin{bmatrix} 1 \\ 1 \end{bmatrix}$$

These quantities are an indication of the sensitivity of the eigensolution to changes in the matrix K. To see this, substitute the preceding expressions into the expansions for $\lambda(\alpha)$ and $\mathbf{u}_i(\alpha)$ given by Equations (3.87) and (3.88). This yields

$$\lambda_1(\alpha) = 2 + 0.5\alpha, \qquad \lambda_2(\alpha) = 4 + 0.5\alpha,$$

$$\mathbf{u}_1(\alpha) = 0.707\begin{bmatrix} 1 \\ 1 \end{bmatrix} + 0.177\alpha\begin{bmatrix} -1 \\ 1 \end{bmatrix}, \mathbf{u}_2(\alpha) = 0.707\begin{bmatrix} -1 \\ 1 \end{bmatrix} - 0.177\alpha\begin{bmatrix} 1 \\ 1 \end{bmatrix}$$

This last set of expressions allows the eigenvalues and eigenvectors to be evaluated for any given parameter change α without having to resolve the eigenvalue problem. These formulas constitute an approximate reanalysis of the system.

It is interesting to note this sensitivity in terms of a percentage. Define the percent change in λ_1 by

$$\frac{\lambda_1(\alpha) - \lambda_1}{\lambda_1} 100\% = \frac{(2 + 0.5\alpha) - 2}{2} 100\% = (25\%)\alpha$$

If the change in the system is small, say $\alpha = 0.1$, then the eigenvalue λ_1 changes by only 2.5%; and the eigenvalue λ_2 changes by 1.25%. On the other hand, the change in the elements of the eigenvector $\mathbf{u}_2$ is 2.5%. Hence, in this case the eigenvector is more sensitive to parameter changes than the eigenvalue is.

By computing higher-order derivatives of λ_i and $\mathbf{u}_i$, more terms of the expansion can be used, and greater accuracy in predicting the eigensolution of the new system results. By using the appropriate matrix computations, the subsequent evaluations of the eigenvalues and eigenvectors as the design is modified can be carried out with substantially less computational effort (reportedly of the order of n^2 multiplications). The sort of calculation provided by eigenvalue and eigenvector derivatives can provide an indication of how changes to an initial design will affect the response of the system. In the above example, the shift in value of the first spring is translated into a percentage change in the eigenvalues and hence in the natural frequencies. If the design of the system is concerned with avoiding resonance, then knowing how the frequencies shift with stiffness is critical.

6.6 Passive and Active Control

In the redesign approach discussed in the previous section, the added structural modification α can be thought of as a passive control. If α represents added stiffness chosen to improve the vibrational response of the system, then it can be thought of as a passive control procedure. As mentioned in Section 1.9, passive control is distinguished from active control by the use of added power or energy in the form of an actuator, required in active control. Control technology of linear systems is a mature discipline with many excellent texts and journals devoted to the topic. There are hundreds of ways to approach control problems and here we focus only on the very basic methods that are useful in applications involving structural vibrations. Two main classification of active control systems are: single-input, single-output methods (SISO) and multiple-input, multiple-output (MIMO) control, depending on the number of sensors and actuators used to implement control. Like design methods, many control designs depend on low-order models of the structure (also called the plant in the controls literature). On the other hand, state space control theory uses matrix theory that is compatible with the vector differential equation commonly used to describe the vibrations of structures.

The words *structure* and *plant* are used interchangeably to describe the vibrating mechanical part or system of interest. The phrase *control system* refers to an actuator (or group of actuators), which is a force-generating device used to apply control forces to the structure, the sensors used to measure the response of the structure (also called the output) and the rule or algorithm that determines how the force is applied. The structure is often called the *open-loop system*, and the structure along with the control system is called the *closed-loop system*. The topic of control has been briefly introduced in Sections 1.9, 2.4, 4.10 and 5.8. The notation of these sections is summarized here as an introduction to the philosophy of active control.

Feedback control requires measurements of the response (by using sensing transducers) and the application of a force to the system (by using force transducers) based on these measurements. The mathematical representation of the method of computing how the force is applied based on the measurements is called the *control law*. As is often the case in modeling physical systems, there are a variety of mathematical representations of feedback control systems. Equation (2.24) represents a method of modeling the use of control forces proportional to position (denoted by $K_p\mathbf{q}$) and velocity (denoted by $K_v\dot{\mathbf{q}}$) used to shape the response of the structure. In the notation of Section 2.4, the closed-loop system is modeled by

$$M\ddot{\mathbf{q}}(t) + [C + G]\dot{\mathbf{q}}(t) + [K + H]\mathbf{q}(t) = -K_p\mathbf{q}(t) - K_v\dot{\mathbf{q}}(t) + \mathbf{f}(t)$$

as given in Equation (2.24) and represents *state variable feedback* (or position and velocity feedback).

Another form of feedback, called *output feedback*, is discussed in Section 4.10 and results if Equation (4.26) is substituted into Equation (4.24) to yield the closed-loop system (with $\mathbf{f} = \mathbf{0}$)

$$M\ddot{\mathbf{q}} + A_2\dot{\mathbf{q}} + K\mathbf{q} = B_f\mathbf{u}$$

In this case, the control vector $\mathbf{u}$ is a function of the response coordinates of interest, denoted by the vector $\mathbf{y}$, i.e. $\mathbf{u}(t) = -G_f\mathbf{y}$. This form of control is called output feedback. The vector $\mathbf{y}$ can be any combination of state variables (i.e. position and velocities), as denoted by the output Equation (4.25), which is

$$\mathbf{y} = C_p\mathbf{q} + C_v\dot{\mathbf{q}}$$

The matrices C_p and C_v denote the locations of and the electronic gains associated with the transducers used to measure the various state variables.

Active control is most often formulated in the state space by

$$\dot{\mathbf{x}} = A\mathbf{x} + B\mathbf{u}$$

as given by Equation (2.25), with output equation

$$\mathbf{y} = C_c\mathbf{x}$$

and feedback control law: $\mathbf{u} = -\boldsymbol{G}_c\mathbf{y}$. The relationships between the physical coordinates (M, A_2, K and B_f) and the state space representation (A, B and C_c) are given in Equation (2.28). Most control results are described in the state space coordinate system. The symbols $\mathbf{y}$ and $\mathbf{u}$ in both the physical coordinate system and the state space coordinate system are the same, because they represent different mathematical models of the same physical control devices. The various relationships between the measurement or output $\mathbf{y}$ and the control input $\mathbf{u}$ determine the various types of control laws, some of which are discussed in the following sections.

The material on isolators and absorbers of Section 6.2 represents two possible methods of passive control. Indeed, the most common passive control device is the vibration absorber. Much of the other work in passive control consists of added layers of damping material applied to various structures to increase the damping ratios of troublesome modes. Adding mass and changing stiffness values are also methods of passive control used to adjust a frequency away from resonance. Damping treatments increase the rate of decay of vibrations, so they are often more popular for vibration suppression.

Active control methods have been introduced in Sections 1.9, 2.4 and 4.10. Here we examine active control as a method for improving the response of a vibrating system. This section introduces the method of *eigenvalue placement* (often called *pole placement*), which is useful in improving the free response of a vibrating system by shifting natural frequencies and damping ratios to desired values.

There are many different methods of approaching the eigenvalue placement problem. The approach taken here is simple. The characteristic equation of the structure is written. Then a feedback law is introduced with undetermined gain coefficients of the form given by Equations (4.24), (4.25) and (4.26). The characteristic equation of the closed-loop system is then written and compared to the characteristic equation of the open-loop system. Equating coefficients of the powers of λ in the two characteristic equations yields algebraic equations in the gain parameters, which are then solved. This yields the control law, which causes the system to have the desired eigenvalues. The procedure is illustrated in the following example.

Example 6.6.1

Consider the undamped conservative system of Example 2.5.4 with $M = I$, $C = 0$, $k_1 = 2$ and $k_2 = 1$. The characteristic equation of the system becomes

$$\lambda^2 - 4\lambda + 2 = 0$$

This has roots $\lambda_1 = 2 - \sqrt{2}$ and $\lambda_2 = 2 + \sqrt{2}$. The system's natural frequencies are then $\sqrt{2 - \sqrt{2}}$ and $\sqrt{2 + \sqrt{2}}$. Suppose now that it is desired to raise the natural frequencies of this system to be $\sqrt{2}$ and $\sqrt{3}$, respectively. Furthermore, assume that the values of k_i and m_i cannot be adjusted, i.e. that passive control is not a design option in this case.

First, consider the control and observation matrices of Section 4.10 and the solution to Example 4.10.1. The obvious choice would be to measure the position $q_1(t)$ and $q_2(t)$, so that $C_v = 0$ and $C_p = I$, and apply forces proportional to their displacements, so that

$$G_f = \begin{bmatrix} g_1 & 0 \\ 0 & g_2 \end{bmatrix}$$

with the actuators placed at x_1 and x_2, respectively. In this case the matrix B_f becomes $B_f = I$. Then the closed-loop system of Equation (4.27) has the characteristic equation

$$\lambda^2 - (4 + g_1 + g_2)\lambda + 2 + g_1 + 3g_2 + g_1 g_2 = 0 \tag{6.27}$$

If it is desired that the natural frequencies of the closed-loop system should be $\sqrt{2}$ and $\sqrt{3}$, then the eigenvalues must be changed to 2 and 3, which means the desired characteristic equation is

$$(\lambda - 3)(\lambda - 2) = \lambda^2 - 5\lambda + 6 = 0 \tag{6.28}$$

By comparing the coefficients of λ and λ^0 (constant) terms of Equations (6.27) and (6.28), it is seen that the gains g_1 and g_2 must satisfy

$$5 = (4 + g_1 + g_2)$$
$$6 = 2 + g_1 + 3g_2 + g_1 g_2$$

which has no real solutions.

From Equation (6.27) it is apparent that in order to achieve the goal of placing the eigenvalues, and hence the natural frequencies, the gains must appear in some different order in the coefficients of Equation (6.27). This condition can be met by reexamining the matrix B_f. In fact, if B_f is chosen to be

$$B_f = \begin{bmatrix} 0 & 0 \\ 1 & 1 \end{bmatrix}$$

the characteristic equation for the closed-loop system becomes

$$\lambda^2 - (4 + g_2)\lambda + 2 + 3g_2 + g_1 = 0 \tag{6.29}$$

Comparison of the coefficients of λ in Equations (6.28) and (6.29) yields values for the gains of $g_1 = 1$ and $g_2 = 1$.

The eigenvalues with these gains can be easily computed as a check to see that the scheme works. They are in fact $\lambda = 2$ and $\lambda = 3$, resulting in the desired natural frequencies.

As illustrated by the preceding example, the procedure is easy to calculate but does not always yield real values or even realistic values of the gains. Note that the ability to choose these matrices is the result of the use of feedback and illustrates the versatility gained by using active control versus using passive control. In passive control, g_1 and g_2 have to correspond to changes in mass or stiffness. In active control, g_1 and g_2 are often electronic settings and hence are easily adjustable within certain bounds (but at other costs).

The use of pole placement assumes that the designer understands, or knows, what eigenvalues are desirable. This knowledge comes from realizing the effect that damping ratios and frequencies, hence the eigenvalues, have on the system response. Often these are interpreted from, or even stated in terms of, design specifications. This is the topic of the next section.

6.7 Controllability and Observability

As pointed out in Example 6.6.1 using pole placement, it is not always possible to find a control law of a given form that causes the eigenvalues of the closed-loop system to have desired values. This inability to find a suitable control law raises the concept of *controllability*. A closed-loop system, meaning a structure and the applied control system, is said to be completely *controllable*, or state controllable, if every state variable (i.e. all positions and velocities) can be affected in such a way as to cause it to reach a particular value within a finite amount of time by some unconstrained (unbounded) control, $\mathbf{u}(t)$. If one state variable cannot be affected in this way, the system is said to be *uncontrollable*. Figures 6.8a and 6.8b illustrate two mechanical oscillators, subject to the same control force $u(t)$ acting on the mass m_2. System (a) of the figure is uncontrollable because m_1 remains unaffected for any choice of $u(t)$. On the other hand, system (b) is controllable, since any nonzero choice of $u(t)$ affects both masses. Note that if a second control force is applied to m_1 of Figure 6.8a, then that system becomes controllable too. Hence, controllability is a function of both the system dynamics and of where and how many control actuators are applied. For instance, the system of Figure 6.8b is controllable with a single actuator, while that of Figure 6.8a requires two actuators to be controllable.

The formal definition of controllability for linear time invariant systems is given in state space rather than in physical coordinates. In particular, consider the first-order system defined as before by

$$\dot{\mathbf{x}}(t) = A\mathbf{x}(t) + B\mathbf{u}(t) \tag{6.30}$$

$$\mathbf{y}(t) = C_c\mathbf{x}(t) \tag{6.31}$$

Recall that $\mathbf{x}(t)$ is the $2n \times 1$ state vector, $\mathbf{u}(t)$ is a $r \times 1$ input vector, $\mathbf{y}(t)$ is a $p \times 1$ output vector, A is the $2n \times 2n$ state matrix, B is a $2n \times r$ input coefficient matrix, and C_c is a $p \times 2n$ output coefficient matrix. The control influence matrix B is determined by the position of control devices (actuators) on the structure. The number of outputs is p, which is the same as $2s$, where s is defined in Equation (4.25) in physical coordinates as the number of sensors. In state space, the state vector includes velocity and position coordinates and hence has twice the size

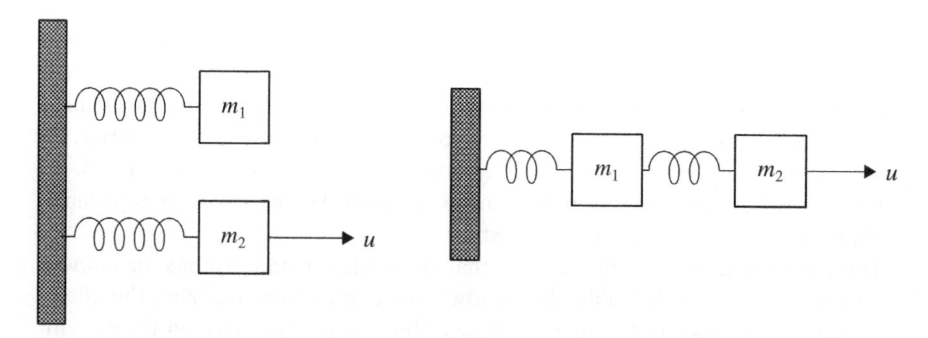

Figure 6.8 Example of (a) an uncontrollable mechanical system and (b) a controllable mechanical system.

($p = 2s$), since the velocities can only be measured at the same locations as the position coordinates. The state, $\mathbf{x}(t)$, is said to be controllable at $t = t_0$ if there exists a piecewise continuous bounded input $\mathbf{u}(t)$ that causes the state vector to move to any final value $\mathbf{x}(t_f)$ in a finite time $t_f > t_0$. If each state $\mathbf{x}(t_0)$ is controllable, the system is said to be *completely state controllable*, which is generally what is implied when a system is said to be controllable (Kailath, 1980; Kuo and Gholoaraghi, 2003).

The standard check for the controllability of a system is a rank test of a certain matrix, similar to the stability conditions used earlier for the asymptotic stability of systems with semidefinite damping. That is, the system of Equation (6.30) is completely state controllable if and only if the $2n \times 2nr$ matrix R, defined by

$$R = [B \quad AB \quad A^2B \cdots A^{2n-1}B] \tag{6.32}$$

has rank $2n$. In this case, the pair of matrices $[A, B]$ is said to be controllable. The matrix R is called the *controllability matrix* for the matrix pair $[A, B]$.

Example 6.7.1

It is easily seen that a damped SDOF system is controllable. In this case $n = 1$, $r = 1$, then

$$A = \begin{bmatrix} 0 & 1 \\ -k/m & -c/m \end{bmatrix}, \quad B = \begin{bmatrix} 0 \\ f/m \end{bmatrix}$$

so that the controllability matrix becomes

$$R = \begin{bmatrix} 0 & f/m \\ f/m & -cf/m^2 \end{bmatrix}$$

which has rank $2 = 2n$. Thus, this system is controllable (even without damping, i.e. even if $c = 0$).

Example 6.7.2

As a second simple example, consider the state matrix of Example 6.7.1 with $k/m = 1$, $c/m = 2$, and a control force applied in such a way as to cause $B = \begin{bmatrix} 1 & -1 \end{bmatrix}^T$. Then the controllability matrix R becomes

$$R = \begin{bmatrix} 1 & -1 \\ -1 & 1 \end{bmatrix}$$

which has rank $1 \neq 2n$ and the system is not controllable. Fortunately, this choice of B is not an obvious physical choice for a control law for this system. In fact, this choice of B causes the applied control force to cancel.

A similar concept to controllability is the idea that every state variable in the system has some effect on the output of the system (response) and is called *observability*. A system is observable if examination of the response (system output) determines information about *each* of the state variables. The linear time-invariant system of Equations (6.30) and (6.31) is completely observable if for each initial state $\mathbf{x}(t_0)$, there exists a finite time $t_f > t_0$ such that knowledge of $\mathbf{u}(t)$, A, C_c, and the output $\mathbf{y}(t)$ is sufficient to determine $\mathbf{x}(t_0)$ for any (unbounded) input $\mathbf{u}(t)$. The test for observability is similar to that for controllability. The system described by Equations (6.30) and (6.31) is completely observable if and only if the $2np \times 2n$ matrix O defined by

$$O = \begin{bmatrix} C_c \\ C_c A \\ \vdots \\ C_c A^{2n-1} \end{bmatrix} \tag{6.33}$$

has rank $2n$. The matrix O is called the *observability matrix*, and the pair of matrices $[A, C_c]$ are said to be observable if the rank of the matrix O is $2n$. The concept of observability is also important to vibration measurement and is discussed in Chapter 12.

Example 6.7.3

Consider again the SDOF system of Example 6.7.1. If just the position is measured, the matrix C_c reduces to the row vector $[1 \quad 0]$ and $\mathbf{y}(t)$ becomes a scalar. The observability matrix becomes

$$O = \begin{bmatrix} 1 & 0 \\ 0 & 1 \end{bmatrix}$$

which has rank 2 $(= 2n)$, and this system is observable. This condition indicates that measurement the displacement $x(t)$ (the output in this case) allows determination of both the displacement and the velocity of the system, i.e. both states are observable.

The amount of effort required to check controllability and observability can be substantially reduced by taking advantage of the physical configuration rather than using the state space formulation of Equations (6.30) and (6.31). For example, Hughes and Skelton (1980) have examined the controllability and observability of conservative systems

$$M\ddot{\mathbf{q}}(t) + K\mathbf{q}(t) = \mathbf{f}_f(t) \tag{6.34}$$

with observations defined by

$$\mathbf{y} = C_p\mathbf{q} + C_v\dot{\mathbf{q}}. \tag{6.35}$$

Here $\mathbf{f}_f(t) = B_f\mathbf{u}(t)$ and $\mathbf{u}(t) = -G_f\mathbf{y}(t)$ defines the input as specified in Section 4.10. In this case, it is convenient to assign $G_f = I$. Then the system is controllable if and only if the $n \times 2n$ matrix R_n defined by

$$R_n = [\tilde{B}_f \; \Lambda_K\tilde{B}_f \ldots \Lambda_K^{n-1}\tilde{B}_f] \tag{6.36}$$

has rank n, where $\tilde{B}_f = S_m^T B_f$ and $\Lambda_K = S_m^T K S_m$. Here S_m is the modal matrix of Equation (6.34), and Λ_K is the diagonal matrix of eigenvalues of Equation (6.34). Thus, controllability for a conservative system can be reduced to checking the rank of a smaller-order matrix rather than the $2n \times 2nr$ matrix R of Equation (6.32).

This condition for controllability can be further reduced to the simple statement that system (6.34) is controllable if and only if the rank of each matrix B_q is n_q, where B_q are the partitions of the matrix $\tilde{B}_f$ according to the multiplicities of the eigenvalues of K. Here, $d \leq n$ denotes the number of distinct eigenvalues of the stiffness matrix and $q = 1, 2, \ldots, d$. The integer n_q refers to the order of a given multiple eigenvalue. The matrix $\tilde{B}_f$ is partitioned into n_q rows. For example, if the first eigenvalue is repeated ($\lambda_1 = \lambda_2$), then $n_1 = 2$ and B_1 consists of the first two rows of the matrix $\tilde{B}_f$. If the stiffness matrix has distinct eigenvalues, the partitions B_q are just the rows of $\tilde{B}_f$. Thus, in particular, if the eigenvalues of K are distinct, then the system is controllable if and only if each row of $\tilde{B}_f$ has at least one nonzero entry.

For systems with repeated roots, this last result can be used to determine the minimum number of actuators required to control the response. Let d denote the number of distinct eigenvalues of K, and let n_q denote the multiplicities of the repeated roots so that $n_1 + n_2 + \cdots + n_q = n$, the number of degrees of freedom of the system, which corresponds to the partitions B_q of $\tilde{B}_f$. Then the minimum number of actuators for the system to be controllable must be greater than or equal to the maximum of the set $\{n_1, n_2, \ldots, n_d\}$. Note that in the case of distinct roots, this test indicates that the system could be controllable with one actuator. Similar results for general asymmetric systems can also be stated. These are discussed by Ahmadian (1985).

As in the rank conditions for stability, if the controllability or observability matrix is square, then the rank check consists of determining if the determinant is nonzero. The usual numerical question then arises concerning how to interpret the determinant having a very small value of say 10^{-6}. This situation raises the concept of "degree of controllability" and "degree of observability". One approach to measuring the degree of controllability is to define a *controllability norm*, denoted by C_q, of

$$C_q = [\det(B_q B_q^T)]^{1/(2n_q)} \tag{6.37}$$

where q again denotes the partitioning of the control matrix $\tilde{B}_f$ according to the repeated eigenvalues of K. According to Equation (6.37), the system is controllable if and only if $C_q > 0$ for all q. In particular, the larger the value of C_q is, the more controllable the modes associated with the qth natural frequency are. Unfortunately, the definition in Equation (6.37) is dependent on the choice of coordinate systems. Another more reliable measure of controllability is given later in Section 6.14, and a modal approach is given in Section 6.15.

Example 6.7.4

Consider the two degrees of freedom system of Figure 6.9. If k_1 is chosen to be unity and $k_2 = 0.01$, two orders of magnitude smaller, then the mass m_2 is weakly coupled to the mass m_1. Because of the increased number of forces acting on the system, a control system that acts on both m_1 and m_2 should be much more controllable than a control system acting just on the mass m_1. The following calculation, based on the controllability norm of Equation (6.37), verifies this

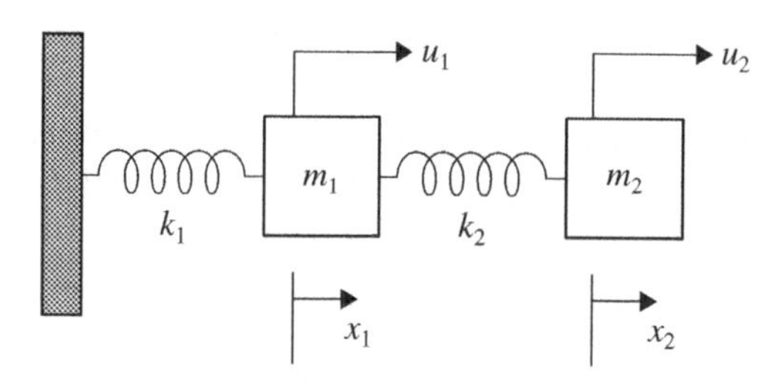

Figure 6.9 A two degrees of freedom structure with two control forces acting on it.

notion. For simplicity, the masses are set at unity, i.e. $m_1 = m_2 = 1$. The equation of motion for the system of Figure 6.9 becomes

$$\begin{bmatrix} 1 & 0 \\ 0 & 1 \end{bmatrix} \ddot{\mathbf{x}} + \begin{bmatrix} 1 & -0.01 \\ -0.01 & 1.01 \end{bmatrix} \mathbf{x} = \begin{bmatrix} 1 & 0 \\ 0 & 1 \end{bmatrix} \begin{bmatrix} u_1 \\ u_2 \end{bmatrix}$$

so that $\mathbf{u} = [u_1 \; u_2]^T$. In this case, $B_f = I$, so that $\tilde{B}_f = S_m^T$, the transpose of the normalized modal matrix of the stiffness matrix K. The matrix K has distinct eigenvalues, so that Equation (6.37) yields $C_1 = C_2 = 1$, and both modes are controllable in agreement with the physical notion that $\mathbf{u}$ affects both x_1 and x_2. Next consider a second control configuration with a single actuator acting on mass m_1 only. In this case, $u_2 = 0$ and B_f becomes the vector $[1 \;\; 0]^T$, since the vector $\mathbf{u}$ collapses to the scalar $u = u_1$. Alternatively, u could still be considered to be a vector, i.e. $\mathbf{u} = [u_1 \; 0]^T$, and B_f would then be the matrix

$$B_f = \begin{bmatrix} 1 & 0 \\ 0 & 0 \end{bmatrix}$$

Using either model, calculation of $\tilde{B}_f$ yields $C_1 = 0.8507$ and $C_2 = 0.5207$. Both of these numbers are smaller then 1, so the controllability measure has decreased from the two-actuator case. In addition, the second mode measure is smaller than the first mode measure ($C_2 < C_1$), so that the second mode is not as controllable as the first with the actuator placed at m_1. In addition, neither mode is as controllable as the two-actuator case. This numerical measure provides quantification of the controllability notion that for the weakly coupled system of this example it would be more difficult to control the response of m_2 (x_2) by applying a control force at m_1 ($C_2 = 0.5207$). The system is still controllable, but not as easily so. Again this is in agreement with the physical notion that pushing on m_1 will affect x_2, but not as easily as pushing on m_2 directly.

Complete controllability results if (but not only if) complete state feedback is used. Complete state feedback results if each of the state variables is used in the feedback law. In the physical coordinates of Equation (2.24), this use of *full state feedback* amounts to nonzero choices of C_p and C_v. In state space, state feedback is obtained by controls of the form

$$\mathbf{u} = -K_f \mathbf{x}, \tag{6.38}$$

where K_f is a feedback gain matrix of appropriate dimension. If the control **u** is a scalar, then K_f is a row vector given by

$$\mathbf{k}_f = \begin{bmatrix} g_1 & g_2 & \cdots & g_{2n} \end{bmatrix} \tag{6.39}$$

and u is a scalar with the form

$$u = g_1 x_1 + g_2 x_2 + \cdots + g_n x_n \tag{6.40}$$

The state Equation (6.30) becomes

$$\dot{\mathbf{x}} = (A - \mathbf{b}\mathbf{k}_f)\mathbf{x} \tag{6.41}$$

where B is reduced to a column vector, **b**. The product $\mathbf{b}\mathbf{k}_f$ is then a matrix (a column vector times a row vector). For complete state feedback, each gain g_i is nonzero, and the gains can be chosen such that the closed-loop system defined by the matrix $(A + \mathbf{b}\mathbf{k}_f)$ has the desired behavior.

By contrast, *output feedback* is defined for the system of Equation (6.30) by

$$\mathbf{u}(t) = -H_f \mathbf{y} \tag{6.42}$$

where H_f is the output feedback gain matrix of dimension $r \times p$. Note that since $\mathbf{y} = C_c \mathbf{x}$, where C_c indicates which states are measured, Equation (6.42) becomes

$$\mathbf{u}(t) = -H_f C_c \mathbf{x} \tag{6.43}$$

This expression appears to be similar to state feedback. The difference between state feedback and output feedback is that unless $H_f C_c = K_f$, output feedback does not use information about each state directly. On the other hand, use of the complete state variable feedback implies that a measurement of each state variable is available and is used in designing the control law. In output feedback, the output **y** is used to determine the control law, whereas in state feedback the state vector **x** is used. In general, the vector **y** is of lower order than **x**. Thus, in state feedback there are more "gains" that can be manipulated to produce the desired effect than there are in output feedback (recall Example 6.6.1). Any control performance achievable by output feedback can also be achieved by complete state feedback, but the converse is not necessarily true.

In the next section, pole placement by output feedback is considered. This task is easier with complete state feedback. Obviously, complete state feedback is the more versatile approach. However, complete state feedback requires knowledge (or measurement) of each state, which is not always possible. In addition, the hardware required to perform full state feedback is much more extensive then that required for output feedback. Section 6.10 discusses how state observers can be used to mimic state feedback when output feedback is not satisfactory for a given application or when hardware issues do not allow for measurement of all of the states.

6.8 Eigenstructure Assignment

Section 6.6 points out a simple method of designing a feedback control system that causes the resulting closed-loop system to have eigenvalues (poles) specified by the designer. In this section, the concept of placing eigenvalues is improved and extended to placing mode shapes as well as natural frequencies. From an examination of the modal expansion for the forced response, it is seen that the mode shapes as well as the eigenvalues have a substantial impact on the form of the response. Hence, by placing both the eigenvalues and eigenvectors, the response of a vibrating system may be more precisely shaped.

First, the procedure of Section 6.6 is formalized into matrix form. To this end, let M_0 and K_0 define a desired mass and stiffness matrix resulting from a redesign process. Therefore the desired system

$$\ddot{\mathbf{q}} + M_0^{-1}K_0\mathbf{q} = \mathbf{0} \tag{6.44}$$

has the *eigenstructure*, i.e. eigenvalues and eigenvectors, that are desired by design. Next, consider the closed-loop system with the existing structure (M and K) as the plant and only position feedback (so that $C_v = 0$). The system is

$$M\ddot{\mathbf{q}} + K\mathbf{q} = B_f\mathbf{u} \tag{6.45}$$

$$\mathbf{y} = C_p\mathbf{q} \tag{6.46}$$

where the various vectors and matrices have the dimensions and definitions stated for Equations (4.24) and (4.25). Recall that the constant matrix C_p represents the placement and instrument gains associated with measurement of the positions, and B_f denotes the position of the force actuators.

The class of control problems considered here uses output feedback. Output feedback uses only the output vector $\mathbf{y}$ rather then the state vector $\mathbf{x}$ in computing the gain and is defined as calculating the matrix G_f such that the control law

$$\mathbf{u}(t) = -G_f\mathbf{y}(t) \tag{6.47}$$

yields the desired response. In this case, it is desired to calculate the gain matrix G_f such that the closed-loop system has the form of Equation (6.44), which has the required eigenvalues and eigenvectors. This procedure is a form of mechanical design.

Proceeding, the closed-loop system Equation (6.45) under output feedback becomes

$$\ddot{\mathbf{q}} + M^{-1}K\mathbf{q} = -M^{-1}B_fG_fC_p\mathbf{q}$$

or

$$\ddot{\mathbf{q}} + M^{-1}(K + B_fG_fC_p)\mathbf{q} = \mathbf{0} \tag{6.48}$$

which has the same form as Equation (4.27) without damping or velocity feedback. If Equation (6.48) is to have the same eigenstructure as the design choice given by the matrices M_0 and K_0, then comparison of Equations (6.48) to (6.44) yields that G_f must satisfy

$$M^{-1}K + M^{-1}B_fG_fC_p = M_0^{-1}K_0 \tag{6.49}$$

Solving this expression for G_f following the rules of matrix algebra yields

$$B_fG_fC_p = M(M_0^{-1}K_0 - M^{-1}K) \tag{6.50}$$

In general, the matrices B_f and C_p are not square matrices (unless each mass has an actuator and sensor attached), so the inverses of these matrices, required to solve for G_f, do not exist. However, a *left generalized inverse*, defined by

$$B_f^I = (B_f^T B_f)^{-1} B_f^T \tag{6.51}$$

and a *right generalized inverse*, defined by

$$C_p^I = C_p^T (C_p C_p^T)^{-1} \tag{6.52}$$

can be used to "solve" Equation (6.50). The matrices C_p^I and B_f^I are called generalized inverses in the sense that

$$B_f^I B_f = (B_f^T B_f)^{-1} B_f^T B_f = I_{r \times r} \tag{6.53}$$

and

$$C_p C_p^I = C_p C_p^T (C_p C_p^T)^{-1} = I_{s \times s} \tag{6.54}$$

where the subscripts on the identity matrices indicate their size. Note that the calculation of the generalized inverses given by Equations (6.51) and (6.52) requires that the matrices $B_f^T B_f$ and $C_p C_p^T$ both be nonsingular. Other solutions can still be found using a variety of methods (Golub and Van Loan, 1996). Generalized inverses are briefly discussed in Appendix B.

Premultiplying Equation (6.50) by Equation (6.51) and postmultiplying Equation (6.50) by Equation (6.52), yields a solution for the $m \times s$ gain matrix G_f to be

$$G_f = (B_f^T B_f)^{-1} B_f^T M (M_0^{-1} K_0 - M^{-1} K) C_p^T (C_p C_p^T)^{-1} \tag{6.55}$$

If this value of the gain matrix G_f is implemented, the resulting closed-loop system will have the eigenstructure and response *approximately* equal to that dictated by the design set M_0 and K_0. Note, if the system is very close to the desired system, the matrix difference $(M_0^{-1}K_0 - M^{-1}K)$ will be small and G_f will be small. However, the matrix G_f depends on where the measurements are made because it is a function of C_p and also depends on the position of the actuators because it is a function of B_f.

Example 6.8.1

Consider the two degrees of freedom system with original design defined by $M = I$ and

$$K = \begin{bmatrix} 2 & -1 \\ -1 & 1 \end{bmatrix}$$

Suppose it is desired to build a control for this system so that the resulting closed-loop system has eigenvalues $\lambda_1 = 2$ and $\lambda_2 = 4$ and eigenvectors given by $\mathbf{v}_1 = [1 \quad -1]^T/\sqrt{2}$ and $\mathbf{v}_2 = [1 \quad 1]^T/\sqrt{2}$, which are normalized. A system with such eigenvalues and eigenvectors can be calculated from Equation (3.21) or

$$M_0^{-1/2} K_0 M_0^{-1/2} = S \,\text{diag}\begin{bmatrix} 2 & 4 \end{bmatrix} S^T$$

This expression can be further simplified if M_0 is taken to be the identity matrix, I. Then

$$K_0 = S\,\text{diag}\begin{bmatrix} 2 & 4 \end{bmatrix} S^T = \begin{bmatrix} 3 & 1 \\ 1 & 3 \end{bmatrix}$$

The pair (I, K_0) then represents the desired system.

Next, based on knowledge of controllability and observability, the matrices C_p and B_f are each chosen to be identity matrices, i.e. each position is measured and each mass has a force applied to it (i.e. an actuator attached to it). This condition ensures that the system is completely controllable and observable and that the controllability and observability norms are large enough. Equation (6.55) for the gain matrix G_f becomes

$$G_f = \left(\begin{bmatrix} 3 & 1 \\ 1 & 3 \end{bmatrix} - \begin{bmatrix} 2 & -1 \\ -1 & 1 \end{bmatrix} \right) = \begin{bmatrix} 1 & 2 \\ 2 & 2 \end{bmatrix}$$

which causes the original system with closed-loop control (i.e. Equation (6.45)) to have the desired eigenstructure. Note that the eigenvalues of K are 0.382 and 2.618 and those of $G_f + K$ are 2 and 4 as desired. In addition, the eigenvectors of $G_f + K$ are computed to be as desired

$$\mathbf{v}_1 = \frac{1}{\sqrt{2}} \begin{bmatrix} 1 \\ -1 \end{bmatrix} \text{ and } \mathbf{v}_2 = \frac{1}{\sqrt{2}} \begin{bmatrix} 1 \\ 1 \end{bmatrix}$$

Although not obvious from the introductory material just presented, Wonham (1967) has shown that all the eigenvalues can be placed if and only if the system is controllable. In case more than one actuator is used, i.e. in the multi-input case, the calculated feedback gain matrix G_f is not uniquely determined by assigning the eigenvalues only (Moore, 1976). Hence, the remaining freedom in the choice of G_f can also be used to place the mode shapes, as was the case in the preceding example. However, only mode shapes that satisfy certain criteria can be placed. These issues are discussed in detail by Andry et al. (1983), who also extended the process to damped and asymmetric systems.

6.9 Optimal Control

One of the most commonly used methods of modern control theory is called optimal control. Like optimal design methods, optimal control involves choosing a cost function or performance index to minimize. Although this method again raises the issue of how to choose the cost function, optimal control remains a powerful method of obtaining a desirable vibration response. Optimal control formulations also allow a more natural consideration of constraints on the state variables as well as consideration for reducing the amount of time, or final time, required for the control to bring the response to a desired level.

Consider again the control system and structural model given by Equations (4.23) and (4.24) in Section 4.10 and its state space representation given in Equations (6.30) and (6.31). The *optimal control problem* is to calculate the control $\mathbf{u}(t)$ that minimizes some specified performance index, denoted by $J = J(\mathbf{q}, \dot{\mathbf{q}}, t, \mathbf{u})$, subject to the constraint that

$$M\ddot{\mathbf{q}} + A_2\dot{\mathbf{q}} + K\mathbf{q} = B_f\mathbf{u}$$

is satisfied, and subject to the given initial conditions $\mathbf{q}(t_0)$ and $\dot{\mathbf{q}}(t_0)$. This last expression is called a differential constraint and is usually written in state space form. The cost function is usually stated in terms of an integral. The design process in optimal control consists of the judicious choice of the function J. The function J must be stated in such a way as to reflect a desired performance. The best, or optimal, $\mathbf{u}$, denoted by $\mathbf{u}^*$, has the property that

$$J(\mathbf{u}^*) < J(\mathbf{u}) \tag{6.56}$$

for any other choice of $\mathbf{u}$. Solving optimal control problems extends the concepts of maximum and minimum from calculus to functionals, J. Designing a vibration control system using optimal control involves deciding on the performance index, J. Once J is chosen, the procedure is systematic.

Before proceeding with the details of calculating an optimal control, $\mathbf{u}^*$, several examples of common choices of the cost function J are given corresponding to various design goals. The *minimum time problem* consists of defining the cost function by

$$J = t_f - t_0 = \int_{t_0}^{t_f} dt$$

which indicates that the state equations take the system from the initial state at time t_0, i.e. $\mathbf{x}(t_{0,})$, to some final state $\mathbf{x}(t_f)$ at time t_f, in a minimum amount of time.

Another common optimal control problem is called the *linear regulator problem*. This problem has specific application in vibration suppression. In particular, the design objective is to return the response (actually, all the states) from the initial state value $\mathbf{x}(t_0)$ to the systems equilibrium position (which is usually $\mathbf{x}_e = \mathbf{0}$ in the case of structural vibrations). The performance index for the linear regulator problem is defined to be

$$J = \frac{1}{2}\int_{t_0}^{t_f} (\mathbf{x}^T Q\mathbf{x} + \mathbf{u}^T R\mathbf{u})dt \tag{6.57}$$

where Q and R are symmetric positive definite *weighting matrices*. The larger Q is, the more emphasis optimal control places on returning the system to zero, since the value of $\mathbf{x}$ corresponding to the minimum of the quadratic form $\mathbf{x}^T Q\mathbf{x}$ is $\mathbf{x} = \mathbf{0}$. On the other hand, increasing R has the effect of reducing the amount, or magnitude of the control effort allowed. Note that positive quadratic forms are chosen so that the functional being minimized has a clear minimum. Using both nonzero Q and R represents a compromise in the sense that, based on a physical argument, making $\mathbf{x}(t_f)$ zero requires $\mathbf{u}(t)$ to be large. The linear regulator problem is an appropriate cost function for control systems that seeks to eliminate, or minimize, transient vibrations in a structure. The need to weight the control effort (R) results from the fact that no solution exists to the variational problem when constraining the control effort. That is when the problem of minimizing J with the inequality constraint $\|\mathbf{u}(t)\| \leq c$, where c is a constant, is not solved. Instead, R is adjusted in the cost function until the control is limited enough to satisfy $\|\mathbf{u}(t)\| \leq c$.

On the other hand, if the goal of the vibration design is to achieve a certain value of the state response, denoted by the state vector $\mathbf{x}_d(t)$, then an appropriate cost function would be

$$J = \int_{t_0}^{t_f} (\mathbf{x} - \mathbf{x}_d)^T Q(\mathbf{x} - \mathbf{x}_d)dt \tag{6.58}$$

where Q is again symmetric and positive definite. This problem is referred to as the *tracking problem*, since it forces the state vector to follow, or track, the vector $\mathbf{x}_d(t)$.

In general, the optimal control problem is difficult to solve and lends itself very few closed-form solutions. With the availability of high-speed computing, the resulting numerical solutions do not present much of a drawback. The following illustrates the problem by analyzing the linear regulator problem.

Consider the linear regulator problem for the state space description of a structure given by Equations (6.30) and (6.31). That is, consider calculating $\mathbf{u}$ such that $J(\mathbf{u})$ given by Equation (6.57) is a minimum subject to the constraint that Equation (6.30) is satisfied. A rigorous derivation of the solution is available in most optimal control texts (Kirk, 1970). Proceeding less formally, assume that the form of the desired optimal control law will be

$$\mathbf{u}^*(t) = -R^{-1}B^T S(t)\mathbf{x}^*(t) \tag{6.59}$$

where $\mathbf{x}^*(t)$ is the solution of the state equation with optimal control $\mathbf{u}^*$ as input and $S(t)$ is a symmetric time-varying $2n \times 2n$ matrix to be determined (not to be confused with the orthogonal matrix of eigenvectors S). Equation (6.59) can be viewed as a statement that the desired optimal control should be in the form of state feedback. With some manipulation (Kirk, 1970), $S(t)$ can be shown to satisfy what is called the *matrix Riccati equation* given by

$$Q - S(t)BR^{-1}B^T S(t) + A^T S(t) + S(t)A + \frac{dS(t)}{dt} = 0 \tag{6.60}$$

subject to the final condition $S(t_f) = 0$. This calculation is a backward-in-time matrix differential equation for the unknown time-varying matrix $S(t)$. The solution for $S(t)$ in turn gives the optimal linear regulator control law (Equation 6.57), causing $J(\mathbf{u})$ to be a minimum. Unfortunately, this calculation requires the solution of $2n(2n + 1)/2$ nonlinear ordinary differential equations simultaneously, backward in time (which forms a difficult numerical problem).

In most practical problems – indeed, even for very simple examples – the Riccati equation must be solved numerically for $S(t)$, which then yields the optimal control law via Equation (6.59).

The Riccati equation, and hence the optimal control problem, becomes simplified if one is interested only in controlling the steady-state vibrational response and controlling the structure over a long time interval. In this case, the final time in the cost function $J(\mathbf{u})$ is set to infinity and the Riccati matrix $S(t)$ is constant for completely controllable, time-invariant systems (Kirk, 1970). Then, $dS(t)/dt$ is zero and the Riccati equation simplifies to

$$Q - SBR^{-1}B^TS + A^TS + SA = 0 \tag{6.61}$$

which is now a nonlinear *algebraic* equation in the constant matrix S. The effect of this method on the vibration response of a simple system is illustrated in the following example.

Example 6.9.1

This example calculates the optimal control for the infinite time linear quadratic regulator problem for an SDOF oscillator of the form

$$\ddot{x}(t) + 4x(t) = f(t)$$

In this case, the cost function is of the form

$$J = \int_0^\infty (\mathbf{x}^TQ\mathbf{x} + \mathbf{u}^TR\mathbf{u})dt$$

which is Equation (6.57) with $t_f = \infty$. A control (state feedback) is sought of the form

$$\mathbf{u} = -R^{-1}B^TS\mathbf{x}(t) = -K_f\mathbf{x}(t)$$

The state equations are

$$\dot{\mathbf{x}} = \begin{bmatrix} 0 & 1 \\ -4 & 0 \end{bmatrix}\mathbf{x} + \begin{bmatrix} 0 & 0 \\ 0 & 1 \end{bmatrix}\mathbf{u}$$

Two cases are considered to illustrate the effect of the arbitrary (but positive definite) weighting matrices Q and R. The system is subject to initial condition $\mathbf{x}(0) = [1 \;\; 1]^T$. In the first case, let $Q = R = I$ and the optimal control is calculated from Equation (6.61) using MATLAB (Moler et al., 1985) to be

$$K_f = \begin{bmatrix} 0 & 0 \\ -0.1231 & -1.1163 \end{bmatrix}$$

The resulting response and control effort are plotted in Figures 6.10 to 6.12. Figure 6.12 is the control law, $\mathbf{u}^*$, calculated by using Equation (6.59); Figures 6.10 and 6.11 illustrate the resulting position and velocity response to initial conditions under the action of the control system.

In case 2, the same problem is solved again with the control weighting matrix set at $R = (10)I$. The result is that the new control law is given by

$$H_f = \begin{bmatrix} 0 & 0 \\ -0.0125 & -0.3535 \end{bmatrix}$$

which is "smaller" than the first case. The resulting position response, velocity response and control effort are plotted in Figures 6.13, 6.14 and 6.15, respectively.

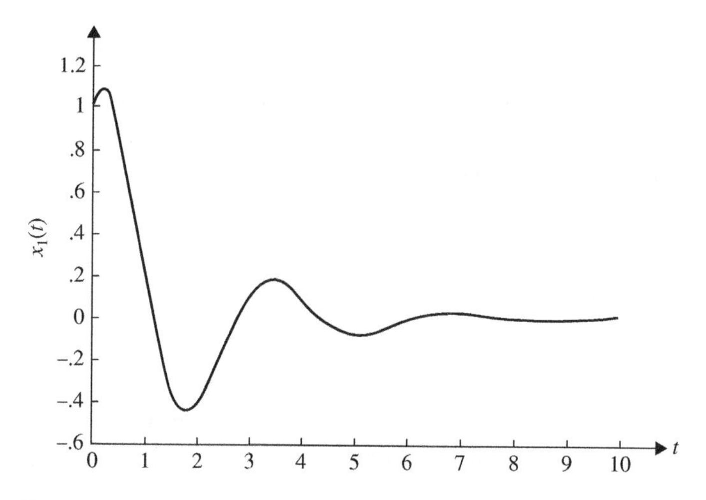

Figure 6.10 Position, $x_1(t)$, versus time for the case $Q = R = I$.

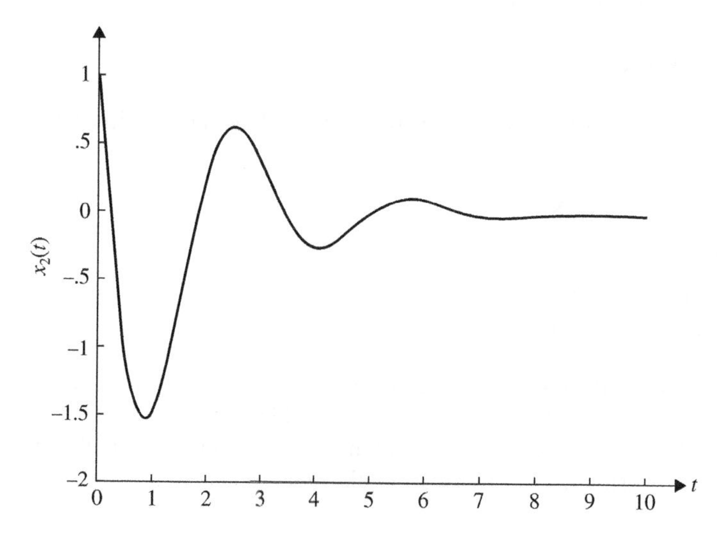

Figure 6.11 Velocity, $x_2(t)$, versus time for the case $Q = R = I$.

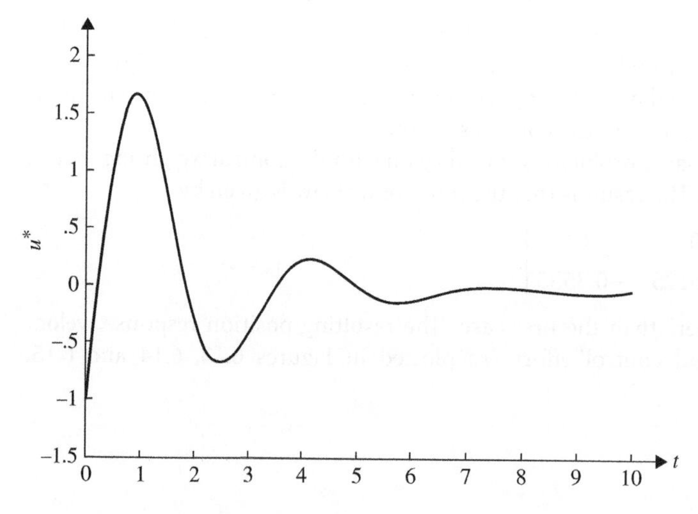

Figure 6.12 Control, $u^*(t)$, versus time for the case $Q = R = I$.

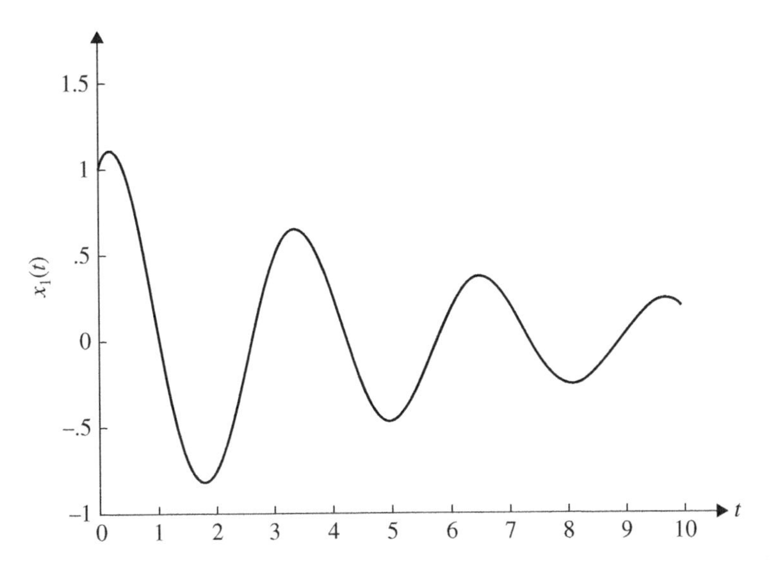

Figure 6.13 Position, $x_1(t)$, versus time for the case $Q = I,\ R = 10I$.

In the second case, with larger values for R, the control effort is initially much more limited, i.e. a maximum value of 0.7 units as opposed to 1.75 units for the first case. In addition, the resulting response is brought to zero faster in the case with more control effort (Figures 6.10, 6.11 and 6.12). These examples illustrate the effect that the weighting matrices have on the response. Note, a designer may have to restrict the control magnitude (hence, use large relative values of R),

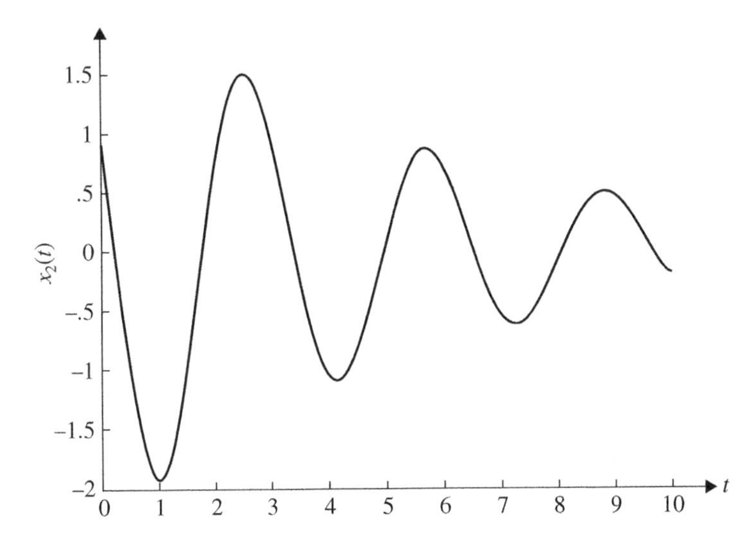

Figure 6.14 Velocity, $x_2(t)$, versus time for the case $Q = I,\ R = 10I$.

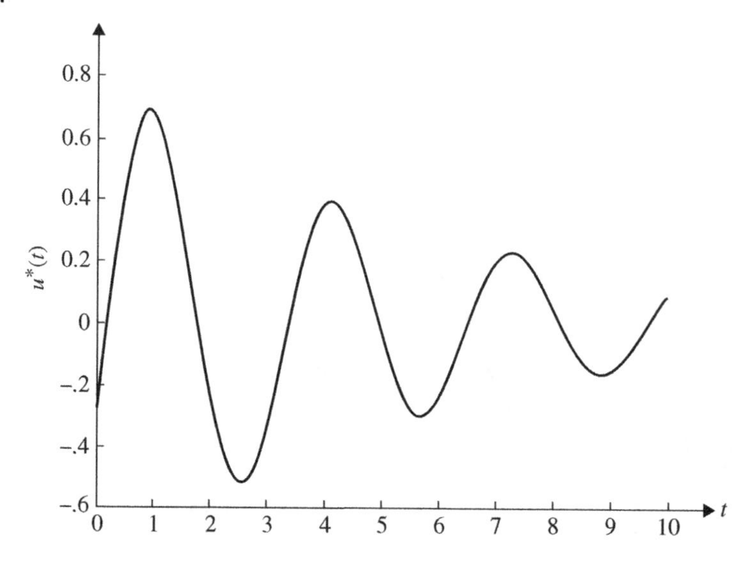

Figure 6.15 Control, $u^*(t)$, versus time for the case $Q = I,\ R = 10.I$.

because the amount of control energy available for a given application is usually limited, even though a better response (shorter time to settle) is obtained with larger values of control effort.

Optimal control has a stabilizing effect on the closed-loop system. Using a quadratic form for the cost function guarantees stability (bounded output) of the closed-loop system. This guarantee can be seen by considering the uncontrolled system subject to Equation (6.58) with $\mathbf{x}_d$ taken as the origin. Asymptotic stability requires that $\mathbf{x}$ approaches zero as time approaches infinity, so that the integral

$$J = \int_0^\infty \mathbf{x}^T Q\mathbf{x}dt$$

converges as long as Q is positive semi-definite. Define the quadratic from $V(t) = \mathbf{x}^T P\mathbf{x}$, where P is the positive definite solution of the Lyapunov Equation (4.29)

$$PA + A^T P = -Q$$

Here, A is the state matrix satisfying Equation (6.30) with $B = 0$. Computing the time derivative of $V(t)$ yields

$$\dot{V}(t) = \frac{d}{dt}(\mathbf{x}^T P\mathbf{x}) = \dot{\mathbf{x}}^T P\mathbf{x} + \mathbf{x}^T P\dot{\mathbf{x}} = \mathbf{x}^T (A^T P + \mathbf{x}PA)\mathbf{x} = -\mathbf{x}^T Q\mathbf{x} < 0$$

Hence, $V(t)$ is positive definite with a negative definite time derivative and is thus a Lyapunov function of the system of Equation (6.30). As a result, the homogenous system is asymptotically stable (recall Section 4.10). Since the homogenous system is asymptotically stable, the closed-loop system (forced response) will be bounded-input, bounded-output stable, as discussed in Section 5.5. Also, note that if $\mathbf{x}_0$ is the initial state, integrating the cost function yields

$$J = \int_0^{\infty} \mathbf{x}^T Q \mathbf{x} dt = \int_0^{\infty} \frac{d}{dt}(-\mathbf{x}^T P \mathbf{x}) dt = \mathbf{x}_0^T P \mathbf{x}_0$$

since asymptotic stability requires $\mathbf{x}$ to approach zero at the upper limit. This calculation indicates that the value of the cost function depends on the initial conditions.

This section is not intended to provide the reader with a working knowledge of optimal control methods. Instead, it is intended to illustrate the use of optimal control as an alternative vibration suppression technique and to encourage the reader to pursue the use of optimal control through one of the references.

6.10 Observers (Estimators)

In designing controllers for vibration suppression, often not all of the velocities and displacements can be conveniently measured. However, if the structure is known and is observable, i.e. if the state matrix A is known and if the measurement matrix C_c is such that O has full rank, one can design a subsystem, called an *observer* (or *estimator*, in the stochastic case) from measurements of the input and of the response of the system. The observer then provides an approximation of the missing measurements.

Consider the state space system given by Equations (6.30) and (6.32), with output feedback as defined by Equation (6.42). To simplify this discussion, consider the case where the control $u(t)$ and the output $y(t)$ are both scalars. This is the SISO case. In this situation the matrix B becomes a column vector, denoted by $\mathbf{b}$, and the matrix C_c becomes a row vector, denoted by $\mathbf{c}^T$.

The output $y(t)$ is proportional to a state variable or linear combination of state variables. Sometimes recovering the state vector from the measurement of $y(t)$ is trivial. For instance, if each state is measured (multiple-output), C_c is square and nonsingular, so that $\mathbf{x} = C_c^{-1}\mathbf{y}$. In this section, it is assumed that the state vector is not directly measured and is not easily determined from the scalar measurement $y(t)$. However, since the quantities, A, $\mathbf{b}$ and $\mathbf{c}^T$, as well as measurements of the input $u(t)$ and the output $y(t)$, are known, the desired state vector can be *estimated*. Constructing this estimated state vector, denoted by $\mathbf{x}_e$, is the topic of this section. State observers can be constructed if the system is completely observable (Chen, 1998).

The simplest observer to implement is the *open-loop estimator*. An open-loop estimator is simply the solution of the state equations with the same initial conditions as the system under consideration. Let $\mathbf{x}_e$ denote the estimated state vector. The integration of

$$\dot{\mathbf{x}}_e = A\mathbf{x}_e + \mathbf{b}u \tag{6.62}$$

yields the desired estimated state vector. The estimated state vector can then be used to perform state feedback or output feedback. Note that integration of

Equation (6.62) requires knowledge of the initial condition, $\mathbf{x}_e(0)$, which is not always available.

Unfortunately, the open-loop estimator does not work well if the original system is unstable (or almost unstable), or if the initial conditions of the unmeasured states are not known precisely. In most situations, the initial state is not known. A better observer can be obtained by taking advantage of the output of the system, $y(t)$, as well as the input. For instance, consider using the difference between the output $y(t)$ of the actual system and the output $y_e(t)$ of the estimator as a correction term in the observer of Equation (6.62). The observer then becomes

$$\dot{\mathbf{x}}_e = A\mathbf{x}_e + \mathbf{r}_e(y - y_e) + \mathbf{b}u \tag{6.63}$$

where $\mathbf{r}_e$ is the gain vector of the observer and is yet to be determined. Equation (6.63) is called an *asymptotic state estimator*, which is designed by choosing the gain vector $\mathbf{r}_e$. The error between the actual state vector $\mathbf{x}$ and the estimated state vector $\mathbf{x}_e$ must satisfy the difference between Equation (6.63) and Equation (6.30). This difference yields

$$(\dot{\mathbf{x}} - \dot{\mathbf{x}}_e) = A(\mathbf{x} - \mathbf{x}_e) + \mathbf{r}_e(y_e - y) \tag{6.64}$$

Since $y_e - y = \mathbf{c}^T(\mathbf{x}_e - \mathbf{x})$, this last expression becomes

$$\dot{\mathbf{e}} = (A - \mathbf{r}_e\mathbf{c}^T)\mathbf{e} \tag{6.65}$$

where the error vector $\mathbf{e}$ is defined to be the difference vector $\mathbf{e} = \mathbf{x} - \mathbf{x}_e$. This expression is the dynamic equation that determines the error between the actual state vector and the estimated state vector. Equation (6.65) describes how the difference between the actual initial condition $\mathbf{x}(0)$ and the assumed condition $\mathbf{x}_e(0)$ evolves in time.

The idea here is to design the observer, i.e. to choose $\mathbf{r}_e$, so that the solution of Equation (6.65) remains as close to zero as possible. For instance, if the eigenvalues of $(A - \mathbf{r}_e\mathbf{c}^T)$ are chosen to have negative real parts that are large enough in absolute value, the vector $\mathbf{e}$ goes to zero quickly, and $\mathbf{x}_e$ approaches the real state $\mathbf{x}$. Thus, any difference between the actual initial conditions and the assumed initial conditions for the observer dies out with time instead of increasing with time, as could be the case with the open-loop observer.

Obviously, there is some difference between using the actual state vector $\mathbf{x}$ and the estimated state vector $\mathbf{x}_e$ in calculating a control law. This difference usually shows up as increased control effort; that is, a feedback control based on $\mathbf{x}_e$ has to exert more energy than one based on the actual state variables $\mathbf{x}$. However, it can be shown that as far as placing eigenvalues are concerned, there is no difference in state feedback between using the actual state and the estimated state (Chen, 1970). Furthermore, the design of the observer and the control can be shown to be equivalent to performing the separate design of a control, assuming that the exact states are available, and the subsequent design of the observer (called the *separation theorem*).

To solve the control problem with state estimation, the state equation with the estimated state vector used as feedback ($u = gy = g\mathbf{c}^T\mathbf{x}_e$, recalling that $\mathbf{c}^T$ is a row vector) must be solved simultaneously with the state estimation Equation (6.65). This solution can be achieved by rewriting the state equation as

$$\dot{\mathbf{x}} = A\mathbf{x} + g\mathbf{b}\mathbf{c}^T\mathbf{x}_e \tag{6.66}$$

and substituting the value for $\mathbf{x}_e$. Then

$$\dot{\mathbf{x}} = A\mathbf{x} + g\mathbf{b}\mathbf{c}^T(\mathbf{x} - \mathbf{e}) \tag{6.67}$$

or upon rearranging

$$\dot{\mathbf{x}} = (A + g\mathbf{b}\mathbf{c}^T)\mathbf{x} - g\mathbf{b}\mathbf{c}^T\mathbf{e}. \tag{6.68}$$

Combining Equations (6.65) and (6.68) yields

$$\begin{bmatrix} \dot{\mathbf{x}} \\ \dot{\mathbf{e}} \end{bmatrix} = \begin{bmatrix} A + g\mathbf{b}\mathbf{c}^T & -g\mathbf{b}\mathbf{c}^T \\ 0 & A - \mathbf{r}_e\mathbf{c}^T \end{bmatrix} \begin{bmatrix} \mathbf{x} \\ \mathbf{e} \end{bmatrix}. \tag{6.69}$$

Here, the zero in the state matrix is a $2n \times 2n$ matrix of zeros. Expression (6.69) is subject to the actual initial conditions of the original state equation augmented by the assumed initial conditions of the estimator. These estimator initial conditions are usually set at zero, so that $[\mathbf{x}^T(0) \quad \mathbf{e}^T(0)]^T = [\mathbf{x}^T(0) \quad \mathbf{0}]^T$. Solution of Equation (6.69) yields the solution to the state feedback problem with the states estimated, rather than directly measured.

The following example illustrates the computation of a state observer as well as the difference between using a state observer and using the actual state in a feedback control problem.

Example 6.10.1

Consider an SDOF **oscillator** ($\omega_n^2 = 4$, $\zeta = 0.25$) with displacement as the measured output. The state space formulation for a single-input, single output systems is

$$\dot{\mathbf{x}} = \begin{bmatrix} 0 & 1 \\ -4 & -1 \end{bmatrix}\mathbf{x} + \mathbf{b}u, \ \mathbf{b} = \begin{bmatrix} 0 \\ 1 \end{bmatrix} \tag{6.70}$$
$$y = [1 \quad 0]\mathbf{x}$$

If output feedback is used, then $u = gy = g\mathbf{c}^T\mathbf{x}$, and the system becomes

$$\dot{\mathbf{x}} = \begin{bmatrix} 0 & 1 \\ -4 & -1 \end{bmatrix}\mathbf{x} + g\begin{bmatrix} 0 & 0 \\ 1 & 0 \end{bmatrix}\mathbf{x}. \tag{6.71}$$

Combining yields

$$\dot{\mathbf{x}} = \begin{bmatrix} 0 & 1 \\ g-4 & -1 \end{bmatrix}\mathbf{x}. \tag{6.72}$$

The asymptotic estimator in this case is given by

$$\dot{\mathbf{x}}_e = A\mathbf{x}_e + \mathbf{r}_e(y - \mathbf{c}^T\mathbf{x}_e) + \mathbf{b}u \tag{6.73}$$

where $\mathbf{r}_e$ is chosen to cause the eigenvalues of the matrix $(A - \mathbf{r}_e\mathbf{c}^T)$ to have negative real parts that are large in absolute value. As mentioned previously, in this case $\mathbf{c}^T = [1 \quad 0]$ is a row vector, so that the product $\mathbf{r}_e\mathbf{c}^T$ is a matrix. Here, the eigenvalues of $(A - \mathbf{r}_e\mathbf{c}^T)$ are chosen to be –6 and –5 (chosen only because they cause the solution of Equation (6.65) to die out quickly; other values could be chosen.) These equations are equivalent to requiring the characteristic equation of $(A - \mathbf{r}_e\, \mathbf{c}^T)$ to be $(\lambda + 6)(\lambda + 5)$, i.e.

$$\begin{aligned}\det(\lambda I - A + \mathbf{r}_e\mathbf{c}^T) &= \lambda^2 + (r_1 + 1)\lambda + (r_1 + r_2 + 4)\\ &= \lambda^2 + 11\lambda + 30\end{aligned} \tag{6.74}$$

Equating coefficients of λ in this expression yields the desired values for $\mathbf{r}_e = [r_1\ r_2]^T$ to be $r_1 = 10$ and $r_2 = 16$. Thus, the estimated state is taken as the solution of

$$\dot{\mathbf{x}}_e = \begin{bmatrix} 0 & 1 \\ -4 & 1 \end{bmatrix} \mathbf{x}_e + \begin{bmatrix} 10 \\ 16 \end{bmatrix} (y - \mathbf{c}^T\mathbf{x}_e) + \mathbf{b}u \tag{6.75}$$

The solution $\mathbf{x}_e$ can now be used as feedback in the original control problem coupled with the state estimation equation. This solution yields (from Equation (6.69) with the control taken as $g = -1$)

$$\begin{bmatrix} \dot{\mathbf{x}} \\ \dot{\mathbf{e}} \end{bmatrix} = \begin{bmatrix} 0 & 1 & 0 & 0 \\ -5 & -1 & 1 & 0 \\ 0 & 0 & -10 & 1 \\ 0 & 0 & -20 & -1 \end{bmatrix} \begin{bmatrix} \mathbf{x} \\ \mathbf{e} \end{bmatrix} \tag{6.76}$$

The plots in Figure 6.17 show a comparison between using the estimated state and the same control with the actual state used. In the figure, the control is fixed to be $g = -1$ and the actual initial conditions are taken to be $\mathbf{x}(0) = [1 \quad 1]^T$. The response for various different initial conditions for the observer $\mathbf{x}_e(0)$ are also plotted.

Note that in both cases the error in the initial conditions of the estimator disappears by about 1.2 time units, lasting just a little longer in Figure 6.17 than in Figure 6.16. This results because the assumed initial conditions for the observer are farther away from the actual initial condition in Figure 6.17 than in

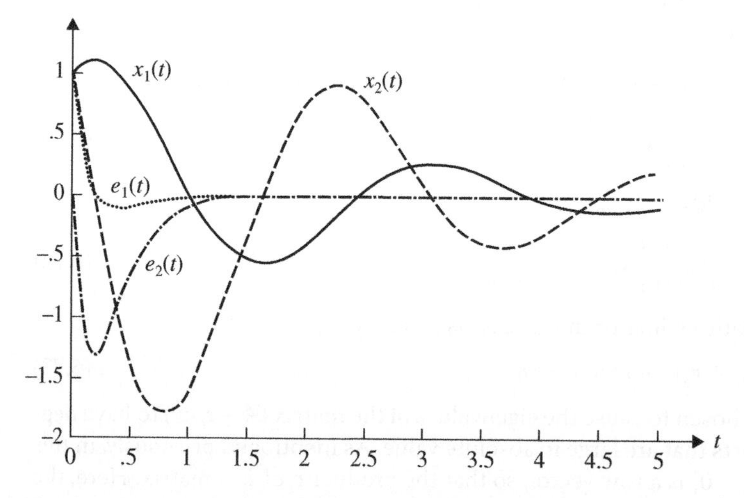

Figure 6.16 The comparison of the error vector $\mathbf{e}(t)$ versus time and the components of the state vector $\mathbf{x}(t)$ versus time for the initial condition $\mathbf{e}(0) = [1 \ \ 0]^T$.

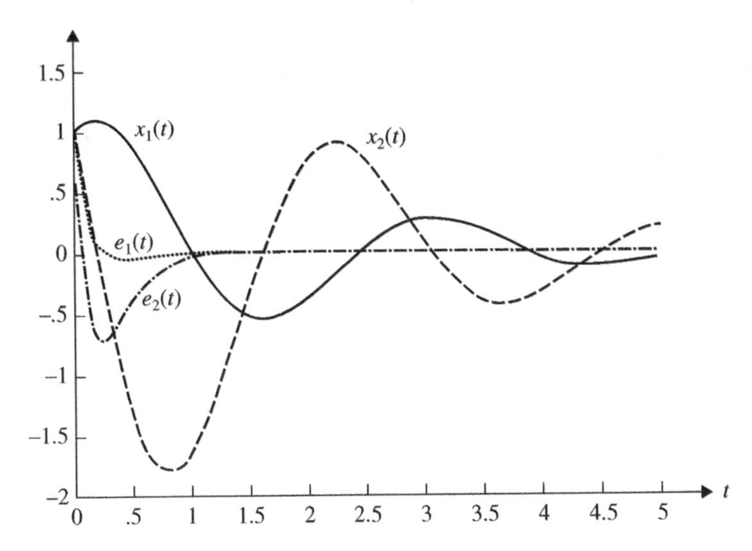

Figure 6.17 The components of the error vector $\mathbf{e}(t)$ versus time and the components of state vector $\mathbf{x}(t)$ versus time for the initial $\mathbf{e}(0) = [1 \;\; 1]^T$.

Figure 6.16. Comparison of the actual response in Figure 6.16 and Figure 6.17 with that of Figure 6.18 shows that the control law calculated using estimated state feedback is only slightly worse (takes slightly longer to decay) than those calculated using actual state feedback for this example.

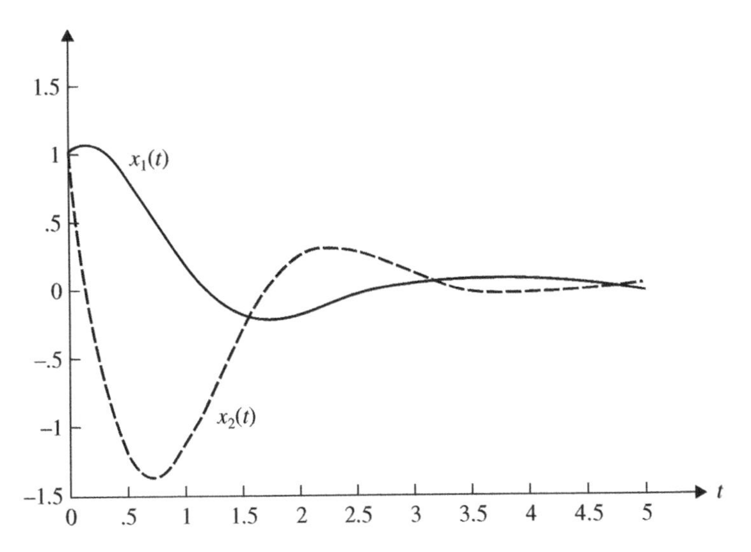

Figure 6.18 The components of the state vector $\mathbf{x}(t)$ versus time for the case of complete state feedback.

In the preceding presentation of state observers, the scenario is that the state vector $\mathbf{x}$ is not available for use in output feedback control. Thus, estimated output feedback control is used instead, i.e. $u(t) = g\mathbf{c}^T\mathbf{x}_e$. A practical alternative use of an observer is to use the estimated state vector $\mathbf{x}_e$ to change a problem that is output feedback control ($g\mathbf{c}^T\mathbf{x}$), because of hardware limitations to one that is augmented by the observer to use complete state variable feedback, i.e. $u(t) = g\mathbf{c}^T\mathbf{x} + \mathbf{k}^T\mathbf{x}_e$. Here the vector $\mathbf{k}$ makes use of each state variable as opposed to $\mathbf{c}$, which uses only some state variables.

If MIMOs are used, this analysis can be extended. The resulting observer is usually called a *Luenberger observer*. If, in addition, a noise signal is added as input to Equation (6.30), the estimation equation can still be developed. In this case they are referred to as *Kalman filters* (Anderson and Moore. 1979).

6.11 Realization

In the preceding section, the state matrix A (and hence, the coefficient matrices M, C and K), along with the input matrix B and the output matrix C_c, are all assumed to be known. In this section, however, the problem is to determine A, B and C_c from the transfer function of the system. This problem was first introduced in Section 1.7, called plant identification, where the scalar coefficients m, c and k were determined from Bode plots. Determining the matrices A, B and C_c from the transfer function of a system is called *system realization.*

Consider again the SISO version of Equations (6.30) and (6.31). Assuming that the initial conditions are all zero, taking the Laplace transform of these two expressions yields

$$s\mathbf{X}(s) = A\mathbf{X}(s) + \mathbf{b}U(s) \tag{6.77}$$

$$y(s) = \mathbf{c}^T\mathbf{X}(s) \tag{6.78}$$

Solving Equation (6.77) for $\mathbf{X}(s)$ and substituting the result into Equation (6.78) yields

$$Y(s) = \mathbf{c}^T(sI - A)^{-1}\mathbf{b}U(s) \tag{6.79}$$

Since y and u are scalars here, the transfer function of the system is just

$$\frac{Y(s)}{U(s)} = G(s) = \mathbf{c}^T(sI - A)^{-1}\mathbf{b} \tag{6.80}$$

Here and in the following, $G(s)$ is assumed to be a rational function of s. If, in addition, $G(s)$ is such that

$$\lim_{s\to\infty} G(s) = 0 \tag{6.81}$$

then $G(s)$ is called a *proper rational function.* These two conditions are always satisfied for the physical models presented in Chapter 2. The triple A, $\mathbf{b}$ and $\mathbf{c}$ is called a *realization* of $G(s)$ if Equation (6.80) holds. The function $G(s)$ can be shown to have a realization if and only if $G(s)$ is a proper rational function. The triple $(A, \mathbf{b}, \mathbf{c})$ of minimum order that satisfies Equation (6.80) is called an *irreducible realization,* or a *minimal realization.* The triple $(A, \mathbf{b}, \mathbf{c})$ can be shown to be an irreducible realization if and only if the triple $(A, \mathbf{b}, \mathbf{c})$ is both controllable and observable (Chen, 1998).

A transfer function $G(s)$ is said to be irreducible if and only if the numerator and denominator of G have no common factor. This statement is true only if the denominator of $G(s)$ is equal to the characteristic polynomial of the matrix A, and only if the degree of the denominator of $G(s)$ is equal to $2n$, the order of the system. While straightforward conditions are available for ensuring the existence of an irreducible realization, the realization is not unique. In fact, if $(A, \mathbf{b}, \mathbf{c})$ is an irreducible realization of the transfer function $G(s)$, then $(A', \mathbf{b}', \mathbf{c}')$ is an irreducible realization if and only if A is similar to A', i.e. there exits a nonsingular matrix P such that $A = PAP^{-1}$, $\mathbf{b} = P\mathbf{b}'$ and $\mathbf{c} = \mathbf{c}'P^{-1}$. Hence, a given transfer function has an infinite number of realizations. This result is very important to remember when studying modal testing in Chapter 12.

There are several ways to calculate realization of a given transfer function. The easiest method is to recall from differential equations the method of writing a $2n$-order differential equation as $2n$ first-order equations. Then, if the transfer function is given as

$$\frac{y(s)}{u(s)} = G(s) = \frac{\beta}{s^{2n} + \alpha_1 s^{2n-1} + \alpha_2 s^{2n-2} + \cdots + \alpha_{2n-1}s + \alpha_{2n}} \tag{6.82}$$

the time domain equivalent is simply obtained by multiplying this out to yield

$$s^{2n}y(s) + \alpha_1 s^{2n-1}y(s) + \cdots = \beta u(s) \tag{6.83}$$

and taking the inverse Laplace transform of Equation (6.83) to get

$$y^{2n} + \alpha_1 y^{(2n-1)} + \cdots + \alpha_{2n-1}y(y) = \beta u(t) \tag{6.84}$$

Here, $y^{(2n)}$ denotes the $2n^{\text{th}}$ time derivative of $y(t)$. Next, define the state variables by the scheme

$$\begin{aligned} x_1(t) &= y(t) \\ x_2(t) &= y^{(1)}(t) \\ &\vdots \\ x_{2n}(t) &= y^{(2n-1)}(t). \end{aligned}$$

The state equations for Equation (6.84) then become

$$\dot{\mathbf{x}}(t) = \begin{bmatrix} 0 & 1 & 0 & \cdots & 0 \\ 0 & 0 & 1 & \cdots & 0 \\ \vdots & \vdots & \vdots & & \vdots \\ 0 & 0 & 0 & \cdots & 1 \\ -\alpha_{2n} & -\alpha_{2n-1} & -\alpha_{2n-2} & \cdots & -\alpha_1 \end{bmatrix} \mathbf{x}(t) + \begin{bmatrix} 0 \\ 0 \\ \vdots \\ 0 \\ \beta \end{bmatrix} u(t) \tag{6.85}$$

and

$$y(t) = [1 \quad 0 \quad 0 \ldots]\mathbf{x}(t) \tag{6.86}$$

The triple $(A, \mathbf{b}, \mathbf{c})$ defined by Equations (6.85) and (6.86) constitutes an irreducible realization of the transfer function given by Equation (6.82).

Realization procedures are also available for MIMO systems (Ho and Kalman, 1965). In Chapter 8, realization methods are used to determine the natural frequencies, damping ratios and mode shapes of a vibrating structure by measuring the transfer function and using it to construct a realization of the test structure.

Example 6.11.1
Consider the transfer function of a simple oscillator, i.e.

$$G(s) = \frac{1}{s^2 + 2\zeta\omega s + \omega^2}$$

Following Equation (6.85), a state space realization of this transfer function is given by

$$A = \begin{bmatrix} 0 & 1 \\ -\omega^2 & -2\zeta\omega \end{bmatrix} \quad \mathbf{b} = \begin{bmatrix} 0 \\ 1 \end{bmatrix} \quad \mathbf{c} = \begin{bmatrix} 1 \\ 0 \end{bmatrix}$$

6.12 Reduced-Order Modeling

Most control methods work best for structures with a small number of degrees of freedom. Many modeling techniques produce structural models of a large order. Hence, it is often necessary to reduce the order of a model before performing a control analysis and designing a control law. This section introduces a systematic method of deciding which degrees of freedom, or coordinate to keep, and which can be removed without sacrificing too much fidelity of the final model.

The approach taken is to reduce the order of a given model based on deleting those coordinates, or modes, that are the least controllable and observable. The idea here is that controllability and observability of a state (coordinate) are indications of the contribution of that state (coordinate) to the response of the structure, as well as the ability of that coordinate to be excited by an external disturbance.

To implement this idea, a measure of the degree of controllability and observability is needed. One such measure of controllability is given by the controllability norm of Equation (6.37). However, an alternative, more useful measure is provided for asymptotically stable systems of the form given by Equations (6.30) and (6.31) by defining the *controllability grammian*, denoted by W_C, by

$$W_C^2 = \int_0^\infty e^{At} BB^T e^{A^T t} dt \tag{6.87}$$

and the *observability grammian*, denoted by W_O, by

$$W_O^2 = \int_0^\infty e^{A^T t} C^T C e^{At} dt \tag{6.88}$$

Here, the matrices A, B and C are defined as in Equations (6.30) and (6.31). The properties of these two matrices provide useful information about the controllability and observability of the closed-loop system. If the system is controllable (or observable), the matrix W_C (or W_O) is nonsingular. These grammians characterize the degree of controllability and observability by quantifying just how far away from being singular the matrices W_C and W_O are. This is equivalent to quantifying rank deficiency. The most reliable way to quantify the rank of a matrix is to examine the singular values of the matrix, which is discussed next.

For any real $m \times n$ matrix A, there exist orthogonal matrices $U_{m \times m}$ and $V_{n \times n}$ such that

$$U^T AV = \text{diag}\begin{bmatrix} \sigma_1 & \sigma_2 & \cdots & \sigma_p \end{bmatrix}, \; p = \min(m, n) \tag{6.89}$$

where the σ_i are real and ordered via

$$\sigma_1 \geq \sigma_2 \geq \cdots \geq \sigma_p \geq 0 \tag{6.90}$$

(Golub and Van Loan, 1996). The numbers σ_i are called the *singular values* of the matrix A. The singular values of the matrix A are the nonnegative square roots of the eigenvalues of the symmetric positive definite matrix A^TA. The vectors $\mathbf{u}_i$ consisting of the columns of the matrix U, are called the *left singular vectors* of A. Likewise, the columns of V, denoted by $\mathbf{v}_i$, are called the *right singular vectors* of A. The process of calculating U, V and diag$[\sigma_1 \ldots \sigma_p]$ is called the *singular value decomposition* (denoted by SVD) of the matrix A. Note that the singular values and vectors satisfy

$$A\mathbf{v}_i = \sigma_i\mathbf{u}_i, \quad i = 1, \ldots, p \tag{6.91}$$

$$A^T\mathbf{u}_i = \sigma_i\mathbf{v}_i, \quad i = 1, \ldots, p \tag{6.92}$$

Note that if A is a square symmetric positive definite matrix, then $U = V$, and the $\mathbf{u}_i$ are the eigenvectors of the matrix A and the singular values of A are identical to the eigenvalues of A^2.

The real square symmetric semi-positive definite matrix A is, of course, singular, or rank deficient, if and only if it has a zero eigenvalue (or singular value). This statement leads naturally to the idea that the size of the singular values of a matrix quantify how close the matrix is to being singular. If the smallest singular value, σ_n, is well away from zero, then the matrix is of full rank and "far away" from being singular (note: unlike frequencies, we order singular values from largest to smallest).

Applying the idea of singular values as a measure of rank deficiency to the controllability and observability grammians yields a systematic model reduction method. The matrices W_O and W_C are symmetric and hence are similar to a diagonal matrix. Moore (1981) showed that there always exists an equivalent system for which these two grammians are both equal and diagonal. Such a system is then called *balanced*.

In addition, Moore showed that W_C and W_O must satisfy the two Liapunov-type equations

$$\begin{aligned} AW_C^2 + W_C^2A^T &= -BB^T \\ A^TW_O^2 + W_O^2A &= -C_c^TC_c \end{aligned} \tag{6.93}$$

for asymptotically stable systems.

Let the matrix P denote a linear similarity transformation, which when applied to Equations (6.30) and (6.31) yields the equivalent system

$$\begin{aligned} \dot{\mathbf{x}}' &= A'\mathbf{x}' + B'\mathbf{u} \\ \mathbf{y} &= C'\mathbf{x}' \end{aligned} \tag{6.94}$$

These two equivalent systems are related by

$$\mathbf{x} = P\mathbf{x}' \tag{6.95}$$

$$A' = P^{-1}AP \tag{6.96}$$

$$B' = P^{-1}B \tag{6.97}$$

$$C_c' = C_cP \tag{6.98}$$

Here, the matrix P can be chosen such that the new grammians defined by

$$W'_C = P^{-1} W_C P \tag{6.99}$$

and

$$W'_O = P^{-1} W_O P \tag{6.100}$$

are equal and diagonal. That is

$$W'_C = W'_O = \Lambda_W = \text{diag}[\sigma_1 \quad \sigma_2 \cdots \sigma_{2n}] \tag{6.101}$$

where the numbers σ_i are the singular values of the grammians and are ordered such that

$$\sigma_i \geq \sigma_{i+1}, \quad i = 1, 2, \dots 2n - 1 \tag{6.102}$$

Under these circumstances, i.e. when Equations (6.101) and (6.102) hold, the system given by Equations (6.94) is said to be *internally balanced*.

Next, let the state variables in the balanced system be partitioned into the form

$$\begin{bmatrix} \dot{\mathbf{x}}'_1 \\ \dot{\mathbf{x}}'_2 \end{bmatrix} = \begin{bmatrix} A_{11} & A_{12} \\ A_{21} & A_{22} \end{bmatrix} \begin{bmatrix} \mathbf{x}'_1 \\ \mathbf{x}'_2 \end{bmatrix} + \begin{bmatrix} B_1 \\ B_2 \end{bmatrix} \mathbf{u} \tag{6.103}$$

$$\mathbf{y} = [C_1 \quad C_2] \begin{bmatrix} \mathbf{x}'_2 \\ \mathbf{x}'_1 \end{bmatrix} \tag{6.104}$$

where A_{11} is a $k \times k$ matrix and $\mathbf{x}'_2$ is the vector containing those states corresponding to the $(2n - k)$ smallest singular values of W_C. It can be shown (Moore, 1981) that the $\mathbf{x}'_2$ part of the state vector for Equations (6.103) and (6.104) affects the output much less than $\mathbf{x}'_1$ does. Thus, if σ_k is much greater than σ_{k+1}, i.e. $\sigma_k \gg \sigma_{k+1}$, the $\mathbf{x}'_2$ part of the state vector does not affect the input output behavior of the system as much as $\mathbf{x}'_1$ does.

The preceding comments suggest that a suitable low-order model of the system of Equations (6.30) and (6.31) is the subsystem given by

$$\dot{\mathbf{x}}'_1 = A_{11}\mathbf{x}'_1 + B_1\mathbf{u} \tag{6.105}$$

$$\mathbf{y} = C_1\mathbf{x}'_1 \tag{6.106}$$

This subsystem is referred to as a *reduced-order model* (often referred to as ROM). Note that, as pointed out by Moore (1981), a realization of Equations (6.30) and (6.31) should yield the reduced-order model of Equations (6.105) and (6.106).

The ROM can be calculated by first calculating an intermediate transformation matrix P_1 based on the controllability grammian. Solving the eigenvalue problem for W_C yields

$$W_C = V_C \Sigma_C^2 V_C^T \tag{6.107}$$

where V_C is the matrix of normalized eigenvectors of W_C and Σ_C^2 is the diagonal matrix of eigenvalues of W_C. The square on Σ_C is a reminder that W_C is positive definite. Based on this decomposition, a nonsingular transformation P_1 is defined by

$$P_1 = V_C \sum_C \tag{6.108}$$

Application of the transformation P_1 to the state equations yields the intermediate state equations defined by

$$A'' = P_1^{-1} A P_1 \tag{6.109}$$

$$B'' = P_1^{-1} B \tag{6.110}$$

$$C_c'' = C_c P_1 \tag{6.111}$$

To complete the balancing algorithm, these intermediate equations are balanced with respect to W_O''. That is, the eigenvalue problem for W_O'' yields the matrices V_O'' and $\Sigma_O''^2$ such that

$$W_O'' = V_O'' \Sigma_O''^2 V_O''^T \tag{6.112}$$

These two matrices are used to define the second part of the balancing transformation, i.e.

$$P_2 = V_O'' \Sigma_O''^{-1/2} \tag{6.113}$$

The balanced version of the original state equations is then given by the product transformation $P = P_1 P_2$ in Equation (6.94). They are

$$A' = P_2^{-1} P_1^{-1} A P_1 P_2 \tag{6.114}$$

$$B' = P_2^{-1} P_1^{-1} B \tag{6.115}$$

$$C' = C P_1 P_2 \tag{6.116}$$

The balanced system is then used to define the ROM of Equations (6.105) and (6.106) by determining the value of k such that $\sigma_k \gg \sigma_{k+1}$. The following example illustrates an internally balanced ROM.

Example 6.12.1

Consider the two degrees of freedom system of Figure 2.6 with $m_1 = m_2 = 1$, $c_1 = 0.2$, $c_2 = 0.1$ and $k_1 = k_2 = 1$. Let an impulse force be applied to m_2 and assume a position measurement of m_2 is available. The state matrix is

$$A = \begin{bmatrix} 0 & 0 & 1 & 0 \\ 0 & 0 & 0 & 1 \\ -2 & 1 & -0.3 & 0.1 \\ 1 & -1 & 0.1 & -0.1 \end{bmatrix}$$

In addition, B becomes the vector $\mathbf{b} = [0 \quad 0 \quad 0 \quad 1]^T$, and the output matrix C_c becomes the vector $\mathbf{c}^T = [0 \quad 1 \quad 0 \quad 0]$.

The controllability and observability grammians can be calculated (for $t \to \infty$) from Equation (6.73) to be

$$W_C = \begin{bmatrix} 4.0569 & 6.3523 & 0.0000 & -0.3114 \\ 6.3523 & 10.5338 & 0.3114 & 0.0000 \\ 0.0000 & 0.3114 & 1.7927 & 2.2642 \\ -0.3114 & 0.0000 & 2.2642 & 4.1504 \end{bmatrix}$$

$$W_O = \begin{bmatrix} 1.8198 & 2.2290 & 0.5819 & 1.1637 \\ 2.2290 & 4.2315 & -0.8919 & 0.4181 \\ 0.5819 & -0.0819 & 4.0569 & 6.3523 \\ 1.1637 & 0.4181 & 6.3523 & 10.5338 \end{bmatrix}$$

Calculation of the matrix P that diagonalizes the two grammians yields

$$P = \begin{bmatrix} 0.4451 & -0.4975 & 0.4962 & -0.4437 \\ 0.7821 & -0.7510 & -0.2369 & 0.3223 \\ 0.2895 & 0.2895 & 0.6827 & 0.8753 \\ 0.4419 & 0.5112 & -0.4632 & -0.4985 \end{bmatrix}$$

The singular values of W_O and W_C are then $\sigma_1 = 9.3836$, $\sigma_2 = 8.4310$, $\sigma_3 = 0.2724$ and $\sigma_4 = 0.2250$. From examination of these singular values, it appears that the coordinates x_1 and x_2 associated with A' of Equation (6.114) are likely candidates for an ROM (i.e. $k = 1$ in this case). Using Equations (6.114), (6.115) and (6.116) yields the balanced system given by

$$A' = \begin{bmatrix} -0.0326 & 0.6166 & 0.0192 & -0.0275 \\ -0.6166 & -0.0334 & -0.0218 & 0.0280 \\ 0.0192 & 0.0218 & -0.1030 & 1.6102 \\ 0.0275 & 0.0280 & 1.6102 & -0.2309 \end{bmatrix}$$

$$B'^T = \begin{bmatrix} 0.7821 & 0.7510 & -0.2369 & -.03223 \end{bmatrix}^T$$

and

$$C' = \begin{bmatrix} 0.7821 & -0.7510 & -0.2369 & 0.3223 \end{bmatrix}$$

Given that $k = 2$ (from examination of the singular values), the coefficients in Equations (6.105) and (6.106) for the ROM become

$$A_{11} = \begin{bmatrix} -0.0326 & 0.6166 \\ -0.6166 & -0.0334 \end{bmatrix}$$

$$B_1 = \begin{bmatrix} 0.7821 & 0.7510 \end{bmatrix}^T$$

and

$$C_1 = \begin{bmatrix} 0.7821 & -0.7510 \end{bmatrix}$$

Plots of the response of the full-order model and the balanced model are given in Figures 6.19 and 6.18, respectively. Note, in Figure 6.20, that the two coordinates

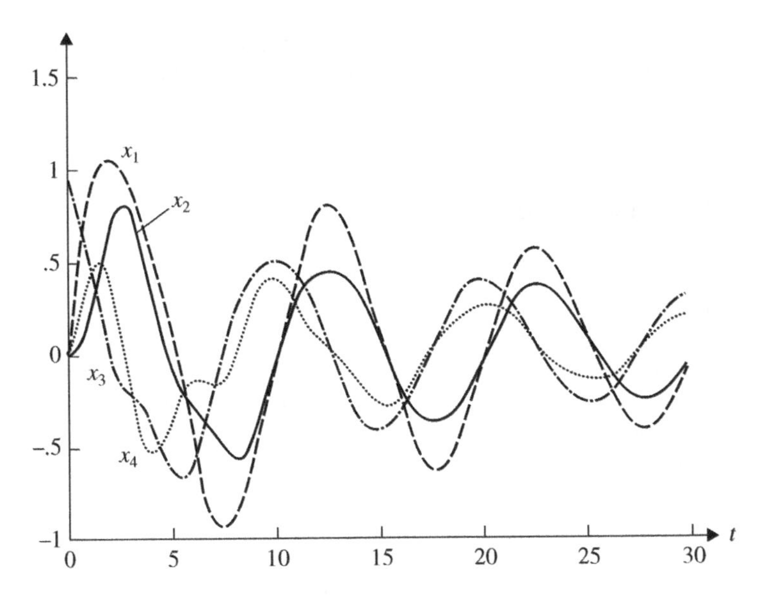

Figure 6.19 Response of the original state variables.

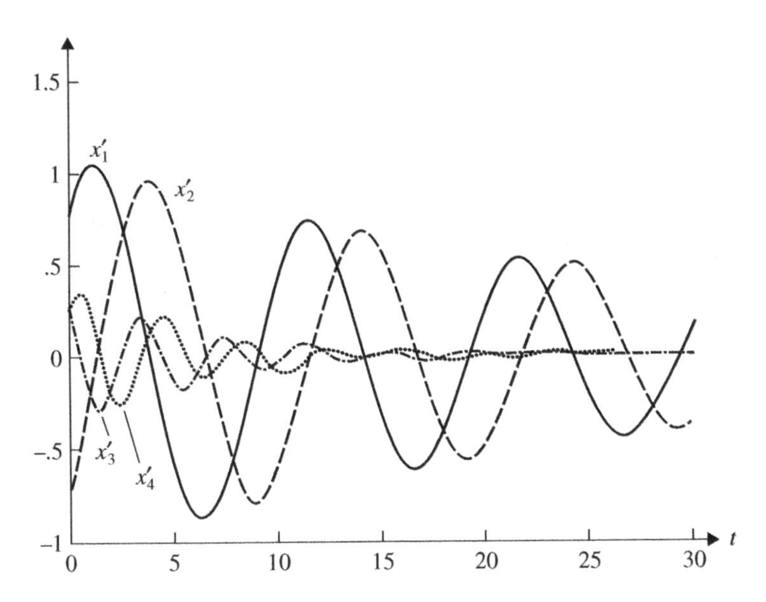

Figure 6.20 Response of the balanced state variables.

(x'_3 and x'_4) neglected in the ROM do not contribute as much to the response and, in fact, die out after about 15 time units. However, all the coordinates in Figure 6.19 and x'_1 and x'_2 are still vibrating after 15 time units.

Note that the balanced ROM will change if different inputs and outputs are considered, as the B and C_c matrices would change, changing the reduction scheme.

6.13 Modal Control in State Space

In general, modal control refers to the procedure of decomposing the dynamic equations of a structure into modal coordinates, such as Equations (5.30) and (5.38), and designing the control system in this modal coordinate system. In broad terms, any control design that employs a modal description of the structure is called a modal control method. Modal control works well for systems in which certain (just a few) modes of the structure dominate the response. That is, modal control works well for systems in which only a few of the modal participation factors (Section 3.3) are large and the rest are relatively small. Modal control can be examined in either state space, via Equations (6.30) and (6.31), or in physical space, via Equations (4.23) to (4.25). This section examines modal control in the state space coordinate system, and the following section examines modal control in the physical coordinate system.

Consider the state space description of Equations (6.30) and (6.31). If the matrix A has a diagonal Jordan form, then, following Section 5.3, there exists a nonsingular matrix U such that

$$U^{-1}AU = \Lambda$$

where Λ is a diagonal matrix of the eigenvalues of the state matrix A. Note that the diagonal elements of Λ will be complex if the system is underdamped. Substituting $\mathbf{x} = U\mathbf{z}$ into Equation (6.30) and premultiplying by the inverse of the nonsingular matrix U yields the diagonal system

$$\dot{\mathbf{z}} = \Lambda\mathbf{z} + U^{-1}B\mathbf{u} \tag{6.117}$$

Here the vector $\mathbf{z}$ is referred to as the *modal coordinate system*. In this form, the controllability problem and the pole placement problem become more obvious.

Consider first the controllability question for the case of a single input (i.e. $\mathbf{u}$ becomes a scalar). Then B is a vector $\mathbf{b}$ and $U^{-1}\mathbf{b}$ is a vector consisting of $2n$ elements denoted by b_i. Clearly, this system is controllable if and only if each $b_i \neq 0$, $i = 1, 2, \ldots, 2n$. If, on the other hand, b_i should happen to be zero for some index i, the system is not controllable. With $b_i = 0$, the ith mode is not controllable, as no feedback law could possibly affect the ith mode. Thus, the vector $U^{-1}\mathbf{b}$ indicates the controllability of the system by inspection.

The form of Equation (6.117) can also be used to perform a quick model reduction. Suppose it is desired to control just the fastest modes or modes in a certain frequency range. Then these modes of interest can be taken as the ROM and the others can be neglected.

Next consider the pole placement problem. In modal form, it seems to be a trivial matter to choose the individual modal control gains, b_i, to place the eigenvalues of the system. For instance, suppose output feedback is used. Then $u = \mathbf{c}^T U\mathbf{z}$, and Equation (6.117), become

$$\dot{\mathbf{z}} = \Lambda\mathbf{z} + U^{-1}\mathbf{b}\mathbf{c}^T U\mathbf{z} = (\Lambda + U^{-1}\mathbf{b}\mathbf{c}^T U)\mathbf{z} \tag{6.118}$$

where $\mathbf{c}^T$ is a row vector. The closed-loop system then has poles determined by the matrix $(\Lambda + U^{-1}\mathbf{b}\mathbf{c}^T U)$. Suppose that the matrix $U^{-1}\mathbf{b}\mathbf{c}^T U$ is also diagonal. In this case the controls are also decoupled, and Equation (6.118) becomes the $2n$ decoupled equations

$$\dot{z}_i = (\lambda_i + u_i)z_i \quad i = 1, 2, \ldots, 2n \tag{6.119}$$

Here u_i denotes the diagonal elements of the diagonal matrix $U^{-1}\mathbf{b}\mathbf{c}^T U$. Note, that the vector $U^{-1}\mathbf{b}\mathbf{c}^T U\mathbf{z}$ in Equation (6.118) is identical to the vector $U^{-1}\mathbf{f}$ in Equation (5.20). The difference between the two equations is simply that in this section the forcing function is the result of a control force manipulated by a designer to achieve a desired response. In the development of Section 5.3, the forcing term represents some external disturbance.

The matrix $U^{-1}\mathbf{b}\mathbf{c}^T U$ in Equation (6.118) may not be diagonal. In this case, Equation (6.118) is not decoupled. The controls in this situation reintroduce coupling into the system. As a result, if it is desired to change a particular modal coordinate, the control force chosen to change this mode will also change the eigenvalues of some of the other modes. Following the arguments in Section 3.5, a necessary and sufficient condition for $U^{-1}\mathbf{b}\mathbf{c}^T U$ to be diagonal is for the matrix $\mathbf{b}\mathbf{c}^T$ to commute with the state matrix A. The remainder of this section is devoted to discussing the coupling introduced by control laws and measurement points in the case that it is desired to control independently a small number of modes of a given structure.

Suppose then that it is desired to control independently a small number of modes. For example, it may be desired to place a small number of troublesome poles while leaving the remaining poles unaffected. Let k denote the number of modes that are to be controlled independently of the remaining $2n - k$ modes. Furthermore, assume that it is the first k modes that are of interest, i.e. the desired modes are the lowest k. Then partition the modal equations with state feedback into the k modal coordinates that are to be controlled and the $2n - k$ modal coordinates that are to be left undisturbed. This yields

$$\begin{bmatrix} \dot{\mathbf{z}}_k \\ \dot{\mathbf{z}}_{2n-k} \end{bmatrix} = \left(\begin{bmatrix} \Lambda_k & 0 \\ 0 & \Lambda_{2n-k} \end{bmatrix} + \begin{bmatrix} \mathbf{b}_k\mathbf{c}_k^T & \mathbf{b}_k\mathbf{c}_{2n-k}^T \\ \mathbf{b}_{2n-k}\mathbf{c}_k^T & \mathbf{b}_{2n-k}\mathbf{c}_{2n-k}^T \end{bmatrix} \right) \begin{bmatrix} \mathbf{z}_k \\ \mathbf{z}_{2n-k} \end{bmatrix} \tag{6.120}$$

Here the matrix $U^{-1}\mathbf{b}\mathbf{c}^T U$ is partitioned into blocks defined by

$$U^{-1}\mathbf{b} = \begin{bmatrix} \mathbf{b}_k \\ \mathbf{b}_{2n-k} \end{bmatrix}$$

$$\mathbf{c}^T U = [\mathbf{c}_k^T \quad \mathbf{c}_{2n-k}^T]$$

where $\mathbf{b}_k$ denotes the first k elements of the vector $\mathbf{b}$, $\mathbf{b}_{2n-k}$ denotes the last $2n - k$ elements, etc. Likewise, the matrices Λ_k and Λ_{2n-k} denote diagonal matrices of the first k eigenvalues and the last $2n - k$ eigenvalues, respectively. Let b_i denote the elements of the vector $U^{-1}\mathbf{b}$ and c_i denote the elements of the vector $\mathbf{c}^T U$. Then $\mathbf{b}_k\mathbf{c}_k^T$ is the $k \times k$ matrix

$$\mathbf{b}_k\mathbf{c}_k^T = \begin{bmatrix} b_1c_1 & b_1c_2 & \cdots & b_1c_k \\ b_2c_1 & b_2c_2 & \cdots & b_2c_k \\ \vdots & \vdots & & \vdots \\ b_kc_1 & b_kc_2 & \cdots & b_kc_k \end{bmatrix} \tag{6.121}$$

Examination of Equation (6.120) illustrates that the first k modes of the system can be controlled independently of the last $(2n - k)$ modes, if and only if the two vectors $\mathbf{b}_{2n\text{-}k}$ and $\mathbf{c}_{2n\text{-}k}$ are both zero. Furthermore, the first k modes can be controlled independently of each other only if Equation (6.121) is diagonal. This, of course, cannot happen, as clearly indicated by setting the off-diagonal terms of Equation (6.121) to zero. Takahashi et al. (1968) discussed decoupled or ideal control in detail. In general, it is difficult to control modes independently, unless a large number of actuators and sensors are used. The vector c_{2n-k}^T indicates the coupling introduced into the system by measurement, and the vector b_{2n-k} indicates coupling due to control action. This phenomenon is known as *observation spillover* and *control spillover* (Balas, 1978).

Example 6.13.1

Consider an overdamped two degrees of freedom structure with a state matrix given by

$$A = \begin{bmatrix} 0 & 0 & 1 & 0 \\ 0 & 0 & 0 & 1 \\ -3 & 1 & -9 & 4 \\ 1 & -1 & 4 & -4 \end{bmatrix}$$

Solving the eigenvalue problem for A yields that the matrices Λ, U and U^{-1} are given by

$$\Lambda = \begin{bmatrix} -10.9074 & 0 & 0 & 0 \\ 0 & -1.2941 & 0 & 0 \\ 0 & 0 & -0.5323 & 0 \\ 0 & 0 & 0 & -0.2662 \end{bmatrix}$$

$$U = \begin{bmatrix} -0.0917 & -0.4629 & 0.7491 & -0.0957 \\ 0.0512 & -0.7727 & 1.0000 & 1.0000 \\ 1.0000 & 0.5990 & -0.3988 & 0.0255 \\ -0.5584 & 1.0000 & 0.5323 & -0.2662 \end{bmatrix}$$

$$U^{-1} = \begin{bmatrix} 0.2560 & -0.1121 & 0.7846 & -0.4381 \\ 0.7382 & 0.3741 & 0.7180 & 1.1986 \\ 1.6791 & 0.3378 & 0.5368 & 0.7167 \\ -1.1218 & 0.9549 & -0.0222 & 0.2320 \end{bmatrix}$$

which constitutes a modal decomposition of the state matrix (note that $A = U\Lambda U^{-1}$). Suppose next that $\mathbf{b} = [0\ 0\ 1\ 0]^T$ and $\mathbf{c}^T = [0\ 0\ 1\ 1]$ are the control and measurement vectors, respectively. In the decoupled coordinate system, the control and measurement vectors become

$$U^{-1}\mathbf{b} = \begin{bmatrix} 0.7846 & 0.7180 & 0.5368 & -0.0222 \end{bmatrix}^T$$

and

$$\mathbf{c}^T U = \begin{bmatrix} 0.44161 & 0.5990 & -0.9311 & -0.2407 \end{bmatrix}$$

Notice that these vectors are fully populated with nonzero elements. This leads to the fully coupled closed-loop system given by Equation (6.118), since the term $U^{-1}\mathbf{b}\mathbf{c}^T U$ becomes

$$U^{-1}\mathbf{b}\mathbf{c}^T U = \begin{bmatrix} 0.3465 & 1.2545 & -0.7305 & -0.1888 \\ 0.3171 & 1.1481 & -0.6685 & -0.1728 \\ 0.2371 & 0.8584 & -0.4999 & -0.1292 \\ 0.0098 & -0.0355 & 0.0207 & 0.0053 \end{bmatrix}$$

This last expression illustrates the recoupling effect caused by the control and measurement locations. Note that the matrix $\Lambda + U^{-1}\mathbf{b}\mathbf{c}^T U$ of Equation (6.118) is not diagonal, but fully populated, recoupling the dynamics.

6.14 Modal Control in Physical Space

As indicated in Section 3.4, the left-hand side of Equation (4.23) can be decoupled into modal coordinates, if and only if $CM^{-1}K = KM^{-1}C$. This was used in Section 5.4 to decouple Equation (5.36) into modal coordinates to solve for the forced response. A similar approach is taken here, except that the input force $S_m^T\mathbf{f}$ of Equation (5.38) becomes the control force $S_m^T B_f\mathbf{u}$. As in the state space, the difficulty lies in the fact that the transformation, S_m, of the equations of motion into modal coordinates, $\mathbf{z}$, does not necessarily decouple the control input, $S_m^T B_f\mathbf{u}$. Note that in this section $\mathbf{z}$ is an $n \times 1$ vector of modal positions, whereas in the previous section, $\mathbf{z}$ is a $2n \times 1$ vector of modal state variables. The control problem in second-order physical coordinates transformed into modal coordinates is, in the notation of Sections 4.10 and 5.4,

$$I\ddot{\mathbf{z}} + \Lambda_D\dot{\mathbf{z}} + \Lambda_K\mathbf{z} = S_m^T B_f\mathbf{u} \tag{6.122}$$

$$\mathbf{y} = C_p S_m\mathbf{z} + C_v S_m\dot{\mathbf{z}} \tag{6.123}$$

Here, the dimensions of the matrices C_p and C_v are as indicated in Equation (4.25). Note that in Equation (6.122) the relative magnitude of the elements $(S_m^T B_f\mathbf{u})_i$ are an indication of the degree of controllability for the ith mode. If $(S_m^T B_f\mathbf{u})_i$ is very small, the ith mode is hard to control. If it happens to be zero, then the ith mode is not controllable. If on the other hand $(S_m^T B_f\mathbf{u})_i$ is relatively large, the ith mode is very controllable.

If output feedback is used of the form suggested in Equation (4.26), then $\mathbf{u} = -G_f\mathbf{y}$. Combining Equation (6.122) with Equation (6.123) and the control law, the system equation becomes

$$I\ddot{\mathbf{z}} + \Lambda_D\dot{\mathbf{z}} + \Lambda_K\mathbf{z} = -S_m^TB_fK_pS_m\mathbf{z} - S_m^TB_fK_vS_m\dot{\mathbf{z}} \tag{6.124}$$

where $K_p = G_fC_p$ and $K_v = G_fC_v$ represent measurement positions and B_f is taken to represent the control actuator locations. Note that by this choice of $\mathbf{u}$, the discussion is now restricted to state variable feedback, i.e. position and velocity feedback. Equation (6.124) can be rewritten as

$$I\ddot{\mathbf{z}} + (\Lambda_D + S_m^TB_fK_vS_m)\dot{\mathbf{z}} + (\Lambda_K + S_m^TB_fK_pS_m)\mathbf{z} = \mathbf{0} \tag{6.125}$$

Now, Equation (6.125) is posed for modal pole placement control.

To cause the closed-loop system of Equation (6.125) to have the desired eigenvalues and hence a desired response, the control gain matrix B_f and the measurement matrices K_v and K_p must be chosen (a design process) appropriately. If, in addition, it is desired to control each mode independently, i.e. each z_i in Equation (6.125), then further restrictions must be satisfied. Namely, it can be seen from Equation (6.125) that an independent control of each mode is possible, if and only if the matrices $S_m^TB_fK_vS_m$ and $S_m^TB_fK_pS_m$ are both diagonal. In the event that the matrix B_fK_v and the matrix B_fK_p are both symmetric, this will be true if and only if

$$B_fK_vM^{-1}C = CM^{-1}B_fK_v \tag{6.126}$$

and

$$B_fK_pM^{-1}K = KM^{-1}B_fK_p \tag{6.127}$$

Unfortunately, this puts *very* stringent requirements on the location and number of sensors and actuators. If, however, the conditions given by Equations (6.126) and (6.127) are satisfied, Equation (6.125) reduces to the n SDOF control problems of the form

$$\ddot{z} + (2\zeta_i\omega_i + \alpha_i)\dot{z}_i + (\omega_i^2 + \beta_i)z_i = 0 \tag{6.128}$$

where α_i and β_i are the diagonal elements of the matrices $S_m^TB_fK_vS_m$ and $S_m^TB_fK_pS_m$, respectively. This last expression represents an independent set of equations that can be solved for α_i and β_i, given desired modal response information. That is, if it is desired that the closed-loop system have a first-mode natural frequency of 10, for instance, then Equation (6.128) requires that $\omega_1^2 + \beta_1 = 10$, or $\beta_1 = 10 - \omega_1^2$. If the known open-loop system has $\omega_1^2 = 6$, then $\beta_1 = 4$ is the desired control gain.

While the modal control equations of Equation (6.128) appear simple, the problem of independent control remains complicated and requires a larger number of sensor and actuator connections. This happens because the α_i's and β_i's of Equation (6.128), while independent, are not always capable of being independently implemented. The design problem is to choose actuator and sensor locations as well as gains such that B, K_v and K_p satisfy Equations (6.126) and (6.127).

The choice of these gains is not independent but rather coupled through the equations

$$S_m^T B_f K_v S_m = \text{diag}[\alpha_i] \tag{6.129}$$

and

$$S_m^T B_f K_p S_m = \text{diag}[\beta_i] \tag{6.130}$$

The modal control of a system in physical coordinates as well as the problem of performing independent modal control are illustrated in the following example.

Example 6.14.1

Consider the system of Figure 2.6 with coefficient values (dimensionless) of $m_1 = 9, m_2 = 1, c_1 = 8, c_2 = 1, k_1 = 24$ and $k_2 = 3$. This produces a system equivalent to the one in Example 3.5.1, where it is shown that $CM^{-1}K = KM^{-1}C$, so that the system decouples. The matrix S_m is given by

$$S_m = \frac{1}{\sqrt{2}}\begin{bmatrix} \frac{1}{3} & -\frac{1}{3} \\ 1 & 1 \end{bmatrix}$$

The decoupled coordinates, listed in Example 3.5.1, yield the modal frequencies and damping ratios as

$$\begin{aligned} \zeta_1 &= 0.2357 \quad \omega_1 = \sqrt{2} = 1.414 \text{ rad/s} \\ \zeta_2 &= 0.3333 \quad \omega_2 = 2.0 \text{ rad/s} \end{aligned}$$

Consider the control problem of calculating a feedback law that will cause the closed-loop system in Equation (6.124) to have a response with a modal damping ratio of 0.4 in the first mode. Three cases of different sensor and actuator placements are considered.

In the first case, consider an SISO system with non-collocated control. Suppose one actuator is used to achieve the desired control and it is connected to only one mass, m_2, and one sensor is used to measure the velocity of m_1. Then

$$K_v = [g_1 \quad 0],\ K_p = \mathbf{0}$$

$$B_f = \begin{bmatrix} 0 \\ g_2 \end{bmatrix}$$

since u is a scalar in this case. Calculating the control and measurement quantities from Equation (6.126) yields

$$S_m^T B_f K_p S_m = 0$$

$$S_m^T B_f K_v S_m = \frac{g_1 g_2}{6}\begin{bmatrix} 1 & -1 \\ 1 & -1 \end{bmatrix}$$

It should be clear from this last expression that no choice of g_1 and g_2 will allow just the first mode damping to be changed, i.e. $S_m^T B_f K_v S_n$ cannot be diagonal.

For the second case, consider using two actuators, one at each mass, and two sensors measuring the two velocities. This is an MIMO system with velocity feedback. Then

$$K_v = \begin{bmatrix} g_1 & 0 \\ 0 & g_2 \end{bmatrix}$$

and

$$B_f = \begin{bmatrix} g_3 & 0 \\ 0 & g_4 \end{bmatrix}$$

Equation (6.126) then yields

$$S_m^T B_f K_v S_m = \frac{1}{18} \begin{bmatrix} g_1 g_3 + 9g_2 g_4 & 9g_2 g_4 - g_1 g_3 \\ 9g_2 g_4 - g_1 g_3 & g_1 g_3 + 9g_2 g_4 \end{bmatrix}$$

An obvious choice for decoupled control is $9g_2g_4 = g_1g_3$, as this makes the above matrix diagonal. Then

$$S_m^T B_f K_v S_m = \begin{bmatrix} g_2 g_4 & 0 \\ 0 & g_2 g_4 \end{bmatrix}$$

Comparing the desired first mode damping ratio yields

$$2\zeta_1\omega_1 + g_2g_4 = 2(0.4)\sqrt{2}$$

Solving this equation for the two unknowns $a = g_2g_4$ yields $g_2g_4 = 0.164$. This choice of the gain keeps the equations of motion decoupled and assigns the first damping ration the desired value of 0.4. Unfortunately, because of the way the gains appear in the closed-loop system, the effect on the second mode velocity coefficient is

$$2\zeta_2\omega_2 + g_2g_4 = 2\hat{\zeta}_2\omega_2$$

This results in a new damping ratio for mode 2 of $\hat{\zeta}_2 = 0.449$. Hence, although a decoupled control has been found, it is still not possible to independently change the damping ratio of mode 1 without affecting mode 2. This would require $g_1g_3 + 9g_2g_4$ to be zero, which of course would also not allow ζ_1 to be controlled at all.

For the third case then, consider an MIMO controller with the velocity signal at m_1 to be fed back to mass m_2 and vice versa. This means that K_v now has the form

$$K_v = \begin{bmatrix} g_1 & g_5 \\ g_6 & g_2 \end{bmatrix}$$

In this case, calculating the velocity feedback coefficient yields

$$S_m^T B_f K_v S_m = \frac{1}{18} \begin{bmatrix} g_3g_1 + 3g_3g_5 + 9g_2g_4 + 3g_4g_6 & 9g_2g_4 + 3g_3g_5 - 3g_4g_6 - g_3g_1 \\ 9g_2g_4 - 3g_3g_5 + 3g_4g_6 - g_3g_1 & 9g_2g_4 + g_3g_1 - 3g_4g_6 - 3g_3g_5 \end{bmatrix}$$

Examining this new feedback matrix shows that *independent* modal control will be possible if and only if the off diagonals are set to zero, decoupling the system, and the element in the (2, 2) position is zero leaving the second mode unchanged. To change the first mode-damping ratio to 0.4, the first entry in the matrix above requires that

$$2\zeta_1\omega_1 + \frac{g_1g_3 + 3g_3g_5 + 9g_2g_4 + 3g_4g_6}{18} = 2(0.4)\omega_1$$

To insure that the controller does not re-couple the equations of motion, the off diagonal elements are set to zero, resulting in the two equations

$$9g_2g_4 + 3g_3g_5 - 3g_4g_6 - g_3g_1 = 0 \quad \text{and} \quad 9g_2g_4 - 3g_3g_5 + 3g_4g_6 - g_3g_1 = 0$$

To insure that the second mode damping is not changed, the element in the (2, 2) position is also set to zero, resulting in the additional condition

$$9g_2g_4 + g_3g_1 - 3g_4g_6 - 3g_3g_5 = 0$$

This last set of equations can be recognized as four linear equations in the four unknowns

$$a = g_1g_3, \quad b = g_2g_4, \quad c = g_3g_5, \quad d = g_4g_6$$

These equations in matrix form are

$$-a + 9b + 3c + 3d = 0$$
$$-a + 9b - 3c + 3d = 0$$
$$a + 9b - 3c - 3d = 0$$
$$\frac{a}{18} + \frac{1}{2}b + \frac{1}{6}c + \frac{1}{6}d = 2(0.4)\omega_1 - 2\zeta_1\omega_1$$

Solving this set of equations with the open-loop values of ω_1 and ζ_1 given above yields

$$a = 2.09 \quad b = 0.232 \quad c = 0.697 \quad d = 0.697$$

Substitution of these values along with the free choice of $g_3 = g_4 = 1$ into the feedback matrix yields

$$S_M^T B_f K_v S_M = \begin{bmatrix} 0.464 & 0 \\ 0 & 0 \end{bmatrix}$$

The resulting closed-loop system yields a new system with the desired first mode damping ratio of 0.4 and the second mode damping ratio unchanged. Furthermore, the closed-loop system remains decoupled. Hence, independent mode control is achieved.

Note that of the six gain values, only four can be determined. In fact, B_f could have been chosen as the identity matrix from the start with the same result. However, in practice, each sensor and actuator will have a coefficient that needs to be accounted for. In many cases, each sensor and actuator will have significant dynamics, ignored here, that could change the plant by adding additional poles and zeros.

Example 6.14.1 illustrates how many control actuators and sensors must be used in order to accomplish an independent control of one mode of a simple two degrees of freedom system. In the case of the last example, as many sensors and actuators were required as degrees of freedom in the system. Hence, it is important in practical control design to consider the placement of actuators and sensors and not just the relationships among the various coefficient matrices for the system.

6.15 Robustness

The concept of robust control systems, or robust systems, has been defined in many ways, not all of which are consistent. However, the basic idea behind the concept of robustness is an attempt to measure just how stable a given system is in the presence of some uncertainty or perturbation in the system. That is, if a system is stable, is it still stable after some changes have been made in the physical or control parameters of the system? This is called *stability robustness*. This same question can be asked with respect to the performance of the system. In this latter case, the question is asked in terms of a given level of acceptability of a specific performance criteria such as overshoot or settling time. For example, if it is required that a given control system has overshot by less than 10% and there is an uncertainty in the control gain, does the overshoot still remain at less than 10% in the presence of that uncertainty? This is called *performance robustness*.

An example of performance robustness is given by Davison (1976). The steady-state error of a control system is defined to be the difference between the response of the system and the desired response of the system as t gets larger in the regulator problem of Section 6.9. A logical measure of control system performance is then whether or not the steady-state error is zero. A given system is then said to be *robust* if there exists a control that regulates the system with zero steady-state error when subjected to perturbations in any of the matrices A, B or C in Equations (6.30) and (6.31).

In the remainder of this section, only stability robustness is discussed. The approach presented here follows that of Patel and Toda (1980). As an example of stability robustness, consider a closed-loop system under state variable feedback of the form

$$\dot{\mathbf{x}} = (A + BK_f)\mathbf{x} = A'\mathbf{x} \tag{6.131}$$

The matrix A' contains the measurement and control matrices and is such that the closed-loop system is asymptotically stable. Let the matrix E_e denote the uncertainty in the system parameters (A) and gains (BK_f). Then, rather than having Equation (6.131), the system may be of the form

$$\dot{\mathbf{x}} = (A' + E_e)\mathbf{x} \tag{6.132}$$

In this equation, the matrix E_e is not known (recall the discussion following Equation (3.93)), but rather only bounds on its elements are known. In particular, it is assumed here that each element of the uncertainty is bounded in absolute value by the same number, ε, so that

$$\left|(E_e)_{ij}\right| < \varepsilon \tag{6.133}$$

for each value of the indices i and j. It was shown by Patel and Toda (1980) that the system with the uncertainty given by Equation (6.133) is asymptotically stable if

$$\varepsilon < \mu_\varepsilon$$

where

$$\mu_\varepsilon = \frac{1}{2n}\left(\frac{1}{\sigma_{max}[F]}\right) \tag{6.134}$$

Here, F is a solution of the Lyapunov matrix Equation (4.29) in the form

$$A'^T F + FA' = -2I \tag{6.135}$$

where I is the $2n \times 2n$ identity matrix. The notation $\sigma_{\max}[F]$ refers to the largest singular value of the matrix F.

Example 6.15.1

Consider a closed-loop system defined by the augmented state matrix given by

$$A' = \begin{bmatrix} 0 & 0 & 1 & 0 \\ 0 & 0 & 0 & 1 \\ -3 & 1 & -9 & 4 \\ 1 & -1 & 4 & -4 \end{bmatrix}$$

the solution of the Lyapunov Equation (6.135) yields

$$F = \begin{bmatrix} 3.5495 & -0.6593 & 0.3626 & 0.0879 \\ 0.6593 & 4.6593 & 0.6374 & 1.6374 \\ 0.3626 & 0.6374 & 0.3956 & 0.5495 \\ 0.0879 & 1.6374 & 0.5495 & 1.2088 \end{bmatrix}$$

The singular values of F are calculated as

$$\sigma_1 = 5.5728, \ \sigma_2 = 3.5202, \ \sigma_3 = 0.6288, \ \sigma_4 = 0.0914$$

From Equation (6.134), the value of μ_ε becomes ($n = 2$)

$$\mu_\varepsilon = \left(\frac{1}{4}\right)\left(\frac{1}{5.5728}\right)$$

Hence, as long as the parameters are not changed by more than 1.3676, i.e. as long as

$$\left|(E_e)_{ij}\right| < 1.3676$$

the system defined by $(A' + E_e)$ will be asymptotically stable.

Many other formulations and indices can be used to discuss stability robustness. For instance, Rew and Junkins (1986) use the sensitivity (Section 6.5) of the closed-loop matrix A' in Equation (6.131) to discuss robustness in terms of the condition number of A'. Kissel and Hegg (1986) have discussed stability robustness with respect to neglected dynamics in control system design. They examine how stability of a closed-loop system is affected if an ROM is used in designing the feedback law. Zhou and Doyle (1997) provide a complete account of the use of robustness principles in control design for both stability and performance.

6.16 Positive Position Feedback Control

A popular modal control method with experimentalists and structural control engineers is called *positive position feedback* (PPF) (Goh and Caughy, 1985), which adds additional dynamics to the system through the control law. A unique feature of the PPF approach is that it can be designed around an experimental transfer function of the structure and does not require an analytical model of the system or plant to be controlled. Goh and Caughy proposed using a special dynamic feedback law designed specifically for use in the second-order form, compatible with Newton's formulation of the equations of motion of a structure. These PPF control circuits are designed to roll off at higher frequency and hence are able to avoid exciting residual modes and introducing spillover, as discussed in Section 6.13.

To illustrate the PPF formulation, consider the SDOF system (or alternately, a single mode of the system)

$$\ddot{x} + 2\zeta\,\omega_n\,\dot{x} + \omega_n^2 x = b\,u \tag{6.136}$$

where ζ and ω_n are the damping ratio and natural frequency of the structure, and b is the input coefficient that determines the level of force applied to the mode of interest. The PPF control is implemented using an auxiliary dynamic system (compensator) defined by

$$\begin{aligned} \ddot{\eta} + 2\zeta_f\,\omega_f\,\dot{\eta} + \omega_f^2\,\eta &= g\omega_f^2 x \\ u &= \frac{g}{b}\,\omega_f^2\,\eta \end{aligned} \tag{6.137}$$

Here ζ_f and ω_f are the damping ratio and natural frequency of the controller and g is a constant. The particular form of Equation (6.137) is that of a second-order system much like a damped vibration absorber. The idea is to choose the PPF frequency and damping ratio so that the response of the structural mode has the desired damping. Combining Equations (6.136) and (6.137) gives the equations of motion in their usual second-order form, which are, assuming no external force

$$\begin{bmatrix} \ddot{x} \\ \ddot{\eta} \end{bmatrix} + \begin{bmatrix} 2\zeta\omega_n & 0 \\ 0 & 2\zeta_f\omega_f \end{bmatrix} \begin{bmatrix} \dot{x} \\ \dot{\eta} \end{bmatrix} + \begin{bmatrix} \omega_n^2 & -g\omega_f^2 \\ -g\omega_f^2 & \omega_f^2 \end{bmatrix} \begin{bmatrix} x \\ \eta \end{bmatrix} = \begin{bmatrix} 0 \\ 0 \end{bmatrix} \tag{6.138}$$

Since the stiffness matrix couples the two coordinates, increasing the filter damping, ζ_f, will effectively add damping to the structural mode. Note also that this is a stable closed-loop system if the symmetric "stiffness" matrix is positive definite for appropriate choices of g and ω_f. That is, if the determinant of displacement coefficient matrix is positive, which happens if

$$g^2\omega_f^2 < \omega_n^2 \tag{6.139}$$

Notice that the stability condition only depends on the natural frequency of the structure, and not on the damping or mode shapes. This is significant in practice because when building an experiment, the frequencies of the structure are usually available with a reasonable accuracy, while mode shapes and damping ratios are much less reliable. The design of the controller then consists of choosing g and ω_f that satisfies Equation (6.139) and choosing ζ_f large enough to add significant damping to the structural mode. Note that the gains of the controller, ζ_f, g and ω_f, are chosen electronically.

The stability property of PPF is also important, because it can be applied to an entire structure eliminating spillover by rolling off at higher frequencies. That is, the frequency response of the PPF controller has the characteristics of a low pass filter. The transfer function of the controller is

$$\frac{\eta(s)}{X(s)} = \frac{g\omega_f^2}{s^2 + 2\zeta_f\omega_f s + \omega_f^2} \tag{6.140}$$

illustrating that it rolls off quickly at high frequencies. Thus the approach is well suited to controlling a mode of a structure with frequencies that are well separated, as the controller is insensitive to the unmodeled high frequency dynamics. If the problem is cast in the state space, the term $\mathbf{b}_{2n\text{-}k}$ in Equation (6.120) is zero, and no spillover results.

The positive position terminology in the term *positive position feedback* comes from the fact that the position coordinate of the structure equation is positively fed to the filter, and the position coordinate of the compensator equation is positively fed back to the structure.

Next, suppose a multi-degree of freedom analytical model of the structure is available. Following the formulation of output feedback discussed in Section 4.10 for an SISO system with no applied force yields

$$M\ddot{\mathbf{q}} + C\dot{\mathbf{q}} + K\mathbf{q} = B_f u \tag{6.141}$$

Coupling this with the PPF controller in the form given in Equation (6.137) written as

$$\begin{aligned} \ddot{\eta} + 2\zeta_f\,\omega_f\,\dot{\eta} + \omega_f^2\,\eta &= g\omega_f^2\,B_f^T\,\mathbf{q} \\ u &= g\,\omega_f^2\,\eta \end{aligned} \tag{6.142}$$

yields

$$\begin{bmatrix} M & 0 \\ 0 & 1 \end{bmatrix}\begin{bmatrix} \ddot{\mathbf{q}} \\ \ddot{\eta} \end{bmatrix} + \begin{bmatrix} C & 0 \\ 0 & 2\zeta_f\omega_f \end{bmatrix}\begin{bmatrix} \dot{\mathbf{q}} \\ \dot{\eta} \end{bmatrix} + \begin{bmatrix} K & -g\omega_f B_f \\ -g\omega_f B_f^T & \omega_f^2 \end{bmatrix}\begin{bmatrix} \mathbf{q} \\ \eta \end{bmatrix} = \begin{bmatrix} 0 \\ 0 \end{bmatrix} \tag{6.143}$$

The system is SISO so that B_f is a vector since u is a scalar. The augmented mass matrix in Equation (6.143) is symmetric and positive definite, the augmented damping matrix is symmetric and positive semi-definite, so the closed-loop stability will depend on the definiteness of the augmented stiffness matrix.

Consider then the definiteness of the augmented stiffness matrix defined by

$$\hat{K} = \begin{bmatrix} K & -g\omega_f B_f \\ -g\omega_f B_f^T & \omega_f^2 \end{bmatrix} \tag{6.144}$$

Let $\mathbf{x}$ be an arbitrary vector partitioned according to $\hat{K}$ and compute

$$\mathbf{x}^T\hat{K}\mathbf{x} = \begin{bmatrix}\mathbf{x}_1^T & \mathbf{x}_2^T\end{bmatrix}\begin{bmatrix} K & -g\omega_f B_f \\ -g\omega_f B_f^T & \omega_f^2 \end{bmatrix}\begin{bmatrix}\mathbf{x}_1^T \\ \mathbf{x}_2^T\end{bmatrix}$$
$$= \mathbf{x}_1^T K\mathbf{x}_1 - g\omega_f\mathbf{x}_1^T B_f\mathbf{x}_2 - g\omega_f\mathbf{x}_2^T B_f^T\mathbf{x}_1 + \omega_f^2\mathbf{x}_2^T\mathbf{x}_2$$

Completing the square and factoring yields

$$\mathbf{x}^T\hat{K}\mathbf{x} = \mathbf{x}_1^T\left(K - g^2B_fB_f^T\right)\mathbf{x}_1 + \left(gB_f^T\mathbf{x}_1 - \omega_f\mathbf{x}_2\right)^T\left(gB_f^T\mathbf{x}_1 - \omega_f\mathbf{x}_2\right)$$

This is of the form

$$\mathbf{x}^T\hat{K}\mathbf{x} = \mathbf{x}_1^T\left(K - g^2B_fB_f^T\right)\mathbf{x}_1 + \mathbf{y}^T\mathbf{y}$$

for arbitrary $\mathbf{x}_1$ and $\mathbf{y}$. Since $\mathbf{y}^T\mathbf{y}$ is always non-negative, $\hat{K}$ will be positive definite if the matrix $K - g^2B_fB_f^T$ is positive definite. If g is chosen, such that $K - g^2B_fB_f^T$ is positive definite, then the closed-loop system will be stable (in fact, asymptotically stable via the discussion in Section 4.5, since the coefficient matrices do not commute).

Example 6.16.1

Consider the two degrees of freedom system of Example 6.14.1 and design a PPF controller to add damping to the first mode of the system without affecting the second mode.

For a single actuator at the location of the first mass, the input matrix becomes

$$B_f = \begin{bmatrix}1\\0\end{bmatrix}$$

The augmented mass and stiffness matrices of Equation (6.143) become

$$\hat{M} = \begin{bmatrix}9 & 0 & 0\\0 & 1 & 0\\0 & 0 & 1\end{bmatrix},\quad \hat{C} = \begin{bmatrix}9 & -1 & 0\\-1 & 1 & 0\\0 & 0 & 2\zeta_f\omega_f\end{bmatrix},\quad \hat{K} = \begin{bmatrix}27 & -3 & -g\omega_f\\-3 & 3 & 0\\-g\omega_f & 0 & \omega_f^2\end{bmatrix}$$

Following the constraint given by inequality in Equation (6.139) for controlling the first mode $g^2\omega_f^2 < \omega_1^2 = 2$, one free choice is $g = \omega_f = 1$. Choosing the PPF damping of $\zeta_f = 0.5$ results in the following closed-loop damping ratios and frequencies

$$\omega_1 = 1.43 \quad \zeta_1 = 0.237$$
$$\omega_2 = 2.00 \quad \zeta_2 = 0.332$$
$$\omega_f = 0.966 \quad \zeta_f = 0.531$$

as computed from the corresponding state matrix. Note that damping is added to the first mode (from 0.235 to 0.237), while mode two damping is only slightly changed and the frequency is not changed at all. All the modes change slightly, because the system is coupled by the filter and, as was shown in Example 6.14.1, multiple sensors and actuators are required to affect only one mode.

6.17 MATLAB Commands for Control Calculations

Most of the calculations in this chapter are easily made in MATLAB. MATLAB contains a "toolbox" just for controls, called the Control System Toolbox (Grace et al., 1992). This is a series of algorithms expressed in M-files, which implements common control design, analysis and modeling methods. Many websites are devoted to using and understanding the control system toolbox. Table 6.1 lists some common commands useful in implementing calculations for active control. The control commands in MATLAB assume the control problem is a linear time invariant system of the form:

$$\dot{\mathbf{x}} = A\mathbf{x} + B\mathbf{u}, \quad \mathbf{y} = C_c\mathbf{x} + D\mathbf{u} \tag{6.145}$$

The developments in this chapter assume $D = 0$.

The follow examples illustrate the use of some of the commands listed in Table 6.1, to perform some of the computations developed in the previous sections.

Table 6.1 Matlab commands for control.

`ss2tf`	converts the state space model to the system transfer function
`tf2ss`	converts a system transfer function to a state space model
`Step`	computes the step response of a system
`Initial`	computes the response of a system to initial conditions
`Impulse`	computes the response of a system to a unit impulse
`Ctrb`	computes the controllability matrix
`Obsv`	computes the observability matrix
`gram`	computes the controllability and observability grammians
`Balreal`	computes the balanced realization
`Place`	computes the pole placement gain matrix
`Lqr`	computes the linear quadratic regulator solution
`Modred`	computes the ROM

Example 6.17.1

The function `initial` can be used to plot the response of the system of Equation (6.145) to a given initial condition. Consider the system in state space form of an SDOF system given by

$$\begin{bmatrix} \dot{x}_1 \\ \dot{x}_2 \end{bmatrix} = \begin{bmatrix} 0 & 1 \\ -4 & -2 \end{bmatrix} \begin{bmatrix} x_1 \\ x_2 \end{bmatrix} + \begin{bmatrix} 1 \\ 0 \end{bmatrix} u, \; y = \begin{bmatrix} 1 & 0 \end{bmatrix} \begin{bmatrix} x_1 \\ x_2 \end{bmatrix}, \; \begin{bmatrix} x_1(0) \\ x_2(0) \end{bmatrix} = \begin{bmatrix} 1 \\ 0 \end{bmatrix}$$

and plot the response using `initial`. Type the following in the command window:

```
>>clear all
>>A=[0 1;-4 -2];
```

```
>>b=[1;0];
>>c=[1 0];
>>d=[0];
>>x0=[1 0];
>>t=0:0.1:6;
>>initial(A,b,c,d,x0,t);
```

This results in the plot given in Figure 6.21.

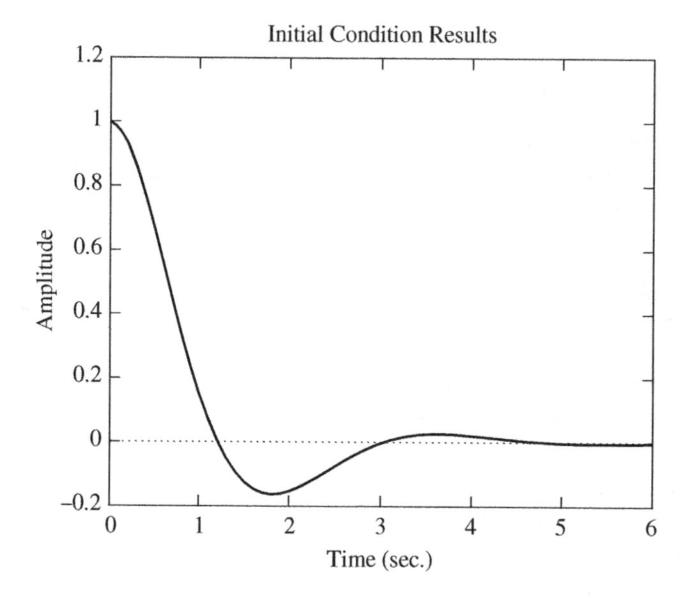

Figure 6.21 The response computed and plotted using the `initial` command.

Next consider using MATLAB to solve the pole placement problem. MATLAB again works with the state space model of Equation (6.145) and uses full-state feedback ($\mathbf{u} = -K\mathbf{x}$) to find the gain matrix K that causes the closed-loop system ($A - \mathbf{b}K$) to have the poles specified in the vector `p`. The following example illustrates the use of the `place` command.

Example 6.17.2

Use the system of Example 6.17.1 and compute the gain that causes the closed-loop system to have two real poles: –2 and –4. Use the code in Figure 6.21 to enter the state space model (`A, b, c, d`), then type the following in the command window:

```
>>p=[ -2 -4];
>>K=place(A,b,p)
place: ndigits= 15
K =
   4   1
```

Thus, the proper gain matrix is a vector in this case. To check that the result works, type:

```
>>eig(A-b*K)
ans =
  -2
  -4
>>eig(A)
ans =
 -1.0000+ 1.7321i
 -1.0000- 1.7321i
```

This shows that the computed gain matrix causes the system to move its poles from the two complex values $-1 \pm 1.7321j$ to the two real values -2 and -4.

Next consider solving the optimal control problem using the linear quadratic regulator problem defined by Equation (6.57). The MATLAB command `lqr` computes the solution of the matrix Ricatta Equation (6.61), calculates the gain matrix K and then the eigenvalues of the resulting closed-loop system (A-BK) for a given choice of the weighting matrices Q and R defined in Section 6.9. In this case, the output matrix B must have the same number of columns as the matrix R. The following example illustrates the procedure.

Example 6.17.3

Compute an optimal control for the system of Example 6.17.1. Here we chose Q to be the identity matrix and R to be $2Q$. The following is typed into the command window:

```
>>A=[0 1;-4 -2];
>>B=eye(2);% the 2x2 identity matrix
>>c=[1 0];
>>d=[0];
>>Q=eye(2);R= 2*Q;% weighting matrices in the cost function
>>[K,S,E]=lqr(A,B,Q,R)
K =
   0.5791   0.0205
   0.0205   0.1309
S =
   1.1582   0.0410
   0.0410   0.2617
E =
  -1.3550+ 1.8265i
  -1.3550- 1.8265i
```

The vector `E` contains new (closed-loop) eigenvalues of the system with the gain computed (`K`) from the solution of the Ricatta equation (`S`). Note that the closed-loop eigenvalues are more heavily damped then the open-loop eigenvalues ($-1 \pm 1.7321j$).

Next consider the model reduction problem of Section 6.12. The balanced realization can be obtained from the command `balreal` and the corresponding ROM can be obtained from the `modred` command. The procedure is illustrated in the next example.

Example 6.17.4
Compute the balanced realization of the system

$$\dot{\mathbf{x}} = \begin{bmatrix} 0 & 0 & 1 & 0 \\ 0 & 0 & 0 & 1 \\ -2 & 1 & -0.3 & 0.1 \\ 1 & -1 & 0.1 & -0.1 \end{bmatrix} \mathbf{x} + \begin{bmatrix} 0 \\ 0 \\ 0 \\ 1 \end{bmatrix} u, \; y = \begin{bmatrix} 0 & 1 & 0 & 0 \end{bmatrix} \mathbf{x}$$

Use the balance realization to reduce the order of the model based on its singular values. First, the state space model is typed into the command window. Then a balanced realization is formed and the singular values computed. The singular values (denoted `g` in the code that follows) are then used with the model reduction routine to compute a ROM. Type the following into the command window:

```
>>A=[0 0 1 0;0 0 0 1;-2 1 -0.3 0.1;1 -1 0.1 -0.1];
>>b=[0;0;0;1]; c=[0 1 0 0]; d=[0];
>>[ab,bb,cb,g]=balreal(A,b,c) % computes the balanced version
and singular values
ab =
   -0.0326   0.6166   0.0192   -0.0275
   -0.6166   -0.0334   -0.0218   0.0280
   0.0192   0.0218   -0.1030   1.6102
   0.0275   0.0280   -1.6102   -0.2309
bb =
   0.7821
   0.7510
   -0.2369
   -0.3223
cb =
   0.7821   -0.7510   -0.2369   0.3223
g =
   9.3836
   8.4310
   0.2724
   0.2250
>>elim=find(g<1); %eliminate states with singular values
less then 1
>>[ar,br,cr]=modred(ab,bb,cb,d,elim) % com-
pute the reduced order
model based on eliminating the states with g<1
```

```
ar =
   -0.0319   0.6173
   -0.6173  -0.0342
br =
   0.7743
   0.7595
cr =
   0.7743  -0.7595
```

The matrix `ar` and the vectors `br` and `cr` are the reduced order state matrix, input and output vectors and form the ROM as defined in Equations (6.105) and (6.106).

There are other commands in the Control System Toolbox that can be used to simplify the calculation of control laws, models and to perform analysis of control systems. Texts in control and the web should be consulted for additional samples and commands.

Chapter Notes

A vast amount of literature is available on methods of vibration isolation and absorption. In particular, the books by Balandin et al. (2001), Rivin (2003) and Korenev and Reznikov (1993) should be consulted to augment the information in Section 6.2. The absorber optimization problem discussed in Section 6.3 is directly from the paper of Soom and Lee (1983). Haug and Arora (1976) provide an excellent account of optimal design methods. More on the use of damping materials can be found in the book by Nashif et al. (1985). Section 6.4 on metastructures is largely from the papers of Reichl and Inman (2016) and motivated by Sun et al. (2010) and Baravelli and Ruzzene (2013). The material in Section 6.5 comes from Whitesell (1980), which was motivated by the work of Fox and Kapoor (1968). More advanced approaches to eigensystem derivatives can be found in Adhikari and Friswell (2001). Next, some control concepts useful in vibration analysis and design are presented, starting in Section 6.6. Linear control engineering is, of course, a topic in its own right, and hence many texts have been written on the subject. The reader is referred to the excellent text by Kuo and Gholoaraghi (2003) for an introduction to control topics and Kailath (1980) for a more advanced treatment of linear systems and control topics. Section 6.7 on controllability and observability stresses the use of these concepts in physical coordinate systems. However, literally hundreds of papers have been written on the topic of the state space formulation. In addition to those references listed in the text, useful conditions for controllability and observability in second-order models are given by Laub and Arnolds (1984) and Bender and Laub (1985). The early important papers on linear control and computation are presented in Patel et al. (1994).

Section 6.8 on eigenstructure assignment extends the concept of pole placement to include placing eigenvectors as well as eigenvalues. Again, hundreds of papers appear in the literature on pole placement methods in the state space. The most comprehensive treatment of eigenstructure assignment with structures in mind is that by Andry et al. (1983), which motivated the presentation in Section 6.8. An excellent text on eigenstructure assignment including MATLAB codes is the one by Liu and Patton (1998).

Section 6.9 on optimal control should be viewed by the reader as a very brief introduction. Again, optimal control is a topic that should be studied in detail by using one of the many texts written on the subject (Kirk. 1970). In addition, optimal control is an intense area of current research, with several journals devoted to the topic. Optimal control is an extremely useful tool in controlling unwanted vibration. As pointed out by Example 6.9.1, however, an element of art remains in optimal control by virtue of the freedom to choose a cost function. Section 6.10 on observers and estimators follows the development given by Chen (1970, 1998) and Kailath (1980). Section 6.11 on realization also closely follows the development in Chen (1998). The purpose of introducing realization is to set the stage for its use in vibration testing in Section 12.6. This illustrates a recent application of a standard control topic (realization) to a vibration problem (testing) and hence should motivate serious vibration engineers to pay more attention to the topic of linear systems and control theory as a source of alternative solutions.

Section 6.12 presents model reduction in the state space. The balancing method presented is improved numerically by Laub (1980). This is again an important area of vibration research and design that benefits from developments that have taken place in linear system theory. There are many other approaches to model reduction, notably the work of Hyland and Bernstein (1985) and Skelton et al. (1982). A recent review of projection-based methods is provided by Benner et al. (2015).

Section 6.14 on modal control in the state space follows the development given by Takahashi et al. (1970). Modal control is popular and appears as a tool in a large number of papers. The section on modal control in physical coordinates is an attempt to extend the idea of modal coordinates to second-order systems. This is again a popular method in vibration control problems (Meirovitch and Baruh, 1982). The section on robustness is intended to provide a brief introduction to a very popular topic in control. The development follows that used by Ridgely and Banda (1986). Note the similarity between robustness and design sensitivity discussed in Section 6.5. The large amount of activity surrounding the topic of robust control and space prevented the inclusion of much material in this area. Instead, the reader is referred to Zhou and Doyle (1997). The material on PPF control in Section 6.16 comes from the article by Friswell and Inman (1999), which contains additional references to PPF formulations in first-order form and connects PPF to standard output feedback and optimal control laws.

References

Adhikari, S. and Friswell, M. I. (2001) Eigenderivative analysis of asymmetric non-conservative systems. *International Journal for Numerical Methods in Engineering*, 51(6), 709–733.

Ahmadian, M. (1985) Controllability and observability of general lumped parameter systems. *AIAA Journal of Guidance, Control and Dynamics*, 8(5), 669–671.

Anderson, B. D. and Moore, J. B. (1979) *Optimal Filtering*. Prentice Hall, Englewood Cliffs, NJ.

Andry, A. N., Jr., Shapiro, E. Y. and Chung, J. C. (1983) Eigenstructure assignment for linear systems. *IEEE Transactions on Aerospace and Electronic Systems*, AES-19(5), 713–729.

Antoulas, A. C. (2005) *Approximation of Large Scale Dynamical Systems*. SIAM, Philadelphia.

Balas, M. J. (1978) Feedback control of flexible structures. *IEEE Transactions on Automatic Control*, AC-23(4), 673–679.

Balandin, D. V., Bolotnik, N. N. and Pilkey, W. D. (2001) *Optimal Protection from Impact, Shock and Vibration*. Gordon & Breach Science Publishers, Amsterdam.

Baravelli, E. and Ruzzene M. (2013) Internally resonating lattices for bandgap generation and low-frequency vibration control. *Journal of Sound and Vibration*, 332, 6562–6579.

Bender, D. J. and Laub, A. J. (1985) Controllability and observability at infinity of multivariable linear second-order models. *IEEE Transaction on Automatic Control*, AC-30(12), 1234–1237.

Benner, P., Gugercin, S. and Willcox, K. (2015) A survey of projection-based model reduction methods for parametric dynamical systems. *SIAM Review*, 57(4), 483–531.

Chen, C.T. (1970) *Introduction to Linear System Theory*. Holt, Rinehart & Winston, New York.

Chen, C. T. (1998) *Linear System Theory and Design*, 3rd Edition. Oxford University Press, Oxford, UK.

Davison, E. J. (1976) Multivariable tuning regulators: the feedforward and robust control of a general servomechanism problem. *IEEE Transactions on Automatic Control*, AC-21, 35–47.

Friswell, M. I. and Inman, D. J. (1999) The relationship between positive position feedback and output feedback controllers. *Smart Materials and Structures*, 8, 285–291.

Fox, R. I. and Kapoor, M. B. H. (1968) Rates of change of eigenvalues and eigenvectors. *AIAA Journal*, 6, 2426–2429.

Fox, R. I. (1971) *Optimization Methods for Engineering Design*. Addison-Wesley, Reading, MA.

Gill, P. E., Murray, W. and Wright, M. (1981) *Practical Optimization*. Academic Press, Orlando, FL.

Goh, C. J. and Caughey, T. K. (1985) On the stability problem caused by finite actuator dynamics in the control of large space Structures. *International Journal of Control*, 41, 787–802.

Golub, G. H. and Van Loan, C. F. (1996) *Matrix Computations*, 3rd Edition. John Hopkins University Press, Baltimore, MD.

Grace, A., Laub, A. J., Little, J. N. and Thompson, C. M. (1992) *Control System Toolbox*. The Mathworks, Inc., Natick, NY.

Haug, E. J. and Arora, J. S. (1976.) *Applied Optimal Design*. John Wiley & Sons, New York.

Ho, B. L. and Kalman, R. E. (1965) Effective construction of linear state variable models from input/output data. *Proceedings of the Third Allerton Conference*, pp. 449–459.

Hyland, D. C. and Bernstein, D. S. (1985) The optimal projection equations for model reduction and the relationship among the methods of Wilson, Skelton and Moore. *IEEE Transactions on Automatic Control*, AC-30(12), 1201–1211.

Hughes, P. C. and Skelton, R. E. (1980) Controllability and observability of linear matrix second-order systems. *ASME Journal of Applied Mechanics*, 47, 415–420.

Kailath, T. (1980) *Linear Systems*. Prentice Hall, Englewood Cliffs, NJ.

Kirk, D. E. (1970) *Optimal Control Theory – An Introduction*. Prentice Hall, Englewood Cliffs, NJ.

Kissel, G. J. and Hegg, D. R. (1986) Stability enhancement for control of flexible space structures. *IEEE Control Systems Magazine*, 6(3), 19–26.

Kuo, B. C. and Gholoaraghi, F. (2003) *Automatic Control Systems*, 8th Edition. John Wiley & Sons, New York.

Korenev, B. G. and Reznikov, L. M. (1993) *Dynamic Vibration Absorbers: Theory and applications*. John Wiley & Sons, New York.

Laub, A. J. (1980) Computation of balancing transformation. *Proceedings of Joint Automatic Control Conference*. No. FA8-E.

Laub, A. J. and Arnold, W. F. (1984) Controllability and observability criteria for multivariable linear second-order models. *IEEE Transactions on Automatic Control*. AC-29(2), 163–165.

Liu, G. P. and Patton, E. J. (1998) *Eigenstructure Assignment for Control System Design*. John Wiley & Sons, Chichester, UK.

Meirovitch, L. and Baruh, H. (1982) Control of self adjoint distributed parameter systems. *AIAA Journal of Guidance Control and Dynamics*, 15(2), 60–66.

Moler, C., Little, J., Banger, S. and Kleiman, S. (1985) *PC-MATLAB User's Guide*. The Math Works, Inc., Portola Valley, CA.

Moore, B. C. (1976) On the flexibility offered by state feedback in multivariable systems beyond closed-loop eigenvalue assignment. *IEEE Transactions on Automatic Control*, AC-21, 689–692.

Moore, B. C. (1981) Principal component analysis in linear systems: controllability, observability and model reduction. *IEEE Transactions on Automatic Control*, AC-26(1), 17–32.

Patel, R. V., Laub, A. J. and Van Dooren, P. M. (eds) (1994) *Numerical Linear Algebra Techniques for Systems and Control*. Institute of Electrical and Electronic Engineers, Inc., New York.

Patel, R. V. and Toda, M. (1980) Quantative measure of robustness for multivariable systems. *Proceedings of the Joint Automatic Control Conference*, TD8-A.

Rivin, E. I.. (2003) *Passive Vibration Isolation*. ASME Press, New York.

Reichl, K. K. and Inman, D. J. (2016) Finite element modelling of longitudinal metastructures for passive vibration suppression. *Proceedings AIAA SciTech 2016*, 4–8 January, San Diego, CA. Paper No. AIAA-2016-1477.

Rew, D. W. and Junkins, J. L. (1986) Multi-criterion approach to optimization of linear regulators. *Proceedings of the AIAA Guidance, Navigation and Control Conference*.

Ridgely, D. B. and Banda, S. S. (1986) *Introduction to Robust Multivariable Control*. Air Force Wright Aeronautical Laboratories Report AFWAL-TR-85-3102.

Skelton, R. E., Hughes, P. C. and Hablani, H. B. (1982) Order reduction for models of space structures using modal cost analysis. *AIAA Journal of Guidance Control and Dynamics*, 5(4), 351–357.

Soom, A. and Lee, M. L. (1983) Optimal design of linear and nonlinear vibration absorbers for damped systems. *ASME Journal of Vibration, Acoustics, Stress, and Reliability in Design*, 105(1), 112–119.

Sun, H., Du, X. and Pai, P. F. (2010) Theory of metamaterial beams for broadband vibration absorption. *Journal of Intelligent Material Systems and Structures*, 21, 1085–1101.

Takahashi, Y., Thal-Larsen, H., Goldenberg, E., Loscutoff, W. V. and Ragetly, P. R. (1968) Mode oriented design viewpoint for linear lumped-parameter multivariable control systems. *ASME Journal of Basic Engineering*, 90, 222–230.

Takahashi, Y., Rabins, M. J. and Auslander, D. M. (1970) *Control and Dynamic Systems*. Addison-Wesley, Reading, MA.

Vakakis, A. F. and Paipetis, S. A. (1986) The effect of viscously damped dynamical absorber on a linear multi-degree-of-freedom system. *Journal of Sound and Vibration*, 105(1), 49–60.

Whitesell, J. E. (1980) Design sensitivity in dynamical systems. PhD thesis, Michigan State University.

Wonham, W. M. (1967) On pole assignment in multi-input controllable linear systems. *IEEE Transactions on Automatic Control*, AC-12, 660–665.

Zhou, K. and Doyle, J. C. (1997) *Essential of Robust Control*. Prentice Hall, Upper Saddle River, NJ.

Problems

6.1 Calculate the value of the damping ratio required in a vibration isolation design to yield a transmissibility ratio of 0.1, given that the frequency ratio

$$\frac{\omega}{\omega_n}$$

is fixed at 6.

6.2 A single degree of freedom system with a mass of 200 kg is connected to its base by a simple spring. The system is being disturbed harmonically at 2 rad/s. Choose the spring stiffness so that the transmissibility ratio is less then one.

6.3 A spring mass system consisting of a 10 kg mass supported by a 2000 Nm spring is driven harmonically by a force of 20 N at 4 rad/s. Design a vibration absorber for this system and compute the response of the absorber mass.

6.4 Find the minimum and maximum points of the function

$$J(\mathbf{y}) = y_1^3 + 3y_1y_2^2 - 3y_1^2 - 3y_2^2 + 4$$

Which points are actually minimum?

6.5 Calculate the minimum of the cost function

$$J(\mathbf{y}) = y_1 + 2y_2^2 + y_3^2 + y_4^2$$

subject to the equality constraints

$$h_1(\mathbf{y}) = y_1 + 3y_2 - y_3 + y_4 - 2 = 0$$
$$h_2(\mathbf{y}) = 2y_1 + y_2 - y_3 + 2y_4 - 2 = 0$$

6.6 Derive Equations (6.19) and (6.20).

6.7 Derive Equation (6.23). (Hint: First multiply Equation (6.22) by M then differentiate.)

6.8 Consider Example 6.5.1. Calculate the change in the eigenvalues of this system if the mass, m_1, is changed to an unknown amount rather than the stiffness (refer to Example 3.3.2).

6.9 Consider the cost function $J(\mathbf{y})$. The partial derivative J with respect to the elements of the vector $\mathbf{y}$ yield only necessary conditions for a minimum. The second-order condition and sufficient condition is that the matrix of second partial derivatives $[J_{ik}]$ be positive definite. Here, J_{ij} denotes the second partial with respect to y_i and y_k. Apply this second condition to Problem 6.2 and verify this result for that particular example.

6.10 Derive second-order conditions for Example 6.3.1 using a symbolic manipulation program.

6.11 Consider Example 2.4.4 with $M = I$, $c_1 = 2$, $c_2 = 1$, $k_1 = 4$ and $k_2 = 1$. Calculate a control law causing the closed-loop system to have eigenvalues $\lambda_{1,2} = -1 \pm j$ and $\lambda_{3,4} = -2 \pm j$ using the approach in Section 6.6.

6.12 The characteristic equation of a given system is

$$\lambda^3 + 5\lambda^2 + 6\lambda + \eta = 0$$

where η is a design parameter. Calculate the stability margin of this system for $\eta_{op} = 15.1$.

6.13 Consider the vibration absorber designed in Problem 6.3. Use numerical simulation to plot the response of the system to an initial displacement disturbance of 0.01 m of m_1 (zero initial velocity). Discuss your results.

6.14 Consider the system of Example 6.8.1 and determine if the system is controllable with one actuator placed at m_1 and observable with one sensor placed at m_2.

6.15 Recalculate the gain matrix G_f for Example 6.8.1 with one actuator placed at m_2 and one sensor placed at m_1 to measure $q_1(t)$ only. Does the resulting system have the desired eigenstructure?

6.16 Consider the system given by $M = I$

$$C = \text{diag}\begin{bmatrix}0.1 & 0.2\end{bmatrix} \text{ and } K = \text{diag}\begin{bmatrix}1 & 2\end{bmatrix}$$

Is the system controllable for $B = [0 \quad 1]^T$? For $B = [1 \quad 0]^T$?

6.17 Repeat Example 6.8.1 with desired eigenvectors [1 −1] and [1 1], desired eigenvalues $\lambda_1 = 1$ and $\lambda_2 = 2$, and desired mass matrix

$$M_o = \begin{bmatrix} 1 & 0 \\ 0 & 2 \end{bmatrix}$$

Assume the same structures as in Example 6.8.1, and place the sensors and actuators as you see fit.

6.18 Calculate G_f in Example 6.8.1 if the desired eigenvalues are $\lambda_1 = 10$ and $\lambda_2 = 20$, with no specification of the system eigenvectors.

6.19 Show that for $\mathbf{b} = [1 \quad 0 \quad 0 \quad \ldots]^T$ the system of Equation (6.41) can be assigned any eigenvalues by proper choice of the gain g_i.

6.20 Show that the mode participation factors are unique for a given set of initial conditions.

6.21 Show that the control given in Example 6.9.1 for the case $Q = R = I$ satisfies Inequality (6.56) by calculating some other values of $\mathbf{u}$ (i.e. H_f).

6.22 Consider Example 6.10.1. Let the observer poles be placed at −2 and −2 so that the speed of the observer is comparable to that of the system. Calculate $\mathbf{e}(t)$ versus time and $\mathbf{x}(t)$.

6.23 Calculate a realization for the vibration absorber problem in Section 6.2.

6.24 Calculate W_C and W_O for the vibration absorber problem of the damped absorber modeled by

$$\begin{bmatrix} m_1 & \\ & m_a \end{bmatrix} \begin{bmatrix} \ddot{x} \\ \ddot{x}_a \end{bmatrix} + \begin{bmatrix} c_1 + c_a & -c_a \\ -c_a & c_a \end{bmatrix} \begin{bmatrix} \dot{x} \\ \dot{x}_a \end{bmatrix} + \begin{bmatrix} k_1 + k_a & -k_a \\ -k_a & k_a \end{bmatrix} \begin{bmatrix} x_1 \\ x_a \end{bmatrix} = \begin{bmatrix} t \\ 0 \end{bmatrix} \cos \omega t$$

Use $m_a = 1, m_1 = 10, k_1 = 5$ *and* $k_a = 1$.

6.25 Calculate the SVD of the following matrices

$$\begin{bmatrix} 1 & 0 & 1 \\ 2 & 3 & 4 \end{bmatrix}, \begin{bmatrix} 1 & 2 \\ 0 & 3 \end{bmatrix}, \begin{bmatrix} 1 & 6 \\ 5 & -1 \\ 1 & 0 \end{bmatrix}$$

6.26 Consider a two degrees of freedom system similar to that of Figure 2.6. Discuss in physical terms how such a system would have to behave if it could be reduced to an SDOF model very accurately.

6.27 Using available software subroutines (for things like SVD), write a program to perform model reduction using balancing. Then reproduce the reduction illustrated in Example 6.12.1.

6.28 Consider Example 2.5.4. Write the system of Equation (2.35) in state space form. Let $m_1 = 1$, $m_2 = 2$, $c_1 = c_2 = 0.1$, $k_1 = 1$ and $k_2 = 4$. Use the modal control approach of Section 6.13 to design a control (i.e. choose g_1 and g_2) to raise the natural frequencies of the structure above 5 Hz. Add additional controls or measurements as necessary.

6.29 Referring to the statement of Problem 6.28, calculate the minimum number of sensors and actuators required to independently control the lowest frequency. That is, for $k = 2$ in Equation (6.120), what control configuration is required for

$$\mathbf{b}_{2n-k} = \mathbf{c}_{2n-k}^T = \mathbf{0}?$$

6.30 Again, consider the feedback control problem for the example in Figure 2.6. Calculate B_f, K_v and K_p for this system, such that Equation (6.125) is diagonal (decoupled).

6.31 Make a physical interpretation of the calculation in Problem 6.29 (i.e. where are the sensors and actuators located and does it make sense?).

6.32 Prove, or derive, that Equations (6.126) and (6.127) are in fact sufficient conditions for the closed-loop system to decouple.

6.33 Suppose that the commutivity condition assumed in Section 6.14 is not satisfied. Then the coefficient of $\dot{\mathbf{z}}$ in Equation (6.122) is *not* diagonal. Calculate conditions on the matrices B_f and K_v so that the coefficient $\dot{\mathbf{z}}$ in Equation (6.125) becomes diagonal. Such a control decouples a coupled system.

6.34 Based on your answer to Problem 6.33, discuss the physical implications of the control in terms of, say, Figure 2.6.

6.35 Is the condition in Problem 6.33 robust? Please discuss your answer.

6.36 Repeat Example 6.15.1 for the system in Problem 6.28, assuming that the coefficients c_1 and c_2 are only accurate to within 25%.

6.37 Compare the controllability norm of Equation (6.37) with the controllability grammian of Equation (6.87). Is there a mathematical relationship between them?

6.38 Show that the realization of Example 6.11.1 is both controllable and observable. In general, show that Equations (6.85) and (6.86) are controllable and observable.

6.39 Show, using the criteria given in Equation (6.32), that the system in Figure 6.8a is uncontrollable and that the system in Figure 6.8b is controllable.

6.40 Show that Equations (6.85) and (6.86) reduce to Equation (6.82) by using the definition of a transfer function given in Equation (6.80).

6.41 One method of performing a model reduction (Section 6.12) is to represent the open-loop system by the first few modes of the structure. Try this by using the first two (state space) modes of Example 6.12.1 as the ROM of the system. Then compare the response of this ROM to the ROM obtained in Example 6.12.1 by the balancing method.

6.42 Design a PPF controller for the mode $\ddot{r}_i + 0.6\dot{r}_i + 9r_i = 0$ that increases the modal damping ratio to 0.3.

7

Distributed Parameter Models

7.1 Introduction

This chapter presents an informal introduction to the vibrations of systems having distributed mass and stiffness, often called distributed parameter systems. For lumped parameter systems, the single-degree-of-freedom system served as a familiar building block upon which more complicated multiple-degree-of-freedom structures can be modeled. Similarly, an examination of the vibrations of simple models of strings, beams and plates provides a set of "building blocks" for understanding the vibrations of systems with distributed mass, stiffness and damping parameters. Such systems are referred to as distributed parameter systems, elastic systems, continuous systems or flexible systems.

This chapter focuses on the basic methods used to solve the vibration problem of flexible systems. A short introduction to deriving the equations of motion for distributed mass systems is given, but the main part of the chapter is focused on listing the equations governing the linear vibrations of several distributed parameter structures, listing the assumptions under which they are valid, and determining the solutions. For rigorous derivations, the works of Meirovitch (2001) and Magrab (1979) should be consulted. These solution methods are made mathematically rigorous and discussed in more detail in Chapters 8 and 9.

7.2 Equations of Motion

As in the case of lumped mass systems, there are several approaches to deriving the equations that govern the vibrations of distributed mass systems. The main approach is to sum forces and moments. However, since distributed mass systems result in partial differential equations, which require boundary conditions as well as initial conditions to solve, the defining relations become more complicated due to the need to determine the boundary conditions. Hence an alternative approach using variational methods is often used to determine the equations of motion as this approach, based on the Lagrangian or energy of the system, reveals the boundary conditions in a more systematic way.

Vibration with Control, Second Edition. Daniel John Inman.

Companion Website: www.wiley.com/go/inmanvibrationcontrol2e

First consider finding the equation of motion of a string fixed between two points and use a force balance approach to find the governing equation. This is illustrated in the following example.

Example 7.2.1

A string of mass density ρ and cross-sectional area A, fixed at both ends and under a tension denoted by τ is illustrated in Figure 7.1. The string moves up and down in the y-direction. The motion at any point on the string is a function of both the time t and the position along the string, x, so that the deflection of the string is denoted by $w(x, t)$.

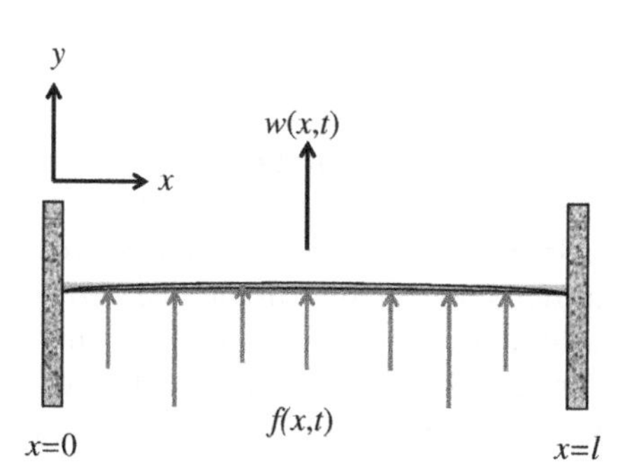

Figure 7.1 A string fixed at each end under tension τ subject to a distributed force $f(x, t)$.

A free body diagram of an infinitesimal element of length Δx of the displaced string is illustrated in Figure 7.2. Let $f(x, t)$ be an external force per unit length also distributed along the string. The subscripts 1 and 2 in the figure refer to values at the left and right end points of the infinitesimal element respectively.

Summing the forces acting on the infinitesimal element in the y-direction and setting the sum equal to the inertial force in the y-direction, $\rho A \Delta x(\partial^2 w/\partial t^2)$, yields

$$-\tau_1 \sin\theta_1 + \tau_2 \sin\theta_2 + f(x, t)\Delta x = \rho A \Delta x \frac{\partial^2 w(x, t)}{\partial t^2} \tag{7.1}$$

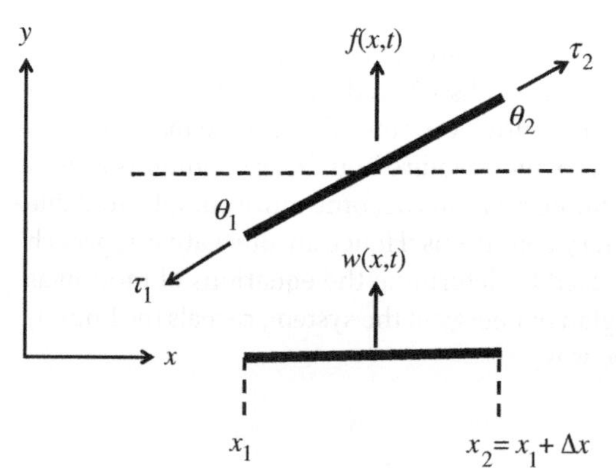

Figure 7.2 A free body diagram of an infinitesimal element of the string in Figure 7.1.

Note that the acceleration is stated in terms of partial derivatives because w is a function of two variables, x and t. The expressions in Equation (7.1) can be approximated in the case of small deflections so that θ_1 and θ_2 are small. In this case, τ_1 and τ_2 can easily be related to the initial tension in the string τ by noting that the horizontal component of the deflected string tension is $\tau_1 \cos\theta_1$ at end 1, and $\tau_2 \cos\theta_2$ at end 2. In the small-angle approximation, $\cos\theta_1 \simeq 1$ and $\cos\theta_2 \simeq 1$, so it is reasonable to set $\tau_1 = \tau_2 = \tau$. Also, for small θ_1

$$\sin\theta_1 \simeq \tan\theta_1 = \left.\frac{\partial w(x,\ t)}{\partial x}\right|_{x_1} \tag{7.2}$$

and

$$\sin\theta_2 \simeq \tan\theta_2 = \left.\frac{\partial w(x,\ t)}{\partial x}\right|_{x_2} \tag{7.3}$$

where $(\partial w/\partial x)|_{x_1}$ is the slope of the string at point x_1 and $(\partial w/\partial x)|_{x_2}$ is the slope of the string at point $x_2 = x_1 + \Delta x$. With these approximations, Equation (7.1) now becomes

$$\left.\left(\tau\frac{\partial w(x,t)}{\partial x}\right)\right|_{x_2} - \left.\left(\tau\frac{\partial w(x,t)}{\partial x}\right)\right|_{x_1} + f(x,t)\Delta x = \rho A\frac{\partial^2 w(x,t)}{\partial t^2}\Delta x \tag{7.4}$$

The slopes in Equation (7.4) are further evaluated by recalling the Taylor series expansion for the function $\tau\,(\partial w/\partial x)$ around the point x_1 from calculus. This yields

$$\left.\left(\tau\frac{\partial w}{\partial x}\right)\right|_{x_2} = \left.\left(\tau\frac{\partial w}{\partial x}\right)\right|_{x_1} + \Delta x\frac{\partial}{\partial x}\left.\left(\tau\frac{\partial w}{\partial x}\right)\right|_{x_1} + O\left(\Delta x^2\right) \tag{7.5}$$

where $O(\Delta x^2)$ denotes terms of order Δx^2 and higher in the Taylor series, which are very small and can be neglected. Substitution of Equation (7.5) into Equation (7.4) yields

$$\frac{\partial}{\partial x}\left.\left(\tau\frac{\partial w}{\partial x}\right)\right|_{x_1}\Delta x + f\,(x,t)\,\Delta x = \rho A\frac{\partial^2 w(x,\ t)}{\partial t^2}\Delta x \tag{7.6}$$

Dividing by Δx and realizing that since Δx is infinitesimal, the designation of point 1 becomes unnecessary, and the equation of motion for the string becomes

$$\frac{\partial}{\partial x}\left(\tau\frac{\partial w\,(x,\ t)}{\partial x}\right) + f\,(x,\ t) = \rho A\frac{\partial^2 w\,(x,\ t)}{\partial t^2} \tag{7.7}$$

Since the tension τ is constant, and if the external force is zero, this becomes

$$c^2 w_{xx}\,(x,\ t) = w_{tt}\,(x,\ t) \quad \text{or} \quad w_{xx}\,(x,\ t) = \frac{1}{c^2}\,w_{tt}\,(x,\ t) \tag{7.8}$$

Here subscripts are used to denote partial derivatives, so $w_t(x, t)$ is the velocity and $w_{tt}(x, t)$ is the acceleration of the string at any point x and time t. The constant $c = \sqrt{\tau/\rho A}$ depends only on the physical properties of the string and is called the *wave speed* and Equation (7.8) is called the *string equation*. The quantity $\tau w_{xx}(x, t)$ is the restoring force of the string and $w_x(x, t)$ is the slope of the string.

There are four derivatives in Equations (7.8), two time derivatives that require two initial conditions and two spatial derivatives requiring two boundary conditions. The initial conditions on position and velocity are spatially dependent and take on the form

$$w(x,0) = w_0(x) \text{ and } w_t(x,0) = \dot{w}_0(x)$$

Here the subscript t is an alternative notation for the partial derivative $\partial/\partial t$, $w_0(x)$ is the initial displacement distribution, and $\dot{w}_0(x)$ the initial velocity distribution of the string. The boundary conditions are determined by the geometric constraints on the string being fixed at the boundary, so that the deflection $w(x, t)$ must be zero at these points for all time

$$w(0,t) = w(l,t) = 0 \quad t > 0 \tag{7.9}$$

These four conditions determine the four constants of integration required to solve for the resulting deflection $w(x, t)$.

Recall Equations (2.12) and (2.13) that defined Lagrange's method for deriving equations of motion of lumped mass systems by working with the kinetic and potential energy of the system rather than balancing forces. A similar approach arises through variational methods and the use of virtual displacements, virtual work and D'Alembert's principle used to define Hamilton's principle for deriving equations of motion. For the distributed parameter systems considered in this chapter, Hamilton's method also prescribes the relevant boundary conditions in a manner sometimes easier than using the force balance and geometry at the boundary to determine such conditions. The procedure examines the motion of a structure between two instants of time and results in a scalar integral.

The method depends on the concept of a virtual displacement, denoted δ. Virtual displacements are infinitesimal displacements that are compatible with all geometric constraints and occur with no change in time. The virtual displacements behave like derivatives following the same rules of calculus. In simple terms, Hamilton's principle states

$$\delta \int_{t_1}^{t_2} L dt = 0, \quad \delta q_k(t_1) = 0 \quad \text{and} \quad \delta q_k(t_2) = 0, \quad \text{for all indices } k \tag{7.10}$$

where L is the Lagrangian given in Section 2.2 as the difference between the kinetic and potential energy. Here the q_k represents all the different displacement variables. Their virtual displacement must be zero at times t_1 and t_2, the limits of integration. A complete derivation can be found in many texts on advanced dynamics (e.g. Meirovitch, 1970). The following example illustrates the method.

Example 7.2.2
To illustrate the use of Hamilton's principle, consider the longitudinal vibration of a thin rod or bar of length l illustrated in Figure 7.3. From solid mechanics

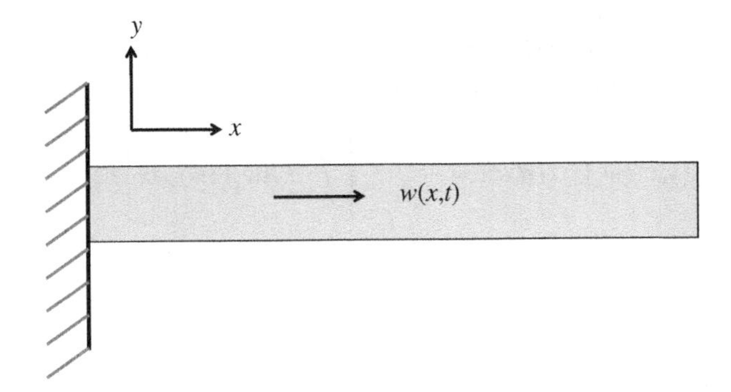

Figure 7.3 A schematic of a rod or bar indicating the direction of longitudinal vibration.

(Dym and Shames, 1973) the strain energy $V(t)$ in a rod of elastic modulus E and cross-sectional area $A(x)$ is given by

$$V(t) = \frac{1}{2}\int_0^l EA(x)[w_x(x,t)]^2 dx \tag{7.11}$$

where $w(x, t)$ denotes the longitudinal displacement and its derivative $w_x(x, t)$ is the strain. The quantity $EA(x)$ is the stiffness in the longitudinal or x-direction. The kinetic energy of the bar is

$$T(t) = \frac{1}{2}\int_0^l \rho A(x)[w_t(x,t)]^2 dx \tag{7.12}$$

Here, ρ is the density per unit cross-section and $w_t(x, t)$ is the velocity. Substitution of these energy expressions into Equation (7.10) yields

$$\begin{aligned} &\delta\int_{t_i}^{t_2}(T-V)dt = \int_{t_1}^{t_2}(\delta T - \delta V)dt = \\ &\int_{t_1}^{t_2}\int_0^l \rho A(x)\tfrac{1}{2}\delta([w_t(x,t)]^2)\,dxdt - \int_{t_1}^{t_2} EA(x)\tfrac{1}{2}\delta([w_x(x,t)]^2)\,dxdt \end{aligned} \tag{7.13}$$

Performing the indicated variations following the rules of calculus yields

$$\begin{aligned} \delta\int_{t_1}^{t_2}(T-V)dt = &\int_{t_1}^{t_2}\int_0^l \rho A(x)w_t\delta w_t dxdt \\ &- \int_{t_1}^{t_2}\int_0^l EA(x)w_x(x,t)\delta w_x(x,t)dxdt \end{aligned} \tag{7.14}$$

Next interchange the time and space integrations and use integration by parts on t to evaluate each of these integrals. The kinetic energy term becomes

$$\begin{aligned} \int_{t_1}^{t_2}\int_0^l \rho A(x)w_t\delta w_t dxdt &= \int_0^l \left[\rho A(x)w_t\delta\, w|_{t_1}^{t_2} - \int_{t_1}^{t_2}\rho A(x)w_{tt}\delta w dt\right]dx \\ &= -\int_{t_1}^{t_2}\int_0^l \rho A(x)w_{tt}\delta w dxdt \end{aligned} \tag{7.15}$$

Note that the first term in the integration by parts is zero because the δw vanishes at t_1 and t_2. Interchanging the variation and derivative operations and integrating by parts on x, the potential energy term becomes

$$-\int_{t_1}^{t_2}\int_0^l EA(x)w_x(x,t)\delta w_x(x,t)\,dxdt = -\int_{t_1}^{t_2}\left[\int_0^l EAw_x(\delta w)_x dx\right]dt$$

$$= -\int_{t_1}^{t_2}\left[EAw_x\delta w|_0^l - \int_0^l (EAw_x)_x \delta w dx\right]dt \tag{7.16}$$

Combining the results of Equations (7.15) and (7.16) into Equation (7.10) yields

$$\delta\int_{t_1}^{t_1}(T-V)dt = 0$$
$$\Rightarrow -\int_{t_1}^{t_2}\left[\int_0^l \rho A(x)w_{tt}\delta w dx - EAw_x\ \delta w|_0^l + \int_0^l (EAw_x)_x\delta w dx\right]dt \tag{7.17}$$

Multiplying and rearranging yields

$$\int_{t_1}^{t_2}\left[\int_0^l \underbrace{\left[(EAw_x)_x - \rho A(x)w_{tt}\right]\delta w dx}_{1} - \underbrace{EAw_x\ \delta w|_0^l}_{2}\right]dt = 0 \tag{7.18}$$

As δw is arbitrary, each of the terms multiplying δw must vanish. Thus the expression labeled 2 above yields the boundary conditions so that

$$EA\frac{\partial(x,t)}{\partial x}\bigg|_0^l = 0 \tag{7.19}$$

In Equation (7.18), δw is arbitrary which means the integral is only zero if the expression given by 1 is zero, so that the equation of motion becomes

$$\frac{\partial}{\partial x}\left(EA(x)\frac{\partial w(x,t)}{\partial x}\right) = \rho A(x)\frac{\partial^2 w(x,t)}{\partial t^2} \tag{7.20}$$

The equation of motion Equation (7.20) and boundary conditions of Equation (7.19) along with the initial conditions of $w(x, t)$ determine the vibration response of the bar.

The two methods of deriving the equations of motion and boundary conditions for distributed parameter systems illustrated here by example are equivalent and produce the same unique result. These alternative approaches provide a choice when considering a vibrating system. Sometimes the energy approach may be easier to use and for other problems summing moments and forces may be more straightforward.

7.3 Vibration of Strings

Figure 7.1 depicts a string fixed at both ends and displaced slightly from its equilibrium position. As noted in Example 7.2.1, the lateral position of the string is denoted by w, which depends not only on the time t but also on the position along the string x. This spatial dependency is the essential difference between lumped parameter systems and distributed parameter systems. The function $w(x,t)$ that satisfies Equation (7.8) also must satisfy two initial conditions because of the second-order time derivatives. The second-order spatial derivative implies that two spatial conditions must also be satisfied. In the cases of interest here, the value of x will vary over a finite range. Physically, if the string is fixed at both ends, then $w(x, t)$ would have to be zero at $x = 0$ and again at $x = l$. Thus the problem of finding the lateral vibrations of a string fixed at both ends can then be summarized as: find a function $w(x, t)$ such that

$$\begin{aligned} w_{xx}(x,t) &= \frac{1}{c^2} w_{tt}(x,t), \quad x \in (0,l) \qquad &&\text{for } t > 0 \\ w(0,t) &= w(l,t) = 0, && t > 0 \\ w(x,0) &= w_0(x), \text{ and } w_t(x,0) = \dot{w}_0(x) \text{ at } t = 0 \end{aligned} \tag{7.21}$$

where $w_0(x)$ and $\dot{w}_0(x)$ are specified time-invariant functions representing the initial (at $t = 0$) displacement and velocity distribution of the string. The notation $x \in (0, l)$ means that the equation holds for values of x in the interval $(0, l)$.

One approach used to solve the system given by Equation (7.21) is to assume that the solution has the form $w(x,t) = X(x)T(t)$. This approach is called the method of *separation of variables* (Boyce and DiPrima, 2012). This method proceeds by substitution of this assumed separated form into Equation (7.21), which yields the set of equations

$$\begin{aligned} X''(x)T(t) &= \frac{1}{c^2} X(x)\ddot{T}(t) \\ X(0)T(t) &= 0, \text{ and } X(l)T(t) = 0 \\ X(x)T(0) &= w_0(x), \text{ and } X(x)\dot{T}(0) = \dot{w}_0(x) \end{aligned} \tag{7.22}$$

where the primes denote total derivatives with respect to x and the over dots indicate total time differentiation. Rearranging the first equation in Equation (7.22) yields

$$\frac{X''(x)}{X(x)} = \frac{1}{c^2} \frac{\ddot{T}(t)}{T(t)} \tag{7.23}$$

Differentiating this expression with respect to x yields

$$\frac{d}{dx}\left(\frac{X''(x)}{X(x)}\right) = 0$$

or that

$$\frac{X''(x)}{X(x)} = \sigma \tag{7.24}$$

where σ is a constant of integration (independent of t or x). The next step is to solve Equation (7.24), i.e.

$$X''(x) - \sigma X(x) = 0 \tag{7.25}$$

subject to the two boundary conditions, which become

$$X(0) = 0, \text{ and } X(l) = 0 \tag{7.26}$$

since $T(t) \neq 0$, for most values of t.

The nature of the constant σ needs to be determined next. There are three possible choices for σ, it can be positive, negative or zero. If $\sigma = 0$, the solution of Equation (7.25) subject to Equation (7.26) becomes $X(x) = 0$, which does not satisfy the condition of a nontrivial solution. If $\sigma > 0$, then the solution to Equation (7.25) is of the form $X(x) = A_1 \cosh(\sigma x) + A_2 \sinh(\sigma x)$. Applying the boundary conditions to this form of the solution then yields

$$0 = A_1$$

$$0 = A_2 \sinh(l\sigma)$$

so that again the only possible solution is the trivial solution, $w(x, t) = 0$.

Thus, the only nontrivial choice for σ is that it must have a negative value. To indicate this, $\sigma = -\lambda^2$ is used. Equations (7.25) and (7.26) become

$$\begin{aligned} &X''(x) + \lambda^2 X(x) = 0 \\ &X(0) = X(l) = 0 \end{aligned} \tag{7.27}$$

Equation (7.27) is called a *boundary value problem*, as the values of the solution are specified at the boundaries. The form of the solution of Equation (7.27) is

$$X(x) = A_1 \sin \lambda x + A_2 \cos \lambda x \tag{7.28}$$

Applying the boundary conditions to the solution in Equation (7.28) indicates that the constants A_1 and A_2 must satisfy

$$A_2 = 0, \text{ and } A_1 \sin \lambda l = 0 \tag{7.29}$$

Examination of Equation (7.29) shows that the only nontrivial solutions occur if

$$\sin \lambda l = 0 \tag{7.30}$$

or when λ has the value $n\pi/l$, where n is any integer value from 1 to infinity, denoted by

$$\lambda_n = \frac{n\pi}{l}, n = 1, 2, \ldots \infty$$

Note the $n = 0$, $\lambda = 0$ is omitted in this case, because it results in zero solution. The solution given in Equation (7.28) is thus the infinite set of functions denoted by

$$X_n(x) = A_n \sin\left(\frac{n\pi}{l} x\right) \tag{7.31}$$

where n is any positive integer. Equation (7.30) resulting from applying the boundary conditions is called the *characteristic equation* and the values λ_n are called *characteristics values.* The functions X_n are *mode shapes* of the system. An important difference between lumped parameter systems and distributed parameter systems is that there are a finite number of solutions to the characteristic equation for a lumped mass system and an infinite number of solutions to the characteristic equation for a distributed mass system.

Substitution of Equation (7.31) into Equation (7.23) then shows that the temporal coefficient $T(t)$ must satisfy the infinite number of equations

$$\ddot{T}_n(t) + \left(\frac{n\pi}{l}\right)^2 c^2 T_n(t) = 0 \tag{7.32}$$

The solution of these equations, one for each n, is given by

$$T_n(t) = A_n^1 \sin\left(\frac{n\pi c}{l}t\right) + A_n^2 \cos\left(\frac{n\pi c}{l}t\right) \tag{7.33}$$

where A_n^1 and A_n^2 are the required constants of integration and the subscript n has been added to indicate that there are an infinite number of solutions of this form. Thus, the solutions of Equation (7.20), also infinite in number, are of the form

$$w_n(x,t) = a_n \sin\left(\frac{n\pi c}{l}t\right)\sin\left(\frac{n\pi}{l}x\right) + b_n \cos\left(\frac{n\pi c}{l}t\right)\sin\left(\frac{n\pi}{l}x\right) \tag{7.34}$$

Here a_n and b_n are constants representing the product of A_n^1 and A_n^2 of Equation (7.33) and the constants of Equation (7.31).

Since Equation (7.34) is a linear system, the sum of all of these solutions is also a solution, so that

$$w(x,t) = \sum_{n=1}^{\infty}\left[a_n \sin\left(\frac{n\pi c}{l}t\right) + b_n \cos\left(\frac{n\pi c}{l}t\right)\right]\sin\left(\frac{n\pi c}{l}t\right) \tag{7.35}$$

This infinite sum may or may not converge, as is discussed in Chapter 9. Next, the constants a_n and b_n need to be calculated. These constants come from the initial conditions. Applying the displacement initial condition to Equation (7.35) yields

$$w(x,0) = w_0(x)\sum_{n=1}^{\infty} b_n \sin\left(\frac{n\pi}{l}x\right) \tag{7.36}$$

This equation can be solved for the constants b_n by using the (orthogonality) property of the functions $\sin(n\pi/l)x$

$$\int_0^l \sin\left(\frac{n\pi}{l}x\right)\sin\left(\frac{m\pi}{l}x\right)dx = \delta_{mn} = \begin{cases} \frac{l}{2} & m = n \\ 0 & m \neq n \end{cases} \tag{7.37}$$

Thus, multiplying Equation (7.36) by $\sin(m\pi x/l)$ and integrating yields the desired constants

$$b_n = \frac{2}{l}\int_0^l w_0(x) sin\left(\frac{n\pi}{l}x\right)dx \tag{7.38}$$

since the sum vanishes for each $n \neq m$. Likewise, if Equation (7.35) is differentiated with respect to time, multiplied by $\sin(m\pi x/l)$ and integrated, the constants a_n are found from

$$a_n = \frac{2}{n\pi c} \int_0^l w_0(x) \sin\left(\frac{n\pi}{l}x\right) dx \tag{7.39}$$

The problem of solving the most basic distributed parameter system is much more complicated than solving for the free response of a simple one degree of freedom lumped parameter system. Also, note that the solution just described essentially yields the theoretical modal analysis solution established for lumped parameter systems, as developed in Section 3.3. The functions $\sin(n\pi x/l)$ serve in the same capacity as the eigenvectors of a matrix in calculating a solution. The major difference between the two developments is that the sum in Equation (3.42) is finite, and hence always converges, whereas the sum in Equation (7.35) is infinite and may or may not converge.

Physically, the functions $\sin(n\pi x/l)$ describe the configuration of the string for a fixed time and hence are referred to as the system's *natural modes of vibration.* Likewise, the numbers $(n\pi c/l)$ are referred to as the *natural frequencies of vibration,* since they describe the motion's periodicity in time. Mathematically, the characteristic values $(n\pi/l)$ are also called the *eigenvalues* of the system, and sin $(n\pi x/l)$ are called the *eigenfunctions* of the system and form an analogy to what is known about lumped parameter systems. These quantities are defined more precisely in Section 8.2. For now, note that the eigenvalues and eigenfunctions defined here serve the same purpose as eigenvalues and eigenvectors defined for matrices. Just as eigenvectors for lumped mass systems are related to their mode shapes, the eigenfuntions define the mode shapes for a distributed mass system.

The basic method of separation of variables combined with the infinite sums and orthogonality as used here forms the basic approach for solving the response of distributed parameter systems. This approach, also called modal analysis, is used numerous times in the following chapters to solve a variety of vibration and control problems.

In the case of lumped parameter systems, a lot was gained by looking at the eigenvalue and eigenvector problem. This information allowed the calculation of the solution of both the free and forced response by using the properties of the eigenstructures. In the following, the same approach (modal analysis) is further developed for distributed parameter systems.

The fact that the solution Equation (7.35) is a series of sine functions should not be a surprise. In fact, Fourier's theorem states that every function $f(x)$ that is piecewise continuous and bounded on the interval $[a, b]$ can be represented as a Fourier series of the form

$$f(x) = \frac{a_0}{2} \sum_{n=1}^{\infty} \left[a_n \cos\left(\frac{n\pi}{l}x\right) + b_n \sin\left(\frac{n\pi}{l}x\right)\right] \tag{7.40}$$

where $l = b - a$. This fact is used extensively in Chapter 12 on vibration testing.

Recall that a function f is said to be *bounded* on the interval $[a, b]$ if there exists a finite constant M such that $|f(x)| < M$ for all x in $[a, b]$. Furthermore, a function $f(x)$ is *continuous* on the interval $[a, b]$, if for every x_1 in $[a, b]$, and for every $\varepsilon > 0$, there exists a number $\delta = \delta(\varepsilon) > 0$ such that $|x - x_1| < \delta$ implies $|f(x) - f(x_1)| < \varepsilon$. A function is *piecewise continuous* on $[a, b]$ if it is continuous on every subinterval of $[a, b]$ (note here that the square brackets indicate that the endpoints of the interval are included in the interval).

In many cases, either all of the coefficients a_n or all of the coefficients b_n are zero. Also, note that many other functions $\varnothing_n(x)$, besides the functions $\sin(n\pi x/l)$ and $\cos(n\pi x/l)$, have the property that arbitrary functions of a certain class can be expanded in terms of an infinite series of such functions, i.e. that

$$f(x) = \sum_{n=1}^{\infty} a_n \varnothing_n(x). \tag{7.41}$$

This property is called *completeness* and is related to the idea of completeness used with orthogonal eigenvectors (Section 3.3). This concept is discussed in detail in Chapter 9.

Note that Equation (7.41) really means that the sequence of partial sums

$$\left\{ a_1\varnothing_1, a_1\varnothing_1 + a_2\varnothing_2, \dots, \sum_{i=1}^{m} a_i\varnothing_i, \dots \right\} \tag{7.42}$$

converges to the function $f(x)$, i.e. that

$$\lim_{m\to\infty} \left(\sum_{n=1}^{m} a_n\varnothing_n \right) = f(x) \tag{7.43}$$

as defined in most introductory calculus texts.

Example 7.3.1

Now that the formal solution of the string has been examined and the eigenvalues and eigenfunctions have been identified, it is important to realize that these quantities are the physical notions of mode shapes and natural frequencies. To this end, suppose that the string is given the following initial conditions

$$w(x, 0) = \sin\left(\frac{\pi}{l}x\right)\ w_t(x, 0) = 0 \tag{7.44}$$

Calculation of the expansion coefficients yields (from Equations (7.38) and (7.39))

$$a_n = \frac{2}{n\pi c}\int_0^l w^t(x, 0)\sin\left(\frac{n\pi}{l}x\right)dx = 0 \tag{7.45}$$

and

$$b_n = \frac{2}{l}\int_0^l \sin\left(\frac{\pi}{l}x\right)\sin\left(\frac{n\pi}{l}x\right)dx = \begin{cases} 1 & n = 1 \\ 0 & n \geq 2 \end{cases} \tag{7.46}$$

The solution thus becomes

$$w(x, t) = \sin\left(\frac{\pi x}{l}\right)\cos\left(\frac{\pi ct}{l}\right) \tag{7.47}$$

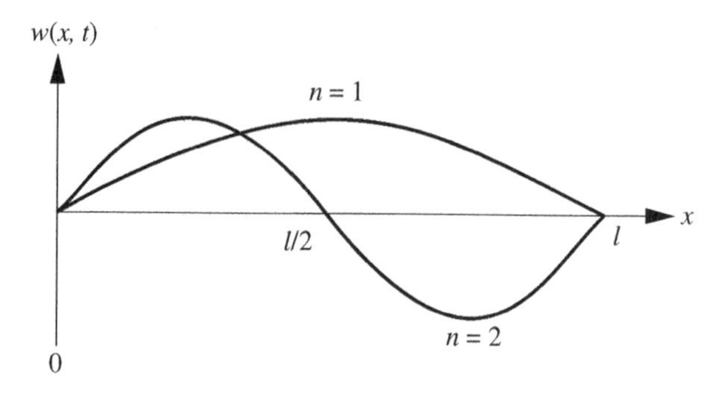

Figure 7.4 The first two mode shapes of a vibrating string fixed at both ends.

In Figure 7.4, this solution is plotted versus x for a fixed value of t. This plot is the *shape* that the string takes if it were viewed by a stroboscope blinking at a frequency of $\pi c/l$. One of the two curves in Figure 7.4 is the plot of $w(x, t)$ that would result if the string were given an initial displacement of $w(x,0) = \sin(\pi x/l)$, the first eigenfunction. The other curve, which takes on negative values, results from an initial condition of $w(x,0) = \sin(2\pi x/l)$, the second eigenfunction. Note that all the eigenfunctions $\sin(n\pi x/l)$ can be generated in this fashion by choosing the appropriate initial conditions, reinforcing the idea that eigenfunctions form the set of *mode shapes* of vibration of the string. These correspond to the mode shapes defined for lumped parameter systems and are the quantities measured in the modal tests described in Chapter 12.

7.4 Rods and Bars

Next, consider the longitudinal vibration of a bar, that is, the vibration of a long slender structure in the direction of its longest axis, as indicated in Figure 7.3. The equation of motion as derived in Example 7.2.2 and given in Equation (7.20) of a bar of length l can be written as

$$[EA(x)w_x(x,t)]_x = \rho(x)A(x)w_{tt}(x,t) \quad x \in (0,l) \tag{7.48}$$

where $A(x)$ is the variable cross-sectional area, $\rho(x)$ represents the variable mass distribution per unit area, E is the elastic modulus, and $w(x,t)$ is the axial displacement (in the x-direction).

The form of Equation (7.48) is the same as that of the string given in Equation (7.8) if the cross-sectional area of the bar is constant. In fact, the "stiffness" *operator* (Section 8.2) in both cases has the form

$$-\alpha\frac{\partial^2}{\partial x^2} \tag{7.49}$$

where α is a constant. Hence, the eigenvalues and eigenfunctions are expected to have the same mathematical form as those of the string, and the solution will be similar. The main difference between these two systems is physical. In Equation (7.48), the function $w(x, t)$ denotes displacements along the long *axis* of the rod, whereas in Equation (7.20), $w(x, t)$ denotes displacements *perpendicular to the axis* of the string.

Several different ways of supporting a rod (and the string) lead to several different sets of boundary conditions associated with Equation (7.48), which can be derived from Equation (7.19). Some are stated in terms of the displacement $w(x,t)$ and others are given in terms of the strain $w_x(x, t)$ (recall strain is the change in length per unit length).

Free Boundary: If the bar is free or unsupported at a boundary, then the stress at the boundary must be zero, i.e. no force should be present at that boundary, or

$$EA(x)\left.\frac{\partial w(x,t)}{\partial x}\right|_{x=0} = 0, \text{ or } EA(x)\left.\frac{\partial w(x,t)}{\partial x}\right|_{x=l} = 0 \tag{7.50}$$

Note that if $A(l) \neq 0$, the strain $w_x(l,t)$ must also be zero. The vertical bar in Equation (7.50) denotes that the function is to be evaluated at $x = 0$ or l after the derivative is taken and indicates the location of the boundary condition.

Clamped Boundary: If the boundary is rigidly fixed, or clamped, then the displacement must be zero at that point or

$$w(x,t)\,|_{x=0} = 0\,, \quad \text{or} \quad w(x,t)\,|_{x=l} = 0 \tag{7.51}$$

Appended Boundary: If the boundary is fastened to a lumped element, such as a spring of stiffness k, the boundary condition becomes

$$EA(x)\frac{\partial w(x,t)}{\partial x}|_{x=0} = kw(x,t)|_{x=0}, \text{ or } EA(x)\frac{\partial w(x,t)}{\partial x}|_{x=l} = -kw(x,t)|_{x=l} \tag{7.52}$$

which expresses a force balance at the boundary. In addition, if the bar has a lumped mass at the end, the boundary condition becomes

$$\begin{aligned} EA(x)w_x(x,t)\big|_{x=0} &= mw_{tt}(x,t)\big|_{x=0}, \quad \text{or} \quad EA(x)w_x(x,t)\big|_{x=l} \\ &= -mw_{tt}(x,t)\big|_{x=l} \end{aligned} \tag{7.53}$$

which also represents a force balance.

These types of problems are discussed in detail in Section 10.4. They represent a large class of applications and are also referred to as structures with time-dependent boundary conditions, constrained structures or combined dynamical systems.

As noted, the equation of motion is mathematically the same for both the string and the bar, so that further discussion of the method of solution for the bar is not given. A third problem again has the same mathematical model: the torsional vibration of circular shafts (rods). The derivation of the equation of motion is very similar, comes from a force balance or from the energy approach (Inman, 2014; Timoshenko et al., 1974). If G represents the shear modulus of elasticity of the shaft of length l, I_ρ is the polar moment of inertia of the shaft, ρ is the mass per unit area, the function $\theta(x, t)$ is the angular displacement of the shaft from its neutral position, x is the distance measured along the shaft, and the equation governing the torsional vibration of the shaft is

$$\theta_{tt}(x,t) = \frac{G}{\rho}\theta_{xx}(x,t)\ x \in (0,l) \tag{7.54}$$

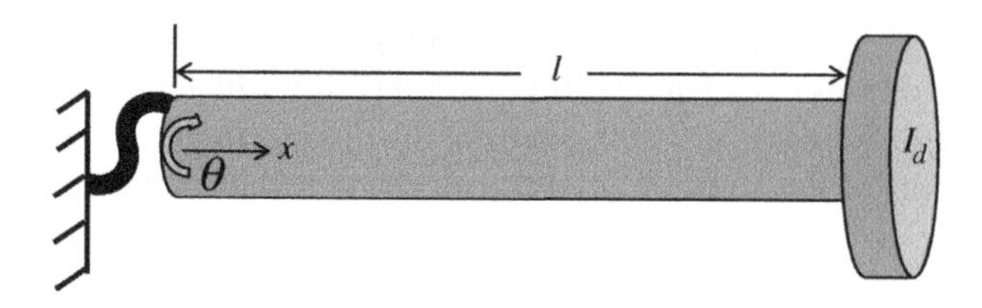

Figure 7.5 A shaft connected to a rotational spring on one end and a disc on the other end.

As in the case of the rod or bar, the shaft can be subjected to a variety of boundary conditions, some of which are described in the following

Free Boundary: If the boundaries of the shaft are not attached to any device, there cannot be any torque acting on the shaft at that point, so that

$$GI_p\theta_x\,|_{x=0} = 0\,, \quad \text{or} \quad GI_p\theta_x\,|_{x=l} = 0 \tag{7.55}$$

at that boundary. If G and I_p are constant, then Equation (9.36) becomes simply

$$\theta_x(x,t)\,|_{x=0} = 0\,, \quad \text{or} \quad \theta_x(x,t)\,|_{x=l} = 0 \tag{7.56}$$

Clamped Boundary: If a boundary is clamped, then no movement of the shaft at that position can occur, so that the boundary condition becomes

$$\theta(x,t)\,|_{x=0} = 0\,, \quad \text{or} \quad \theta(x,t)\,|_{x=l} = 0 \tag{7.57}$$

Appended Boundaries: If a torsional spring of stiffness k is attached at the right end of the shaft (say at $x = l$), the spring force (torque) must balance the internal bending moment. The boundary condition becomes

$$GI_p\theta(x,t)\,|_{x=l} = k\theta(x,t)\,|_{x=l} \tag{7.58}$$

At the left end this becomes (Figure 7.5)

$$GI_p\theta_x(x,t)\,|_{x=0} = -k\theta(x,t)\,|_{x=0} \tag{7.59}$$

Quite often a shaft is connected to a disc at one end or the other. If the disc has mass polar moment of inertia I_d at the right end, the boundary condition becomes

$$GI_p\theta_x(x,t)\,|_{x=l} = -I_d\theta_{tt}(x,t)\,|_{x=l} \tag{7.60}$$

or

$$GI_p\theta_x(x,t)\,|_{x=0} = I_d\theta_{tt}(x,t)\,|_{x=0} \tag{7.61}$$

if the disc is placed at the left end.

The shaft could also have both a spring and a mass at one end, in which case the boundary condition, obtained by summing forces, is

$$GI_p\theta_x(x,t)\,|_{x=0} = -I_d\theta_{tt}(x,t)\,|_{x=0} - k\theta(x,t)\,|_{x=0} \tag{7.62}$$

As illustrated in the examples and exercises, the boundary conditions affect the natural frequencies and mode shapes. These quantities have been tabulated for many common boundary conditions (Blevins, 2001; Gorman, 1975).

Example 7.4.1
Consider the vibration of a shaft that is fixed at the left end ($x = 0$) and has a disc attached to the right end ($x = l$). Let G, I_p, and ρ all have unit values and calculate the eigenvalues and eigenfunctions of the system.

Following the separation of variables procedure used in the solution of the string problems, a solution of the form $\theta(x, t) = \Theta(x)T(t)$ is assumed and substituted into the equation of motion and boundary conditions resulting in

$$\Theta''(x) + \lambda^2\Theta(x) = 0$$
$$\Theta(0) = 0$$
$$GI_p\Theta'(l)T(t) = I_d\Theta(l)\ddot{T}(t)$$

Recall that $T(t)$ is harmonic, so that $\ddot{T}(t) = -\lambda^2 T(t)$, and the last boundary condition can be written as

$$GI_p\Theta'(l) = -\lambda^2 I_d\Theta(l)$$

which removes the time dependence. The general spatial solution is

$$\Theta(x) = A_1 \sin \lambda x + A_2 \cos \lambda x$$

Application of the boundary condition at $x = 0$ yields

$$A_2 = 0 \text{ and } \Theta(x) = A_1 \sin \lambda x$$

The second boundary condition yields (for the case $G = I_p = 1$)

$$\lambda A_1 \cos \lambda l = -\lambda^2 I_d A_1 \sin \lambda l$$

which is satisfied for all values of λ such that

$$\tan(\lambda l) = -\frac{l}{I_d}\frac{1}{\lambda l} \tag{7.63}$$

Equation (7.63) is a transcendental equation for the values of λ, the eigenvalues, and has an infinite number of solutions denoted by λ_n, calculated either graphically or numerically from the points of intersection given in Figure 7.6. The values of λ correspond to the values of (λl) at the intersections of the two curves in Figure 7.6. Note that the effect of the disc inertia, I_d, is to shift the

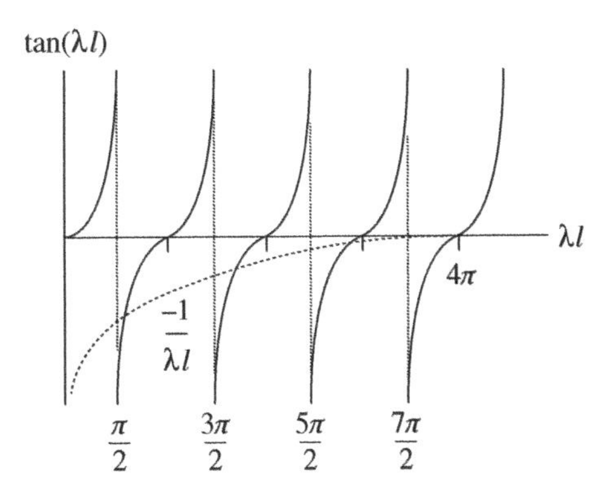

Figure 7.6 The graphical solution of the transcendental equation for tan $\lambda l = -1/\lambda l$.

points of intersection of the two curves. The transcendental equation of Equation (7.53) is solved numerically near each crossing, using the plot to obtain an initial guess for the numerical procedure. For values of n greater then 3, the crossing points in the plot approach the zeros of the tangent, and hence the sine or $n\pi$.

7.5 Vibration of Beams

In this section, the transverse vibration of a beam is examined. A beam is represented in Figure 7.7. The beam has mass density $\rho(x)$, cross-sectional area $A(x)$, and moment of inertia $I(x)$ about its neutral axis. The deflection $w(x, t)$ of the beam is the result of two effects, bending and shear. The derivation of equations and boundary conditions for the vibration of the beam can be found in several texts (Timoshenko et al., 1974).

A beam is a transversely loaded (along the x-axis in Figure 7.7), prismatic structural element with length l, which is large in value when compared to the magnitude of the beam's cross-sectional thickness (h_y) and width (h_z). Let $A(x)$ denote the cross-sectional area in the y–z plane. In the previous section, vibration of such a structure in the x-direction was considered and referred to as the longitudinal vibration of a bar. In this section, the *transverse vibration* of the beam, i.e. vibration in the y-direction, perpendicular to the long axis of the beam, is considered.

A fundamental structural element considered in many vibration problems is the Euler-Bernoulli beam, characterized by being uniform, linear, homogenous, long and slender (length greater than 10 times the other dimensions), in plain strain (no deflections in the y–z plane), so that the sides of an infinitesimal element remain parallel as it vibrates. An infinitesimal element of the beam in Figure 7.7 as it deflects and its free-body diagram is illustrated in Figure 7.8. Note that the sides at x and $x + dx$ are parallel. The deflection $w(x, t)$ is in the y-direction, $V(x, t)$ denotes the shear force and $M(x, t)$ denotes the bending moment as the element deforms. Inherent in this is the assumption that deflections are small.

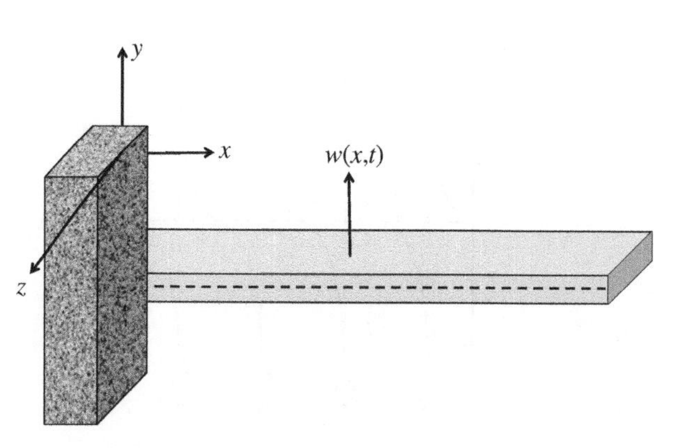

Figure 7.7 A beam indicating the deflection defining transverse vibration (motion in the y-direction).

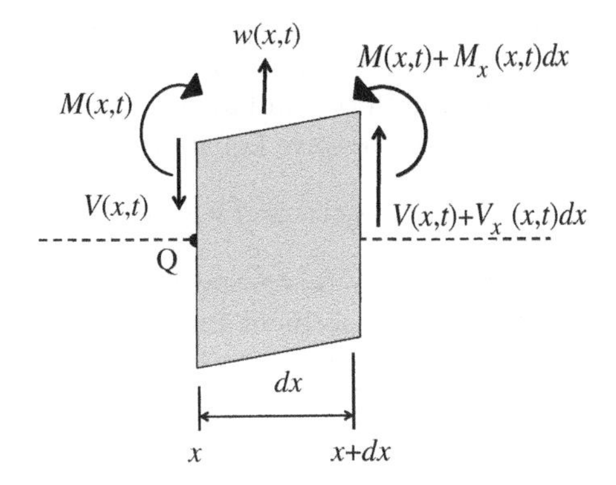

Figure 7.8 The free-body diagram of an infinitesimal element of a slender beam of Figure 7.7 undergoing small deflections.

Summing the forces in Figure 7.8 in the vertical direction yields

$$V(x,t) + \frac{\partial V(x,t)}{\partial x}dx - V(x,t) = \rho A(x)\,dx\frac{\partial^2 w(x,t)}{\partial t^2} \tag{7.64}$$

The term on the right is the inertia of the element with mass given as the product of the material density ρ , the cross-sectional area A and the length dx. The summation of moments acting on the element dx about point Q in Figure 7.8 yields (assuming zero rotational inertia)

$$M(x,t) + \frac{\partial M(x,t)}{\partial x}dx - M(x,t) + \left(V(x,t) + \frac{\partial V(x,t)}{\partial x}dx \right) dx = 0 \tag{7.65}$$

Simplifying this last equation, dropping the term multiplied $(dx)^2$ and setting the coefficient of dx equal to zero yields that the shear force and moment are related by

$$V(x,t) = -\frac{\partial M(x,t)}{\partial x} \tag{7.66}$$

However, from solid mechanics, the bending moment is related to the displacement by

$$M(x,t) = EI(x)\frac{\partial^2 w(x,t)}{\partial x^2} \tag{7.67}$$

where $I(x)$ is the cross-sectional area moment of inertial about the z-axis, E is Young's elastic modulus and the second derivative of the displacement is related to the radius of curvature for small displacements. Combining Equations (7.64), (7.66) and (7.67) yields the Euler-Bernoulli beam equation for transverse vibrations

$$\rho A(x)\frac{\partial^2 w(x,t)}{\partial t^2} = -\frac{\partial^2}{\partial x^2}\left(EI(x)\frac{\partial^2 w(x,t)}{\partial x^2} \right) \tag{7.68}$$

For constant cross-sectional area A and stiffness EI, the equation can be written as

$$\frac{\partial^2 w(x,t)}{\partial t^2} + c^2\frac{\partial^4 w(x,t)}{\partial x^4} = 0, \quad c = \sqrt{\frac{EI}{\rho A}} \tag{7.69}$$

The constant c is again called a wave speed. These equations hold on the domain of $w(x, t)$, which in this case is all values of $x \in (0, l)$. The two time derivatives require two initial conditions and the four spatial derivatives require four boundary conditions in order to solve Equation (7.69). The initial conditions are of the form

$$w(x, 0) = w_0(x), \text{ and } w_t(x, 0) = w_0(x) \text{ at } t = 0$$

where $w_0(x)$ and $\dot{w}_0(x)$ are specified time-invariant functions representing the initial (at $t = 0$) displacement and velocity distribution of the beam.

The boundary conditions required to solve Equation (7.69) are obtained by examining the deflection $w(a, t)$, the slope of the deflection

$$\left.\frac{\partial w\,(x,\ t)}{\partial x}\right|_{x=a}$$

the bending moment

$$EI(x)\left.\frac{\partial^2 w(x, t)}{\partial x^2}\right|_{x=a}$$

or the shear force

$$\left.\frac{\partial}{\partial x}\left(EI(x)\frac{\partial^2 w(x, t)}{\partial x^2}\right)\right|_{x=a}$$

at each end of the beam (i.e. $a = 0$ or l). A common configuration is *clamped–free* or *cantilevered*, as illustrated in Figure 7.7. In addition to a boundary being clamped or free, the end of a beam could be resting on a support restrained from bending or deflecting. The situation is called *simply supported* or *pinned*. A *sliding* boundary is one in which displacement is allowed but rotation is not. The shear load at a sliding boundary is zero.

If a beam in transverse vibration is free at the end $x = a$, the deflection and slope at that end are unrestricted, but the bending moment and shear force must vanish

$$\text{bending moment} = EI(x)\left.\frac{\partial^2 w(x, t)}{\partial x^2}\right|_{x=a} = 0 \tag{7.70}$$

$$\text{shear force} = \frac{\partial}{\partial x}\left.\left[EI\frac{\partial^2 w(x, t)}{\partial x^2}\right]\right|_{x=a} = 0$$

If, on the other hand, a beam is clamped (or fixed) at $x = a$, the bending moment and shear force are unrestricted, but the deflection and slope must vanish at that end

$$\text{deflection} = w(x, t)|_{x=a} = 0$$

$$\text{slope} = \left.\frac{\partial w(x, t)}{\partial x}\right|_{x=a} = 0 \tag{7.71}$$

At a simply supported or pinned end, the slope and shear force are unrestricted and the deflection and bending moment must vanish

$$\text{deflection} = w(x,t)|_{x=a} = 0$$
$$\text{bending moment} = EI(x)\left.\frac{\partial^2 w(x,t)}{\partial x^2}\right|_{x=a} = 0 \tag{7.72}$$

At a sliding end, the slope or rotation is zero and no shear force is allowed. On the other hand, the deflection and bending moment are unrestricted. Hence, at a sliding boundary

$$\text{slope} = \left.\frac{\partial w(x,t)}{\partial x}\right|_{x=a} = 0 \tag{7.73}$$

$$\text{shear force} = \left.\frac{\partial}{\partial x}\left(EI\frac{\partial^2 w(x,t)}{\partial x^2}\right)\right|_{x=a} = 0$$

Other boundary conditions are possible by connecting the ends of a beam to a variety of devices such as lumped masses, springs, etc. These boundary conditions can be determined by force and moment balances or derived from Hamilton's principle. These are also listed in tabular form in Inman (2014) and Blevins (2001), among others.

To correct for the effect of cross-sectional dimensions which become significant when the beam is no longer slender and/or used at high frequencies, the rotational inertia is included as well as the shear deformation (Elishakoff et al., 2015). For small angles, the rotational inertia is inertia of the beam related to the slope of the deflection curve, which is denoted $w_x(x,t) = \partial w(x,t)/\partial x$ and corresponds to an acceleration of

$$\frac{\partial^2}{\partial t^2}\left(\frac{\partial w}{\partial x}\right) = \frac{\partial^3 w(x,t)}{\partial t^2 \partial x} \tag{7.74}$$

Including the inertia term corresponding to this rotational acceleration in the moment equation about point Q in Figure 7.8 becomes

$$\frac{\partial M(x,t)}{\partial x} + V(x,t) = \rho I(x)\frac{\partial^3 w(x,t)}{\partial t^2 \partial x} \tag{7.75}$$

Solving Equation (7.75) for the shear force, $V(x,t)$, changes Equation (7.66) to become

$$V(x,t) = -\frac{\partial M(x,t)}{\partial x} + \rho I(x)\frac{\partial^3 w(x,t)}{\partial t^2 \partial x} \tag{7.76}$$

Summing the forces in the y-direction yields (as before)

$$\frac{\partial V(x,t)}{\partial x}dx = \rho A(x)\,dx\frac{\partial^2 w(x,t)}{\partial t^2} \tag{7.77}$$

Substitution of Equation (7.76) into Equation (7.77) yields

$$\frac{\partial}{\partial x}\left(-\frac{\partial M(x,t)}{\partial x} + \rho I(x)\frac{\partial^3 w(x,t)}{\partial t^2 \partial x}\right) = \rho A(x)\frac{\partial^2 w(x,t)}{\partial t^2} \tag{7.78}$$

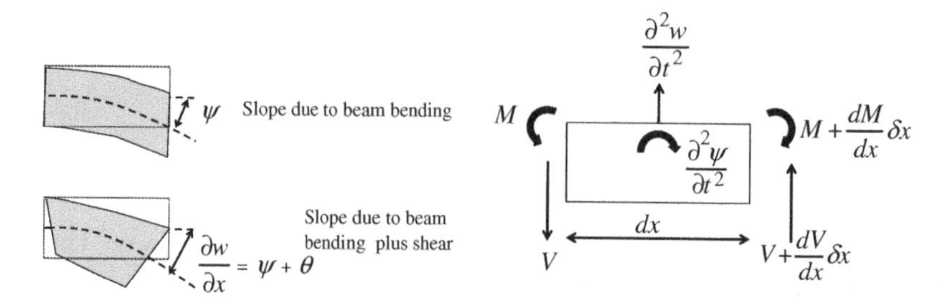

Figure 7.9 (left) The angles used to represent the effects of shear deformation. (right) Free body diagram of a differential element showing the forces and moments.

Using the value of the moment, $M(x, t)$ given by Equation (7.67) results in

$$\frac{\partial^2}{\partial x^2}\left(EI(x)\frac{\partial^2 w(x,t)}{\partial x^2}\right) = \rho I(x)\frac{\partial^3 w(x,t)}{\partial t^2 \partial x} - \rho A(x)\frac{\partial^2 w(x,t)}{\partial t^2} \tag{7.77}$$

Equation (7.77) is the equation of motion for a slender beam, including the effects of rotary inertia. This is called the *Rayleigh beam model* (Han et al., 1999).

The deflection due to shear introduces an angle of shear, θ, made with the tangent to the neutral axis so that the total slope now becomes

$$\frac{\partial w(x,t)}{\partial x} = \psi(x,t) + \theta(x,t) \tag{7.78}$$

Figure 7.9 illustrates these various angles. The slope $\psi(x, t)$ now becomes the relevant function to consider and introduces a second equation of motion coupled with the equation for the deflection $w(x, t)$. Note that if the shear is neglected, θ is zero and the slope due to bending is identical to $w_x(x, t)$. Recall from the study of solid mechanics and taking into account the slope due to bending, that

$$M(x,t) = EI\frac{\partial \psi(x,t)}{\partial x}, \text{and } V(x,t) = -\kappa^2 AG\theta = -\kappa^2 AG\left(\frac{\partial w(x,t)}{\partial x} - \psi(x,t)\right) \tag{7.79}$$

Here κ^2 is called the shear correction factor or shear coefficient and depends on the shape of the cross-section A. The constant G is the shear modulus. See Han et al. (1999) or Cowper (1966) for values of κ^2 for various shapes.

Summing moments from Figure 7.9, considering the shear angle yields

$$-V(x,t)\,dx + \frac{\partial M(x,t)}{\partial x}dx - \rho I(x)\frac{\partial^2 \psi(x,t)}{\partial t^2}dx = 0 \tag{7.80}$$

Substitution of Equations (7.79) into Equation (7.80) and setting the coefficient of dx to zero yields

$$\kappa^2 AG\left(\frac{\partial w(x,t)}{\partial x} - \psi(x,t)\right) + \frac{\partial}{\partial x}\left[EI\frac{\partial \psi(x,t)}{\partial x}\right] = \rho I(x)\frac{\partial^2 \psi(x,t)}{\partial t^2} \tag{7.81}$$

Summing forces in the vertical direction in Figure 7.9 yields

$$-\frac{\partial V(x,t)}{\partial x} - \rho A(x)\frac{\partial^2 w(x,t)}{\partial t^2} = 0 \tag{7.82}$$

This is of the same form as Equation (7.64) for the Euler-Bernoulli beam, except the interpretation of V is now based on Equation (7.79) so that the force balance becomes

$$\kappa^2 A(x)G\frac{\partial}{\partial x}\left(\frac{\partial w(x,t)}{\partial x} - \psi(x,t)\right) - \rho A(x)\frac{\partial^2 w(x,t)}{\partial t^2} = 0 \tag{7.83}$$

Equations (7.81) and (7.83) represent two coupled partial differential equations defining the vibrations of a beam, including the effect of both shear deformation and rotary inertia, which is called a *Timoshenko beam* model.

The Timoshenko beam equations can be combined and written as one equation in the displacement $w(x, t)$, the quantity that is easiest to measure and defines the response. Performing the derivative indicated in Equation (7.83) and solving for $\psi_x(x,t)$ yields

$$\frac{\partial \psi(x,t)}{\partial x} = \frac{\partial^2 w(x,t)}{\partial x^2} - \frac{\rho}{\kappa^2 G}\frac{\partial^2 w(x,t)}{\partial t^2} = 0 \tag{7.84}$$

Taking the partial derivative of Equation (7.81) and using Equation (7.84) to eliminate $\psi_x(x,t)$yields (suppressing the x and t dependence and assuming E and I constant)

$$EI\frac{\partial^4 w}{\partial x^4} + \rho A\frac{\partial^2 w}{\partial t^2} - \rho I\left(1 + \frac{E}{\kappa^2 G}\right)\frac{\partial^4 w}{\partial t^2 \partial x^2} + \frac{\rho I}{\kappa^2 G}\frac{\partial^4 w}{\partial x^4} = 0 \tag{7.85}$$

The boundary conditions for the Timoshenko beam equation are necessarily more complex. These are stated in terms of both the slope $\psi(x,t)$ and the deflection $w(x, t)$. For a clamped end, the boundary conditions become (at $x = 0$, say)

$$\psi(0,t) = w(0,t) = 0 \tag{7.86}$$

At a simply supported end, they become

$$EI\frac{\partial \psi\,(0,\ t)}{\partial x} = w\,(0,\ t) = 0 \tag{7.87}$$

and at a free end, they become

$$\kappa^2 AG\left(\frac{\partial w}{\partial x} - \psi\right) = EI\frac{\partial \psi}{\partial x} = 0 \tag{7.88}$$

In order to solve the Timoshenko beam equations, four initial conditions are also needed

$$\begin{aligned}
\psi(x,0) &= \psi_0(x)\\
\psi_t(x,0) &= \dot{\psi}_0(x)\\
w(x,0) &= w_0(x)\\
w_t(x,0) &= \dot{w}_0(x)
\end{aligned} \tag{7.89}$$

where the functions on the left are assumed to be given.

Example 7.5.1

Consider a cantilevered Euler-Bernoulli beam (clamped at $x = 0$ end free at the $x = l$) and compute the natural frequencies and mode shapes using separation of variables as described previously. Assume E, I, ρ and l are constant. The method is basically the same as used for a string, rod and bar, but the resulting boundary value problem is slightly more complex. The clamped-free boundary conditions are

$$w(0,t) = w_x(0,t) = 0 \quad \text{and} \quad w_{xx}(l,t) = w_{xxx}(l,t) = 0$$

The equation of motion is

$$\frac{\partial^2 w}{\partial t^2} + \left(\frac{EI}{\rho A}\right)\frac{\partial^4 w}{\partial x^4} = 0$$

Using the method of separation of variables, assume the solution is of the form: $w(x,t) = X(x)T(t)$ to obtain

$$\left(\frac{EI}{\rho A}\right)\frac{X''''}{X} = -\frac{\ddot{T}(t)}{T(t)} = \omega^2$$

The spatial equation becomes

$$X''''(x) - \left(\frac{\rho A}{EI}\right)\omega^2 X(x) = 0$$

Next define: $\beta^4 = \dfrac{\rho A \omega^2}{EI}$ so that the equation of motion becomes $X'''' - \beta^4 X = 0$, which has the solution

$$X(x) = C_1 \sin\beta x + C_2 \cos\beta x + C_3 \sinh\beta x + C_4 \cosh\beta x$$

Applying the boundary conditions in separated form

$$X(0) = X'(0) = 0 \qquad \text{and} \qquad X''(l) = X'''(l) = 0$$

yields four equations for unknown coefficients C_i. These four equations are written in matrix form as

$$\begin{bmatrix} 0 & 1 & 0 & 1 \\ 1 & 0 & 1 & 0 \\ -\sin\beta l & -\cos\beta l & \sinh\beta l & \cosh\beta l \\ -\cos\beta l & \sin\beta l & \cosh\beta l & \sinh\beta l \end{bmatrix}\begin{bmatrix} C_1 \\ C_2 \\ C_3 \\ C_4 \end{bmatrix} = 0$$

For a nonzero solution for the coefficients C_i, the matrix determinant must be zero. The determinant yields the characteristic equation

$$\begin{aligned} &(-\sin\beta l - \sinh\beta l)(\sin\beta l - \sinh\beta l) \\ &\quad -(-\cos\beta l - \cosh\beta l)(-\cos\beta l - \cosh\beta l) = 0 \end{aligned}$$

Simplifying, the characteristic equation becomes: $\cos\beta l \cosh\beta l = -1$, or

$$\cos\beta_n l = -\frac{1}{\cosh\beta_n l}$$

This last expression is solved numerically for the values βl, which are now indexed with the subscript $n = 1, 2, 3 \ldots\infty$ to indicate the many solutions. The frequencies are then

$$\omega_n = \sqrt{\frac{\beta_n^4 EI}{\rho A}}, \; n = 1, 2, 3 \cdots \infty$$

The mode shapes given by the solution for each β_n are

$$X_n = C_{1n} \sin\beta_n x + C_{2n} \cos\beta_n x + C_{3n} \sinh\beta_n x + C_{4n} \cosh\beta_n x$$

Using the boundary condition information that $C_4 = -C_2$ and $C_3 = -C_1$ yields

$$-C_1(\sin\beta l + \sinh\beta l) = C_2(\cos\beta l + \cosh\beta l)$$

so that

$$C_1 = -C_2\left(\frac{\cos\beta l + \cosh\beta l}{\sin\beta l + \sinh\beta l}\right)$$

The mode shapes can then be expressed as

$$X_n = -C_{2n}\left[-\left(\frac{\cos\beta_n l + \cosh\beta_n l}{\sin\beta_n l + \sinh\beta_n l}\right)\sin\beta_n x + \cos\beta_n x\right.$$
$$\left. + \left(\frac{\cos\beta_n l + \cosh\beta_n l}{\sin\beta_n l + \sinh\beta_n l}\right)\sinh\beta_n x - \cosh\beta_n x\right]$$

Additional examples of the solution to the beam equation can be found in the next chapter, which discusses formal methods of solution.

7.6 Coupled Effects

Beams are often the structure of choice when trying to investigate new vibration problems. In this section a basic beam is used to understand the coupling between several different effects: coupled bending and torsion, fluid structure interaction, and piezoelectric excitation and sensing. These topics are only introduced here. Each of these topics deserves a lot more attention than provided and the references should be consulted for more details (Erturk and Inman, 2011).

Obviously beams can vibrate in more than one direction. Here we consider a slender beam structure that is coupled in bending and torsion as it vibrates. Such coupled vibration can occur in beams of any geometry; however, here we consider a U-shaped beam because its geometry is easier to visualize. Some examples of common situations that result in coupled bending and torsion motion are aircraft wings, helicopter blades, beams made of composites, fan blades, bladed discs, rotating shafts, etc. In airfoil shapes, the classic issue is flutter instability induced by the coalescence of bending and torsional modes. Consider the beam of Figure 7.10. Because of the geometry, the shear center (s) and center of gravity

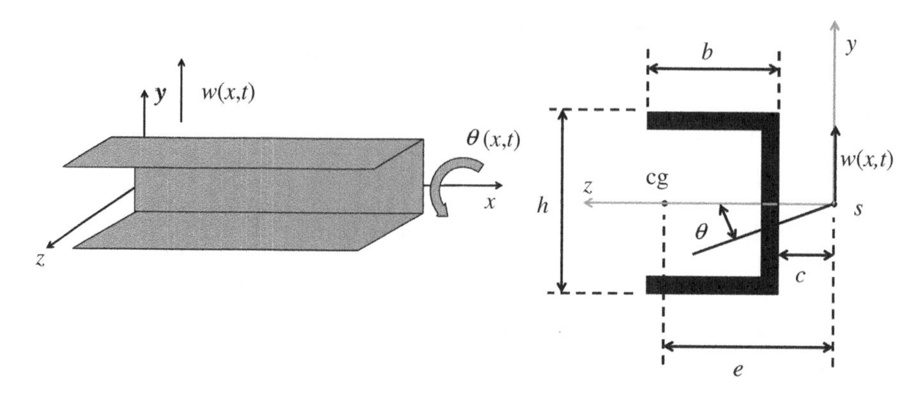

Figure 7.10 Dimensions and parameters for a U beam. The figure on the right shows a cross-section.

(cg) no longer coincide, causing the coupling between bending vibration and torsional vibration.

The equations of motion of the beam in Figure 7.10 are derived from Weaver *et al.* (1990). They are

$$\begin{aligned} EI_z\frac{\partial^4 w(x,t)}{\partial x^4} &= -\rho A\frac{\partial^2}{\partial t^2}[w(x,t) - c\theta(x,t)] \\ R\frac{\partial^2\theta(x,t)}{\partial x^2} - R_1\frac{\partial^4\theta(x,t)}{\partial x^4} &= -\rho Ac\frac{\partial^2}{\partial t^2}[w(x,t) - c\theta(x,t)] + \rho I_p\frac{\partial^2\theta(x,t)}{\partial t^2} \end{aligned} \tag{7.90}$$

Here w and θ are the transverse deflection and rotation respectively about the x-axis, as indicated in the Figure 7.10. The constant R is the torsional rigidity, R_1 is the warping rigidity, ρ is the mass density, I_p is the centroidal polar of moment, A is the cross-sectional area, EI_z is flexural rigidity and c and e are geometric constants defined by

$$c = e + \frac{b^2}{2b+h}, \qquad e = \frac{b^2h^2t_1}{4I_z} \tag{7.91}$$

where b and h are defined in Figure 7.10 and t_1 is the thickness of the material defining the beam. As in most cases, the easiest set of boundary conditions to use are pinned-pinned boundary conditions. They are for a beam of length l

$$\begin{aligned} &w(0,t) = 0,\ w_{xx}(0,t) = 0,\ \theta(0,t) = 0,\ \theta_{xx}(0,t) = 0 \\ &w(l,t) = 0,\ w_{xx}(l,t) = 0,\ \theta(l,t) = 0,\ \theta_{xx}(l,t) = 0 \end{aligned} \tag{7.92}$$

The most common case useful for wings and rotor blades is the clamped free case, which has boundary conditions (clamped at 0 and free at l)

$$\begin{aligned} &w(0,t) = 0, \quad w_x(0,t) = 0, \quad \theta(0,t) = 0, \quad \theta_x(0,t) = 0 \\ &w_{xx}(l,t) = 0, \quad w_{xx}(l,t) = 0, \quad \theta_{xx}(l,t) = 0, \quad -EI_p\theta_x(l,t) + R\theta_x(l,t) = 0 \end{aligned} \tag{7.93}$$

Using the appropriate boundary conditions and the equations of motion, the frequencies can be determined by separation of variables or by the approximate methods indicated in Chapter 11. In order to solve for displacement and rotation, four initial conditions are needed, two for w and two for θ (displacements and velocities). Equations (7.90) effectively represent the coupling of Equation (7.68) for the Euler-Bernoulli beam and Equation (7.54) for the torsional vibration of a shaft.

Example 7.6.1

Using pinned-pinned boundary conditions, find expressions for the natural frequencies of the beam of Figure 7.10.

Solution: Assume a separation of variables solution of the form

$$w(x,t) = W(x)T(t) \quad \text{and} \quad \theta(x,t) = \Theta(x)T(t)$$

Substitution into Equations (7.90) and dividing by $T(t)$ yields

$$\begin{aligned} EI_z W''''(x) &= \{-\rho A W(x) + c\rho A\Theta(x)\}\frac{\ddot{T}(t)}{T(t)} \\ R\Theta''(x) - R_1\Theta''''(x) &= \{-\rho A c W(x) + \rho A c^2\Theta(x) + \rho I_p\Theta(x)\}\frac{\ddot{T}(t)}{T(t)} \end{aligned} \tag{7.94}$$

Assuming harmonic motion in time, the frequencies of vibration will be of the form

$$-\omega_n^2 = \frac{\ddot{T}(t)}{T(t)}$$

Furthermore, note that the functions $W(x)$ and $\Theta(x)$ satisfy the simply supported boundary conditions if they have the form

$$W(x) = a\sin\frac{n\pi x}{l}, \quad \Theta(x) = b\sin\frac{n\pi x}{l}$$

Substitution of these last expressions into Equations (7.94) and manipulating yields

$$EI_z\frac{n^4\pi^4}{l^4}a = -\{-\rho A a + c\rho A b\}\omega_n^2$$

$$\left\{R\frac{n^2\pi^2}{l^2} - R\frac{n^4\pi^4}{l^4}\right\}b = -\{-\rho A c a + (\rho A c^2 + \rho I_p)b\}\omega_n^2$$

Solving the first equation for a in terms of b, substituting into the second equation above and cancelling a yields the frequency equation

$$\begin{aligned} &\omega_n^4 - \omega_{bn}^2\omega_n^2 + \omega_{tn}^2\omega_n^2 + \omega_{bn}^2\omega_{tn}^2 = \lambda\omega_n^4 \\ &\Rightarrow (1-\lambda)\omega_n^4 + \left(\omega_{tn}^2 - \omega_{bn}^2\right)\omega_n^2 + \omega_{bn}^2\omega_{tn}^2 = 0 \end{aligned} \tag{7.95}$$

where

$$\lambda = \frac{Ac^2}{I_p + Ac^2}, \quad \omega_{bn}^2 = \frac{E_z n^4\pi^4}{L^4\rho A}, \quad \omega_{tn}^2 = \frac{(Rn^2\pi^2L^2 + R_1 n^4\pi^4)}{L^4\rho(I_p + Ac^2)}$$

Solving Equation (7.95) using the quadratic formula yields

$$\omega_n^2 = \frac{\left(\omega_{tn}^2 + \omega_{bn}^2\right) \pm \sqrt{\left(\omega_{tn}^2 - \omega_{bn}^2\right)^2 + 4\lambda\omega_{bn}^2\omega_{tn}^2}}{2(1-\lambda)} \tag{7.96}$$

Note that if $c = 0$ so that shear center coincides with the centroid, then the equations decouple and the frequencies become those of transverse vibration and torsional vibration separately. For other boundary conditions, the frequencies are computed numerically.

Next consider a thin walled pipe carrying a fluid, as indicated in Figure 7.11.

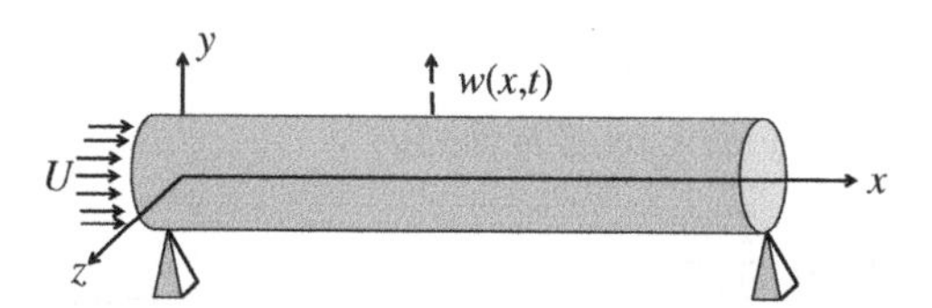

Figure 7.11 Vibration model of a pipe carrying a fluid in transverse vibration $w(x, t)$.

The flow in the pipe of velocity U and density μ is modeled by a force per unit length given by

$$F(x,t) = \mu dx\left(U^2\frac{\partial^2 w(x,t)}{\partial x^2} + 2U\frac{\partial^2 w(x,t)}{\partial x\partial t} + \frac{\partial^2 w(x,t)}{\partial t^2}\right) + \rho dx\frac{\partial^2 w(w,t)}{\partial t^2}$$

Here ρ is the density of the pipe material of stiffness EI. Following Euler-Bournoulli beam assumptions, the equation of vibration becomes (Ashley and Haviland, 1950)

$$\frac{\partial^2}{\partial x^2}\left(EI\frac{\partial^2 w(x,t)}{\partial x^2}\right) + \mu U^2\frac{\partial^2 w(x,t)}{\partial x^2} + 2U\frac{\partial^2 w(x,t)}{\partial x\partial t} + (\mu+\rho)\frac{\partial^2 w(x,t)}{\partial t^2} = 0 \tag{7.97}$$

As another example of a coupled system, consider a Euler-Bournoulli beam with a piezoceramic patch attached to it, as illustrated in Figure 7.12. Such systems are discussed in more detail in Section 7.10.

There are numerous approaches to modeling a beam with a piezoelectric patch, either as a sensor or actuation device, or both. One simple approach is to model

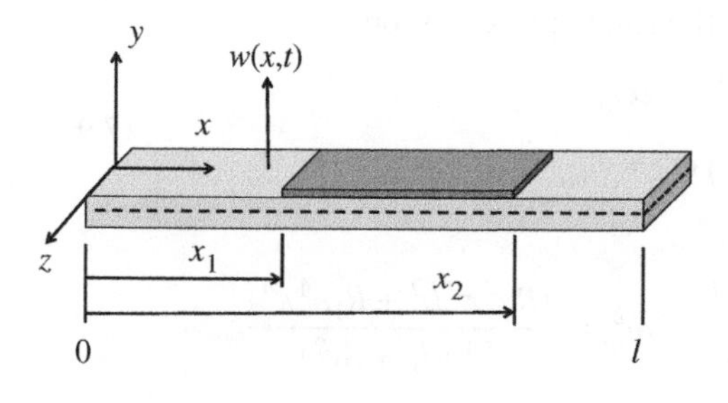

Figure 7.12 A beam with a piezoelectric patch layered between points x_1 and x_2.

the patch as applying a moment to the beam over the length of the patch as the result of an applied voltage denoted $V(t)$ given by

$$M(x) = aV[\Phi(x - x_1) - \Phi(x - x_2)] \tag{7.98}$$

Here Φ denotes the Heaviside step function and a is a constant, depending on the geometry and various physical parameters of the piezoelectric material. Using Hamilton's method of Section 7.2, the strain and kinetic energy expressions are

$$V = \frac{1}{2}\int_0^L EI\left(\frac{\partial^2 w}{\partial x^2}\right)^2 dx, \quad \text{and} \quad T = \frac{1}{2}\int_0^L \rho\left(\frac{\partial^2 w}{\partial t^2}\right)^2 dx$$

The applied moment defines the non-conservative work term given by

$$W = \frac{1}{2}\int_0^L M(x)\frac{\partial^2 w}{\partial x^2} dx$$

By substitution of these values into the Equation (7.10), the resulting equation of motion is

$$\rho A(x)\frac{\partial^2 w(x,t)}{\partial t^2} + \frac{\partial^2}{\partial x^2}\left(EI(x)\frac{\partial^2 w(x,t)}{\partial x^2}\right) = \frac{\partial^2 M(x,t)}{\partial x^2} \tag{7.99}$$

Given the appropriate boundary and initial conditions, solving Equation (7.99) is a bit tricky because of the indicated derivatives of the Heaviside functions indicated in the moment expression and the fact that the density and stiffness are not constants but also depend upon x. A more detailed model is taken up in Section 7.10, where piezoelectric patches bonded to structures are used in several applications.

7.7 Membranes and Plates

In this section, the equations for linear vibrations of membranes and plates are discussed. These objects are two-dimensional versions of the strings and beams discussed in the preceding sections. They occupy plane regions in space. The membrane represents a two-dimensional version of a string, and a plate can be thought of as a membrane with bending stiffness.

First, consider the equations of motion for a membrane. A membrane is basically a two-dimensional system that lies in a plane when in equilibrium. A common example is a drum head. The structure itself provides no resistance to bending, so that the restoring force is due only to the tension in the membrane. Thus, a membrane is similar to a string and, as was mentioned, is a two-dimensional version of a string. The reader is referred to Timoshenko et al. (1974) for the derivation of the membrane equation. Let $w(x,y,t)$ represent the displacement in the z-direction of a membrane, lying in the x–y plane at the point (x,y) and time t. The displacement is assumed to be small, with small slopes, and is perpendicular to the x–y plane. Let T be the tensile force per unit length of the membrane assumed to be the same in all directions and ρ be the mass per unit area of the membrane. Then the equation for free vibration is given by

$$T\nabla^2 w(x,y,t) = \rho w_{tt}(x,y,t) \quad x,y \in \Omega \tag{7.100}$$

where Ω denotes the region in the x–y plane occupied by the membrane. Here ∇^2 is the *Laplace operator*. In rectangular coordinates, this operator has the form

$$\nabla^2 = \frac{\partial^2}{\partial x^2} + \frac{\partial^2}{\partial y^2}. \tag{7.101}$$

The boundary conditions for the membrane must be specified along the shape of the boundary, not just at points, as in the case of the string. If the membrane is fixed or clamped at a segment of the boundary, then the deflection must be zero along that segment. If $\partial\Omega$ is the curve in the x–y plane corresponding to the edge of the membrane, i.e. the boundary of Ω, then the clamped boundary condition is denoted by

$$w(x,y,t) = 0 \quad x,y \in \partial\Omega \tag{7.102}$$

If, for some segment of $\partial\Omega$, denoted by $\partial\Omega_1$, the membrane is free to deflect transversely, then there can be no force component in the transverse direction, and the boundary condition becomes

$$\frac{\partial w(x,y,t)}{\partial n} = 0 \quad x,y \in \partial\Omega_1 \tag{7.103}$$

Here, $\partial w/\partial n$ denotes the derivative of $w(x,y,t)$ normal to the boundary in the reference plane of the membrane.

Example 7.7.1

Consider the vibration of a square membrane, as indicated in Figure 7.13, clamped at all of the edges. With $c^2 = T/\rho$, the equation of motion (7.100) becomes

$$c^2\left[\frac{\partial^2 w}{\partial x^2} + \frac{\partial^2 w}{\partial y^2}\right] = \frac{\partial^2 w}{\partial t^2} \quad x,y \in \Omega \tag{7.104}$$

Assuming that the solution separates, i.e. that $w(x,y,t) = X(x)Y(y)T(t)$, Equation (7.104) becomes

$$\frac{1}{c^2}\frac{\ddot{T}}{T} = \frac{X''}{X} + \frac{Y''}{Y} \tag{7.105}$$

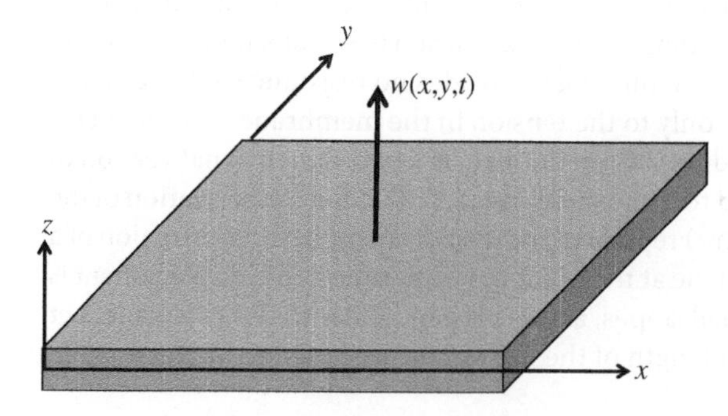

Figure 7.13 A square membrane or plate illustrating vibration perpendicular to its surface.

Equation (7.105) implies that $\ddot{T}/(Tc^2)$ is a constant (recall the argument used in Section 7.3). Denote the constant by ω^2, so that

$$\frac{\ddot{T}}{(Tc^2)} = -\omega^2 \tag{7.106}$$

Then Equation (7.105) implies that

$$\frac{X''}{X} = -\omega^2 - \frac{Y''}{Y} \tag{7.107}$$

By the same argument used before, both X''/X and Y''/Y must be constant (that is independent of t and x or y). Hence

$$\frac{X''}{X} = -\alpha^2 \tag{7.108}$$

and

$$\frac{Y''}{Y} = -\gamma^2 \tag{7.109}$$

where α^2 and γ^2 are constants. Equation (7.107) then yields

$$\omega^2 = \alpha^2 + \gamma^2 \tag{7.110}$$

These expressions result in two spatial equations to be solved

$$X'' + \alpha^2 X = 0 \tag{7.111}$$

which has a solution (A and B constants of integration) of the form

$$X(x) = A \sin\ ax + B\ \cos\ ax \tag{7.112}$$

and

$$Y'' + \gamma^2 Y = 0 \tag{7.113}$$

which yields (C and D constant of integration) a solution of the form

$$Y(y) = C \sin \gamma y + D \cos \gamma y \tag{7.114}$$

The total spatial solution is the product $X(x)Y(y)$, or

$$\begin{aligned} X(x)Y(y) = {} & A_1 \sin \alpha x \sin \gamma y + A_2 \sin \alpha x \cos \gamma y \\ & + A_3 \cos \alpha x \sin \gamma y + A_4 \cos \alpha x \cos \gamma y \end{aligned} \tag{7.115}$$

Here the constants A_i consist of the products of the constants in Equations (7.112) and (7.114) and are to be determined by the boundary and initial conditions.

Equation (7.115) can now be used with the boundary conditions to calculate the eigenvalues and eigenfunctions of the system. The clamped boundary condition, along $x = 0$ in Figure 7.13, yields

$$T(t)X(0)Y(y) = T(t)B(A_3 \sin \gamma y + A_4 \cos \gamma y) = 0$$

or

$$A_3 \sin \gamma y + A_4 \cos \gamma y = 0 \tag{7.116}$$

Now Equation (7.116) must hold for any value of y. Thus, as long as γ is not zero (a reasonable assumption, since if it is zero the system has a rigid body motion), A_3 and A_4 must be zero. Hence the spatial solution must have the form

$$X(x)Y(y) = A_1 \sin \alpha x \sin \gamma y + A_2 \sin \alpha x \cos \gamma y \tag{7.117}$$

Next, application of the boundary condition $w = 0$ along the line $x = 1$ yields

$$A_1 \sin \alpha \sin \gamma y + A_2 \sin \alpha \cos \gamma y = 0 \tag{7.118}$$

Factoring this expression yields

$$\sin \alpha (A_1 \sin \gamma y + A_2 \cos \gamma y) = 0 \tag{7.119}$$

Now either $\sin \alpha = 0$ or, by the preceding argument, A_1 and A_2 must be zero. However, if A_1 and A_2 are both zero, the solution is zero. Hence, in order for a nontrivial solution to exist, $\sin \alpha = 0$, which yields

$$\alpha = n\pi, \; n = 1, 2, \ldots \infty \tag{7.120}$$

Using the boundary condition $w = 0$ along the line $y = 1$ results in a similar procedure and yields

$$\gamma = m\pi, \quad m = 1, 2, \ldots \infty \tag{7.121}$$

Note that the possibility of $\gamma = \alpha = 0$ is not used because it was necessary to assume $\gamma \neq 0$ in order to derive Equation (7.117). Equation (7.110) yields that the constant ω in the temporal equation must have the form

$$\begin{aligned} \omega_{mn} &= \sqrt{\alpha_n^2 + \gamma_m^2} \\ &= \pi \sqrt{m^2 + n^2} \quad m, n = 1, 2, 3, \ldots \infty \end{aligned} \tag{7.122}$$

Thus the eigenvalues and eigenfunctions for the clamped membrane are Equation (7.122) and $\{\sin n\pi x$ and $\sin m\pi y\}$, respectively. The solution of Equation (7.104) becomes

$$w(x, y, t) = \sum_{m=1}^{\infty} \sum_{n=1}^{\infty} (\sin m\pi x \sin n\pi y)[A_{mn} \sin(\omega_{mn} c\pi t) + B_{mn} \cos(\omega_{nm} c\pi t)] \tag{7.123}$$

where A_{mn} and B_{mn} are determined by the initial conditions and the orthogonality of the eigenfunctions.

In progressing from the vibration of a string to considering the transverse vibration of a beam, the beam equation allowed for bending stiffness. In the same manner, a plate differs from a membrane because plates have bending stiffness. The reader is referred to Reismann (1988), Reismann and Pawlik (1974) or Sodel (1993) for a more detailed explanation and a precise derivation of the plate equation. Basically, the plate, like the membrane, is defined in a plane (x–y) with the deflection $w(x,y,t)$ taking place along the z-axis perpendicular to the x–y plane. The basic assumption is again small deflections with respect to the thickness, h. Thus, the plane running through the middle of the plate is assumed not to deform during bending (called a *neutral plane*). In addition, normal stresses in the direction transverse to the plate are assumed to be negligible. Again, there is no thickness stretch. The displacement equation of motion for the free vibration of the plate is

$$-D_E\nabla^4 w(x,y,t) = \rho w_{tt}(x,y,t), \quad x,y \in \Omega \tag{7.124}$$

where ρ is the mass density (per unit area), and the constant D_E, the plate flexural rigidity, is defined in terms of Poisson's ratio ν, and the plate thickness h and E the elastic modulus as

$$D_E = \frac{Eh^3}{12(1-\nu^2)}. \tag{7.125}$$

The operator ∇^4, called the *biharmonic operator*, is a fourth-order operator, the exact form of which depends on the choice of coordinate systems. In rectangular coordinates, the biharmonic operator becomes

$$\nabla^4 = \frac{\partial^4}{\partial x^4} + 2\frac{\partial^4}{\partial x^2 \partial y^2} + \frac{\partial^4}{\partial y^4} \tag{7.126}$$

The boundary conditions for a plate are a little more difficult to write, as their form, in some cases, also depends on the coordinate system in use. The following are example boundary conditions for a plate in a rectangular coordinate system.

Clamped Edge: For a clamped edge the deflection and normal derivative $\partial/\partial n$ are both zero along the edge

$$w(x,y,t) = 0, \quad \text{and} \quad \frac{\partial w(x,y,t)}{\partial n} = 0, \quad x,y \in \partial\Omega \tag{7.127}$$

Here the normal derivative is the derivative of w normal to the neutral plane.

Simply Supported: For a rectangular plate, the simply supported boundary conditions become

$$\begin{aligned} w(x,y,t) &= 0 \quad \text{along all edges} \\ \frac{\partial^2 w(x,y,t)}{\partial x^2} &= 0 \quad \text{along the edges } x = 0,\ x = l_1 \end{aligned} \tag{7.128}$$

$$\frac{\partial^2 w(x,y,t)}{\partial y^2} = 0 \quad \text{along the edges } y = 0,\ y = l_2 \tag{7.129}$$

where l_1 and l_2 are the lengths of the plate edges and the second partial derivatives reflect the normal strains along these edges.

7.8 Layered Materials

The use of layered materials and composites in the design of modern structures has become very popular because of increased strength to weight ratios. The theory of vibration of layered materials is not as developed, but does offer some interesting design flexibility.

The transverse vibration of a three-layer beam consisting of a core between two faceplates, as indicated in Figure 7.14, is considered here. The layered beam consists of two faceplates of thickness h_1 and h_3, which are sandwiched around a core beam of thickness h_2. The distance between the center lines of the two faceplates is denoted by d. The displacement equation of vibration becomes (Sun and Lu, 1995)

$$\frac{\partial^6 w(x,t)}{\partial x^6} - g(1+\beta)\frac{\partial^4 w(x,t)}{\partial x^4} + \frac{\rho}{D_e}\left[\frac{\partial^4 w}{\partial x^2 \partial t^2} - g\frac{\partial^2 w}{\partial t^2}\right] = 0 \quad x \in (0,l) \tag{7.130}$$

where

$$g = \frac{G}{h_2}\left[\frac{1}{E_1 h_1} + \frac{1}{E_3 h_3}\right]$$

G = shear modulus of the core

E_i = Young's modulus of the ith face plate

$$D_e = \frac{(E_1 h_1^3 + E_3 h_3^3)}{12}$$

$$\beta = \frac{d^2 E_1 h_1 E_3 h_3^3 (E_1 h_1 + E_3 h_3)}{D_e}$$

$$d = h_2 + \frac{(h_1 + h_3)}{2}$$

ρ = mass per unit length of the entire structure

The boundary conditions are again not as straightforward as those of a simple beam. In fact, since the equation for free vibration contains six derivatives, there

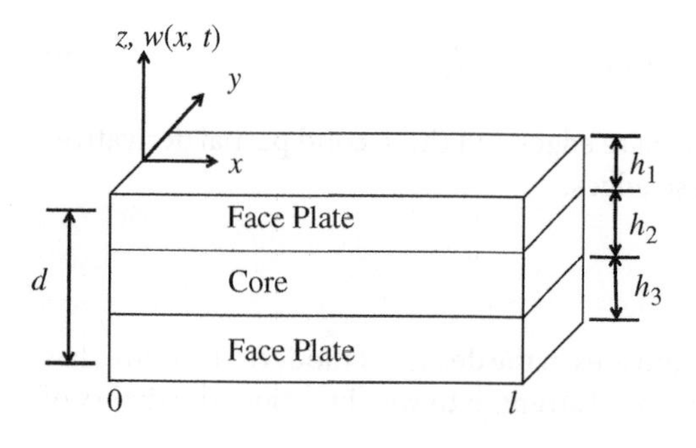

Figure 7.14 The dimensions of a three-layer beam for transverse vibration analysis.

are six boundary conditions that must be specified. If the beam has both ends clamped, the boundary conditions are

$$\begin{aligned}
&w(0,t) = w(l,t) = 0 \quad \text{(zero displacement)}\\
&w_x(0,t) = w_x(l,t) = 0 \quad \text{(zero rotation)}\\
&w_{xxxx}(0,t) - g(1+\beta)w_{xx}(0,t) - \frac{\rho}{D_e}w_{tt}(0,t) = 0\\
&w_{xxxx}(l,t) - g(1+\beta)w_{xx}(l,t) - \frac{\rho}{D_e}w_{tt}(l,t) = 0 \quad \text{(zero bending moments)}
\end{aligned} \tag{7.131}$$

Note that the form of this equation is different from all the other structures considered in this chapter. In all the previous cases, the equation for linear vibration can be written in the form

$$\begin{aligned}
w_{tt}(\mathbf{x},t) + L_2 w(\mathbf{x},t) &= 0 \quad \mathbf{x} \in \Omega\\
Bw(\mathbf{x},t) &= 0 \quad \mathbf{x} \in \partial\Omega
\end{aligned} \tag{7.132}$$

plus initial conditions, where Ω is a region in three-dimensional space bounded by $\partial\Omega$. Here L_2 is a linear operator in the spatial variables only, $\mathbf{x}$ is a vector consisting of the spatial coordinates x, y and z, and B represents the boundary conditions. As long as the vibration of a structure fits into the form of Equation (7.132) and the operator L_2 satisfies certain conditions (specified in Chapter 9), separation of variables can be used to solve the problem. However, Equation (7.130) is of the form

$$\begin{aligned}
L_0 w_{tt}(\mathbf{x},t) + L_2 w(\mathbf{x},t) &= 0 \quad \mathbf{x} \in \Omega\\
Bw(\mathbf{x},t) &= 0 \quad \mathbf{x} \in \partial\Omega
\end{aligned} \tag{7.133}$$

where both L_0 and L_2 are linear operators in the spatial variables. Because of the presence of two operators, it is not clear if separation of variables will work as a solution technique. The circumstances and assumptions required for separation of variables to work is the topic of the next chapter. In addition, classifications and further discussion of the operators are contained in Chapter 9.

The boundary conditions associated with the operator B in Equation (7.132) can be separated into two classes. Boundary conditions that arise clearly as the result of the geometry of the structure, such as Equation (7.71), are called *geometric boundary conditions*. Boundary conditions that arise by requiring a force (or moment) balance at the boundary are called *natural boundary conditions*. Equation (7.70) represents an example of a natural boundary condition.

7.9 Damping Models

Viscous damping, introduced in Section 1.3, is most commonly used. The reason for this consideration is that viscous damping lends itself to analytical solutions for transient as well as steady-state response by relatively simple techniques. While there is significant evidence indicating the inaccuracies of viscous damping models (Snowden, 1968), modeling dissipation as viscous damping represents a significant improvement over the conservative models given by Equations (7.132) or (7.133). In this section, several distributed parameter models of damping are introduced.

First, consider the transverse free vibration of a membrane in a surrounding medium (such as air) furnishing resistance to the motion that is proportional to the velocity (i.e. viscous damping). The equation of motion is Equation (7.100) with the addition of a damping force. The resulting equation is

$$\rho w_{tt}(x, y, t) + \gamma w_t(x, y, t) - T\nabla^2 w(x, y, t) = 0, \; x, y \in \Omega \tag{7.134}$$

where ρ, T and $\nabla^2$are as defined for Equation (7.100) and γ is the viscous damping coefficient. The positive constant γ reflects the proportional resistance to velocity. This system is subject to the same boundary conditions as discussed in Section 7.7.

The solution method (separation of variables) outlined in Example 7.7.1 works equally well for solving the damped membrane equation, Equation (7.134). The only change in the solution is that the temporal function $T(t)$ becomes an exponentially decaying sinusoid rather than a constant amplitude sinusoid, depending on the relative size of γ.

External damping of the viscous type can also be applied to the model of the free flexural vibration of a plate. This problem has been considered by Murthy and Sherbourne (1972). They modeled the damped plate by

$$\rho w_{tt}(x, y, t) + \gamma w_t(x, y, t) + D_E \nabla^4 w(x, y, t) = 0 \quad x, y \in \Omega \tag{7.135}$$

subject to the same boundary conditions as Equation (7.134). Here D_E, ρ and ∇^4 are as defined for Equation (7.104), and γ again represents the constant viscous damping parameter. The plate boundary conditions given in Section 7.7 also apply to the damped plate Equation (7.135)

Consider next the longitudinal vibration of a bar subject to both internal and external damping. In this case, the equation of vibration for the bar in Figure 7.3 becomes

$$w_{tt}(x, t) + 2\left(\gamma - \beta\frac{\partial^2}{\partial x^2}\right) w_t(x, t) - \frac{EA}{\rho} w_{xx}(x, t) = 0 \quad x \in \Omega \tag{7.136}$$

subject to the boundary conditions discussed in Section 7.4. The quantities E, A and ρ are taken to be constant versions of the like quantities defined in Equation (7.48). The constant γ is again a viscous damping factor derived from an external influence, whereas the constant β is a viscous damping factor representing an internal damping mechanism.

Note that the internal model is slightly more complicated than the other viscous damping models considered in this section because of the inclusion of the second spatial derivative. Damping models involving spatial derivatives can cause difficulties in computing analytical solutions and are discussed in Sections 8.4 and 9.6. In addition, damping terms can alter the boundary conditions.

Umland et al. (1991) proposed a viscous damping model for bending vibrations in a thin beam and experimentally determined the damping constants, c_1 and c_2, in the following model

$$\rho A(x)\frac{\partial^2 w(x, t)}{\partial t^2} + c_1\frac{\partial w(x, t)}{\partial t} + c_2 I\frac{\partial^5 w(x, t)}{\partial x^4 \partial t} + EI\frac{\partial^4 w(x, t)}{\partial x^4} = F(x, t) \tag{7.137}$$

Here $F(x, t)$ is an applied force. The boundary conditions for a cantilever configuration (fixed at 0 and free at L) become

$$\begin{aligned} &w(0,t) = w_x(0,t) = 0 \\ &Elw_{xx}(L,t) + c_2 Iw_{txx}(L,t)Elw_{xxx}(L,t) + c_2 Iw_{txxx}(L,t) = 0 \end{aligned} \tag{7.138}$$

Thus the damping coefficients c_1 and c_2 also show up in the boundary conditions.

Banks and Inman (1991) further investigated damping models for beams by discussing a Kelvin-Voigt (also called *strain rate damping*) term in Equation (7.137) defined by the operator

$$L = c_2 I \frac{\partial^5 w(x,t)}{\partial x^4 \partial t}$$

They showed the need to expand the boundary conditions to include the damping moment

$$M = c_2 I \frac{\partial^3 w(x,t)}{\partial x^2 \partial t}$$

These various viscous damping forms are often successful in modeling damping in metals and structures subject to air damping (or other fluid influences).

Another form of damping present in soft materials such as polymers, adhesives and rubbers is called *hysteretic damping,* because it introduces a hysteresis effect to the vibration response. Some composites also exhibit hysteresis. There are several mathematical models used to represent this effect. A common one is to use stress proportional to strain plus, the past history of strain by introducing an integral term of the form

$$\int_{-r}^{0} g(s) w_{xx}(x, t+s) ds, \quad g(s) = \frac{\alpha}{\sqrt{-s}} e^{\beta s}$$

Here α and β are constants determined by measurements. The equation of an Euler-Bernoulli beam with hysteretic damping becomes

$$\rho A \frac{\partial^2 w(x,t)}{\partial t^2} + \frac{\partial^2}{\partial x^2}\left[EI \frac{\partial^2 w(x,t)}{\partial x^2} - \int_{-r}^{0} g(s) w_{xx}(x, t+s) ds \right] = 0 \tag{7.139}$$

Because the hysteretic term affects the strain, the boundary conditions for a cantilever configuration also change and become

$$w(0,t) = w_x(0,t) = 0$$

at the clamped end and

$$\begin{aligned} &EI \frac{\partial^2 w(x,t)}{\partial x^2} - \int_{-r}^{0} g(s) w_{xx}(x, t+s) ds = 0 \\ &\frac{\partial}{\partial x}\left[EI \frac{\partial^2 w(x,t)}{\partial x^2} - \int_{-r}^{0} g(s) w_{xx}(x, t+s) ds \right] = 0 \end{aligned}$$

at the free end. The above forms are discussed in the context of using experiments to determine α, β, c_1 and c_2 in Banks and Inman (1991). They also discuss a concept of spatial hysteresis, but this form has not received much use.

The general model for vibration of distributed parameter systems with viscous damping can be written as

$$\begin{aligned} L_0 w_{tt}(\mathbf{x}, t) + L_1 w_t(\mathbf{x}, t) + L_2 w(\mathbf{x}, t) &= 0 \quad \mathbf{x} \in \Omega \\ Bw(\mathbf{x}, t) &= 0 \quad \mathbf{x} \in \partial\Omega \end{aligned} \tag{7.140}$$

plus appropriate initial conditions. Here the operators B, L_0 and L_2 are as defined for Equation (7.133), and the operator L_1 is exemplified by the models illustrated in this section. The operator L_0 is called the mass operator, the operator L_1 is called the damping operator, and the operator L_2 is called the stiffness operator. As illustrated by the examples, the operator L_0 is often the identity operator. The properties of these operators, the nature of the solutions of Equations (7.132) and (7.140), and their relationship to the vibration problem, are topics of the next three chapters.

7.10 Modeling Piezoelectric Wafers

Piezoelectric materials briefly introduced in Section 7.6 have numerous uses in vibration related topics. The piezoelectric effect is evident in a number of different types of materials, including both ceramic based and polymer based. The piezoelectric effect basically couples mechanical strain in a material with its electric displacement. The resulting coupling means that straining a piezoelectric material generates a voltage across the material and applying a voltage to the material causes the material to strain. Thus a piezoelectric patch can be used as a sensor and/or an actuator (Leo, 2007). In one dimension, the equations describing the coupling due to the piezoelectric effect are

$$\begin{aligned} S &= s^E T + dE \\ D &= dT + \varepsilon^T E \end{aligned} \tag{7.141}$$

Here S is the mechanical strain, T is the mechanical stress, E is the electric field strength, D is the dielectric displacement, s^E is the compliance measured at zero electric field, and ε^T is the dielectric constant measured at zero stress. The piezoelectric constant, d, determines the coupling between mechanical and electrical fields. Since the piezoelectric effect is three-dimensional, Equation (7.141) is really a full tensor relationship. This unusual notation for stress and strain, and the use of E for electric field, rather than modulus, results from electrical engineers being the first to model the piezoelectric effect. In most circumstances, the piezoelectric material is in the shape of a thin plate with the electric field perpendicular to the resulting strain with the respective coupling being denoted by d_{31} (see Leo, 2007, for values).

First consider using a piezoelectric layer to actuate a beam. The schematic in Figure 7.15 shows of a typical "bimorph" configuration consisting of an identical layer of material on the top and bottom and glued onto the beam with an adhesive. As a voltage is applied to the piezoelectric patch perpendicular to the x-axis, the patch will stretch or shrink in the x-direction. If each patch is activated with an opposite voltage, one patch will stretch and one will shrink causing a bending moment on the beam.

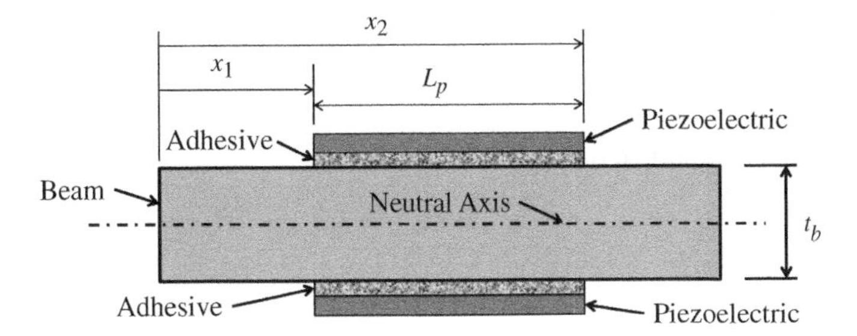

Figure 7.15 Dimensions of a piezoelectric "bimorph" beam, with adhesive of thickness t_a and piezoelectric wafer of thickness t_p.

Ignoring the adhesive effects, the moment applied by to the beam is a function of the geometry of the patch (of width b), its modulus, E_p, the moment arm $(t_a + t_b)$ and the applied voltage, $v(t)$, and is give by

$$M(x,t) = bE_p d_{31}(t_a + t_b)v(t)[\Phi(x - x_2) - \Phi(x - x_1)] \tag{7.142}$$

As in Equation (7.98), Φ denotes the Heaviside Step Function and the location of the patch. Usually the piezoelectric material is chosen to be a piezoceramic because of its higher value of d_{31} compared with the piezoelectric polymers. Typical values for a piezoceramic material are $E_p = 63$ GPa and $d_{31} = 120 \times 10^{-12}$ m/V.

The equations of motion can be derived using the Hamiltonian and results in the same form as Equation (7.99) for the single patch (called a unimorph configuration). The only differences between the bimorph and unimoph equations of motion are the value of the coefficient in Equations (7.98) and (7.142) and the functional form of the modulus and density.

The strain energy $V(t)$ in the transverse direction of a beam is given by

$$V(t) = \frac{1}{2}\int_0^L EI(x)\left[\frac{\partial^2 w(x,t)}{\partial x^2}\right]^2 dx \tag{7.143}$$

where the term in brackets is derived from the radius of curvature of the beam assuming small strain and $EI(x)$ is the beam stiffness. The kinetic energy of a differential element of the beam is

$$T(t) = \frac{1}{2}\int_0^L \rho A(x)\left[\frac{\partial w(x,t)}{\partial t}\right]^2 dx \tag{7.144}$$

where $\rho(x)$ is the density of the beam. Next consider the non-conservative work due to the presence of the piezoelectric layer

$$W(t) = \int_0^L M(x)\left[\frac{\partial^2 w(x,t)}{\partial x^2}\right] dx \tag{7.145}$$

Again the term in brackets is derived from the radius of curvature of the beam. The Lagrangian modified to account for non-conservative forces is

$$L = T - V + W \tag{7.146}$$

Application of Hamilton's principle becomes

$$\delta \int_{t_1}^{t_2} (T - V + W)dt = 0 \tag{7.147}$$

Here the variational operator δ causes an infinitesimal change in a function for a fixed value of x, so that $\delta x = 0$, and δ is commutative with respect to the differentiation and integration (Reddy, 1984). The following computes the variations of each term independently to avoid long expressions. First, consider the variation of the kinetic energy term

$$\begin{aligned}\delta \int_{t_1}^{t_2} T dt &= \frac{1}{2}\delta \int_{t_1}^{t_2}\int_0^L \rho A(x)\left[\frac{\partial w(x,t)}{\partial t}\right]^2 dxdt \\ &= \frac{1}{2}\int_{t_1}^{t_2}\int_0^L \rho A(x) 2 w_t \delta w_t dxdt = \int_{t_1}^{t_2}\int_0^L \rho A(x) w_t \delta w_t dxdt\end{aligned}$$

Here the variational operator δ passes through $\rho(x)$ because the variation is taken for a fixed value of x. To shorten the expressions, the x and t dependence has been suppressed and the index notation sub t replaces the partial derivative symbol. Integrating by parts yields (Equation (7.15)

$$\delta \int_{t_1}^{t_2} T dt = -\int_{t_1}^{t_2}\int_0^l \rho A(x) w_{tt} \delta w dxdt \tag{7.148}$$

Next consider the variation in the potential energy term

$$\begin{aligned}\delta \int_{t_1}^{t_2} V dt &= \frac{1}{2}\delta \int_{t_1}^{t_2}\int_0^L EI(x)\left[\frac{\partial^2 w(x,t)}{\partial x^2}\right]^2 dxdt \\ &= \frac{1}{2}\int_{t_1}^{t_2}\int_0^L EI(x) 2(w_{xx})\delta(w_{xx})\, dxdt \\ &= \int_{t_1}^{t_2}\int_0^L \underbrace{EI(x)w_{xx}}_{u} \underbrace{\frac{\partial}{\partial x}(\delta w_x)dxdt}_{dv} \\ &= \int_{t_1}^{t_2}\left\{ \underbrace{EI(x)w_{xx}}_{u} \underbrace{\delta w_x}_{v}\Bigg|_0^L - \int_0^L \underbrace{\delta w_x}_{v} \underbrace{\frac{\partial}{\partial x(EI(x)w_{xx}dx}}_{du}\right\} dt\end{aligned}$$

Here integration of parts was used ($\int udv = uv - \int vdu$) and again the variational operator δ treats $EI(x)$ as a constant as δ operates over a constant displacement. Integrating the x integral again, where this time $u = (EIw_{xx})_x$ and $dv = \delta w_x$, the variation of the potential becomes

$$\delta \int_{t_1}^{t_2} Vdt = \int_{t_1}^{t_2} \int_0^l \left\{ EI(x)\frac{\partial^2 w}{\partial x^2}\delta\frac{\partial w}{\partial x}\bigg|_0^L - \frac{\partial}{\partial x}\left(EI(x)\frac{\partial^2 w}{\partial x^2}\right)\delta w\bigg|_0^L + \int_0^L \frac{\partial^2}{\partial x^2}\left(EI(x)\frac{\partial^2 w}{\partial x^2}\right)\delta w dx \right\} dt \tag{7.149}$$

Repeating this procedure for the piezoelectric layer and realizing that $M(x)$ vanishes at the $x = 0$ and L yields

$$\begin{aligned}\delta W &= \delta \int_{t_1}^{t_2}\int_0^L M(x)\left[\frac{\partial^2 w(x,t)}{\partial x^2}\right] dxdt \int_{t_1}^{t_2}\int_0^L \underbrace{m(x)\delta}_{v} \underbrace{\left[\frac{\partial^2 w(x,t)}{\partial x^2}\right]}_{dv} dxdt \\ &= \int_{t_1}^{t_2}\left\{ M(x)\frac{\partial^2 w(x,t)}{\partial t^2}\bigg|_0^L - \int_0^L \frac{\partial M}{\partial x}\frac{\partial}{\partial x}(\delta x)\, dx \right\} dt = \int_{t_1}^{t_2} \frac{\partial^2 M}{\partial x^2}\delta w dt\end{aligned} \tag{7.150}$$

The boundary terms all vanish and integration by parts was used again in the second line of Equation (7.150). Gathering up all the terms in the Lagrangian and substituting into Equation (7.147), using pinned-pinned boundary conditions and realizing that the coefficients of δw must vanish, yields the following equation of motion and boundary conditions for a pinned-pinned beam

$$\begin{aligned}&\rho A(x)\frac{\partial^2 w(x,t)}{\partial t^2} + \frac{\partial^2}{\partial x^2}\left[EI(x)\frac{\partial^2 w(x,t)}{\partial x^2}\right] = \\ &\quad v(t)bE_p d_{31}(t_a + t_b)\frac{\partial^2}{\partial x^2}[\Phi(x - x_2) - \Phi(x - x_1)]\end{aligned} \tag{7.151}$$

$$\text{deflection} = w(x,t)|_{x=a} = 0$$

$$\text{bending moment} = EI(x)\frac{\partial^2 w(x,t)}{\partial x^2}\bigg|_{x=a} = 0$$

The constant a in Equation (7.151) takes on the value of either 0 or L. Here the control input is the voltage $v(t)$ applied to the piezoelectric wafers. Equation (7.151) allows the design and implementation of an open loop control for shaping the response of the beam. However, if feedback is desired, the sensing aspect of the piezoelectric wafer must be considered as discussed next.

As the beam in Figure 7.15 bends, it induces a charge in the piezoelectric wafer due to the stress created in the wafer. The stress at the centroid at a point on the centroidal axis, denoted as y_c, is

$$T(x) = \frac{y_c E_s}{r} = y_c E_s \frac{\partial^2 w(x,t)}{\partial x^2} \tag{7.152}$$

Here r is the radius of curvature for small deflections as before. The electric displacement caused by the strain (from Equation (7.141) with zero applied electric field) is

$$D = d_{31}T(x, y) = y_c d_{31} E_s \frac{\partial^2 w(x, t)}{\partial x^2} \tag{7.153}$$

Here E_s denotes the modulus of the piezoelectric sensor wafer (same as E_p). The charge induced by the electric displacement is

$$q = \int_A \mathbf{D}\cdot\mathbf{n}da = \int_{x_2}^{x_2} y_c d_{31} E_s \frac{\partial^2 w(x, t)}{\partial x^2} dx = y_c d_{31} E_s b[w_x(x_2, t) - w_x(x_1, t)] \tag{7.154}$$

where $\mathbf{n}$ is the unit normal vector to the area A. Both $\mathbf{D}$ and $\mathbf{n}$ are scalars in this case. The induced voltage $v_s(t)$ is just the charge divided by the capacitance of the piezoelectric waver, C_a, so that

$$v_s(t) = K_s[w_x(x_2, t) - w_x(x_1, t)], \;\; K_s = \frac{y_c d_{31} E_s b}{c_a} \tag{7.155}$$

The constant K_s is called the sensor coefficient.

Equations (7.151) and (7.155) form an actuator and sensor system for use in a number of applications. Sensing can be accomplished by solving Equation (7.151) for a given set of initial conditions and substituting the solution into Equation (7.155) to compute a voltage representative of the motion of the beam (proportional to strain). Open loop control can be accomplished by specifying the voltage in Equation (7.151). Combining both the sensor function and control function allows the implementation of closed loop control. Typically, one piezoelectric wafer is used for actuation using Equation (7.151), a separate wafer is used as a sensor computed from Equation (7.155) and a control law is devised to perform active vibration suppression and/or shape control. However, one wafer can be used to perform both functions taking into consideration the coupling indicated in Equation (7.141); for details see Dosch et al (2001). Fanson and Caughey (1990) use these actuation and sensing equations for implementing positive position feedback control on a beam with piezoceramic actuation and sensing.

By adding a resistor across the piezoelectric wafer, the resistive heating of the induced sensor voltage across the resister extracts energy from the beam as it vibrates forming a damping effect, called a *shunt damper* (Lesieutre, 1998). Shunt dampers are effective compared to viscoelastic damping treatments in some situations, because they are less affected by temperature differences. However, in order to capture the effect of shunt damping, Equations (7.151) and (7.155) must be coupled to the relevant electric circuit through the piezoelectric constitutive laws given in Equation (7.141). Details of the solutions of the above-mentioned uses of these equations are given in Section 10.9.

Chapter Notes

This chapter presents a brief introduction to the linear vibration of distributed parameter systems without regard for mathematical rigor or a proper derivation of the governing equations. The chapter is intended to review and familiarize the reader with some of the basic structural elements used as examples in the study of distributed parameter systems and points out the easiest and most commonly used method of solution, separation of variables.

Section 7.3 introduces the classic vibrating string. The derivation and solution of the string equation can be found in almost any text on vibration, partial differential equations, or applied mathematics. The vibration of bars covered in Section 7.4 is almost as common and can again be found in almost any vibration text. An excellent detailed derivation of most of the equations can be found in Magrab (1979) and more basic derivations can be found in Inman (2014) or any of the other excellent introductory texts (Rao, 2007, 2011). The material on membranes and plates in Section 7.7 is also standard and can be found in most advanced texts, such as Meirovitch (1967, 1997, 2001). Ventsel and Krauthammer (2001) present a complete derivation of the thin plate equations used here. A classic reference for plates is Sodel (1993). The material on layered structures of Section 7.8 is non-standard, but such materials have made a significant impact in engineering design and should be considered. Sun and Lu's (1995) book gives a list of useful papers in this area. Blevins (2001) is an excellent reference and tabulates natural frequencies and mode shapes for a variety of basic elements (strings, bars, beams, plates in various configurations). Elishakoff (2005) gives the solutions of several unusual configurations.

Section 7.9 introduces some simple viscous damping models for distributed parameter systems. Information on such models in the context of transient vibration analysis is difficult to come by. The majority of work on damping models centers on the steady-state forced response of such systems and presents a very difficult problem. Snowden (1968) and Nashif et al. (1985) present an alternative view on damping and should be consulted for further reading. Banks et al. (1994) discuss damping in beams.

As indicated, most of the material in this chapter is standard. However, this chapter does include some unusual boundary conditions representing lumped parameter elements appended to distributed parameter structures. These configurations are very important in vibration design and control. The text by Gorman (1975) and of Blevins (2001) tabulate the natural frequencies of such systems. Rao (2007) devotes an entire book to the vibration of distributed parameter, or continuous structures. The modeling presented in Section 7.10 is after Fanson and Caughey (1990), which follows the original works collected in Das et al. (2001).

References

Ashley, H. and Haviland, G.. (1950) Bending vibrations of a pipeline containing flowing fluid. *Journal of Applied Mechanics*, 7, 229–232.

Banks, H. T. and Inman, D. J. (1991) On damping mechanisms in beams. *ASME Journal of Applied Mechanics*, 58(3), 716–723.

Banks, H. T., Wang, Y. and Inman, D. J. (1994) Bending and shear damping in beams: frequency domain estimation techniques. *Journal of Vibration and Acoustics*, 116(2), 188–197.

Blevins, R. D. (2001) *Formulas for Mode Shapes and Natural Frequencies*. Krieger Publishing Company, Malabar, FL.

Boyce, W. E. and DiPrima, R. C. (2012) *Elementary Differential Equations and Boundary Value Problems*, 10th Edition. John Wiley & Sons, Hoboken, NJ.

Cowper, G. R. (1966) The shear coefficient in Timoshenko's beam theory. *ASME Journal of Applied Mechanics*, 3, 335–340.

Das, A and Wada, B, (eds) (2001) *Selected Papers on Smart Structures for Spacecraft*. SPIE Milestones Series, vol. MS167, The International Society for Optical Engineering, Bellingham, WA.

Dosch, J. J., Inman, D. J. and Garcia, E. (2001) A self sensing piezoelectric actuator for collocated control, in *Selected Papers on Smart Structures for Spacecraft*, Das, A. and Wada, B. (eds), *SPIE Milestones Series*, MS167, 319–338. (A reprint of the article originally appearing in *Journal of Intelligent Material Systems and Structures*, 1992, 3(1), 166–185.)

Dym, C. L. and Shames, I. H. (1973) *Solid Mechanics: A variational approach*. McGraw-Hill, New York.

Elishakoff, I. (2005) *Eigenvalues of Inhomogeneous Structures*. CRC Press, Boca Raton, FL.

Elishakoff, I., Kaplunov, J. and Nolde, E. (2015) Celebrating the centenary of Timoshenko's study of effects of shear deformation and rotary inertia. *Applied Mechanics Reviews*, 67(6), 060802.

Erturk, A. and Inman, D. J. (2011) *Piezoelectric Energy Harvesting*. John Wiley & Sons, Ltd., Chichester, UK, 416 p.

Fanson, J. L. and Caughey, T. K. (1990) Positive position feedback control for large space structures. *AIAA Journal*, 28(4), 717–724.

Gorman, D. J. (1975) *Free Vibration Analysis of Beams and Shafts*. John Wiley & Sons, New York.

Han, S. M., Benaroya, H. and Wei, T. (1999) Dynamics of transversely vibrating beams using four engineering theories, *Journal of Sound and Vibration*, 225(5), 935–988.

Inman, D. J. (2014) *Engineering Vibration*, 4th Edition. Pearson Education, Upper Saddle River, NJ.

Leo, D. J. (2007) *Smart Material Systems: Analysis, design, and control*. John Wiley & Sons, Hoboken, NJ.

Lesieutre, G. A. (1998) Vibration damping and control using shunted piezoelectric materials. *The Shock and Vibration Digest*, 30(3), 187–195.

Meirovitch, L. (1967) *Analytical Methods in Vibration*. Macmillan, New York.

Meirovitch, L. (1970) *Methods of Analytical Dynamics*. McGraw-Hill, New York.

Meirovitch, L. (1997) *Principles and Techniques of Vibration*. Prentice Hall, Upper Saddle River, NJ.

Meirovitch, L. (2001) *Fundamentals of Vibration*. McGraw-Hill Higher Education, New York.

Magrab, E. B. (1979) *Vibrations of Structural Members*. Sijthoff &Noordhoff, Alphen aan den Rijn, The Netherlands.

Murthy, D. N. S. and Sherbourne, A. N. (1972) Free flexural vibrations of damped plates. *ASME Journal of Applied Mechanics*, 39, 298–300.
Nashif, A. D., Jones, D. I. G. and Henderson, J. P. (1985) *Vibration Damping*. John Wiley & Sons, New York.
Reddy, J. N. (1984) *Energy and Variational Methods*. John Wiley & Sons, New York.
Reismann, H. (1988) *Elastic Plates: Theory and application*. John Wiley & Sons, New York.
Reismann, H. and Pawlik, P. S. (1974) *Elastokinetics, An Introduction to the Dynamics of Elastic Systems*. West Publishing, St. Paul, MN.
Snowden, J. C. (1968) *Vibration and Shock in Damped Mechanical Systems*. John Wiley & Sons, New York.
Sodel, W. (1993) *Vibration of Shells and Plates*, 2nd Edition. Marcel Decker, New York.
Rao, S. S. (2011) *Mechanical Vibrations*, 5th Edition. Pearson, Upper Saddle River, NJ.
Rao, S. S. (2007) *Vibration of Continuous Systems*. John Wiley & Sons, Hoboken, NJ.
Timoshenko, S. P., Young, D. H. and Weaver, W., Jr. (1974) *Vibration Problems in Engineering*, 4th Edition. John Wiley & Sons, New York.
Sun, C. T. and Lu, Y. P. (1995,) *Vibration Damping of Structural Elements*. Prentice Hall, Englewood Cliffs, NJ.
Umland, J. W., Inman, D. J. and Banks, H. T. (1991) Damping in coupled rotation and bending – an experiment. *Proceedings of the 1991 Joint Automatic Control Conference* (Invited) May 1991, III, 2994–2999.
Ventsel E. and Krauthammer, T. (2001) *Thin Plates and Shells*. Marcel Decker, New York.
Weaver, W., Jr., Timoshenko, S. P. and Young, D. H. (1990) *Vibration Problems in Engineering*, 5th Edition. John Wiley & Sons, New York.

Problems

7.1 Provide the details of integration by parts suggested to derive Equations (7.15) and (7.16), and hence verify that those equations are correct.

7.2 A string is fixed at one end, and held taught by a mass sliding in a channel connected to two springs, as illustrated in Figure 7.16. Use Hamilton's principle to derive the equations of motion and boundary conditions.

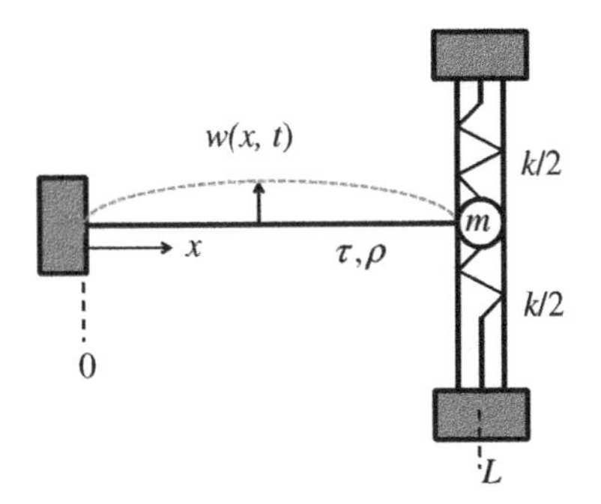

Figure 7.16 A string fixed at one end and connected to a spring and mass at the other end.

7.3 A beam is pinned at one end, and supported by a spring at the other end, as illustrated in Figure 7.17. Use Hamilton's principle to derive the equations of motion and boundary conditions.

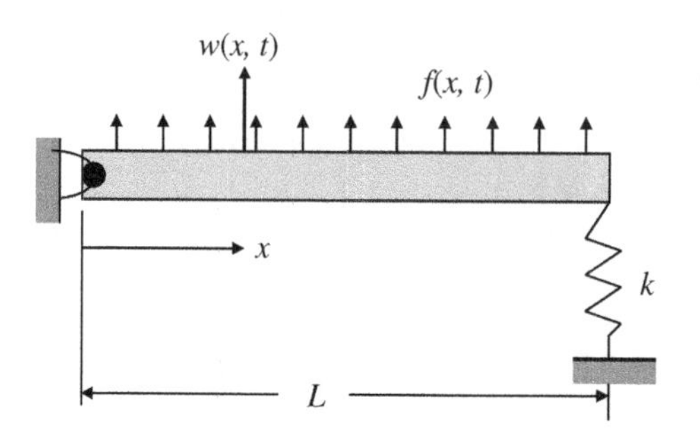

Figure 7.17 A beam fixed at one end and connected to a spring at the other end.

7.4 Consider the bar of Section 7.4. Using the method of separation of variables, calculate the natural frequencies if the bar is clamped at one end and connected to a linear spring of stiffness k at the other end.

7.5 Consider the string equation of Section 7.3, with clamped boundaries at each end. At the midpoint, $x = l/2$, the density changes from ρ_1 to ρ_2, but the tension T remains constant. Derive the characteristic equation.

7.6 Calculate the natural frequencies of vibration of the Euler-Bernoulli beam clamped at both ends.

7.7 Consider a non-uniform bar described by

$$\frac{\partial}{\partial x}[EA(x)u_x] = m(x)u_{tt} \text{ on } (0, l)$$

with

$$EA(a) = 2EA_0\left(1 - \frac{x}{l}\right) \text{ and } m(x) = 2m_0\left(1 - \frac{x}{l}\right)$$

fixed at 0 and free at l. Calculate the free vibration. What are the first three eigenvalues?

7.8 Derive the Timoshenko beam equation, assuming that it is subject to an applied force $q(x, t)$.

7.9 Verify the orthogonality condition for the set of functions $\left\{\sin \frac{n\pi x}{l}\right\}$ on the interval $[0, l]$.

7.10 Calculate the natural frequencies of the system in Figure 7.3 and compare them to those of Example 7.4.1.

7.11 Calculate the solution of the internally and externally damped bar given by Equation (7.136) with a clamped boundary at $x = 0$ and a free boundary at $x = l$.

7.12 Are there values of the parameters γ and ρ in Equation (7.136) for which the damping for one or more of the terms $T_n(t)$ is zero? A minimum? Use the clamped boundary conditions of Problem 7.11.

7.13 Compute the natural frequencies of a beam clamped at the right end (0) with a tip mass of value M at the left end.

7.14 Compute the characteristic equation of a shaft with a disc of inertia I at each end.

7.15 Consider Example 7.6.1 and show that the form of the functions $W(x)$ and $\Theta(x)$ satisfy the pinned-pinned boundary conditions given in Equation (7.92).

7.16 Compute the characteristic equation of a shaft with a disc of inertia J at each end.

8

Formal Methods of Solutions

8.1 Introduction

This chapter examines various methods of solving for the vibrational response of the distributed parameter systems introduced in the previous chapter. As in finite-dimensional systems, the response of a given system is made up of two parts: the transient, or free response, and the steady state, or forced response. In general, the steady state response is easier to calculate and in many cases the steady state is all that is necessary. The focus in this chapter is the free response. The forced response is discussed in more detail in Chapter 10.

Several approaches to solving distributed parameter vibration problems are considered. The formal notion of an operator and the eigenvalue problem associated with the operator are introduced. The traditional separation of variables method used in Chapter 7 is compared with the eigenvalue problem. The eigenfunction expansion method is introduced and examined for systems including damping. Transform methods and integral formulations in the form of Green's functions are also introduced in less detail.

8.2 Boundary Value Problems and Eigenfunctions

As discussed in Section 7.8, a general formulation of the undamped boundary value problems presented in Equation (7.132) can be written as (the subscript on L is dropped here for notational ease)

$$\begin{aligned} w_{tt}(\mathbf{x},t) + Lw(\mathbf{x},t) &= 0 \qquad && \mathbf{x} \in \Omega \qquad && \text{for } t > 0 \\ Bw &= 0 && \mathbf{x} \in \partial\Omega && \text{for } t > 0 \\ w(\mathbf{x},0) &= w_0(\mathbf{x}), && w_t(\mathbf{x},0) = \dot{w}_0(\mathbf{x}), && \text{at } t = 0 \end{aligned} \tag{8.1}$$

where $w(\mathbf{x}, t)$ is the deflection, $\mathbf{x}$ is a three-dimensional vector of spatial variables, and Ω is a bounded region in three-dimensional space with boundary $\partial\Omega$. The

Vibration with Control, Second Edition. Daniel John Inman.

Companion Website: www.wiley.com/go/inmanvibrationcontrol2e

operator L is a differential operator of spatial variables only. For example, for the longitudinal vibration of a bar (string or rod), the operator L has the form

$$L = -\alpha \frac{\partial^2}{\partial x^2} \tag{8.2}$$

where α is a constant. An *operator*, or *transformation*, is a rule that assigns to each function $w(\mathbf{x}, t)$ belonging to a certain class, another function ($-\alpha w_{xx}$ in the case of the string operator) belonging to another, perhaps different, class of functions. Note that a matrix satisfies this definition. The notation B is an operator representing the boundary conditions as given, for example, by Equation (7.9). As indicated previously, Equations (8.1) define a boundary value problem. A common method of solving Equations (8.1) is to use separation of variables, as illustrated by the examples in Chapter 7. As long as the operator L does not depend on time and if L satisfies certain other conditions (discussed in the next chapter), this method will work.

In many situations, the separation-of-variables approach yields an infinite set of functions of the form $\phi_n(\mathbf{x})a_n(t)$ that are solutions of Equation (8.1). The most general solution is then the sum, i.e.

$$w(\mathbf{x}, t) = \sum_{n=1}^{\infty} a_n(t)\phi_n(\mathbf{x}) \tag{8.3}$$

A related method, modal analysis, also uses these functions and is described in the next section.

Similar to the eigenvectors of a matrix, some operators have eigenfunctions. A nonzero function $\phi(\mathbf{x})$ that satisfies the relationships

$$L\phi(\mathbf{x}) = \lambda\phi(\mathbf{x}) \quad \mathbf{x} \in \Omega$$
$$B\phi(\mathbf{x}) = 0 \qquad \mathbf{x} \in \partial\Omega$$

is called an *eigenfunction* of the operator L with boundary condition B. The scalar λ (possibly complex) is called an *eigenvalue* of the operator L with respect to the boundary conditions B. In some cases the boundary conditions are not present, as in the case of a matrix, and in some cases the boundary conditions are contained in the domain of the operator L. The *domain* of the operator L, denoted by $D(L)$, is the set of all functions $u(x)$ for which Lu is defined and of interest.

To see the connection between separation of variables and eigenfunctions, consider substitution of the assumed separated form $w(\mathbf{x}, t) = a(t)\phi(\mathbf{x})$ into Equations (8.1). This yields

$$\frac{\ddot{a}(t)}{a(t)} = \frac{L\phi(\mathbf{x})}{\phi(\mathbf{x})} \quad \mathbf{x} \in \Omega \quad t > 0 \tag{8.4}$$

$$a(t)B\phi(\mathbf{x}) = 0, \quad \mathbf{x} \in \partial\Omega, \quad t > 0 \tag{8.5}$$

$$a(0)\phi(\mathbf{x}) = w_0(\mathbf{x}), \quad \dot{a}(0)\phi(\mathbf{x}) = \dot{w}_0(\mathbf{x}) \tag{8.6}$$

As before, Equation (8.4) implies that each side is constant, so that

$$L\phi(\mathbf{x}) = \lambda\phi(\mathbf{x}), \quad \mathbf{x} \in \Omega \tag{8.7}$$

where λ is a scalar. In addition, note that since $a(t) \neq 0$ for all t, Equation (8.5) implies that

$$B\phi(\mathbf{x}) = 0, \quad x \in \partial\Omega, \quad \text{and} \quad t > 0 \tag{8.8}$$

Equations (8.7) and (8.8) are a statement that $\phi(\mathbf{x})$ and λ constitute an eigenfunction and eigenvalue of the operator L.

Example 8.2.1

Consider the operator formulation of the longitudinal bar equation, presented in Section 7.3. The form of the beam operator is

$$L = -\alpha \frac{\partial^2}{\partial x^2} \quad x \in (0, l)$$

with boundary conditions ($B = 1$ at $x = 0$ and $B = \partial/\partial x$ at $x = l$)

$$\phi(0) = 0 \text{ (clamped end)}, \text{ and } \phi_x(l) = 0 \text{ (free end)}$$

and where $\partial\Omega$ consists of the points $x = 0$ and $x = l$. Here, the constant a represents the physical parameters of the beam, i.e. $\alpha = EA/\rho$. The eigenvalue problem $L\phi = \lambda\phi$ becomes

$$-\alpha\phi_{xx} = \lambda\phi$$

or

$$\phi_{xx} + \frac{\lambda}{\alpha}\phi = 0$$

This last expression is identical to Equation (9.6) and the solution is

$$\phi(x) = A_1 \sin\left(\sqrt{\frac{\lambda}{\alpha}}x\right) + A_2 \cos\left(\sqrt{\frac{\lambda}{\alpha}}x\right)$$

where A_1 and A_2 are constants of integration. Using the boundary conditions yields

$$0 = \phi(0) = A_2 \quad \text{and} \quad 0 = \phi_x(l) = A_1\sqrt{\frac{\lambda}{\alpha}} \cos\sqrt{\frac{\lambda}{\alpha}}l$$

This requires that $A_2 = 0$ and

$$A_1 \cos\sqrt{\frac{\lambda}{\alpha}}l = 0$$

Since A_1 cannot be zero

$$\sqrt{\frac{\lambda}{\alpha}}l = \frac{n\pi}{2}$$

for all odd integers n. Thus λ is depends on n and

$$\lambda_n = \frac{\alpha n^2\pi^2}{4l^2}, \quad n = 1, 3, 5, \ldots \infty$$

Thus, there are many eigenvalues λ, denoted now by λ_n, and many eigenfunctions ϕ, denoted by ϕ_n. The eigenfunctions and eigenvalues of the operator L are given by the sets

$$\{\phi_n(x)\} = \left\{A_n \sin\frac{n\pi x}{2l}\right\}, \quad \text{and} \quad \{\lambda_n\} = \left\{\frac{\alpha n^2\pi^2}{4l^2}\right\}, \quad n \text{ odd}$$

respectively. Note that, as in the case of a matrix eigenvector, eigenfunctions are determined only to within a multiplicative constant (A_n in this case).

Comparing the eigenfunctions of the operator for the beam with the spatial functions calculated in Chapter 7, shows that the eigenfunctions of the operator correspond to the mode shapes of the structure. This correspondence is exactly analogous to the situation for the eigenvectors of a matrix.

8.3 Modal Analysis of the Free Response

The eigenfunctions associated with the string equation are shown in Section 7.3 to be the mode shapes of the string. Also, by using the linearity of the equations of motion, the solution is given as a summation of mode shapes. This summation of mode shapes, or eigenfunctions, given in Equation (7.35), constitutes the *eigenfunction expansion* or *modal analysis* of the solution and provides an alternative point of view to the separation-of-variables technique.

First, as in the case of eigenvectors of a matrix, eigenfunctions are conveniently normalized to fix a value for the arbitrary constant. To this end, let the eigenfunctions of interest be denoted by $A_n\phi_n(\mathbf{x})$. If the constants A_n are chosen such that

$$\int_\Omega A_n^2\phi_n(\mathbf{x})\phi_n(\mathbf{x})d\Omega = 1 \tag{8.9}$$

then the eigenfunctions $\emptyset_n = A_n\phi_n$ are said to be *normalized*, or normal. If, in addition, they satisfy

$$\int_\Omega \emptyset_n\emptyset_m d\Omega = \delta_{mn} \tag{8.10}$$

the eigenfunctions are said to be *orthonormal*, exactly analogous to the eigenvector case. The method of modal analysis assumes that the solution of Equation (8.1) can be represented as the series

$$w(\mathbf{x}, t) = \sum_{n=1}^{\infty} a_n(t)\emptyset_n(\mathbf{x}) \tag{8.11}$$

where $\emptyset_n(\mathbf{x})$ are the normalized eigenfunctions of the operator L. Substitution of Equation (8.11) into Equation (8.1), multiplying by $\emptyset_m(\mathbf{x})$, and integrating (assuming uniform convergence) over the domain Ω reduces Equations (8.1) to an infinite set of uncoupled ordinary differential equations of the form

$$\ddot{a}_n(t) + \lambda_n a_n(t) = 0, \quad n = 1, 2, \ldots, \infty \tag{8.12}$$

Equation (8.12) can then be used along with the appropriate initial conditions to solve for each of the temporal functions. Here, λ_n is the eigenvalue associated with the nth mode, so that

$$\int_\Omega \emptyset_n L \emptyset_n d\Omega = \lambda_n \int_\Omega \emptyset_n \emptyset_n d\Omega = \lambda_n \tag{8.13}$$

where Equations (8.7) and (8.10) are used to evaluate the integral.

Example 8.3.1
Consider the transverse vibration of a Euler-Bernoulli beam with hinged boundary conditions. Calculate the eigenvalues and eigenfunctions for the associated operator.

The stiffness operator for constant mass, cross-sectional area, and area moment of inertia is given by (see Equation 7.69)

$$L = \frac{EI}{m}\frac{\partial^4}{\partial x^4} = \beta \frac{\partial^4}{\partial x^4}$$
$$\emptyset(0) = \emptyset_{xx}(0) = 0$$
$$\emptyset(l) = \emptyset_{xx}(l) = 0$$

The eigenvalue problem $Lu = \lambda u$ then becomes

$$\beta \emptyset_{xxxx} = \lambda \emptyset$$

which has a solution of the form

$$\emptyset(x) = C_1 \sin \mu x + C_2 \cos \mu x + C_3 \sinh \mu x + C_4 \cosh \mu x$$

where $\mu^4 = \lambda/\beta$. Applying the four boundary conditions to this expression yields the four simultaneous equations

$$\emptyset(0) = C_2 + C_4 = 0$$
$$\emptyset_{xx}(0) = -C_2 + C_4 = 0$$
$$\emptyset(L) = C_1 \sin \mu l + C_2 \cos \mu l + C_3 \sinh \mu l + C_4 \cosh \mu l = 0$$
$$\emptyset_{xx}(l) = -C_1 \sin \mu l - C_2 \cos \mu l + C_3 \sinh \mu l + C_4 \cosh \mu l = 0$$

These four equations in the four unknown constants C_i can be solved by examining the matrix equation

$$\begin{bmatrix} 0 & 1 & 0 & 1 \\ 0 & -1 & 0 & 1 \\ \sin \mu l & \cos \mu l & \sinh \mu l & \cosh \mu l \\ -\sin \mu l & -\cos \mu l & \sinh \mu l & \cosh \mu l \end{bmatrix} \begin{bmatrix} C_1 \\ C_2 \\ C_3 \\ C_4 \end{bmatrix} = \begin{bmatrix} 0 \\ 0 \\ 0 \\ 0 \end{bmatrix}$$

Recall from Chapter 3 that in order for a nontrivial vector $\mathbf{c} = [C_1\ C_2\ C_3\ C_4]^T$ to exist, the coefficient matrix must be singular. Thus, the determinant of the coefficient matrix must be zero. Setting the determinant equal to zero yields the characteristic equation

$$4 \sin(\mu l) \sinh(\mu l) = 0$$

This, of course, can be true only if $\sin(\mu l) = 0$, leading to

$$\mu = \frac{n\pi}{l}, \quad n = 1, 2, \ldots \infty$$

Here $n = 0$ is excluded because it results in the trivial solution. In terms of the physical parameters of the structure, the eigenvalues become (here n is an integer and m is the mass per unit length of the beam)

$$\lambda_n = \frac{n^4\pi^4 EI}{ml^4}$$

Solving for the four constants C_i yields $C_2 = C_3 = C_4 = 0$ and that C_1 is arbitrary. Hence, the eigenfunctions are of the form

$$\left[A_n \sin\left(\frac{n\pi x}{l}\right)\right]$$

The arbitrary constants A_n can be fixed by normalizing the eigenfunctions

$$\int_0^l A_n^2 \sin^2\left(\frac{n\pi x}{l}\right) dx = 1$$

so that $A_n^2 l/2 = 1$, or $A_n = \sqrt{2/l}$. Thus, the normalized eigenfunctions are the set

$$\{\emptyset_n\} = \left\{\sqrt{\frac{2}{l}} \sin\left(\frac{n\pi x}{l}\right)\right\}_{n=1}^{\infty}$$

Hence the temporal coefficient in the series expansion of the solution in Equation (8.11) will be determined from the initial conditions and the finite number of equations

$$\ddot{a}_n(t) + \frac{n^4\pi^4 EI}{ml^4} a_n(t) = 0, \quad n = 1, \ldots, \infty$$

Equation (8.11) then yields the total solution.

8.4 Modal Analysis in Damped Systems

As in the matrix case for lumped parameter systems, the method of modal analysis (and separation of variables) can still be used for certain types of viscous damping modeled in a distributed structure. Systems that can be modeled by partial differential equations of the form

$$w_{tt}(\mathbf{x}, t) + L_1 w_t(\mathbf{x}, t) + L_2 w(\mathbf{x}, t) = 0 \quad \mathbf{x} \in \Omega \tag{8.14}$$

where L_1 and L_2 are operators, with the similar properties to L and such that L_1 and L_2 have the same eigenfunctions, can be solved by the method of modal analysis illustrated in Equation (8.11) and Example 8.3.1. Section 7.9 lists some examples of different damping models.

To see this solution method, let L_1 have eigenvalues $\lambda_n^{(1)}$ and L_2 have eigenvalues $\lambda_n^{(2)}$. Substitution of Equation (8.11) into Equation (8.14) then yields (assuming convergence)

$$\sum_{n=1}^{\infty}\left[\ddot{a}_n \varnothing_n(\mathbf{x}) + \lambda_n^{(1)}\dot{a}_n \varnothing_n(\mathbf{x}) + \lambda_n^{(2)} a_n \varnothing_n(\mathbf{x})\right] = 0 \tag{8.15}$$

Multiplying by $\varnothing_n(\mathbf{x})$, integrating over Ω, and using the orthogonality conditions in Equation (8.10), yields the decoupled set of n ordinary differential equations

$$\ddot{a}_n(t) + \lambda_n^{(1)}\dot{a}_n(t) + \lambda_n^{(2)} a_n(t) = 0, \quad n = 1, 2, \ldots, \infty \tag{8.16}$$

subject to the appropriate initial conditions.

The actual form of damping in distributed parameter systems is not always clearly known. In fact, the form of L_1 is an elusive topic of current research and several texts (Nashif et al., 1985; Sun and Lu, 1995). Often the damping is modeled as being proportional, i.e. $L_1 = \alpha I + \beta L_2$, where α and β are arbitrary scalars and L_1 satisfies the same boundary conditions as L_2. In this case, the eigenfunctions of L_1 are the same as those of L_2. Damping is often estimated using equivalent viscous proportional damping of this form as an approximation. A more advanced approach to estimating damping is given by Banks et al. (1994)

Example 8.4.1

As an example of a proportionally damped system, consider the transverse free vibration of a membrane in a surrounding medium, such as a fluid, providing resistance to the motion that is proportional to the velocity. The equation of motion given by Equation (7.100) is Equation (8.14), with

$$L_1 = 2\frac{\gamma}{\rho}$$

$$L_2 = -\frac{T}{\rho}\nabla^2$$

where T, ρ and ∇^2 are as defined for Equation (7.100) and γ is a constant defined by the damping provided by the fluid. The position $\mathbf{x}$ in this case is the vector $[x \; y]$ in two-dimensional space. If $\lambda_1^{(2)}$ is the first eigenvalue of L_2, then the solutions to Equation (8.16) are of the form

$$a_n(t) = e^{-\frac{\gamma}{\rho}t}\left[A_n \sin\sqrt{\lambda_n^{(2)} - \frac{\gamma^2}{\rho^2}}\,t + B_n \cos\sqrt{\lambda_n^{(2)} - \frac{\gamma^2}{\rho^2}}\,t\right]$$

where A_n and B_n are determined by the initial conditions as done in Equations (7.38) and (7.39).

Not all damped systems have this type of damping. Systems that have proportional damping are called *normal mode systems,* since the eigenfunctions of the operator L_2 serve to "decouple" the system. Decouple, as used here, refers to the fact that Equation (8.16) depends only on n and not on any other index. This topic is considered theoretically in Section 9.6.

8.5 Transform Methods

An alternative to using separation of variables and modal analysis is to use a transform to solve for the vibrational response. As with the Laplace transform method used on the temporal variable in state space analysis for lumped parameter systems, a Laplace transform can also be used in solving Equation (8.1). In addition, a Fourier transform can be used on the spatial variable to calculate the solution. These methods are briefly mentioned here. The reader is referred to a text such as Churchill (1972) for a rigorous development.

The Laplace transform taken on the temporal variable of a partial differential equation can be used to solve for the free or forced response of Equations (8.1) and (8.14). The transform approach is best explained by considering an example.

Consider the vibrations of a bar with constant force F_0 applied to one end and fixed at the other. Recall that the equation for longitudinal vibration is

$$w_{tt}(x,t) = \alpha^2 w_{xx}(x,t) \tag{8.17}$$

with boundary conditions

$$w(0,t) = 0, \quad EAw_x(l,t) = F_0\delta(t) \tag{8.18}$$

Here $\alpha^2 = EA/\rho$, as defined in Section 7.4. Assuming that the initial conditions are zero, the Laplace transform of Equation (8.17) yields

$$s^2 W(x,s) - \alpha^2 W_{xx}(x,s) = 0 \tag{8.19}$$

and of Equation (8.18) yields

$$\begin{aligned} W_x(l,s) &= \frac{F_0}{EAs} \\ W(0,s) &= 0 \end{aligned} \tag{8.20}$$

Here W denotes the Laplace transform of w. The solution of Equation (8.19) is of the form

$$W(x,s) = A_1 \sinh\frac{sx}{\alpha} + A_2 \cosh\frac{sx}{\alpha}$$

Applying the boundary condition at $x = 0$, gives $A_2 = 0$. Differentiating with respect to x and taking the Laplace transform yields the boundary condition at $x = l$. The constant A_1 is then determined as

$$A_1 = \left(\frac{\alpha F_0}{EA}\right)\left(\frac{1}{s^2\cosh(sl/\alpha)}\right)$$

The solution in terms of the transform variable s then becomes

$$W(x,s) = \frac{\alpha F_0 \sinh(sx/\alpha)}{EAs^2\cosh(sl/\alpha)} \tag{8.21}$$

Taking the inverse Laplace transform of Equation (8.21) and using residue theory, the solution in the time domain is obtained. The inverse is given by Churchill (1972) to be

$$w(x,t) = \frac{F_0}{E}x + \left(\frac{8l}{\pi}\right)^2 \sum_{n=1}^{\infty} \frac{(-1)^n}{(2n-1)^2} \sin\frac{(2n-1)\pi x}{2l} \cos\frac{(2n-1)\pi a t}{2l} \tag{8.22}$$

A text on transforms should be consulted for the details. Basically, the expansion comes from the zeros in the complex plane of $s^2\cosh(sl/a)$, i.e. the poles of $W(x,s)$.

This same solution can also be obtained by taking the finite Fourier sine transform of Equations (8.17) and (8.18) on the spatial variable x, rather than the Laplace transform of the temporal variable (Meirovitch, 1967). Usually transforming the spatial variable is more productive because the time dependence is a simple initial value problem.

When boundary conditions have even-order derivatives, a finite sine transformation (Fourier transform) is appropriate. The sine transform is defined by

$$W(n,t) = \int_0^l w(x,t)\sin\frac{n\pi x}{l}dx \tag{8.23}$$

Note here that the transform in this case is over the spatial variable.

Again, the method is explained by example. To that end, consider the vibration of a string clamped at each end subject to nonzero initial velocity and displacement, i.e.

$$\begin{aligned} w_{xx} &= \frac{1}{c^2}w_{tt}(x,t) \\ w(0,t) &= w(l,t) = 0, \quad w(x,0) = f(x), \quad w_t(x,0) = g(x) \end{aligned} \tag{8.24}$$

The finite sine transform of the second derivative is

$$W_{xx}(n,t) = \frac{n\pi}{l}\left|(-1)^{n+1}W(l,t) + W(0,t)\right| - \left(\frac{n\pi}{l}\right)^2 W(n,t) \tag{8.25}$$

which is calculated from integration by parts of Equation (8.23). Substitution of the boundary conditions yields the transformed string equation

$$W_{tt}(n,t) + \left(\frac{n\pi}{l}\right)^2 W(n,t) = 0 \tag{8.26}$$

This equation is subject to the transform of the initial conditions, which are

$$W(n,0) = \int_0^l f(x)\sin\frac{n\pi x}{l}dx \tag{8.27}$$

and

$$W_t(n,0) = \int_0^l g(x)\sin\frac{n\pi x}{l}dx$$

Thus

$$w(n,t) = w(n,0)\cos\frac{n\pi ct}{l} + w_t(n,0)\frac{l}{n\pi c}\sin\frac{n\pi ct}{l} \tag{8.28}$$

Again, Equation (8.28) has to be inverted to get the solution $w(x,t)$. The inverse finite Fourier transform is given by

$$w(x,t) = \frac{2}{l}\sum_{n=1}^{\infty} W(n,t)\sin\left(\frac{n\pi x}{l}\right) \tag{8.29}$$

so that

$$w(x,t)=\frac{2}{l}\sum_{n=1}^{\infty}\left[\left\{W(n,0)\cos\left(\frac{n\pi ct}{l}\right)+\frac{W_t(n,0)l}{n\pi c}\sin\left(\frac{n\pi ct}{l}\right)\right\}\sin\frac{n\pi x}{l}\right] \tag{8.30}$$

Transform methods are attractive for problems defined over infinite domains and for problems with odd boundary conditions. The transform methods yield a quick "solution" in terms of the transformed variable. However, the inversion back into the physical variable can be difficult and may require as much work as using separation of variables or modal analysis. However, in some instances, the only requirement may be to examine the solution in its transformed state (Section 12.5).

8.6 Green's Functions

Yet another approach to solving the free vibration problem is to use the integral formulation of the equations of motion. The basic idea here is that the free response is related to the eigenvalue problem

$$\begin{aligned} Lw &= \lambda w \\ Bw &= 0 \end{aligned} \tag{8.31}$$

where L is a differential operator and B represents the boundary conditions. The inverse of this operator will also yield information about the free vibrational response of the structure. If the inverse of L exists, Equation (8.31) can be written as

$$L^{-1}w=\frac{1}{\lambda}w \tag{8.32}$$

where L^{-1} is the inverse of the differential operator or an integral operator.

The problem of solving for the free vibration of a string fixed at both ends by working essentially with the inverse operator is approached in this section. This approach is done by introducing the concept of a Green's function. To this end, consider again the problem of a string fixed at both ends and deformed from its equilibrium position. This time, however, instead of looking directly at the vibration problem, the problem of determining the static deflection of the string due to a transverse load concentrated at a point is first examined. This related problem is called the *auxiliary problem*. In particular, if the string is subject to a point load of unit value at x_0, which is somewhere in the interval (0,1), the equation of the deflection $w(x)$ for a string of tension T is

$$-T\frac{d^2w(x)}{dx^2}=\delta(x-x_0) \tag{8.33}$$

where $\delta(x-x_0)$ is the Dirac delta function. The delta function is defined by

$$\delta(x-x_0)=\begin{cases} 0 & x\neq x_0 \\ \infty & x=x_0 \end{cases} \tag{8.34}$$

and

$$\int_0^1 \delta(x - x_0)dx = \begin{cases} 0 & \text{if } x_0 \text{ is not in } (0,1) \\ 1 & \text{if } x_0 \text{ is in } (0,1) \end{cases} \tag{8.35}$$

If $f(x)$ is a continuous function, then it can be shown that

$$\int_0^1 f(x)\delta(x - x_0)dx = f(x_0) \tag{8.36}$$

for x_0 in (0,1). Note that the Dirac delta function is not really a function in the strict mathematical sense (Stakgold (1979)).

Equation (8.33) can be viewed as expressing the fact that the force causing the deflection is applied only at the point x_0. Equation (8.33) plus boundary conditions is now viewed as the auxiliary problem of finding a function $g(x,x_0)$, known as the *Green's function* for the operator $L = -T d^2/dx^2$, with boundary conditions $g(0, x_0) = 0$ and $g(1, x_0) = 0$. In more physical terms, $g(x,x_0)$ represents the deflection of the string from its equilibrium position at point x due to a unit force applied at point x_0. The Green's function thus defined is also referred to as an *influence function.* The following example is intended to clarify the procedure for calculating a Green's function.

Example 8.6.1

Calculate the Green's function for the string of Figure 8.1. The Green's function is calculated by solving the equation on each side of the point x_0 and then matching up the two solutions. Thus, since $g'' = 0$ for all x not equal to x_0, integrating yields

$$g(x, x_0) = \begin{cases} Ax + B & 0 \le x < x_0 \\ Cx + D & x_0 < x \le 1 \end{cases}$$

where A, B, C and D are constants of integration. Applying the boundary condition at $x = 0$ yields

$$g(0, x_0) = 0 = B$$

and the boundary condition at 1 yields

$$g(1, x_0) = 0 = C + D$$

Hence, the Green's function becomes

$$g(x, x_0) = \begin{cases} Ax & 0 \le x < x_0 \\ C(x - 1) & x_0 < x \le 1 \end{cases}$$

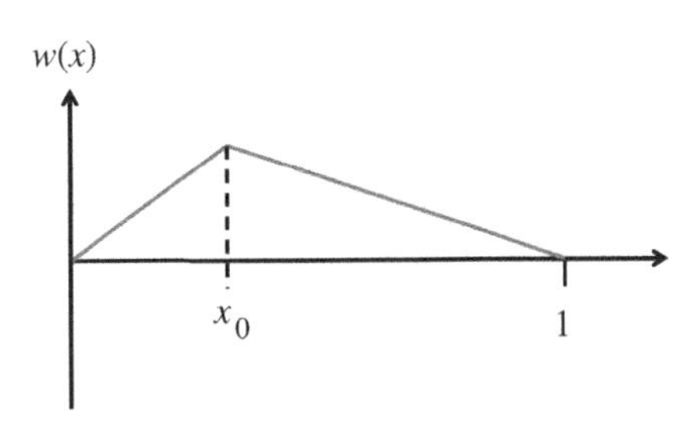

Figure 8.1 A statically deflected string fixed at both ends.

Since the string does not break at x_0, $g(x,x_0)$ must be continuous at x_0 and this allows determination of one more constant. In particular, this continuity condition requires

$$Ax_0 = C(x_0 - 1)$$

The Green's function now becomes

$$g(x, x_0) = \begin{cases} Ax & 0 \le x < x_0 \\ A\dfrac{x_0(x-1)}{(x_0-1)} & x_0 < x \le 1 \end{cases}$$

The remaining constant, A, can be evaluated by considering the magnitude of the applied force required to produce the deflection. In this case, a unit force is applied, so that integration of Equation (8.33) along a small interval containing x_0, say $x_0 - \varepsilon \le x \le x_0 + \varepsilon$, yields

$$\int_{x_0-\varepsilon}^{x_0+\varepsilon} \frac{d^2g}{dx^2} dx = -\frac{1}{T}\int_{x_0-\varepsilon}^{x_0+\varepsilon} \delta(x - x_0)dx$$

or

$$\left.\frac{dg}{dx}\right|_{x_0-\varepsilon}^{x_0+\varepsilon} = -\frac{1}{T}$$

Denoting the derivative by a subscript and expanding yields

$$g_x(x_0 + \varepsilon, x_0) - g_x(x_0 - \varepsilon, x_0) = -\frac{1}{T}$$

This last expression is called a *jump discontinuity* in the derivative. Upon evaluating the derivative, the above expression becomes

$$A\frac{x_0}{(x_0-1)} - A = -\frac{1}{T}$$

Solving this for the value of A yields

$$A = \frac{1 - x_0}{T}$$

Green's function, with all the constants of integration evaluated, is thus

$$g(x, x_0) = \begin{cases} \dfrac{(1-x_0)x}{T} & 0 \le x < x_0 \\ \dfrac{(1-x)x_0}{T} & x_0 < x \le 1 \end{cases}$$

The Green's function actually defines the inverse operator (when it exists) of the differential operator L and can be used to solve for the forced response of the string. Consider the equations (for the string operator of the example)

$$\begin{aligned} Lu &= f(x) \\ Bu &= 0 \end{aligned} \tag{8.37}$$

where $f(x)$ is a piecewise continuous function, and L is a differential operator that has an inverse. Let G denote the operator defined by

$$Gf(x) = \int_0^1 g(x, x_0)f(x_0)dx_0$$

The operator G defined in this way is called an *integral operator*. Note that the function

$$u(x) = \int_0^1 g(x, x_0)f(x_0)dx_0 \tag{8.38}$$

satisfies Equation (8.37), including the boundary conditions, which follows from a straightforward calculation. Equation (8.38) can also be written

$$u(x) = \int_0^{x-\varepsilon} g(x, x_0)f(x_0)dx_0 + \int_{x+\varepsilon}^1 g(x, x_0)f(x_0)dx_0$$

where the integration has been split over two separate intervals for the purpose of treating the discontinuity in g_x. Using the rules for differentiating an integral (Leibnitz's rule) applied to this expression yields

$$\begin{aligned} u_x(x) &= \int_0^{x-\varepsilon} g_x(x, x_0)f(x_0)dx_0 + g(x, x-\varepsilon)f(x-\varepsilon) \\ &\quad + \int_{x+\varepsilon}^1 g_x(x, x_0)f(x_0)dx_0 - g(x, x+\varepsilon)f(x+\varepsilon) \\ &= \int_0^{x-\varepsilon} g_x(x, x_0)f(x_0)dx_0 + \int_{x+\varepsilon}^1 g_x(x, x_0)f(x_0)dx_0 \end{aligned}$$

Taking the derivative of this expression for u_x yields

$$\begin{aligned} u_{xx}(x) &= \int_0^{x-\varepsilon} g_{xx}(x, x_0)f(x_0)dx_0 + g_x(x, x-\varepsilon)f(x-\varepsilon) \\ &\quad + \int_{x+\varepsilon}^1 g_{xx}(x, x_0)f(x_0)dx_0 - g_x(x, x+\varepsilon)f(x+\varepsilon) \end{aligned}$$

The discontinuity in the first derivative yields that

$$g_x(x, x-\varepsilon)f(x-\varepsilon) - g_x(x, x+\varepsilon)f(x+\varepsilon) = -\frac{f(x)}{T}$$

Hence

$$u_{xx} = \int_0^{x-\varepsilon} g_{xx}(x, x_0)f(x_0)dx_0 + \int_{x+\varepsilon}^1 g_{xx}(x, x_0)f(x_0)dx_0 - \frac{f(x)}{T} \tag{8.39}$$

But, $g_{xx} = 0$ in the intervals specified in the two integrals. Thus, this last expression is just the equation $Lu = f$. The function $u(x)$ then becomes

$$u(x) = \int_0^1 g(x, y)f(y)dy$$

which satisfies Equation (8.37) as well as the boundary condition.

Now note that $Gf = u$, so that G applied to $Lu = f$ yields $G(Lu) = Gf = u$, and hence $GLu = u$. Also, L applied to $Gf = u$ yields $LGf = Lu = f$, so that $LGf = f$. Thus, the operator G is clearly the inverse of the operator L.

In the same way, Green's function can also be used to express the eigenvalue problem for the string. In fact

$$\int_0^1 g(x, x_0)\theta(x_0)dx_0 = \mu\theta(x) \tag{8.40}$$

yields the eigenfunctions $\theta(x)$ for the operator L as defined in Equation (8.2), where $\mu = 1/\lambda$ and Equation (8.32) is defined by G.

To summarize. consider the slightly more general operator given by

$$Lw = a_0(x)w_{xx}(x) + a_1(x)w_x(x) + a_2(x)w(x) = 0 \tag{8.41}$$

with boundary conditions given by

$$B_1 w\,|_{x=0} = 0 \text{ and } B_2 w\,|_{x=1} = 0 \tag{8.42}$$

The Green's function for the operator given by Equations (8.41) and (8.42) is defined to be the function $g(x,x_0)$ such that

1) $0 < x \leq 1, 0 \leq x_0 < 1$;
2) $g(x, x_0)$ is continuous for any fixed value of x_0 and satisfies the boundary conditions in Equation (8.42);
3) $g_x(x, x_0)$ is continuous, except at $x = x_0$;
4) As a function of x, $Lg = 0$ everywhere except at $x = x_0$;
5) The jump discontinuity $g_x(x, x_0 + \varepsilon) - g_x(x, x_0 - \varepsilon) = 1/a_0(x)$ holds.

The Green's function defines the inverse of the operators defined in Equations (8.41) and (8.42). Furthermore, the eigenvalue problem associated with the vibration problem can be recast as an integral equation, as given by Equation (8.40). The Green's function approach can be extended to other operators. Both of these concepts are capitalized on in the following chapters.

Chapter Notes

The majority of this chapter is common material found in a variety of texts, only some of which are mentioned here. Section 8.2 introduces eigenfunctions and makes the connection between eigenfunctions and separation of variables as methods of solving boundary value problems arising in vibration analysis. The method of separation of variables is discussed in most texts on vibration as well as those on differential equations, such as the text by Boyce and DiPrima (2000). Eigenfunctions are also discussed in texts on operator theory, such as Naylor and Sell (1982). Few texts make an explicit connection between the two methods. The procedure is placed on a firm mathematical base in Chapter 9.

Section 8.3 examines, in an informal way, the method of modal analysis, a procedure made popular by the excellent text by Meirovitch (1967, 2001). Here, however, the method is more directly related to eigenfunction expansions. Section 8.4 introduces damping as a simple velocity-proportional operator commonly used as a first attempt, as described in Section 7.9. Damping models represent a discipline by themselves. Here a model of mathematical convenience is used. A brief look at using transform methods is provided in Section 8.5 for the sake of completeness. Transform methods have been developed by Yang (1992). Most transform methods are explained in detail in the basic text by Churchill (1972). The last section on Green's functions follows closely the development in Stakgold (1979, 2000). Green's function methods provide a strong basis for the theory to follow in the next chapter. Most texts on applied mathematics in engineering discuss Green's functions.

References

Banks, H. T., Wang, Y. and Inman, D. J. (1994) Bending and shear damping in beams: frequency domain estimation techniques. *Journal of Vibration and Acoustics*, 116(2), 188–197.

Boyce, W. E. and DiPrima, R. C. (2000) *Elementary Differential Equations and Boundary Value Problems*, 7th Edition. John Wiley & Son, New York.

Churchill, R. V. (1972) *Operational Mathematics*, 3rd Edition. McGraw-Hill, New York.

Meirovitch, L. (1967) *Analytical Methods in Vibration*. Macmillan, New York.

Meirovitch, L. (2001) *Fundamentals of Vibration*. McGraw-Hill Higher Education, New York.

Nashif, A. D., Jones, D. I. G. and Henderson, J. P. (1985) *Vibration Damping*. John Wiley & Sons, New York.

Naylor, A. W. and Sell, G. R. (1982) *Linear Operator Theory in Engineering and Science*, 2nd Edition. Springer-Verlag, New York.

Stakgold, I. (1979) *Green's Functions and Boundary Value Problems*. John Wiley & Sons, New York.

Stakgold, I. (2000) *Boundary Value Problems of Mathematical Physics*, vol. 1, Philadelphia: Society for Industrial and Applied Mathematics (Classics in Applied Mathematics, vol. 29).

Sun, C. T. and Lu, Y. P. (1995) *Vibration Damping of Structural Elements*. Prentice Hall, Englewood Cliffs, NJ.

Yang, B. (1992) Transfer functions of constrained/combined one-dimensional continuous dynamic systems, *Journal of Sound and Vibration*, 156(3), 425–443.

Problems

8.1 Compute the eigenvalues and eigenfunctions for the operator

$$L = -\frac{d^2}{dx^2}$$

with boundary conditions $u(0) = 0$, $u_x(1) + u(1) = 0$.

8.2 Normalize the eigenfunctions of Problem 8.1

8.3 Show that $u_{nm}(x,y) = A_{nm} \sin n\pi x \sin m\pi y$ is an eigenfunction of the operator

$$L = \frac{\partial^2}{\partial x^2} + \frac{\partial^2}{\partial y^2}$$

with boundary conditions

$$u(x,0) = u(x,1) = u(1,y) = u(0,y) = 0$$

This is the membrane operator for a unit square.

8.4 Normalize the eigenfunctions of a membrane (clamped) of Problem 8.3 and show that they are orthonormal.

8.5 Calculate the temporal coefficients, $a_n(t)$, for the problem of Example 8.3.1.

8.6 Calculate the initial conditions required in order for $a_n(t)$ to have the form given in Example 8.4.1.

8.7 Rewrite Equation (8.16) for the case that the eigenfunctions of L_1 are not the same as those of L_2.

8.8 Solve for the free longitudinal vibrations of a clamped-free bar in the special case that the damping is approximated by the operator $L_1 = 0.1I$, $EA = \rho = 1$, and the initial conditions are $w_t(x,0) = 0$ and $w(x,0) = 10^{-2}$.

8.9 Calculate Green's function for the operator given by $L = 10^6 \times \partial^2/\partial x^2$, $u(0) = 0$, $u_x(l) = 0$. This corresponds to a clamped bar.

8.10 Calculate Green's function for the operator $L = \partial^4/\partial x^4$, with boundary conditions $u(0) = u(1) = u_{xx}(1) = 0$. (*Hint*: The jump condition is

$$g_{xxx}(x, x+\varepsilon) - g_{xxx}(x, x-\varepsilon) = 1)$$

8.11 Normalize the eigenfunctions of a shaft fixed at one end and attached to a disc at the other end, as computed in Example 7.4.1 and discuss the orthogonality conditions.

9

Operators and the Free Response

9.1 Introduction

Just as knowledge of linear algebra and matrix theory is helpful in the study of lumped parameter vibration problems, a working knowledge of functional analysis and operator theory is useful in the study of the vibrations of distributed parameter systems. A complete introduction to these topics requires a sound background in mathematics and is not possible within the limited space here. However, this chapter introduces some of the topics of relevance in vibration analysis. One of the main concerns of this section is to consider the convergence of the series expansions of eigenfunctions used in the separation of variables and modal analysis methods introduced in the previous chapters. The intent of this chapter is similar to that of Chapter 3, which introduced linear algebra as needed for discussing the free response of lumped mass systems (also called finite-dimensional systems). The goal of this chapter is to provide a mathematical analysis of the methods used in Chapter 8.

In addition, some results are presented that examine the qualitative nature of the solution of linear vibration problems. In many instances, the describing differential equations cannot be solved in closed form. In these situations, knowing the nature of the solution rather than its details may be satisfactory. For example, knowing that the first natural frequency of a structure is bounded away from a driving frequency, rather than knowing the exact numerical value of the natural frequency, may be sufficient. The qualitative behavior of solutions is discussed in this chapter. As indicated in Chapter 7, qualitative results are very useful in design situations.

It should be noted that the mathematical properties of an operator may not always be apparent, but physical considerations demand that the equations of motion for structures admit operator descriptions that have the property of possessing a full set of natural frequencies and mode shapes. It should thus be a matter of writing the various operators in the proper form.

It is well known in the structural dynamics community that the backbone of testing for validation is centered on measuring natural frequencies and mode shapes computed from both time response and frequency response data. All structures admit these properties when tested. From the analyst point of view, analytical

Vibration with Control, Second Edition. Daniel John Inman.

Companion Website: www.wiley.com/go/inmanvibrationcontrol2e

mode shapes and natural frequencies are derived from eigenvalues and eigenfunctions determined from the defining operators. Unfortunately, not all operators admit eigenvalues and eigenfunctions, even though they model structures that admit such properties. The operators used to describe the linear vibrations of common distributed parameter structures (beams, plates and shells) are unbounded, which leaves them open for not having eigenvalues. However, the differential operators for strings, beams, membranes, plates and so on, all have inverses defined by integral operators, which are bounded. This connection of vibration equations to a bounded operator is significant in verifying the existence of an eigenstructure and convergent eigenfunction expansions compatible with testing, i.e. the analytical properties needed to match experimental modal analysis results.

9.2 Hilbert Spaces

The definition of integration familiar to most engineers from introductory calculus is *Riemann integration*. Riemann integration can be defined on the interval $[a, b]$ as

$$\int_a^b f(x)dx = \lim_{\substack{n \to \infty \\ \Delta x \to 0}} \sum_{i=1}^{n} f(x_i)\Delta x_i \tag{9.1}$$

where Δx_i is a small interval of the segment $b - a$, for instance, $(b - a)/n$. In most engineering applications, this type of integration is adequate. However, often a sequence of functions $\{f_n(x)\}$, defined for values of x in the interval $[a, b]$, converges to a function $f(x)$, i.e.

$$\{f_n(x)\} \to f(x) \tag{9.2}$$

as n approaches infinity. Then being able to conclude that

$$\lim_{n\to\infty}\left(\int_a^b f_n(x)dx\right) \to \int_a^b f(x)dx \tag{9.3}$$

is important. For example, when using modal analysis, this property is required. Unfortunately, Equation (9.3) is not always true for Riemann integration. The *Lebesgue Integral* was developed to force Equation (9.3) to be true.

Lebesgue integration was developed using the concept of measurable sets, which is beyond the scope of this text. Thus, rather than defining Lebesgue integration directly, the properties are listed:

1) If $f(x)$ is Riemann integrable, then it is Lebesgue integrable, and the two integrations yield the same value (the reverse is *not* true).
2) If α is a constant and $f(x)$ and $g(x)$ are Lebesgue integrable, then $\alpha f(x)$ and $g(x) + f(x)$ are Lebesgue integrable.
3) If $f^2(x)$ is Lebesgue integrable, then $\int f^2(x)dx = 0$, if and only if $f(x) = 0$ *almost everywhere* (i.e. everywhere, except at a few points).
4) If $f(x) = g(x)$ almost everywhere, then $\int f^2(x)dx = \int g^2(x)dx$.

5) If $f_n(x)$ is a sequence of functions that are Lebesgue integrable over the interval $[a, b]$, if the sequence $\{f_n(x)\}$ converges to $f(x)$, and if for sufficiently large n there exists a function $F(x)$ that is Lebesgue integrable and $|f_n(x)| < F(x)$, then
   a) $f(x)$ is Lebesgue integrable
   b) $\lim_{n\to\infty} \left\{ \int_a^b f_n(x)dx \right\} = \int_a^b f(x)dx$

Any function $f(\mathbf{x})$, $\mathbf{x} \in \Omega$, such that $f^2(\mathbf{x})$ is Lebesgue integrable is called square integrable in the Lebesgue sense, denoted by $f \in L_2^R(\Omega)$ and, as will be illustrated, defines an important class of functions. In fact, the set of functions that are $L_2^R(\Omega)$ make up a linear space (see Appendix B or Naylor and Sell, 1982). In this notation, the subscript 2 denotes square integrable, the superscript R denotes the functions are all real, and Ω denotes the domain of integration.

Another important concept is that of linearly independent sets of functions. An arbitrary set of functions is said to be *linearly independent* if every finite subset of elements is linearly independent. Note that this definition is consistent with the notion of linear independence introduced in Chapter 3 and, in fact, is based on that definition.

The concept of a linear space, or vector space, is sufficient for the study of matrices and vectors. However, more mathematical structure is required to discuss operators. In particular, a linear space is defined to be an *inner product* space if, with every pair of elements u and v in the space, there exists a unique complex number (u, v), called the inner product of u and v, such that

1) $(u, v) = (v, u)^*$, the asterisk indicates the complex conjugate;
2) $(\alpha u_1 + \beta u_2, v) = \alpha(u_1, v) + \beta(u_2, v)$;
3) $(u, u) \geq 0$;
4) $(u, u) = 0$, if and only if $u = 0$.

The inner product most often used in vibrations is that defined by the Lebesgue integral given in the form

$$(u, v) = \int_\Omega u(\mathbf{x})v^*(\mathbf{x})d\Omega \tag{9.4}$$

Next, a few more definitions and results are required in order to mimic the structure used in linear algebra to analyze vibration problems. The *norm* of an element in a linear space is denoted by $\|u\|$ and is defined, in the case of interest here, by

$$\|u(\mathbf{x})\| = \sqrt{(u(\mathbf{x}), u(\mathbf{x}))} \tag{9.5}$$

Note that $\|u(\mathbf{x})\| = 0$ if and only if $u(\mathbf{x}) = 0$ almost everywhere. In addition, the norm satisfies the following conditions:

1) $\|u\| > 0$;
2) $\|\alpha u\| \leq |\alpha| \, \|u\|$, where α is a scalar;
3) $\|u + v\| < \|u\| + \|v\|$ with equality, if and only if $v = \alpha u > 0$ (referred to as the *triangle inequality*).

This last set of conditions introduces even more structure on a linear space. A linear space with such a norm is called a *normed linear space*. The set $L_2^R(\Omega)$ can be shown to form an inner product space with the preceding definition of an inner product and a normed linear space with the preceding definition of a norm.

Note that in the case of vectors, the scalar product $\mathbf{x}^T\mathbf{x}$ satisfies the preceding definition of inner product. An important property of sets of elements is that of orthogonality. Just as in the case of vectors, two elements, u and v, of an inner product space are said to be *orthogonal* if $(u,v) = 0$. Orthogonality is used extensively in Chapter 8 with respect to modal analysis.

Based on the definitions of convergence used in calculus for scalars, a definition of convergence for elements of a normed linear space can be stated. Let $\{u_n(x)\}$ be a set of elements in a normed linear space, denoted by V. The sequence $\{u_n(x)\}$ *converges* to the element $u(x)$ in V if for all $\varepsilon > 0$, there exists a number $N(\varepsilon)$ such that

$$\|u_n(x) - u(x)\| < \varepsilon$$

whenever $n > N(\varepsilon)$. This form of convergence is denoted by any of the following

$$u_n \to u \text{ as } n \to \infty$$

$$\lim_{n\to\infty} u_n(x) = u(x)$$

$$\lim_{n\to\infty} \|u_n(x) - u(x)\| = 0$$

In particular, the notation

$$u(x) = \sum_{n=1}^{\infty} u_n(x)$$

implies that the sequence of partial sums converges to $u(x)$, i.e.

$$\lim_{k\to\infty} \sum_{n=1}^{k} u_n(x) = u(x)$$

Convergence defined this way is referred to as *strong convergence*, or *norm convergence*. As pointed out briefly in Section 8.4, this convergence is required when writing the modal expansion of Equation (8.11).

In the case of vectors, writing a series of weighted sums of eigenvectors is sufficient. The resulting sum is always finite, since the sum in the modal expansion contains a finite number of terms. However, since the sums in general normed linear spaces may have an infinite number of terms, convergence to some finite valued function is not obvious.

As an aid in considering the convergence of a sequence of functions, a Cauchy sequence is used. A sequence $\{u_n(x)\}$ is defined to be a *Cauchy sequence* if for all numbers $\varepsilon > 0$, there exists a number $N(\varepsilon)$, such that $\|u_n(x) - u_m(x)\| < \varepsilon$ for every index m and n larger than $N(\varepsilon)$. An immediate consequence of this definition, and the triangle inequality, is that every convergent sequence is a Cauchy sequence. However, in general, not every Cauchy sequence converges. Requiring Cauchy sequences to converge leads to the concept of completeness. A normed linear space is defined to be *complete* if every Cauchy sequence in that space converges in that space.

Note that in a complete normed linear space, convergent sequences and Cauchy sequences are identical. The concept of completeness means that the limits of all of the sequences in the space that converge are also in that space. Comparing the set of real numbers with the set of rational numbers is analogous to the concept of completeness. The set of rational numbers is not complete, since one can construct a sequence of rational numbers that converges to an irrational number (such as the square root of two), which is not rational and hence is not in the set.

A complete normed linear space is called a *Banach space.* The set of all vectors of dimension n with real elements and with norm defined by $\|\mathbf{x}\|^2 = \mathbf{x}^T\mathbf{x}$ is a familiar example of a Banach space. The major difference in working with lumped parameter vibration problems versus working with distributed parameter vibration problems is based on the fact that every finite-dimensional normed linear space is complete and many common infinite-dimensional spaces are not. This possibility requires some concern over issues of convergence when using mode summation methods.

Example 9.2.1

Show by example that the space defined by the set of all continuous functions defined on the interval $[-1,1]$ with norm

$$\|u\|^2 = \int_{-1}^{1} |u|^2 dx$$

is not complete and also that a Cauchy sequence in the space does not converge to something in the space. Consider the sequence of continuous functions $u_n(x)$, where

$$u_n(x) = \begin{cases} 0 & -1 \le x \le 0 \\ nx & 0 < x \le \frac{1}{n} \\ 1 & \frac{1}{n} < x \le 1 \end{cases}$$

A quick computation verifies that the sequence is a Cauchy sequence, but the sequence $\{u_n\}$ converges to the function $u(x)$ given by

$$u(x) = \begin{cases} 0 & -1 \le x \le 0 \\ 1 & 0 < x \le 1 \end{cases}$$

which is *discontinuous* and hence is not an element in the space defined on the set of continuous functions.

In this last example, note that both $u_n(x)$ and $u(x)$ are square integrable in the Lebesgue sense. In other words, if instead of requiring the linear space in the example to be the set of continuous functions, the set of square integrable functions in which the Lebesgue sense are used, the space *is* a complete normed linear space. Since using modal analysis is desirable, mimicking the procedure used in finite dimensions, the natural choice of linear spaces to work in is $L_2^R(\Omega)$ or, when appropriate, $L_2^C(\Omega)$.

Again, motivated by the method of modal analysis, it is desirable to equip the linear space with an inner product. Gathering all this mathematical structure together yields the class of functions most useful in vibration analysis. This space is called a *Hilbert Space,* denoted by $\mathcal{H}$. A Hilbert Space is defined as a complete inner product space. Again, the set of real vectors with inner product $\mathbf{x}^T\mathbf{x}$ is an example of a Hilbert Space. The Hilbert Spaces of interest here are $L_2^R(\Omega)$ and $L_2^C(\Omega)$.

A further requirement placed on Hilbert Spaces used in vibration applications is the assumption that the space is separable. A *separable Hilbert Space* is a Hilbert Space that contains a countable set of elements $\{f_n(x)\}$, such that for any element f in the space and any positive real number ε, there exists an index N and a set of constants $\{\alpha_i\}$ such that

$$\left\| f(x) - \sum_{n=1}^{N} \alpha_n f_n(x) \right\| < \varepsilon \tag{9.6}$$

Here, the set $\{f_n\}$ is called a *spanning set* for the Hilbert Space. These concepts and their significance are discussed in the next section. However, note again that finite-dimensional vector spaces are separable and that this property is useful in writing modal expansions. Both spaces $L_2^R(\Omega)$ and $L_2^C(\Omega)$ are separable Hilbert Spaces. The symbol H is used to denote a general separable Hilbert Space.

9.3 Expansion Theorems

In this section, the expansion theorem, or modal expansion, used informally in the last chapter is placed in a more rigorous setting by generalizing the Fourier series. Let $\{\phi_k\}$ be an infinite set of orthonormal functions in $\mathcal{H}$, i.e. $(\phi_k,\phi_i) = \delta_{ki}$. For any element $u \in \mathcal{H}$, the scalar (u,ϕ_k) is called the *Fourier coefficient* of u relative to the set $\{\phi_k\}$. Furthermore, the sum

$$\sum_{k=1}^{\infty} (u, \phi_k)\phi_k(x)$$

is called the *Fourier series* of $u(x)$ relative to the set $\{\phi_k\}$.

The following results (stated without proof) are useful in extending the Fourier theorem and in understanding how to use modal expansions in applications. Let $\{\phi_n\}$ again be an orthonormal set of elements in $\mathcal{H}$, let u also be an element in $\mathcal{H}$, and let $\{\lambda_k\}$ denote a set of complex scalars. Then the following relationships hold

$$\left\| u - \sum_{k=1}^{n} \lambda_k \phi_k \right\| \geq \left\| u - \sum_{k=1}^{n} (u, \phi_k)\phi_k \right\| \tag{9.7}$$

$$\sum_{k=1}^{\infty} \left|(u, \phi_k)\right|^2 < \|u(x)\|^2 \text{ (called Bessel's inequality)} \tag{9.8}$$

and

$$\lim_{k\to\infty} (u, \phi_k) = 0 \tag{9.9}$$

This last result is referred to as the *Riemann-Lebesgue lemma.*

Furthermore, a famous result, the *Riesz-Fischer theorem*, relates the convergence of a modal expansion to the convergence of the coefficients in that expansion:

1) If $\sum_{k=1}^{\infty} |\lambda_k|^2 < \infty$, then $\sum_{k=1}^{\infty} \lambda_k \phi_k(x)$ converges to an element, u in $\mathcal{H}$ and $\lambda_k = (u, \phi_k)$, i.e. the λ_k are the Fourier coefficients of the function $u(x)$.
2) If $\sum_{k=1}^{\infty} |\lambda_k|^2$ diverges, then so does the expansion $\sum_{k=1}^{\infty} \lambda_k \phi_k$.

A set of orthonormal functions satisfying (1) is called a *complete set* of functions. Recall that a complete space is a space in which Cauchy sequences converge. A complete set, on the other hand, is a set of elements in a space such that every element in the space can be represented as a series expansion of elements of the set. In general, an orthonormal set of functions $\{\phi_k\}$ in $\mathcal{H}$ is complete if any of the following relationships hold

$$u = \sum_{k=1}^{\infty} (u, \phi_k)\phi_k \text{ for each in } \mathcal{H} \tag{9.10}$$

$$\|u\|^2 = \sum_{k=1}^{\infty} |(u, \phi_k)|^2 \text{ for each continuous } u \text{ in } \mathcal{H} \tag{9.11}$$

$$\text{if } (u, \phi_n) = 0 \text{ for each index } n, \text{then } u = 0 \tag{9.12}$$

Equation (9.11) in this list is referred to as *Parseval's equality.*

The preceding theorems generalize the concept of a Fourier series expansion of a function and provide a framework for modal expansions.

9.4 Linear Operators

The idea of a linear operator was introduced in Chapter 8 as related to vibration problems. Here, the definition of a linear operator is formalized, and some properties of a linear operator are developed. The eigenfunctions and eigenvalues of operators are also discussed in detail and the concept of adjoint is introduced.

A *subspace* of a linear space is a subset of elements of that space that again has the structure of a linear space. A linear operator is briefly defined in Section 8.2 as a mapping from one set of functions to another. Then the subspace $D(L)$ of the space $\mathcal{H}$ denotes the *domain* of the operator L and is the space of elements in $\mathcal{H}$ that the operator L is defined to act on. A rule L is defined to be an operator if, for each $u \in D(L)$, there is a uniquely determined element Lu that lies in $\mathcal{H}$. An operator L is linear if for every complex scalar α and β, as well as for u and v in $D(L)$, the following is true

$$L(\alpha u + \beta v) = \alpha L u + \beta L v \tag{9.13}$$

The operator L defines two other spaces of interest. The first is the *range space*, or *range*, of the operator L, denoted by $R(L)$; it is defined to be the set of all functions $\{Lu\}$ where $u \in D(L)$. The domain and range of an operator are exactly analogous to the domain and range of a function. Another very important space associated with an operator is the *null space*. The null space of an operator L is the set of all functions $u \in D(L)$, such that $Lu = 0$. This space is denoted by $N(L)$ and corresponds to rigid body modes in a structure. The spaces $R(L)$ and $N(L)$ are in fact subspaces of $\mathcal{H}$.

Obvious examples of these various spaces associated with an operator are the transformation matrices of Chapter 3. In fact, linear operator theory is an attempt to generalize the theory of matrices and linear algebra.

An important difference between matrices and operators can be pointed out by defining *equality* of two operators. Two operators L_1 and L_2 are equal if and only if

$$D(L_1) = D(L_2) \text{ and } L_1 u = L_2 u \text{ for all } u(x) \in D(L_1)$$

Example 9.4.1

Consider the linear operator associated with the string equation with fixed ends. Define this operator's domain and null space. The operator is

$$L = -\frac{\partial^2}{\partial x^2}$$

The domain, $D(L)$, consists of all functions in $L_2^R(0, l)$ having two derivatives and satisfying the boundary conditions $u(0) = u(l) = 0$. The null space of L is the set of all functions u in $D(L)$, such that $u'' = 0$. Integrating $u'' = 0$ requires that $u = ax + b$, where a and b are constants of integration. However, elements in $D(L)$ must also satisfy $a(0) + b = 0$ and $al + b = 0$, so that $a = b = 0$. Thus the null space of this operator consists of only the zero function.

An operator L is called *one-to-one* if $Lu_1 = Lu_2$ holds, if and only if $u_1 = u_2$. A linear operator L has an *inverse*, denoted by L^{-1} and defined by $L^{-1}u = v$, if and only if $u = Lv$. L is said to be nonsingular in this case and singular if L^{-1} does not exist. Note that for nonsingular operators

$$D(L^{-1}) = R(L)$$
$$R(L^{-1}) = D(L)$$
$$LL^{-1}u = u \qquad \text{for all } u \text{ in } D(L^{-1}) = R(L)$$

and

$$L^{-1}Lu = u \qquad \text{for all } u \text{ in } D(L) = R(L^{-1})$$

Also, L^{-1} is one-to-one. The inverse operator, as is shown in Section 8.6, of a differential operator often turns out to be an integral operator defined by a Green's function.

The concept of null space and inverse are also closely related to the eigenvalues of an operator. In particular, suppose zero is an eigenvalue of L_1. Then $L_1 u = 0$ for $u \neq 0$ (since eigenfunctions are never zero). Thus, there exists a nonzero function u in the null space of the operator L. Actually, a stronger statement holds true. If L is a one-to-one operator, L has an inverse, and $N(L)$ contains only zero if and only if zero is not an eigenvalue of the operator L.

An operator is bounded if there exists a finite constant $c > 0$ such that

$$\|Lu\| < c\,\|u\|$$

for each function $u(x)$ in $D(L)$. A related operator property is continuity. An operator L is *continuous* at $u \in D(L)$, if whenever $\{u_n\}$ is a sequence of functions in $D(L)$ with limit u, then $Lu_n \to Lu$. This definition of continuity is often abbreviated as $u_n \to u \Rightarrow Lu_n \to Lu$. For linear operators, this definition is equivalent to requiring L to be continuous at every element in $D(L)$. In addition, a linear operator is continuous if and only if it is a bounded operator.

Example 9.4.2

Differential operators are not bounded. Consider $L = d/dx$ with $D(L)$ consisting of all functions in $L_2^R(0,1)$, such that Lu is in $L_2^R(0,1)$. This operator is not bounded since the element $u(x) = \sin(n\pi x) \in L_2^R(0,1)$. Then $\|u\|$ = constant = a, independent of n, and

$$\|Lu\| = n\left(\int \cos^2 n\pi x dx\right)^{1/2} = nb,$$

where b is a constant. Hence, there is a function $u \in D(L)$ such that $\|Lu\| > \|u\|$, and the operator cannot satisfy the definition of a bounded operator because n may be arbitrarily large.

Example 9.4.3

Integral operators are bounded. Let $u \in L_2^R(0,1)$ and define the operator L by

$$Lu(x) = \int_0^x u(s)ds = f(x)$$

Next consider $f(x)$ as defined earlier and square its modulus to get

$$|f(x)|^2 = \left|\int_0^x u(s)ds\right|^2$$

But, the Cauchy-Schwarz inequality yields that

$$\left|\int_0^x u(s)ds\right|^2 \leq \int_0^1 |1|^2 ds \int_0^1 |u(s)|^2 ds = \|u\|$$

Hence $\|Lu\| < \|u\|$, and this operator is bounded.

The operators used to describe the linear vibrations of common distributed parameter structures are unbounded. However, the differential operators for strings, beams, membranes, and so on, all have inverses defined by Green's functions. Green's functions define integral operators, which are bounded. This connection of vibration equations to a bounded operator is significant in verifying convergence and eigenfunction expansions.

Another important operator is called the *adjoint operator*, which is defined in the following paragraphs and is basically the generalization of the transpose of a real matrix. First, consider a linear operator T, which maps elements in $\mathcal{H}$ into a set of complex numbers. Such an operator is called a *linear functional*. In particular, consider the linear functional defined by an inner product. Let the linear functional T_v be defined as

$$T_v u = (u, v) = \int_{\Omega} uv^* d\Omega \tag{9.14}$$

where $u \in \mathcal{H}$, and v is a fixed element in $\mathcal{H}$. The action of T_v defines a bounded operator.

The following result, called the *Riesz representation theorem*, allows the adjoint operator for a bounded operator to be defined. Let T be a bounded linear functional defined on $\mathcal{H}$. Then there exists a unique element, f, in $\mathcal{H}$ such that

$$Tu = (u, f). \tag{9.15}$$

The significance of the Riesz representation theorem is as follows. Suppose that the operator L is bounded, and u and v are in $\mathcal{H}$. Then

$$Tu = (Lu, v) \tag{9.16}$$

defines a bounded linear functional. However, via the Riesz representation theorem, there exists a unique element f in $\mathcal{H}$ such that

$$Tu = (u, f) \tag{9.17}$$

Hence, Equation (9.16) yields that

$$(Lu, v) = (u, f) \tag{9.18}$$

where $f \in \mathcal{H}$ is unique. The unique element f is denoted as $f = L^* v$, where L^* is defined to be the *adjoint* of the operator L. Then, Equation (9.18) becomes

$$(Lu, v) = (u, L^* v) \tag{9.19}$$

The adjoint defined this way is linear and unique. The adjoint operator L^* is also bounded if L is bounded.

Unfortunately, the Riesz representation theorem and the preceding discussion of adjoints only hold for bounded operators. The equations of interest in vibrations, however, yield unbounded operators. Since the inverses of these unbounded operators are in fact bounded, the idea of an adjoint operator for unbounded operators is still used. However, to denote that a formal proof of existence of the adjoint operator for unbounded operators does not exist, the adjoint of an unbounded operator is often referred to as a *formal adjoint*.

Let L be an unbounded differential operator with domain $D(L)$. The operator L^* defined on a domain $D(L^*)$ is the formal adjoint of L if

$$(Lu, v) = (u, L^*v), \quad u \in D(L), \quad v \in D(L^*) \tag{9.20}$$

Note that $D(L)$ is characterized by certain boundary conditions. The domain $D(L^*)$ will also be characterized by possibly different boundary conditions called *adjoint boundary conditions.*

An operator (possibly unbounded) is defined to be *formally self-adjoint* if

$$D(L) = D(L^*) \quad \text{and} \quad Lu = L^*u \quad \text{for} \quad u \in D(L) \tag{9.21}$$

The prefix *formal* is not always used, however. It is important to note that if L is formally self-adjoint, then for all $u, v \in D(L)$

$$(Lu,\ v) = (u,\ Lv) \tag{9.22}$$

which follows from Equation (9.21). This last relationship is used to define a related class of operators. An operator L is *symmetric* if Equation (9.22) holds for all pairs u and v in $D(L)$.

Note that the difference between the definitions of symmetric and formally self-adjoint is the domain requirement of Equation (9.21). That is, if the form of the operator and its adjoint are the same, then the operator is symmetric. If in addition, both the operator and the adjoint operator have the same boundary conditions, then the operator is self adjoint. If L is a bounded operator, then $D(L) = \mathcal{H}$, and formally self-adjoint and symmetric mean the same. If L is unbounded, however, an operator may be symmetric but not self-adjoint (i.e. $D(L) \neq D(L^*)$). A formally self-adjoint operator, however, is a symmetric operator.

Example 9.4.4

Consider the operator for a fixed-fixed string and determine its adjoint. Here L is of the form

$$L = -\alpha \frac{\partial^2}{\partial x^2}$$

and its domain is defined as

$$D(L) = \left\{u | u(0) = u(l) = 0, u, u', u'' \in L_2^R(0, l)\right\}$$

The right-hand side of this expression denotes the set of all functions $u(x)$ such that the boundary conditions are satisfied (i.e. $u(0) = \mathrm{u}(l) = 0$) and the functions u, u' and u'' belong in the set $L_2^R(0, l)$. Calculating the linear product (Lu,v) for $u \in D(L)$ and $v \in D(L^*)$ yields

$$(Lu, v) = -\alpha \int_0^l uvdx$$

Integrating by parts twice yields

$$(Lu, v) = \left[-\alpha u'(l)v(l) + \alpha u'(0)v(0) + \alpha u(l)v'(l) - \alpha u(0)v'(0)\right] - \alpha \int_0^l uv''dx$$

Since $u \in D(L)$, the above reduces to

$$(Lu, v) = \left[-\alpha u'(l)v(l) + \alpha u'(0)v(0)\right] + (u, Lv) \tag{9.23}$$

Now if both u and v are in $D(L)$, then $v(l) = v(0) = 0$ and Equation (9.23) shows that L is a symmetric operator. To see that L is in fact formally self-adjoint, however, consider Equation (9.23) with $v \in D(L^*)$. Since $v \in D(L^*)$, the integration by parts requires that v, v' and v'' are in $L_2^R(0, l)$. In order for $(Lu,v) = (u,Lv)$ to be satisfied, the term in brackets in Equation (9.23) must be zero, i.e.

$$-\alpha u'(l)v(l) + \alpha u'(0)v(0) = 0$$

But since $u'(l)$ and $u'(0)$ are arbitrary (i.e. there is no restriction on these values in $D(L)$), $v(l)$ and $v(0)$ must both be zero. Thus

$$D(L^*) = \{v|v(0) = v(l) = 0, v, v', v'' \in L_2^R(0, l)\} = D(L)$$

and this operator is in fact formally self-adjoint. The boundary conditions, $v(0) = v(l) = 0$, are the adjoint boundary conditions.

In order for a differential operator to be symmetric, it must be of even order (Problem 9.20). Most of the physical structures examined so far are self-adjoint with respect to most boundary conditions. The following example illustrates a symmetric, nonself-adjoint-operator.

Example 9.4.5

Consider the operator $L = \partial^2/\partial x^2$ on the domain

$$D(L) = \{u|u(0) = u'(0) = 0, u, u', u'' \in_2^R (0, l)\}$$

Equation (9.23) still holds with $a = -1$. However, the term in brackets is now

$$[u'(l)v(l) - u(l)v'(l)] = 0$$

Since $u'(l)$ and $u(l)$ are arbitrary, this last expression requires that

$$v(l) = v'(l) = 0$$

Hence

$$D(L^*) = \{v|v, v', v'' \in L_2^R(0, l) \text{ and } v(l) = v'(l) = 0\} \neq D(L)$$

and the operator is not self-adjoint. However, this operator is symmetric.

Symmetric operators (and thus so do self-adjoint operators) have special properties, as do symmetric matrices, which are useful in the study of vibrations. If L is symmetric, the following are true:

1) (Lu, u) is real.
2) If L has eigenvalues, they are all real.
3) Eigenfunctions corresponding to distinct eigenvalues are orthogonal.
4) Furthermore, if L is self-adjoint, it is called *positive definite* if $(Lu,u) > 0$ for all nonzero $u \in D(L)$ and *positive semidefinite* if $(Lu,u) \geq 0$ for all nonzero $u \in D(L)$.

Example 9.4.6
Consider the operator of Example 9.4.4, and show that it is positive definite. Start by calculating (Lu,u) for an arbitrary element $u \in D(L)$. Integration by parts yields

$$(Lu, u) = -(u', u)\Big|_0^l + \int_0^l (u')^2 dx$$
$$= \int_0^l (u')^2 dx \geq 0$$

where the boundary conditions eliminate the constant terms. This calculation shows that the operator L is positive semidefinite. To see that the operator is, in fact, strictly positive definite, note that if $u' = 0$, then $u = c$, a constant. However, $u(0) = u(l) = 0$ must be satisfied, so that u must be zero, contradicting the semidefinite condition. Thus, $(Lu,u) > 0$ and L is in fact positive definite.

In the case of matrices, a symmetric matrix is positive definite if and only if its eigenvalues are all positive real numbers. For an operator, a weaker version of this statement holds. A positive definite operator has positive eigenvalues (assuming it has eigenvalues). With further assumptions, which are clarified in the next section, a stronger statement can be made that is more in line with the matrix result.

9.5 Compact Operators

In this section, the last major mathematical requirement for eigenfunction expansions is considered. Compact operators are defined and some of their principal properties examined. Self-adjoint compact operators have essentially the same eigenstructure as symmetric matrices, and hence, are ideal for modal analysis.

The notion of a compact operator is based on restricting a bounded operator. Let $\{u_n\}$ be a uniformly bounded, infinite sequence in $\mathcal{H}$. A *uniformly bounded sequence* is one for which $\|u_n\| \leq M$ for all values of n (i.e. M does not depend on n). Let L be a bounded linear operator defined on $\mathcal{H}$. The operator L is defined to be *compact* if, from the sequence $\{Lu_n\}$, one can extract a subsequence, denoted by $\{(Lu_n)_k\}$, that is a Cauchy sequence. Another way to describe a compact operator is to note that a compact operator maps bounded sets into compact sets. A compact set is a set such that each sequence of elements in the set contains a convergent subsequence. In this way, the notion of a compact operator is related to a compact set.

The idea of a compact operator is stronger than that of a bounded operator. In fact, if an operator is compact, it is also bounded. However, bounded operators are not necessarily compact. Since bounded operators are continuous operators and compactness requires more of the operator, compact operators are also called *completely continuous* operators.

The identity operator is an example of a bounded operator, i.e. $\|Iu\| \leq a\|u\|$, which is not necessarily compact. Every bounded operator defined on a finite-dimensional Hilbert space is compact, however. Bounded operators are compact because every set containing a finite number of elements is compact.

Consider next the integral operator defined by Green's function. Such operators are compact. In fact, if $g(x, y)$ is any continuous function where x and y are both in the interval $(0, l)$, then the integral operator defined by

$$Lu = \int_0^l g(x, y)u(y)dy \tag{9.24}$$

is a compact operator on $L_2^R(0, l)$.

Self-adjoint compact operators have the desired expansion property, namely, if L is compact and self-adjoint on $\mathcal{H}$, then L can be shown to have nonzero eigenvalues $\{\mu_n\}$ and orthonormal eigenfunctions $\{\emptyset_n\}$. Next, let $u(x)$ be any element in H; so that $u(x)$ can then be represented as the generalized Fourier series

$$u(x) = \sum_{n=1}^{\infty} (u, \emptyset_n)\emptyset_n + u_0(x) \tag{9.25}$$

where $u_0(x)$ lies in the null space of L. Furthermore, the function Lu can also be represented as

$$Lu(x) = \sum_{n=1}^{\infty} \mu_n(u, \emptyset_n)\emptyset_n \tag{9.26}$$

Note that if L is nonsingular, $u_0(x) = 0$. Also note that by comparing Equation (9.10) with Equation (9.25), the eigenfunctions of a compact self-adjoint operator are complete in $\mathcal{H}$. These two powerful results form the backbone of modal analysis of distributed parameter systems.

Expression (9.26) also allows a more concrete statement about the relationship between the eigenvalues of an operator and the definiteness of an operator. In fact, a compact self-adjoint operator L is positive definite if and only if each of its eigenvalues are positive real numbers.

A key feature of a bounded operator is the concept of compactness. Self-adjoint compact operators have essentially the same eigenstructure as symmetric matrices, and hence, are ideal for modal analysis. An operator L is defined to be *compact* if it maps bounded sets into compact sets. A compact set is a set such that each sequence of elements in the set contains a convergent subsequence. So basically if a differential operator is self adjoint and has a compact inverse, then the desired eigenstructure properties follow. Furthermore, a compact self-adjoint operator L is positive definite if and only if each of its eigenvalues are positive real numbers, lining up well again with measured structural properties.

This concept of having a full set of eigenfunctions and eigenvalues can be extended to differential operators that do not have an inverse, such as those with free-free boundary conditions by defining the *resolvent* operator, which is the inverse of a singular operator shifted by the identity operator. In terms of the notation here, the resolvent of the (singular) differential operator L is

$$(L - \alpha I)^{-1}$$

which always exist. As long as this operator is compact, then all the above-mentioned properties exist. This is why modal analysis works for freely supported structures.

9.6 Theoretical Modal Analysis

In this section, the idea of a compact operator, along with the associated expansion, is applied to a generic model of a differential equation describing the linear vibration of a distributed parameter system. This theory results in the modal analysis of such structures and provides a firm mathematical foundation for the material presented in Sections 8.3 and 8.4.

Consider again Equation (8.1), repeated here for convenience

$$\begin{aligned} w_{tt}(x,t) + L_2 w(x,t) &= 0 \qquad x \in \Omega \qquad t > 0 \\ Bw &= 0 \qquad x \in \partial\Omega \qquad t > 0 \\ w(x,0) &= w_0(x) \quad w_t(x,0) = \dot{w}_0(x) \end{aligned} \tag{9.27}$$

With the additional assumptions that the (nonsingular) operator L_2 is self-adjoint and has a compact inverse, the following shows that the sum of Equation (8.3), i.e. the modal expansion of the solution, converges.

Since L_2^{-1} is compact, the eigenfunctions of L_2^{-1} are complete. As noted before, these eigenfunctions are also those of L_2, so that the eigenfunctions of L_2 are complete. As a result, the solution $w(x,t)$, considered as a function of x defined on the Hilbert space $\mathcal{H}$ for a fixed value of t, can be written as

$$w(x,t) = \sum_{n=1}^{\infty} a_n(t)\varnothing_n(x) \tag{9.28}$$

where it is anticipated that the Fourier coefficient a_n in Equation (9.28) will depend on the fixed parameter t, $t > 0$. From Equation (9.10), the coefficient $a_n(t)$ is

$$a_n(t) = \int_\Omega w(x,t)\varnothing_n d\Omega = (w, \varnothing_n). \tag{9.29}$$

Multiplying Equation (9.27) by $\varnothing_n(x)$ and integrating over Ω yields

$$\int_\Omega w_{tt}\varnothing_n(x)d\Omega + \int_\Omega L_2 w\varnothing_n(x)d\Omega = 0 \tag{9.30}$$

Equation (9.30) can be rewritten as

$$\frac{\partial^2}{\partial t^2}\left[\left(w, \varnothing_n\right)\right] + \left(L_2 w, \varnothing_n\right) = 0 \tag{9.31}$$

Using the self-adjoint property of L_2 and the fact that $\varnothing_n$ is an eigenfunction of L_2 with corresponding eigenvalue λ_n yields

$$\frac{\partial^2}{\partial t^2}[(w, \varnothing_n)] + \lambda_n(w, \varnothing_n) = 0$$

or, from Equation (9.29)

$$\ddot{a}_n(t) + \lambda_n a_n(t) = 0 \quad t > 0 \tag{9.32}$$

This expression, along with the appropriate initial conditions, can be used to calculate $a_n(t)$. In particular, if L_2 is positive definite, $\lambda_n > 0$ for all n, then

$$a_n(t) = \frac{\dot{a}_n(0)}{\omega_n} \sin \omega_n t + a_n(0) \cos \omega_n t \tag{9.33}$$

where

$$a_n(0) = \int_\Omega w(x,0)\varnothing_n(x)d\Omega$$

$$\dot{a}_n(0) = \int_\Omega w_t(x,0)\varnothing_n(x)d\Omega$$

$$\omega_n = \sqrt{\lambda_n}$$

These formulas are, of course, consistent with those developed formally in Section 8.3. Here, however, the convergence of Equation (9.28) is guaranteed by the compactness theorem of Section 9.5.

As indicated in Section 8.4, this procedure can be repeated for damped systems if the operators L_1 and L_2 commute on a common domain. This result was first pointed out formally by Caughey and O'Kelly (1965), but the development did not concern itself with convergence. The approach taken by Caughey and O'Kelley, as well as in Section 8.4, is to substitute the series of Equation (9.28) into Equation (8.14). Unfortunately, this substitution raises the issue of convergence of the derivative of the series. The convergence problem is circumvented by taking the approach described in Equations (9.30) to (9.32).

Repeating this procedure for the damped system, as described by Equation (8.14), requires that L_1 and L_2 have the same eigenfunctions and that L_2 is self-adjoint with compact inverse. Multiplying Equation (8.14) by $\varnothing_n(x)$ and integrating yields

$$\frac{d^2}{dt^2}[(w,\varnothing_n)] + \frac{d}{dt}[(L_1 w,\varnothing_n)] + (L_2 w,\varnothing_n) = 0 \tag{9.34}$$

Using the property that L_1 and L_2 are self-adjoint and denoting the eigenvalues of L_1 by $\lambda_n^{(1)}$ and those of L_2 by $\lambda_n^{(2)}$, Equation (9.34) becomes

$$\ddot{a}_n(t) + \lambda_n^{(1)}\dot{a}_n(t) + \lambda_n^{(2)}a_n(t) = 0 \tag{9.35}$$

Equations (9.32) and (9.35) constitute a theoretical modal analysis of a distributed mass system. Equation (9.34) is solved using the methods of Section 1.3 from initial conditions determined by using mode orthogonality.

9.7 Eigenvalue Estimates

As indicated previously, knowledge of a system's eigenvalues (natural frequencies) provides knowledge of the dynamic response of a structure. Unfortunately, the eigenvalue problem for the operators associated with many structures cannot be solved. Having no solution requires the establishment of estimates and bounds on the eigenvalues of operators. Note, that while this section is an extension of

Section 3.6 on eigenvalue estimates for matrices, the operator problem is more critical because in the matrix case estimates are primarily used as an analytical tool, whereas in the operator case the estimates are used where solutions do not even exist in closed form.

Consider first the conservative vibration problems of Equation (8.1) for the case that the operator L is self-adjoint and positive definite with compact inverse. As noted in Section 9.6, these assumptions guarantee the existence of a countable infinite set of positive real eigenvalues $\{\lambda_i\}$, which can be ordered as

$$0 \le \lambda_1 \le \lambda_2 \le \cdots \le \lambda_n \le \cdots \tag{9.36}$$

with corresponding orthonormal eigenfunctions, $\{\emptyset_n(x)\}$, complete in $L_2^R(\Omega)$.

Next, further assume that the operator L is *coercive*, i.e. that there exists a constant c such that

$$c\|u\|^2 \le (Lu, u) \tag{9.37}$$

for all $u \in D(L)$. Thus

$$\frac{(Lu, u)}{\|u\|^2} \ge c \tag{9.38}$$

for all $u \in D(L)$. Thus, the quantity $(Lu,u)/\|u\|^2$ is bounded below, and therefore there is a *greatest lower bound,* denoted by *glb*, of this ratio. Define the functional $R(u)$ by

$$R(u) = \frac{(Lu, u)}{\|u\|^2} \tag{9.39}$$

The functional $R(u)$ is called the *Rayleigh quotient* of the operator L.

Since the eigenfunctions of L are complete, Equation (9.11) yields

$$\|u\|^2 = \sum_{n=1}^{\infty} |(u, \emptyset_n)|^2$$

and

$$(Lu, u) = \sum_{n=1}^{\infty} \lambda_n |(u, \emptyset_n)|^2 \tag{9.40}$$

Hence, the Rayleigh quotient can be written as

$$R(u) = \frac{\sum \lambda_n |(u, \emptyset_n)|^2}{\sum |(u, \emptyset_n)|^2} \ge \lambda_1 \tag{9.41}$$

since Equation (9.7) implies that

$$\sum \lambda_n |(u, \emptyset_n)|^2 > \lambda_1 \sum |(u, \emptyset_n)|^2 \tag{9.42}$$

Here, the summation limits have been suppressed. Also, note that $R(\emptyset_1) = \lambda_1$, so that

$$\lambda_1 = \min(\lambda_i) = \underset{\substack{u \neq 0 \\ u \in D(L)}}{\text{glb}} [R(u)] \tag{9.43}$$

As in the matrix case, the Rayleigh quotient yields a method of finding an upper bound to the eigenvalues of a system.

Example 9.7.1

Consider the operator $L = -\partial^2/\partial x^2$ defined on the domain $D(L) = \{u \mid u(0) = u(1) = 0, u, u', u'' \in L_2^R(0,1)\}$. Estimate the first natural frequency by using the Rayleigh quotient.

Note that the function $u(x) = x(1-x)$ is in $D(L)$. Calculation of the Rayleigh quotient then yields

$$R(u) = \frac{-(u'', u)}{(u, u)} = \frac{1/3}{1/30} = 10$$

A calculation of the exact value of λ_1 for this operator yields

$$\lambda_1 = (\pi^2) < 10$$

so that the Rayleigh quotient in this case provides an upper bound to the lowest eigenvalue.

Bounds are also available for the other eigenvalues of an operator. In fact, with the previously mentioned assumptions on the operator L, the domain of L can be split into two subspaces, M_k, and $M_k^\perp$ (read "M_k perp") defined by the eigenfunctions of L. Let $M_k = \{\emptyset_1, \emptyset_2, \ldots, \emptyset_k\}$ be the set of the first k eigenfunctions of L and $M_k^\perp$ be the set of remaining eigenfunctions, i.e. $M_k^\perp = \{\emptyset_{k+1}, \emptyset_{k+2}, \ldots\}$. From these considerations (Stakgold, 1967, 2000), the following holds

$$\lambda_k = \min_{u \in M_{k-1}^\perp} [R(u)] = \max_{u \in M_k} [R(u)] \tag{9.44}$$

This formulation of the eigenvalues of an operator as a minimum or a maximum over sets of eigenfunctions can be further extended to extremals over arbitrary subspaces. Equation (11.44) is called the *Courant minimax principle.* Again, with L satisfying the assumptions of this section, let E_k be any k-dimensional subspace of $D(L)$, then

$$\lambda_k = \min_{E_k \in D(L)} \left[\max_{\substack{u \in E_k \\ u \neq 0}} (R(u)) \right] = \max_{E_k \in D(L)} \left[\max_{\substack{u \in E_{k-1}^\perp \\ u \neq 0}} (R(u)) \right] \tag{9.45}$$

where the value of $u(x)$ satisfying Equation (9.45) becomes $\emptyset_k$. The minimum over $E_k \in D(L)$ refers to the minimum over all the subspaces E_k of dimension k contained in $D(L)$. The maximum value of $R(u)$, $u \in E_k$, refers to the maximum value of $R(u)$ for each element u in the set E_k. The difference between Equations (9.44) and (9.45) is that Equation (9.44) is restricted to subspaces generated by the eigenfunctions of L, whereas in Equation (9.45) the subspaces are any subspace of $D(L)$.

9.8 Enclosure Theorems

The Rayleigh quotient and formulations of the eigenvalue problem of the previous section provide a means of estimating an upper bound of an operator's eigenvalues by examining sets of arbitrary functions in the domain of the operator. In this section, lower bounds and enclosure bounds are examined. Furthermore, bounds in terms of related operators are examined. In this way eigenvalues of operators that are difficult or impossible to calculate are estimated in terms of operators with known eigenvalues.

The first two results follow from the definition of the definiteness of an operator. This definition can be used to build a partial ordering of linear operators and to provide an eigenvalue estimate. For two self-adjoint operators L_1 and L_2, the operator inequality denoted by

$$L_1 \leq L_2 \tag{9.46}$$

is defined to mean that

$$\mathcal{H} \supset D(L_1) \supset D(L_2) \tag{9.47}$$

and

$$(L_1 u, u) \leq (L_2 u, u) \text{ for all } u \in D(L_2) \tag{9.48}$$

where Equation (9.47) denotes that $D(L_2)$ is a subset of $D(L_1)$, and so on.

If Equation (9.48) holds with strict equality, i.e. if L_1 and L_2 have the same form, with L_1 defined on a subspace of L_2 and if L_1 and L_2 are positive definite with compact inverses, then

$$\lambda_i^{(1)} \leq \lambda_i^{(2)}, \quad i = 1, 2, \ldots \tag{9.49}$$

Here, $\lambda_i^{(1)}$ denotes the eigenvalues of L_1, and $\lambda_i^{(2)}$ denotes those of the operator L_2. This inequality is called the first *monotonicity principle*.

Example 9.8.1

Consider the two operators defined by

$$L_1 = L_2 = -\partial^2/\partial x^2$$

with domains

$$D(L_1) = \left\{u | u(0) = u(l_1) = 0,\ u,\ u',\ u'',\ \in L_2^R(0, l_1)\right\}$$

and

$$D(L_2) = \{u | u(0) = u(l_2) = 0,\ u,\ u',\ u'' \in L_2^R(0, l_2)\}$$

Consider the case with $l_1 \leq l_2$. Then redefine the domain $D(L_1)$ to be the completion of the set of functions that vanish outside of $D(L_1)$, i.e. for $x > l_1$. Then

$D(L_2) \supset D(L_1)$ and expression (9.47) and inequality (9.49) are satisfied so that $L_2 < L_1$.

Thus, inequality (9.49) yields (note the indices are interchanged)

$$\lambda_i^{(2)} \leq \lambda_i^{(1)}$$

To see that this is true, note that

$$\lambda_i^{(2)} = \sqrt{i\pi/l_2} \leq \sqrt{i\pi/l_1} = \lambda_i^{(1)}.$$

This example shows that shrinking the domain of definition of the problem increases the eigenvalues, as expected.

The trick in using the first monotonicity principle is the ability to extend the domain $D(L_2)$ so that it can be considered as a subspace of $D(L_1)$. Extending the domain works in the example, because the boundary condition is $u = 0$ along the boundary. The method fails, for instance, for the membrane equation with clamped boundaries and a hole removed (Weinberger, 1974).

The preceding example was chosen to illustrate that the principle works. The use of this principle is more interesting, however, in a situation where one of the boundaries is such that the eigenvalues cannot be calculated, i.e. as in the case of an odd-shaped membrane. In this case, the unknown eigenvalues, and hence natural frequencies, can be bracketed by two applications of the first monotonicity theorem.

For instance, if the eigenvalues of a membrane of irregular shape are required, the eigenvalues of inscribed and circumscribed rectangles can be used to provide both upper and lower bounds of the desired eigenvalues. Thus, the monotonicity principle can also be thought of as an *enclosure theorem* (Problem 9.17).

If the operators L_1 and L_2 are positive definite and of different form, i.e. if equality in expression (9.48) does not hold, then inequality (9.49) is known as the *second monotonicity theorem*. The following example illustrates how the monotonicity results can be used to create enclosures for the eigenvalues of certain operators.

Example 9.8.2

Consider the membrane operator $L = -\nabla^2$, as defined by Equation (7.101). In particular, consider the three operators

$$L_1 = -\nabla^2,\ D(L_1) = \left\{ u \,\middle|\, \frac{\partial u}{\partial n} + k_1 u = 0 \text{ on } \partial\Omega, u, u_x, u_y, u_{xy}, u_{xx}, u_{yy} \in L_2^R(\Omega) \right\}$$

$$L_2 = -\nabla^2,\ D(L_2) = \left\{ u \,\middle|\, \frac{\partial u}{\partial n} + k_2 u = 0 \text{ on } \partial\Omega, u, u_x, u_y, u_{xy}, u_{xx}, u_{yy} \in L_2^R(\Omega) \right\}$$

$$L_3 = -\nabla^2,\ D(L_3) = \left\{ u \,\middle|\, u = 0 \text{ on } \partial\Omega, u, u_x, u_y, u_{xy}, u_{xx}, u_{yy} \in L_2^R(\Omega) \right\}$$

where $k_1 > k_2 > 0$. Compare the eigenvalues of these three operators.

With $k_1 > k_{2,}$ integration yields

$$(u, L_2 u) < (u, L_1 u)$$

For all $u \in D(L_2)$ and using the second monotonicity principle yields

$$\lambda_i^{(2)} < \lambda_i^{(1)}$$

Comparing operators L_3 and L_1, note that $D(L_3) \supset D(L_1)$, so that application of the first monotonicity principle yields

$$\lambda_i^{(2)} \leq \lambda_i^{(1)} \leq \lambda_i^{(3)}$$

Example 9.8.3

Consider a uniform beam, a structure easily understood and having the desired operator properties. Three cases are presented, two with known analytical eigenvalues and one bracketed by these two that does not have analytical solutions in order to illustrate how the enclosure theory works. The three beams are depicted in Figure 9.1. The tapered shaped beam outlined with red dashed lines is an example of a "practical" component of "odd" geometric shape. For a beam with cross-sectional area A, modulus E, area moment of inertia I, and density ρ, the relevant operator is

$$L = \frac{\partial^2}{\partial^2 x}\left[\frac{EI}{\rho A}\frac{\partial^2}{\partial^2 x}\right]$$

The two gray beams that geometrically bracket the "dashed" beam have constant E, I, A and ρ, so that the analytical frequencies are easily calculated to be

$$\lambda_n = \sqrt{\frac{E_j I_j}{\rho_j A_j}}\pi f(n)$$

Here $f(n)$ is function of integers depending on the boundary conditions. The index, $j = 1, 3$, denotes the properties of the two beams under consideration. For this case of constant physical parameters, the eigenvalues are a function of the ratio of the area moment of inertia and the cross-sectional area. This ratio turns out, for the case illustrated, to be just a function of the vertical height of the beam, denoted h (distance along the y-direction from the neutral axis to the surface). Thus, as long as the height of the inscribed red beam lies between the two "analytical" beams, simple integration shows that the three operators satisfy the inequalities

$$L_1 \leq L_2 \leq L_3$$

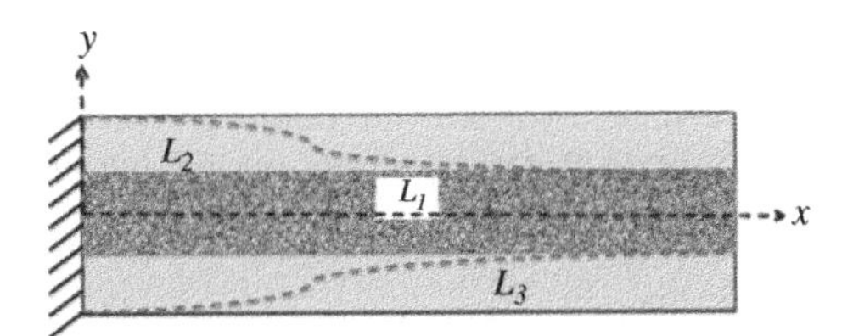

Figure 9.1 A non-uniform beam (dashed outline) with unknown spectrum bracketed by two beams with analytically known spectrum.

because the domains are all the same. This yields

$$\sqrt{\frac{EI_1}{\rho A_1}}f(n) \le \lambda_n^{(2)} \le \sqrt{\frac{EI_3}{\rho A_3}}f(n)$$

providing analytical bounds on the eigenvalues and hence natural frequencies of an oddly-shaped structure.

Thus, the idea is to find structures with closed form eigenvalues that bracket the unknown structure that would normally have to be solved numerically.

Chapter Notes

The material of this chapter is a much condensed, and somewhat oversimplified, version of the contents of a course in applied functional analysis. Several texts are recommended and were used to develop the material here. The books by Stakgold (1967, 1968) present most of the material here in two volumes. This text, republished in 2000 (Stakgold, 2000a,b), is recommended because it makes a very useful comparison between finite-dimensional systems (matrices) and infinite-dimensional systems. The book by Hocstadt (1973) presents an introduction to Hilbert spaces and compact operators in fairly short order. A good fundamental text on functional analysis, such as Bachman and Narici (1966), or on operator theory, such as Naylor and Sell (1982), presents the information of Sections 9.2, 9.3, 9.4 and 9.5 in rigorous detail. The material of Section 9.5 is usually not presented, except formally in vibrations texts. MacCluer (1994) presents conditions for the existence of convergent eigenfuction expansions and connects this to operator properties that result in the successful application of the method of separation of variables.

The eigenvalue estimate methods given in Section 9.7 are the most common eigenvalue estimates and can be found in most texts, including Stakgold (1967, 2000). The literature is full of improvements and variations of these estimates. The short and excellent text by Weinberger (1974) is summarized in Section 9.8 on enclosure theorems. These theorems are useful, but apparently have not been taken advantage of by the engineering community.

References

Bachman, G. and Narici, L. (1966) *Functional Analysis*. Academic Press, New York.

Caughey, T. K. and O'Kelley, M. E. J. (1965) Classical normal modes in damped linear dynamic systems. *ASME Journal of Applied Mechanics*, 32, 583–288.

Hocstadt, H. (1973) *Integral Equations*. John Wiley & Sons, New York.

MacCluer, C. R. (1994) *Boundary Value Problems and Orthogonal Expansions*. IEEE Press, New York.

Naylor, A. W. and Sell, G. R. (1982) *Linear Operator Theory in Engineering and Science*. Springer-Verlag, New York.

Stakgold, I. (1967) *Boundary Value Problems of Mathematical Physics*, vol. 1. Macmillan Publishing Co., New York.
Stakgold, I. (1968) *Boundary Value Problems of Mathematical Physics*, vol. 2. Macmillan Publishing Co., New York.
Stakgold, I. (2000a) *Boundary Value Problems of Mathematical Physics*, vol. 1. Philadelphia Society for Industrial and Applied Mathematics, Philadelpia, PA. (*Classics in Applied Mathematics*, vol. 29).
Stakgold, I. (2000b) *Boundary Value Problems of Mathematical Physics*, vol. 2. Philadelphia Society for Industrial and Applied Mathematics, Philadelpia, PA. (*Classics in Applied Mathematics*, vol. 29).
Weinberger, H. F. (1974) *Variational Methods for Eigenvalue Approximation*. Society for Industrial and Applied Mathematics, Philadelpia, PA.

Problems

9.1 Show that $L_2^R(\Omega)$ is a linear space.

9.2 Show that the sequence of Example 9.2.1 is in fact a Cauchy sequence.

9.3 Show that the sequence of Example 9.2.1 converges to $u(x)$ as claimed.

9.4 Show that the operator of Example 9.4.1 is linear.

9.5 For the operator of Example 9.4.1, show that $N(L)$, $R(L)$ and $D(L)$ are linear spaces and hence, subspaces of $\mathcal{H}$.

9.6 Consider the operator $-\partial^2/\partial x^2$, defined on $L_2^R(0, l)$ such that $u_x(0) = u_x(l) = 0$. Calculate the operator's null space, and show that it is a subspace of $\mathcal{H}$. Is this operator equal to the operator of Example 9.4.1?

9.7 Show that the linear functional defined by Equation (9.14) is in fact bounded.

9.8 Show that the operator defined by the Green's function of Example 8.6.1 is a bounded operator.

9.9 Calculate the adjoint of the operator in Problem 9.6. Is the operator formally self-adjoint? Positive definite?

9.10 Show that the operator $-\partial^4/\partial x^4$ defined on $L_2^R(0, l)$ with boundary conditions $u(0) = u(l) = u'(0) = u'(l) = 0$ is formally self-adjoint.

9.11 Consider the transverse vibrations of a beam with variable stiffness ($EI(x)$) and of dimension compatible with the Euler-Bernoulli assumptions and with cantilevered boundary conditions. Show that the corresponding operator is symmetric, positive definite and self-adjoint.

9.12 Calculate the adjoint of the operator $L = \partial/\partial x$ with boundary conditions $u(0) = \alpha u(1)$, α constant. (Do not forget to calculate $D(L^*)$.)

9.13 Suppose that A is a compact linear operator and B is a bounded linear operator, such that AB is defined. Show that AB is compact.

9.14 Show that if the linear self-adjoint operators L_1 and L_2 commute and if L_2 has a compact inverse and L_1 is nonsingular, then L_1 and L_2 have a common set of eigenfunctions.

9.15 Show that the identity operator is not compact.

9.16 Prove that if L has a compact inverse, it is positive definite if and only if each of its eigenvalues are positive.

9.17 Calculate some estimates of the eigenvalues of the operator for the transverse vibration of a simply supported, non-uniform beam with $EI(x) = (1.1 - x)$. Compare the results of your estimates to the exact values for $EI = 1$.

9.18 Calculate the eigenvalues of a square membrane clamped along its boundary on each of the following:
a) The square defined by the axis and the line $x = 1, y = 1$.
b) The square defined by the axis and the line $x = 2, y = 2$.
c) The square defined by the axis and line $x = 3, y = 3$.
Compare the results of these three operators using the enclosure result of Example 9.8.2 and illustrate that the monotonicity results hold.

9.19 Consider the transverse vibrations of three beams, all with dimensions compatible with the Euler-Bernoulli assumptions and all with cantilevered boundary conditions. Suppose two of the beams have constant stiffness denoted by E_1I_1 and E_2I_2 respectively and that the third beam has a variable stiffness denoted $EI(x)$. Show that if $E_1I_1 < EI(x) < E_2I_2$, then the eigenvalues of the variable stiffness beam fall in between those of the constant stiffness beams.

9.20 Show that a differential operator must be of even order for it to be symmetric. (*Hint*: Use integration of parts to show that an odd-order differential operator is not symmetric.)

10

Forced Response and Control

10.1 Introduction

This chapter considers the response of distributed parameter structures that are under some external influence. This includes consideration of the response of distributed mass structures to applied external forces, the response of distributed mass structures connected to lumped mass elements, and the response of distributed mass structures under the influence of both passive and active control devices.

If the equations of motion can be decoupled, then many of the results used for lumped mass systems described in Chapter 5 can be repeated for the distributed mass case. However, because of the infinite dimensional nature of distributed mass systems, convergence of solutions occasionally preempts the use of these methods. Convergence issues are especially complicated if the structure is subjected to control forces or unknown disturbances.

10.2 Response by Modal Analysis

This section considers the forced response of damped distributed parameter systems of Equation (8.14) of the form

$$w_{tt}(x,t) + L_1 w_t(x,t) + L_2 w(x,t) = f(x,t) \quad x \in \Omega \tag{10.1}$$

with appropriate boundary and initial conditions. Here the operators L_1 and L_2 are self-adjoint, positive definite operators; L_2 has a compact inverse, and L_1 shares the set of eigenfunctions, $\{\emptyset_n(x)\}$, with L_2 (i.e. L_1 and L_2 commute). For the moment, the only assumption made of $f(x, t)$ is that it lies in $L_2^R(\Omega)$.

Since $f(x, t) \in L_2^R(\Omega)$, Equation (10.1) can be multiplied by the function $\emptyset_n(x)$ and then integrated over Ω. This integration yields

$$(w_{tt}, \emptyset_n) + (L_1 w_t, \emptyset_n) + (L_2 w, \emptyset_n) = (f, \emptyset_n) \tag{10.2}$$

The left-hand side of this equation is identical to Equation (10.16). Applying the analysis of Section 9.6, Equation (10.2) becomes

$$\ddot{a}_n(t) + \lambda_n^{(1)} \dot{a}_n(t) + \lambda_n^{(2)} a_n(t) = f_n(t) \quad n = 1, 2, 3 \ldots \infty \tag{10.3}$$

Vibration with Control, Second Edition. Daniel John Inman.

Companion Website: www.wiley.com/go/inmanvibrationcontrol2e

where $f_n(t)$ has the form

$$f_n(t) = \int_{\Omega} f(x,t)\varnothing_n(x)d\Omega \quad n = 1,2,3\ldots\infty \tag{10.4}$$

This scalar equation in the function $a_n(t)$ can be solved and analyzed using the single degree of freedom (SDOF) model of Section 1.4.

Equation (10.3) is essentially the same as Equation (5.39), and the solution is thus given by Equation (5.40). That is, if each mode of the system is assumed to be underdamped, then for zero initial conditions

$$a_n(t) = \frac{1}{\omega_{dn}} \int_0^t e^{-\zeta_n \omega_n \tau} f_n(t-\tau)\sin(\omega_{dn}\tau)d\tau \tag{10.5}$$

where for each value of the index n

$$\omega_n = \sqrt{\lambda_n^{(2)}}, \text{ the } n^{\text{th}} \text{ natural frequency} \tag{10.6}$$

$$\zeta_n = \frac{\lambda_n^{(1)}}{2\sqrt{\lambda_n^{(2)}}}, \text{ the } n^{\text{th}} \text{ modal damping ratio} \tag{10.7}$$

$$\omega_{dn} = \omega_n\sqrt{1-\zeta_n^2}, \text{ the } n^{\text{th}} \text{ damped natural frequency} \tag{10.8}$$

Thus, in the solution where the operators L_1 and L_2 commute. the temporal coefficients in the series solution are determined by using results from SDOF theory discussed in Chapter 1. The solution to Equation (10.1) is the sum

$$w(x,t) = \sum_{n=1}^{\infty} a_n(t)\varnothing_n(x) \tag{10.9}$$

where the $a_n(t)$ are determined by Equation (10.5) for the case that the initial conditions are set to zero and the set $\{\varnothing_n(x)\}$ consists of the eigenfunctions of the operator L_2. Since the set of functions $\{\varnothing_n(x)\}$ are the modes of free vibration, the just-described procedure is referred to as a modal analysis solution of the forced response problem.

Example 10.2.1

Consider the hinged-hinged beam of Example 10.3.1. Assuming the beam is initially at rest ($t = 0$), calculate the response of the system due to a harmonic force of $\sin(t)$ applied at $x = l/2$, where l is the length of the beam. Assume the damping in the beam is of the form $2\alpha w_t(x,t)$, where α is a constant. First note that the operator

$$L_2 = \frac{EI}{m}\frac{\partial^4}{\partial x^4}$$

has a compact inverse and is self-adjoint and positive definite with respect to the given boundary conditions. Furthermore, the eigenfunctions of the operator L_2

serve as eigenfunctions for the operator $L_1 = 2\alpha I$. Thus, the eigenvalues of the operator $4L_2 - L_1^2$ are (for the given boundary conditions)

$$4\left[n^4\pi^4\frac{EI}{(ml^4)} - \alpha^2\right]$$

which are greater than zero for every value of the index n if

$$\frac{\pi^2}{l^2}\sqrt{\frac{EI}{m}} > \alpha$$

Hence, each coefficient, $a_n(t)$, is underdamped according to Inman and Andry (1982). The solution given by Equation (10.5) then applies.

The forcing function for the system is described by

$$f(x,t) = \delta\left(x - \frac{l}{2}\right)\sin t$$

where $\delta(x - l/2)$ is the Dirac delta function. Substitution of this last expression into Equation (10.4), along with the normalized eigenfunctions of Example 8.3.1, yields

$$\begin{aligned} f_n(t) &= \sqrt{\frac{2}{l}}\sin t\int_0^l \sin\left(\frac{n\pi x}{l}\right)\delta\left(x - \frac{l}{2}\right)dx \\ &= \sqrt{\frac{2}{l}}\sin t\sin\frac{n\pi}{2} \end{aligned}$$

In addition, the natural frequency, damping ratio and damped natural frequency become

$$\begin{aligned} \omega_n &= \left(\frac{n\pi}{l}\right)^2\sqrt{\frac{EI}{m}} \\ \zeta_n &= \frac{\alpha}{\omega_n} \\ \omega_{dn} &= \omega_n\sqrt{1 - \frac{\alpha^2}{\omega_n^2}} \end{aligned}$$

With these modal damping properties determined, specific computation of Equation (10.5) can be performed. Note that the even modes are not excited in this case, since $f_{2n} = 0$ for each n. Physically, these are zero because the even modes all have nodes at the point of excitation, $x = l/2$.

If the damping in the system is such that the system is overdamped or mixed damped, the solution procedure is the same. The only difference is that the form of $a_n(t)$ given by Equation (10.5) changes. For instance, if there is zero damping in the system, then $\zeta_n \to 0$, $\omega \to \omega_n$, and the solution of Equation (10.5) becomes

$$a_n(t) = \frac{1}{\omega_n}\int_0^t f_n(t-\tau)\sin(\omega_n\tau)d\tau \tag{10.10}$$

10.3 Modal Design Criteria

The previous section indicates how to calculate the forced response of a given structure due to an external disturbance by modal analysis. This modal approach is essentially equivalent to decoupling a partial differential equation into an infinite set of ordinary differential equations. This section examines some of the traditional design formulas for SDOF oscillators applied to the modal coordinates of a distributed parameter structure of the form given in Equation (10.9). This modal design approach assumes that the summation of Equation (10.9) is uniformly convergent and the set of eigenfunctions $\{\varnothing_n(x)\}$ is complete. Hence, there is a value of the index n, say $n = N$, for which the difference between $w(x, t)$ and the partial sum

$$\sum_{n=1}^{N} a_n(t)\varnothing_n(x)$$

is arbitrarily small. Physically, observation of certain distributed mass systems indicates that some key modes seem to dominate the response, $w(x, t)$, of the system. Both the mathematics and the physics in this case encourage the use of these dominant modes in the design criteria.

As an illustration of modal dominance, consider again the problem of Example 10.2.1. With zero initial conditions, the response $a_n(t)$ is of the same form as Equation (1.22) multiplied by 0, 1 or −1, depending on the value of n (i.e. $\sin n\pi/2$). In fact, integration of Equation (10.5) for the case $f_n(t) = f_{n0}\sin \omega t$ yields

$$a_n(t) = X_n \sin(\omega t + \beta_n) \tag{10.11}$$

The coefficient X_n is determined (Section 1.4) to be

$$X_n = \frac{f_{n0}}{\sqrt{\left(\lambda_n^{(2)} - \omega^2\right)^2 + \left(\lambda_n^{(1)}\omega\right)^2}} \tag{10.12}$$

and the phase shift β_n becomes

$$\beta_n = \tan^{-1} \frac{\left(\lambda_n^{(1)}\omega\right)}{\lambda_n^{(2)} - \omega^2} \tag{10.13}$$

The quantity X_n can be thought of as a *modal participation factor* in that it is an indication of how dominant the nth mode is. For a fixed value of the driving frequency, ω, the values of X_n steadily decrease as the index n increases. The modal participation factor decreases unless the driving frequency is close to the square root of one of the eigenvalues of the operator L_2. In this case, the modal participation factor for that index may be a maximum. By examining the modal participation factors or modal amplitudes, the designer can determine which modes are of interest or which modes are most important. The following example illustrates this point.

Example 10.3.1
Calculate the modal amplitudes for the clamped beam of Example 10.2.1. Note that in this case the driving frequency is 1, i.e. $\omega = 1$. For the sake of simplicity, let $EI = m = \alpha = 1$ and $l = 2$, so that the system is underdamped. From Example 10.2.1, $f_{n0} = 1$ for each n. Also, $\lambda_n^{(1)} = 2$ for each n and

$$\lambda_n^{(2)} = 6.088n^4$$

for each value of the index n. In this case, Equation (10.12) yields that

$$X_1 = 0.183, \quad X_2 = 0.010, \quad X_3 = 0.002$$

Note that the modal participation factor, X_n decreases rapidly with increasing n. Next consider the same problem with the same physical parameters, except with a new driving frequency of $\omega = 22$. In this case, the modal participation factors are

$$\begin{aligned} X_1 &= 0.002, & X_2 &= 0.003 \\ X_3 &= 0.022, & X_4 &= 0.0009 \\ X_5 &= 0.0003, & X_6 &= 0.0001 \end{aligned}$$

This example illustrates that if the driving frequency is close to a given mode frequency (X_3 in this case), the corresponding modal amplitude will increase in absolute value.

By examining the solution $w(x, t)$ mode by mode, certain design criteria can be formulated and applied. For example, the magnification curve of Figure 1.9 follows directly from Equation (10.12) on a per-mode basis. Indeed, all the design and response characterizations of Section 1.4, such as bandwidth and overshoot, can be applied per mode. However, all the design procedures become more complicated because of coefficient coupling between each of the mode equations given by Equation (10.11). While the equations for $a_n(t)$ are decoupled in the sense that each $a_n(t)$ can be solved independently of each other, the coefficients in these equations will depend on the same physical parameters (i.e. E, I, ρ, m, etc.). This is illustrated in the following example.

Example 10.3.2
Consider the step response of the clamped beam of Example 10.2.1. A modal time to peak can be defined for such a system by using Equation (1.37) applied to Equation (10.3). With a proper interpretation of ζ_n and ω_n, the *modal time to peak*, denoted by t_{pn}, is

$$t_{pn} = \frac{\pi}{\omega_n\sqrt{1-\zeta_n^2}}$$

where $\omega_n = (n^2\pi^2/l^2)\sqrt{EI/m}$ and $\zeta_n = \alpha/\omega_n$. Examination of this formula shows that if E, I, m and α are chosen so that t_{p2} has a desired value, then t_{p3}, t_{p4},... are fixed. Thus, the peak time, overshoot, etc. of a distributed mass system cannot be independently chosen on a per-mode basis. even though the governing equations decouple.

10.4 Combined Dynamical Systems

Many systems are best modeled by combinations of distributed mass components and lumped mass components. Such systems are called *hybrid systems*, distributed systems with lumped appendages, or *combined dynamical systems*. This section discusses the natural frequencies and mode shapes of such structures and the use of the eigensolution to solve for the forced response of such structures.

As an example of such a system, consider the free vibration of a beam of length l connected to a lumped mass and spring, as illustrated in Figure 10.1. The equation of motion of the beam with the effect of the oscillator modeled as an external force, $f(t)\delta(x - x_1)$, is

$$EIw_{xxxx} + \rho Aw_{tt} = f(t)\delta(x - x_1) \quad x \in (0, l) \tag{10.14}$$

The equation of motion of the appended system is given by

$$m\ddot{z}(t) + kz(t) = -f(t) \tag{10.15}$$

where m is the appended mass and k the associated stiffness. Here the coordinate, $z(t)$, of the appended mass is actually the displacement of the beam at the point of attachment, i.e.

$$z(t) = w(x_1, t) \tag{10.16}$$

Combining Equations (10.14) and (10.15) yields

$$\left[EI\frac{\partial^4}{\partial x^4} + k\delta(x - x_1)\right] w(x, t) + [\rho A + m\delta(x - x_1)]w_{tt}(x, t) = 0 \tag{10.17}$$

The solution $w(x, t)$ is now assumed to separate, i.e. $w(x, t) = u(x)a(t)$. Following the method of separation of variables, substitution of the separated form into Equation (10.17) and rearrangement of terms yields

$$\frac{EIu''''(x) + k\delta(x - x_1)u(x)}{[\rho A + m\delta(x - x_1)]u(x)} = -\frac{\ddot{a}(t)}{a(t)} \tag{10.18}$$

As before (Section 7.3), each side of the equality must be constant. Taking the separation constant to be ω^2, the temporal function $a(t)$ has the harmonic form

$$a(t) = Asin(\omega t) + B\cos(\omega t) \tag{10.19}$$

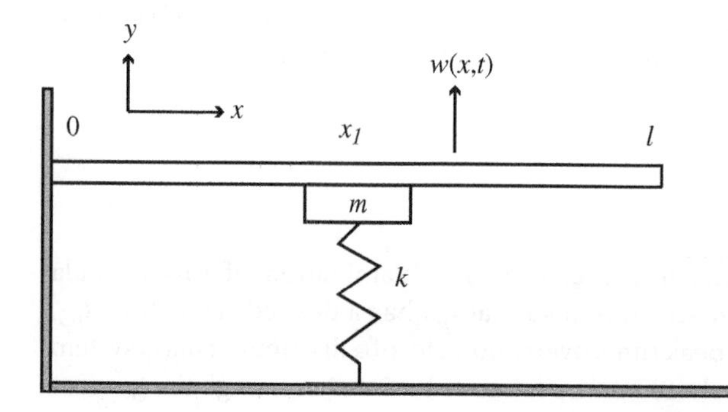

Figure 10.1 A beam with an attached lumped mass-spring system.

where A and B are constants of integration determined by the initial conditions. The spatial equation becomes

$$ELu''''(x) + [(k - m\omega^2)\delta(x - x_1) - \rho A\omega^2]u(x) = 0 \tag{10.20}$$

subject to the appropriate boundary conditions.

Solution of Equation (10.20) yields the generalized eigenfunctions, $\emptyset_n(x)$, and eigenvalues, ω_n^2, for the structure. These are called generalized eigenfunctions because Equation (10.20) does not formally define an operator eigenvalue problem, as specified in Section 8.2. Hence, the procedure and modal analysis are performed formally. Note, however, that if $k/m = \omega_n^2$, i.e. if the appended spring-mass system is tuned to a natural frequency of the beam, the related eigenfunction becomes that of the beam without the appendage. The solution of Equation (10.20) can be constructed by use of a Green's function for the vibrating beam.

The Green's function $g(x, x_1)$ for a beam satisfies

$$g'''' - \beta^4 g = \delta(x - x_1) \tag{10.21}$$

where $\beta^4 = \rho A\omega^2/(EI)$ and g satisfies the appropriate boundary conditions. Following the development of Section 8.6, Equation (10.21) has the solution

$$g(x, x_1) = -\frac{1}{2\beta^3 \sin\beta l \sinh\beta l}\begin{cases} y(x, x_1), & 0 \le x < x_1 \\ y(x_1, x), & x_1 < x \le l \end{cases} \tag{10.22}$$

where the function $y(x, x_1)$ is symmetric in x_1 and x, and has the form

$$\begin{aligned} y(x, x_1) = {} & \sin(\beta l - \beta x_1)\sin(\beta x)\sinh(\beta l) \\ & - \sinh(\beta l - \beta x_1)\sinh(\beta x)\sin(\beta l) \end{aligned} \tag{10.23}$$

In terms of the Green's function just defined, the solution to Equation (10.20) for the simply supported case can be written as (Nicholson and Bergman, 1986)

$$u(x) = \frac{1}{EI}(m\omega^2 - k)g(x, x_1)u(x_1) \tag{10.24}$$

If $u(x_1)$ were known, then Equation (10.24) would specify the eigenfunctions of the system. Fortunately, the function $u(x_1)$ is determined by writing Equation (10.24) for the case $x = x_1$, resulting in

$$\left[EI - \left\{m\left(\omega^2 - \frac{k}{m}\right)\right\}g\left(x_1, x_1\right)\right]u(x_1) = 0 \tag{10.25}$$

which yields the characteristic equation for the system. In order to allow $u(x_1)$ to be nonzero, the coefficient in Equation (10.25) must vanish, yielding an expression for computing the natural frequencies, ω. This transcendental equation in ω contains terms of the form $\sin\beta l$ and hence has an infinite number of roots, denoted by ω_n. Thus, Equation (10.24) yields an infinite number of eigenfunctions, denoted by $u_n(x)$.

Both the free and forced response of a combined dynamical system, such as the one described in Figure 10.1, can be calculated using a modal expansion for a cantilevered beam. Following Section 7.5, the eigenfunctions are, from Equation (10.20), those of a non-appended cantilevered beam, i.e.

$$\emptyset_i(x) = \cosh\beta_i x - \cos\beta_i x - \alpha_i(\sinh\beta_i x - \sin\beta_i x) \tag{10.26}$$

Here the constants α_i are given by

$$\alpha_i = \frac{\cosh \beta_i l + \cos \beta_i l}{\sinh \beta_i l + \sin \beta_i l} \tag{10.27}$$

and the eigenvalues β_i are determined from the transcendental equation

$$1 + \cosh \beta_i l \cos \beta_i l = 0 \tag{10.28}$$

Note that in this case the arguments of Section 9.6 hold and the functions $\emptyset_i(x)$ form a complete orthogonal set of functions. Hence the spatial solution $u(x)$ can be written as

$$u(x) = \sum_{i=1}^{\infty} b_i \emptyset_i(x) \tag{10.29}$$

with the set $\{\emptyset_i\}$ normalized so that

$$(\emptyset_i, \emptyset_j) = l\delta_{ij} \tag{10.30}$$

Substitution of Equation (10.29) for $u(x)$ in Equation (10.20), multiplying by $\emptyset_j(x)$, using the property

$$\emptyset_j''''(x) = \beta_j^4 \emptyset_j(x)$$

and integrating over the interval $(0, l)$ yields

$$\begin{aligned} EIl\beta_i^4 b_i + (k - m\omega^2)\emptyset_i(x_1) b_i \emptyset_i(x_1) - \rho A l \omega^2 b_i = 0, \ \text{for } i = j \\ (k - m\omega^2)\emptyset_j(x_1) b_i \emptyset_i(x_1) = 0, \quad \text{for } i \neq j \end{aligned} \tag{10.31}$$

Dividing this last expression by $\rho A l$ and defining two new scalars, A_{ij} and B_{ij} by

$$A_{ij} = \frac{k\emptyset_i(x_1)\emptyset_j(x_1)}{\rho A l} + \frac{EI\beta_i^4}{\rho A}\delta_{ij} \tag{10.32}$$

$$B_{ij} = \frac{1}{\rho A l} m\emptyset_i(x_1)\emptyset_j(x_1) + \delta_{ij} \tag{10.33}$$

allows Equation (10.31) to be simplified. Equation (10.31) can be rewritten as

$$\sum_{j=1}^{\infty} A_{ij} b_j = \omega^2 \sum_{j=1}^{\infty} B_{ij} b_j \tag{10.34}$$

This last expression is in the form of a generalized infinite matrix eigenvalue problem for ω^2 and the generalized Fourier coefficients, b_j. The elements b_j are the modal participation factors for the modal expansion given by Equation (10.29).

The orthogonality relationship for the eigenfunctions $\{\emptyset_n(x)\}$ is calculated from Equation (10.20) and rearranged in the form

$$\emptyset_n''''(x) - \left(\frac{\rho A \omega^2}{EI}\right)\emptyset_n(x) = \frac{1}{EI}(\omega_n^2 m - k)\delta(x - x_1)\emptyset_n(x) \tag{10.35}$$

Premultiplying Equation (10.35) by $\varnothing_m(x)$ and integrating yields (Problem 10.5)

$$\rho A \int_0^l \varnothing_m(x)\varnothing_n(x)dx = -\int_0^l m\delta(x-x_1)\varnothing_m(x)\varnothing_n(x)dx$$

or

$$\int_0^l \left[1+\frac{m}{A\rho}\delta(x-x_1)\right]\varnothing_m(x)\varnothing_n(x)dx = \delta_{nm} \tag{10.36}$$

The preceding characteristic equation and orthogonality relationship completes the modal analysis of a cantilevered Euler-Bernoulli beam connected to a spring and lumped mass.

Equipped with the eigenvalues, eigenfunctions and the appropriate orthogonality condition, a modal solution for the forced response of a damped structure can be carried out for a proportionally damped beam connected to a lumped spring-mass dashpot arrangement following these procedures. Bergman and Nicholson (1985) showed that the modal equations for a damped cantilevered beam attached to a spring-mass dashpot appendage have the form

$$\ddot{a}_n(t) + \sum_{m=1}^{\infty}\left\{\varepsilon_b\delta_{nm} + \mu\left(\frac{\varepsilon\alpha_m^4\alpha_n^4}{\alpha_0^8} - \varepsilon_b\right)A_mA_n\varnothing_m(x_1)\varnothing_n(x_1)\dot{a}_m(t)\right\}$$
$$+\alpha_n^4 a_n(t) = f_n(t) \tag{10.37}$$

where:

$f_n(t) = \gamma\int_0^L \varnothing_n(x)f(x,t)dx$
$f(x,t) =$ externally applied force
$\varepsilon_b =$ distributed damping coefficient
$\varepsilon =$ lumped damping rate
$\mu =$ lumped mass
$\alpha_n^2 =$ system natural frequencies
$\alpha_0 =$ lumped stiffness
$A_n = \frac{\alpha_0^4}{\alpha_0^4-\alpha_n^4}$

With the given parameters and orthogonality conditions, the combined system has a modal solution given by

$$w(x,t) = \sum_{n=1}^{\infty} a_n(t)\varnothing_n(x) \tag{10.38}$$

where $a_n(t)$ satisfies Equation (10.37) and the appropriate initial conditions. Note from Equation (10.37) that proportional damping results if $\varepsilon\alpha_m^4\alpha_n^4 = \varepsilon_b\alpha_0^8$.

10.5 Passive Control and Design

The lumped appendage attached to a beam in the previous section can be viewed as a passive control device, much in the same way that the absorber of Section 6.2 can be thought of as a passive control element. In addition, the layered materials of Section 7.5 can be thought of as either a passive control method or a redesign method. In either case, the desired result is to choose the system's parameters in such a way that the resulting structure has improved vibration response.

First consider a single absorber added to a cantilevered beam. The equations of motion as discussed in the previous section have a temporal response governed by Equation (10.37). Thus the rate of decay of the transient response is controlled by the damping terms

$$\sum_{m=1}^{\infty} \{\varepsilon_b \delta_{nm} + \mu \frac{\varepsilon \alpha_m^4 \alpha_n^4}{\alpha_0^8} - \varepsilon_b A_m A_n \emptyset_m(x_1) \emptyset_n(x_1)\} \dot{a}_m(t) \tag{10.39}$$

The design problem becomes that of choosing x_1, ε, μ and α_0 so that Equation (10.39) has the desired value. With only four parameters to choose and an infinite number of modes to effect, there are not enough design parameters to solve the problem. In addition, the summation in Equation (10.39) effectively couples the design problem so that passive control cannot be performed on a per-mode basis. However, for specific cases the summation can be truncated, making the design problem more plausible.

Next consider the layered material of Section 7.8. Such materials can be designed to produce both a desired elastic modulus and a desired loss factor (Nashif et al., 1985). Consider the problem of increasing the damping in a beam so that structural vibrations in the beam decay quickly. Researchers in the materials area often approach the problem of characterizing the damping in a material by use of the concept of loss factor, introduced as η in Section 1.4, and the concept of *complex modulus* introduced next.

For a distributed mass structure, it is common practice to introduce damping in materials by simply replacing the elastic modulus for the material, denoted by E, by a complex modulus of the form

$$E(1 + i\eta) \tag{10.40}$$

where η is the experimentally determined loss factor for the material and i is the square root of (−1). The rationale for this approach is based on an assumed temporal solution of the form $Ae^{i\omega t}$. If $Ae^{i\omega t}$ is substituted into the equation of motion of a damped structure, the velocity term yields a coefficient of the form $i\omega$, so that the resulting equation may be viewed as having a complex stiffness. This form of damping is also called the *Kimball-Lovell complex stiffness* (Bert, 1973).

The loss factor for a given structure made of standard metal is usually not large enough to suppress unwanted vibrations in many applications. One approach to designing more highly damped structures is to add a layer of damping material to the structure, as indicated in Figure 10.2. The new structure then has different elastic modulus (frequencies) and loss factor. In this way, the damping material

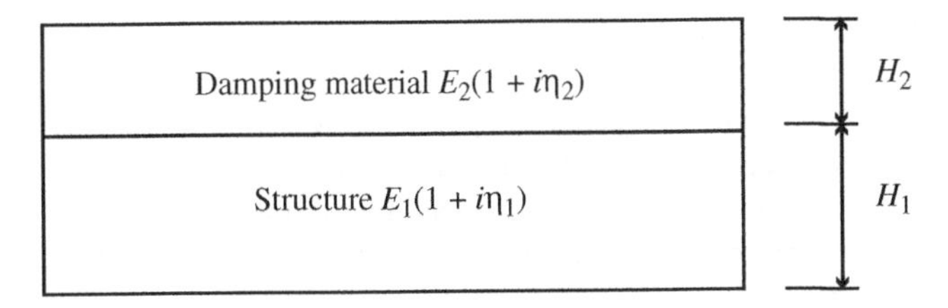

Figure 10.2 Passive vibration control by using a damping layer.

can be thought of as a passive control device used to change the poles of an existing structure to more desirable locations. Such a treatment of structures is called *extensional damping.* Sometimes it is referred to as *unconstrained layer damping,* or *free layer damping.*

Let E and η denote the elastic modulus and loss factor of the combined system of Figure 10.2. Let E_1 and η_1 denote the modulus and loss factor of the original beam and let E_2 and η_2 denote the modulus and loss factor of the added damping material. In addition, let H_2 denote the thickness of the added damping layer and H_1 denote the thickness of the original beam. Let $e_2 = E_2/E_1$ and $h_2 = H_2/H_1$. The design formulas relating the "new" modulus and loss factor to those of the original beam and added damping material are given in Nashif et al. (1985) as

$$\frac{EI}{E_1 I_1} = \frac{1 + 4e_2h_2 + 6e_2h_2^2 + 4e_2h_2^3 + e_2^2h_2^4}{1 + e_2h_2} \tag{10.41}$$

and

$$\frac{\eta}{\eta_1} = \frac{e_2h_2(3 + 6h_2 + 4h_2^2 + 2e_2h_2^3 + e_2^2h_2^4}{(1 + e_2h_2)(1 + 4e_2h_2 + 6e_2h_2^2 + 4e_2h_2^3 + e_2^2h_2^4)} \tag{10.42}$$

where $(e_2h_2)^2$ is assumed to be much smaller than e_2h_2.

Equations (10.41) and (10.42) can be used to choose an appropriate damping material to achieve a desired response.

The preceding complex modulus approach can also be used to calculate the response of a layered structure. Note that the response of an undamped uniform beam can be written in the form

$$w(x, t) = \sum_{n=1}^{\infty} g(E)a_n(t, E)\varnothing_n(x, E) \tag{10.43}$$

where $g(E)$ is some function of the modulus E. This functional dependence is usually not explicitly indicated but rather is contained in the eigenvalues of the eigenfunctions $\varnothing_n(x, E)$ and in the temporal coefficients $a_n(t, E)$. One approach used to include the effects of damping in a layered beam is simply to substitute the values of $E + i\eta$ obtained from Equations (10.41) and (10.42) into g, a_n and $\varnothing_n$ in Equation (10.43). Each term of the series in Equation (10.43) is complex and of the form $g(E(1 + i\eta))\, a_n(t,E(1 + i\eta))\varnothing_n(x, E(1 + i\eta))$, so some manipulation is required to calculate the real and imaginary parts. This approach should be treated as an approximation, as it is not rigorous.

10.6 Distributed Modal Control

In this section, the control of systems governed by partial differential equations of the form of Equation (10.1) is considered. The control problem is to find some function $f(x, t)$ such that the response $w(x, t)$ has a desired form. If $f(x, t)$ is a function of the response of the system, then the resulting choice of $f(x, t)$ is called active control. If, on the other hand, $f(x, t)$ is thought of as a change in the design of the structure, it is referred to as passive control (or redesign). Modal control methods can be used in either passive or active control. Any control method that uses the eigenfunctions, or modes, of the system in determining the control law $f(x, t)$ is considered to be a modal control method.

Repeating the analysis of Section (10.2) yields the modal control equations. For the control problem, the functions $f_n(t)$ of Equation (10.4) are thought of as modal controls, or inputs, in the jargon of Chapter 6. As indicated in Chapter 6, there are many possible control techniques to apply to Equation (10.3). Perhaps the simplest and most physically understood is state feedback. Viewed by itself, Equation (10.3) is a two-state model with the states being the generalized velocity, $\dot{a}_n(t)$, and position, $a_n(t)$. If $f_n(t)$ is chosen in this way, Equation (10.3) becomes

$$\ddot{a}_n(t) + \lambda_n^{(1)}\dot{a}_n(t) + \lambda^{(2)}a_n(t) = -c_n^p a_n(t) - c_n^v \dot{a}_n(t) \tag{10.44}$$

where c_n^p and c_n^v are modal position and velocity gains, respectively. Obviously, the choice of the position and velocity feedback gains completely determines the nth temporal coefficient in the free response given by Equation (10.10). In theory, c_n^p and c_n^v can be used to determine such performance criteria as the overshoot, decay rate, speed of response, etc. These coefficients can be chosen as illustrated for the SDOF problem of Example 10.6.1.

The question arises, however, about the convergence of $f(x, t)$. Since

$$f_n(t) = -c_n^p a_n(t) - c_n^v \dot{a}_n(t) = \int_{\Omega} f(x,t)\varnothing_n(x)d\Omega \tag{10.45}$$

the series

$$f(x,t) = \sum_{n=1}^{\infty} \left[-c_n^p a_n(t) - c_n^v \dot{a}_n(t)\right]\varnothing_n(x)d\Omega \tag{10.46}$$

must converge. Furthermore, it must converge to some function $f(x, t)$ that is physically realizable as a control. Such controls $f(x, t)$ are referred to as *distributed controls,* because they are applied along the spatial domain Ω.

Example 10.6.1

Consider the problem of controlling the first mode of a flexible bar. An internally damped bar clamped at both ends has equations of motion given by Equation (10.1) with

$$L_1 = -2b\frac{\partial^2}{\partial x^2},\ L_2 = -\alpha\frac{\partial^2}{\partial x^2}$$

and boundary conditions $w(x, t) = 0$ at $x = 0$ and $x = 1$. Here, b is the constant denoting the rate of internal damping and α denotes a constant representing the stiffness in the bar (EI/ρ). Solution of the eigenvalue problem for L_1 and L_2 and substitution of the appropriate eigenvalues into Equation (10.3) yields

$$\ddot{a}_n(t) + 2bn^2\pi^2\dot{a}_n(t) + \alpha n^2\pi^2 a_n(t) = f_n(t)$$

For the sake of illustration, assume $\alpha = 400\pi^2$ and $b = 1$ in the appropriate units.

Suppose it is desired to control only the lowest mode. Furthermore, suppose it is desired to shift the frequency and damping ratio of the first mode. Note that the equation for the temporal coefficient for the first mode is

$$\ddot{a}_1(t) + 2\pi^2\dot{a}_1(t) + 400\pi^4 a_1(t) = f_1(t)$$

so that the first mode has an undamped natural frequency of $\omega_1 = 20\pi^2$ and a damping ratio of $\zeta_1 = 0.05$.

The control problem is taken to be that of calculating a control law, $f(x, t)$, that raises the natural frequency to $25\pi^2$ and the damping ratio to 0.1. This goal will be achieved if the displacement coefficient, after control is applied, has the value

$$(25\pi^2)^2 = 624\pi^4$$

and the velocity coefficient of the closed loop system has the value

$$2\zeta_1\omega_1 = 2(0.1)(25\omega^2) = 5\omega^2$$

Using Equation (10.44) with $n = 1$ yields

$$\ddot{a}_1(t) + 2\pi^2\dot{a}_1(t) + 400\pi^4 a_1(t) = -c_1^p a_1(t) - c_1^v \dot{a}_1(t)$$

Combining position coefficients and then velocity coefficients yields the following two simple equations for the control gains

$$c_1^p + 400\pi^4 = 625\pi^4$$
$$c_1^v + 2\pi^2 = 5\pi^2$$

Thus, $c_1^p = 225\pi^4$ and $c_1^v = 3\pi^2$ will yield the desired first mode values. The modal control force is thus

$$f_1(t) = -225\pi^4 a_1(t) - 3\pi^2\dot{a}_1(t)$$

In order to apply this control law only to the first mode, the control force must be of the form

$$f(x, t) = f_1(t)\varnothing_1(x)$$

For this choice of $f(x, t)$, the other modal controls, $f_n(t)$, $n > 1$, are all zero, which is very difficult to achieve experimentally because the result requires $f(x, t)$ to be distributed along a single mode.

Unfortunately, designing distributed actuators is difficult in practice. The design of actuators that produce a spatial distribution along a given mode, as required by the example, is even more difficult. Based on the availability of actuators, the more practical approach is to consider actuators that act at a point, or points, in the domain of the structure. The majority of control methods used for distributed parameter structures involve using finite-dimensional models of the structure. Such models are often obtained by truncating the series expansion of the solution of Equation (10.9). The methods in Chapter 6 are then used to design a vibration control system for the structure. The success of such methods is tied to the process of truncation (Gibson, 1981). Truncation is discussed in more detail in Chapter 11.

10.7 Nonmodal Distributed Control

An example of a distributed actuator that provides a nonmodal approach to control is the use of a piezoelectric polymer. Piezoelectric devices offer a convenient source of distributed actuators. One such actuator has been constructed and used for vibration control of a beam (Bailey and Hubbard, 1985) and is presented here.

Consider the transverse vibrations of a cantilevered beam of length l with a piezoelectric polymer bonded to one side of the beam. The result is a two-layer material similar to the beam illustrated in Figure 10.2, presented in Section 7.6, Figure 7.13 and derived in Section 7.10. Bailey and Hubbard (1985) have shown that the equation governing the two-layer system is

$$\frac{\partial^2}{\partial x^2}\left[EI\frac{\partial^2 w}{\partial x^2}\right] + \rho A\frac{\partial^2}{\partial t^2} = 0 \quad x \in \Omega \tag{10.47}$$

with boundary conditions

$$w(0,t) = w_x(0,t) = 0 \tag{10.48}$$

$$EIw_{xx}(l,t) = -cf(t), \text{ and } w_{xxx}(l,t) = 0 \tag{10.49}$$

where it is assumed that the voltage applied by the polymer is distributed evenly along x, i.e. that its spatial dependence is constant. Here EI reflects the modulus and inertia of both the beam and the polymer, ρ is the density, and A is the cross-sectional area. The constant c is the bending moment per volt of the material and $f(t)$ is the voltage applied to the polymer. This distributed actuator behaves mathematically as a *boundary control.*

One approach to solving this control problem is to use a Lyapunov function, $V(t)$, for the system, and choose a control $f(t)$ to minimize the time rate of change of the Lyapunov function. The Lyapunov function is chosen to be

$$V(t) = \frac{1}{2}\int_0^l \left[\left(\frac{\partial^2 w}{\partial x^2}\right)^2 + \left(\frac{\partial w}{\partial t}\right)^2\right] dt \tag{10.50}$$

which is a measure of how far the beam is from its equilibrium position. Minimizing the time derivative of this functional is then equivalent to trying to bring the system to rest (equilibrium) as fast as possible. Differentiating Equation (10.49) and substitution of Equation (10.47) yields

$$\frac{\partial V}{\partial t} = \int_0^l \left[1 - \frac{EI}{\rho A}\right] w_{xxt}w_{xx}dx + \frac{c}{\rho A}f(t)w_{xx}(l,t) \tag{10.51}$$

The voltage $f(t)$ is thus chosen to minimize this last quantity. Bailey and Hubbard (1985) showed that $f(t)$, given by

$$f(t) = -\text{sgn}(cw_{xt}(l, t))f_{\max} \tag{10.52}$$

is used as a minimizing control law. Here sgn denotes the signum function.

Not only is the control law of Equation (10.52) distributed and independent of the structure's modal description, but it also allows the control force to be magnitude limited, i.e. $|f(t)| \leq f_{\max}$. These are both very important practical features. In addition, the control law depends only on feeding back the velocity of the tip of the beam. Hence, this distributed control law requires that a measurement be taken at a single point at the tip ($x = l$).

10.8 State Space Control Analysis

This section examines the control problem for distributed parameter structures cast in the state space formulation. Considering Equation (10.1), define the two-dimensional vector $z(x, t)$ by

$$z(x, t) = [w(x, t) \quad w_t(x, t)]^T \tag{10.53}$$

The state equation for the system of Equation (10.2) then becomes

$$z_t = Az + bu \tag{10.54}$$

where the matrix of operators A is defined by

$$A = \begin{bmatrix} 0 & I \\ -L_2 & -L_1 \end{bmatrix} \tag{10.55}$$

the vector b is defined by $b = [0 \ 1]^T$, and $u = u(x, t)$ is now used to denote the applied force, which in this case is a control. As in the lumped mass case, there needs to be an observation equation, denoted here as

$$y(x, t) = Cz(x, t) \tag{10.56}$$

In addition, the state vector z is subject to boundary conditions and initial conditions and must have the appropriate smoothness (i.e. the elements z belong to a specific function space). Equations (10.55) and (10.56) form the state space equations for the control of distributed mass systems and are a direct generalization of Equations (2.25) and (2.28) for the state space representation of lumped mass systems.

As discussed in the preceding sections, the input or control variable $u(x, t)$ can be either distributed or lumped in nature. In either case, the general assumption is that the function $u(x, t)$ separates in space and time. The most common form that the control $u(x, t)$ takes, describes the situation in which several time-dependent control forces are applied at various points in the domain of the structure. In this case

$$u(x, t) = \sum_{i=1}^{m} \delta(x - x_i)u_i(t) \tag{10.57}$$

where the m control forces of the form $u_i(t)$ are applied to the m locations x_i. Note that this formulation is consistent with the development of combined dynamical systems of Section 10.4.

The output, or measurement, of the system is also subject to the physical constraint that most devices are lumped in nature. Measurements are most often proportional to a state or its derivatives. In this case, $y(x, t)$ takes the form

$$y(x,t) = \sum_{j=1}^{p} c_j \delta(x - x_j) z(x,t) \tag{10.58}$$

where c_j are measurement gains of each of the p sensors located at the p points, x_j, in the domain of the structure.

The concepts of controllability and observability are of course equally as critical for distributed mass systems as they are for lumped mass systems. Unfortunately, a precise definition and appropriate theory is more difficult to develop and hence is not covered here. Intuitively, however, the actuators and sensors should not be placed on nodes of the vibrational modes of the structures. If this practice is adhered to, then the system will be controllable and observable. This line of thought leads to the idea of modal controllability and observability (Goodson and Klein, 1970).

An optimal control problem for a distributed parameter structure can be formulated following the discussion in Section 6.9 by defining various cost functionals. In addition, pole placement and state feedback schemes can be devised for distributed parameter systems, generalizing the approaches used in Chapter 6. Note, however, that not all finite-dimensional control methods have direct analogs in distributed parameter systems. This lack of analogy is largely due to the difference between functional analysis and linear algebra.

10.9 Vibration Suppression using Piezoelectric Materials

Consider the a piezoceramic wafer similar to that of Figure 7.16, where the upper layer is used as a control actuator and the bottom layer is used as a sensor. The sensor equation for a piezoelectric patch placed on a beam between points x_1 and x_2 is given in Equation (7.155) in terms of the voltage generated as

$$v_s(t) = \frac{y_c d_{31} E_s b}{C_a}[w_x(x_2,t) - w_x(x_1,t)] = K[w_x(x_2,t) - w_x(x_1,t)] \tag{10.59}$$

Here the constant, $K = y_c d_{31} b / C_a$, depends on the centroidal axis y_c, the piezoelectric capacitance C_a, the wafer thickness b, the wafer's elastic modulus E_s and the piezoelectric constant d_{31}. The equation of a beam with piezoceramic actuator is given in Equation (7.151) as

$$\begin{aligned} \rho A(x)\frac{\partial^2 w(x,t)}{\partial t^2} + \frac{\partial^2}{\partial x^2}\left[EI(x)\frac{\partial^2 w(x,t)}{\partial x^2}\right] &= \\ v(t) b E_p d_{31}(t_a + t_b)\frac{\partial^2}{\partial x^2}[\Phi(x - x_2) - \Phi(x - x_1)] \end{aligned} \tag{10.60}$$

Next, consider approximating the solution by using separation of variables where the spatial function is an eigenfunction of the beam operator and the stiffness EI and mass ρA are considered constant. Thus, assume the solution $w(x,t) = \phi_n(x)a_n(t)$, substitute into Equation (10.60), multiply by $\phi_m(x)$ and integrate over the length of the beam to get the modal version

$$\ddot{a}_n(t) + \omega_n^2 a_n(t) = v(t)\gamma \int_0^L \phi_n(x)\frac{\partial^2}{\partial x^2}[\Phi(x-x_2) - \Phi(x-x_1)] \tag{10.61}$$

The frequency ω_n is determined from the eigenvalues of the stiffness operator for the given boundary conditions and has the form

$$\omega_n = \beta_n^2\sqrt{\frac{EI}{\rho A}} \tag{10.62}$$

and β_n is determined by the boundary conditions and the solution to the eigenvalue problem. See Example 7.5.1 for computation of β_n. The constant γ is a function of the piezoceramic properties: $\gamma = bE_p d_{31}(t_a + t_b)$. Using separation of variables on the integral term in Equation (10.61) and the properties of the Heaviside function and the delta function, $\Phi'(x-a) = \delta\,(x-a)$, yields

$$\ddot{a}_n(t) + \omega_n^2 a_n(t) = \frac{\gamma}{\rho A}\left[\phi_n'(x_2) - \phi_n'(x_1)\right] v(t) \tag{10.63}$$

A closed loop control law can add additional dynamics by many different techniques (Chapter 6). For example, position feedback can be applied by simply choosing $v(t) = gv_s(t)$, where the gain g is chosen to adjust the frequencies (i.e. shift the stiffness) of the closed loop system.

A popular control law with experimentalists and application engineers is called Positive Position Feedback (Fanson and Caughey, 1990) defined for the lumped parameter systems in Section 6.16, because its closed loop stability and design are straightforward to implement, requiring only knowledge of the open-loop frequencies and not the entire model of the structure to be controlled. To implement this control law, the input voltage, $v(t)$, is formulated as

$$v(t) = \frac{g\rho A\omega_{nf}^2\eta(t)}{\gamma[\phi_n'(x_2) - \phi_n'(x_1)]} \tag{10.64}$$

where g is to be determined control gain. Following Equation (6.137), $\eta(t)$ is the "position" coordinate of an auxiliary dynamic system (called a *compensator*) defined by

$$\ddot{\eta}(t) + 2\zeta_f\omega_{nf}\dot{\eta}(t) + \omega_{nf}^2\eta(t) = g\omega_{nf}^2\frac{v_s(t)}{K[w_x(x_2,t) - w_x(x_1,t)]} = g\omega_{nf}^2 a_n(t) \tag{10.65}$$

Combining Equations (10.63) and (10.65) in matrix form yields

$$\begin{bmatrix}\ddot{a}_n(t)\\ \ddot{\eta}(t)\end{bmatrix} + \begin{bmatrix}0 & 0\\ 0 & 2\zeta_f\omega_{nf}\end{bmatrix}\begin{bmatrix}\dot{a}_n(t)\\ \dot{\eta}(t)\end{bmatrix} + \begin{bmatrix}\omega_n^2 & -g\omega_{nf}^2\\ -g\omega_{nf}^2 & \omega_{nf}^2\end{bmatrix}\begin{bmatrix}a_n(t)\\ \eta(t)\end{bmatrix} = \begin{bmatrix}0\\ 0\end{bmatrix} \tag{10.66}$$

This is the modal version for distributed parameter systems of Equation (6.141) for lumped parameter systems.

Repeating the argument given in Section 6.16, the closed loop system is stable, because the "stiffness" matrix is positive definite as long as

$$g\omega_{nf}^2 < \omega_n^2 \tag{10.67}$$

Notice that the stability condition only depends on the natural frequency of the structure, and not on the damping or mode shapes. This is significant in practice, because when building an experiment, the frequencies of the structure are usually available with a reasonable accuracy, while mode shapes and damping ratios are much less reliable. The modal transfer function (Equation 6.140) illustrates that the response rolls off quickly at high frequencies. Thus the approach is well suited to controlling a mode of a structure with frequencies that are well separated, as the controller is insensitive to the unmodeled high frequency dynamics.

Chapter Notes

This chapter discusses the analysis of the forced response and control of structures with distributed mass. Section 10.2 presents standard, well-known modal analysis of the forced response of a distributed mass system. Such an approach essentially reduces the distributed mass formulation to a system of SDOF models that can be analyzed by the methods in Chapter 1. Section 10.3 examines some design specifications for distributed mass systems in modal coordinates. One cannot assign design criteria to each mode independently, as is sometimes suggested by using modal coordinates.

Section 10.4 examines a method of calculating the response of hybrid systems, i.e. systems composed of both distributed mass elements and lumped mass elements. Several authors have approached this problem over the years. Most recently, Nicholson and Bergman (1986) produced a series of papers (complete with computer code) discussing combined dynamical systems based on beam equations. Their paper is essentially paraphrased in Section 10.4. Banks et al. (1998) clarify the existence of normal modes in combined dynamical systems. The papers by the Bergman group also contain an excellent bibliography of this area. The importance of the analysis of combined dynamical system is indicated in Section 10.5 on passive control. The most common passive vibration suppression technique is to use an absorber or an isolator. The theory by the Bergman group provides an excellent analytical tool for the design of such systems. An alternative and very useful approach to vibration design and passive control is to use layers of material and high damping (loss factor). This approach is discussed extensively in the book by Nashif et al. (1985), which provides a complete bibliography.

Section 10.6 discusses the concept of modal control for distributed mass structures with distributed actuators and sensors. This material is expanded in a paper by Inman (1984). The details of using a modal control method are outlined by Meirovitch and Baruh (1982) and were originally introduced by Gould and Murray-Lasso (1966). Gibson (1981) discusses some of the problems associated with using finite-dimensional state models in designing control laws for distributed mass systems. This result has caused interest to arise in nonmodal control methods, an example of which is discussed in Section 10.7. The material of Section 10.7 is taken from the paper by Bailey and Hubbard (1985).

Section 10.8 presents a very brief introduction to formulating the control problem in the state space. Several books, notably Komkov (1970) and Lions (1972), discuss this topic in more detail. A more practical approach to the control problem is presented in the next chapter. Tzou and Bergman (1998) present a collection of works in the vibration and control of distributed mass systems. Section 10.9 follows Fanson and Caughey (1990).

References

Bailey, T. and Hubbard, J. E. (1985) Distributed piezoelectric polymer active vibration control of a cantilevered beam. *AIAA Journal of Guidance Control and Dynamics*, 8(5), 605–611.

Banks, H. T., Bergman, L. A., Inman, D. J. and Luo, Z. (1998) On the existence of normal modes of damped discrete continuous systems. *Journal of Applied Mechanics*, 65(4), 980–989.

Bergman, L. A. and Nicholson, J. W. (1985) Forced vibration of a damped combined linear system. *ASME Journal of Vibration, Acoustics, Stress and Reliability in Design*, 107, 275–281.

Bert, C. W. (1973) Material damping: an introductory review of mathematical models measures and experimental techniques. *Journal of Sound and Vibration*, 19(2), 129–153.

Fanson, J. L. and Caughey, T. K. (1990) Positive position feedback control of large space structures, *AIAA Journal*, 28(4), 717–724.

Gibson, J. S. (1981) An analysis of optimal modal regulation: convergence and stability. *SIAM Journal of Control and Optimization*, 19, 686–706.

Goodson, R. E. and Klein, R. E. (1970) A definition and some results for distributed system observability. *IEEE Transactions on Automatic Control*, AC-15, 165–174.

Gould, L. A. and Murray-Lasso, M. A. (1966) On the modal control of distributed systems with distributed feedback. *IEEE Transactions on Automatic Control*, AC-11(4), 729–737.

Inman, D. J. and Andry, A. N., Jr. (1982) The nature of temporal solutions of damped distributed parameter systems with classical normal modes, *ASME Journal of Applied Mechanics*, 49, 867–870.

Inman, D. J. (1984) Modal decoupling conditions for distributed control of flexible structures. *AIAA Journal of Guidance, Control and Dynamics*, 7(6), 750–752.

Komkov, V. (1970) *Optimal Control Theory for the Damping of Vibrations of Simple Elastic Systems*. Springer-Verlag Lecture Notes in Mathematics, New York. 153 p.

Lions, J. L. (1972) *Some Aspects of the Optimal Control of Distributed Parameter Systems*. Society of Industrial and Applied Mathematics, Phaladelphia, PA.

Meirovitch, L. and Baruh, H. (1982) Control of self-adjoint distributed parameter systems. *AIAA Journal of Guidance, Control and Dynamics*, 5, 60–66.
Nashif, A. D., Jones, D. I. G. and Henderson, J. P. (1985) *Vibration Damping*. John Wiley & Sons, New York.
Nicholson, J. W. and Bergman, L. A. (1986) Free vibration of combined dynamical systems. *ASCE Journal of Engineering Mechanics*, 112(1), 1–13.
Tzou, H. S. and Bergman, L.A. (eds) (1998) *Dynamics and Control of Distributed Systems*. Cambridge University Press, Cambridge, UK.

Problems

10.1 Calculate the response of the first mode of a clamped membrane of Equation (7.134) subject to zero initial conditions and an applied force of

$$f(x, y, t) = 3\sin t\delta(x - 0.5)\delta(y - 0.5)$$

10.2 Derive the modal response (i.e. $a_n(t)$) for the general system given by Equation (10.1) and associated assumptions if, in addition to $f(x, t)$, the system is subject to initial conditions of the form

$$w(x, 0) = w_0 \quad w_t(x, 0) = w_{t0}$$

Use the notation of Equations (10.3) to (10.8).

10.3 Calculate an expression for the modal participation factor for Problem (10.1).

10.4 Define a modal logarithmic decrement for the system of Equation (10.1) and calculate a formula for it.

10.5 Derive Equation (10.36) from Equation (10.35) by performing the suggested integration. Integrate the term containing $\emptyset_n''''$ four times using the homogeneous boundary conditions and again using Equation (10.35) to evaluate $\emptyset_m''''$.

10.6 Discuss the possibility that the sum in Equation (10.37) can be truncated because

$$\varepsilon\alpha_m^4\alpha_n^4 = \alpha_0^8\varepsilon_b$$

for some choices of m and n.

10.7 Show that the complex stiffness is a consistent representation of the equation of motion by substituting the assumed solution $Ae^{i\omega t}$ into Equation (8.14). What assumption must be made on the operators L_1 and L_2?

10.8 a) Calculate the terms $g(E)$, $a_n(t,E)$ and $\emptyset_n(t,E)$ explicitly in terms of the modulus E for a damped free beam of unit length.
b) Next substitute $e(1 + i\eta)$ for E in your calculation and compare your result with the same beam having a damping operator of $L_1 = 2\eta I$, where I is the identity operator.

10.9 Formulate an observer equation for a beam equation using the state space formulation of Section 10.8.

10.10 Consider the transverse vibration of a beam of length l, modulus E and mass density ρ. Suppose an accelerometer is mounted at the point $x = l/2$. Determine the observability of the first three modes.

10.11 Consider the beam of Example 8.3.1 with the following numerical values: $\rho = 2715\ \text{kg/m}^3$, $l = 0.5128\ \text{m}$, $I = 6.96 \times 10^{10}\ \text{n/m}^2$ and $A = 8.16\ \text{m}^2$. Compute a control law to give the first mode a damping value of $\zeta = 0.1$ using positive position feedback.

11

Approximations of Distributed Parameter Models

11.1 Introduction

This chapter is devoted to examining approximations of distributed parameter systems by lumped parameter models. Since the solutions of distributed parameter systems are often given in terms of an infinite series, and since only a few configurations have closed form solutions, there is a need to cast distributed parameter systems into finite dimensional systems that can easily be solved numerically. In addition, control and design are well developed for lumped parameter systems, providing further motivation to approximate distributed systems with the more easily manipulated lumped systems. From the experimentalist point of view, most common measurement methods only "see" a finite (dimensional) number of points.

In this chapter, several common methods of approximating distributed mass structures by lumped mass models are presented. Most of these methods eliminate the spatial dependence in the solution technique by discretizing the spatial variable in some way, effectively approximating an eigenfunction with an eigenvector. This chapter ends with a discussion of the effects of active control of distributed mass structures and the accuracy of the approximation.

11.2 Modal Truncation

Since the solution of the vibration problem given by

$$w_{tt}(x,t) + L_1 w_t(x,t) + L_2 w(x,t) = f(x,t)\ x \in \Omega \tag{11.1}$$

plus appropriate boundary conditions are of the form

$$w(x,t) = \sum_{n=1}^{\infty} a_n(t)\phi_n(x) \tag{11.2}$$

which converges uniformly, it is possible to approximate the solution by

$$w_N(x,t) = \sum_{n=1}^{N} a_n(t)\phi_n(x) \tag{11.3}$$

Vibration with Control, Second Edition. Daniel John Inman.

Companion Website: www.wiley.com/go/inmanvibrationcontrol2e

where N is finite. This finite sum approximation ignores the sum given by

$$w_R(x,t) = \sum_{n=N+1}^{\infty} a_n(t)\phi_n(x) \tag{11.4}$$

called the *residual*. The modes in this sum are called the *truncated modes*, i.e. the functions $\phi_n(x)$ for values of the index $n = N+1 \to \infty$. The assumption is that the residual solution is small, i.e. that $\|w_R(x,t)\| < \varepsilon$. This assumption is often satisfied by physical structures, giving rise to the statement that structures behave like low-pass filters.

Substitution of Equation (11.3) into Equation (11.1) yields

$$\sum_{n=1}^{N} \left[\ddot{a}_n(t)\phi_n(x) + \dot{a}_n L_1\phi_n(x) + a_n(t)L_2\phi_n(x)\right] = \sum_{n=1}^{N} b_n(t)\phi_n(x) \tag{11.5}$$

where $f(x,t)$ has also been expanded in terms of the functions $\phi_n(x)$ with coefficients $b_n(t)$. Premultiplying Equation (11.5) by $\phi_m(x)$ and integration over Ω yields two possibilities. Note that the sum is now finite, so that convergence is not a problem. First, if $L_1L_2 = L_2L_1$ on the appropriate domain, then Equation (11.5) becomes N decoupled ordinary differential equations of the form

$$\ddot{a}_n(t) + \lambda_n^{(1)}\dot{a}_n(t) + \lambda_n^{(2)}a_n(t) = b_n(t) \tag{11.6}$$

In matrix form, this becomes

$$I\ddot{\mathbf{a}} + \Lambda_C\dot{\mathbf{a}} + \Lambda_K\mathbf{a} = \mathbf{f} \tag{11.7}$$

which can then be analyzed by the methods in Chapter 5. Here Λ_C and Λ_K are diagonal matrices and $\mathbf{a}$ and $\mathbf{f}$ are N-vectors of obvious definition.

If the commutivity condition does not hold, then Equation (11.6) becomes

$$I\ddot{\mathbf{a}} + C\dot{\mathbf{a}} + \Lambda_K\mathbf{a} = \mathbf{f} \tag{11.8}$$

where the elements of C are

$$c_{ij} = \int_{\Omega} \phi_i L_1 \phi_j d\Omega. \tag{11.9}$$

and the functions $\phi_j(x)$ are the eigenfunctions of L_2. The boundary conditions are incorporated in the matrices C, Λ_C and Λ_K automatically by virtue of the integration. The initial conditions on $\mathbf{a}(t)$ are defined by

$$a_i(0) = \int_{\Omega} w(x,0)\phi_i(x)d\Omega, \text{ and } \dot{a}_i(0) = \int_{\Omega} w_t(x,0)\phi_i(x)d\Omega \tag{11.10}$$

In both cases, it is required that $w_R(x,t)$ be as small as possible, i.e. that the higher modes do not contribute much to the solution. In practice, this is often so. For instance, $N = 3$ is often adequate to describe the longitudinal vibration of a simple cantilevered beam (recall Example 10.3.1). Equations (11.6) and (11.8) are finite-dimensional approximations of Equations (11.1), derived by truncating the higher modes of the structure's response (i.e. setting $w_R = 0$) and as such are referred to as a *truncated modal model*. Damping tends to kill off the higher modes, because each term in Equation (11.4) is multiplied by an exponential of the form: $e^{-\zeta_n \omega_n t}$. Since ω_n increases substantially with n, the terms in Equation (11.4) become less significant, contributing little to the response.

11.3 Rayleigh-Ritz-Galerkin Approximations

The Rayleigh quotient was introduced in Section 9.7 as a means of approximating the natural frequencies of a conservative system. Ritz used this concept to calculate an approximate solution for the eigenfunctions (mode shapes) in terms of an assumed series of trial functions. This approach is similar to modal truncation but rather than using the exact mode shapes as the expanding basis, any complete set of basis functions that satisfy the boundary conditions is used. In other words, the Rayleigh-Ritz (as it is usually called) approximation does not require any knowledge of the eigenfunctions. Furthermore, the Rayleigh quotient can be written in terms of energy, rather than in terms of the eigenvalue problem, reducing the number of derivatives and boundary conditions that need to be satisfied by the choice of "trial" functions.

Trial functions are functions which:

a) satisfy the boundary conditions (or at least some of them);
b) are orthogonal to each other; and
c) have enough derivatives to be fit into the equation of motion.

Trial functions are further divided up into those that satisfy all of the boundary conditions (called comparison functions) and those that satisfy only the geometric boundary conditions (called admissible functions). In forming the sum of Equation (11.5), there are three classifications of functions that can be used:

1) *Eigenfunctions*: these satisfy the equation of motion plus all the boundary conditions;
2) *Comparison functions*: these are orthogonal and satisfy all the boundary conditions (but not the equation of motion);
3) *Admissible functions*: these are orthogonal and satisfy only the geometric boundary conditions (i.e. things like displacements and slopes).

Boundary conditions are classified as either:

a) *natural boundary conditions*: those that involve force and moment balances; or
b) *geometric boundary conditions*: those that satisfy displacement and slope conditions at the boundary.

Using trial functions eliminates the need to know the eigenfunctions of the structure before approximating the system. Let $\{\theta_n(x)\}$ be a linearly independent set of basis functions that are complete in a subspace of $D(L_2)$ and satisfy the appropriate boundary conditions. The Nth approximate solution of Equation (11.1) is then given by the expression

$$w_N(x, t) = \sum_{n=1}^{N} a_n(t)\theta_n(x). \tag{11.11}$$

Likewise, $f(x, t)$ is approximated by

$$f_N(x,t) = \sum_{n=1}^{N} b_n(t)\theta_n(x) \tag{11.12}$$

Substitution of Equations (11.9) and (11.10) for $w(x, t)$ and $f(x, t)$ into Equation (11.1), respectively, yields

$$\sum_{n=1}^{N} \left[\ddot{a}(t)\theta_n + \dot{a}_n(t)L_1\theta_n + a_n(t)L_2\theta_n\right] = \sum_{n=1}^{N} b_n(t)\theta_n(x) \tag{11.13}$$

Premultiplying Equation (11.13) by $\theta_m(x)$ and integrating (thus using the boundary conditions) yields the finite-dimensional approximation

$$M\ddot{\mathbf{x}} + C\dot{\mathbf{x}} + K\mathbf{x} = \mathbf{f}(t) \tag{11.14}$$

where the matrices M, C and K and the vector $\mathbf{f}$ are defined by

$$m_{ij} = \int \theta_i\theta_j d\Omega \tag{11.15}$$

$$c_{ij} = \int \theta_i L_1 \theta_j d\Omega \tag{11.16}$$

$$k_{ij} = \int \theta_i L_2 \theta_j d\Omega \tag{11.17}$$

$$f_i = \int f(x,t)\theta_i d\Omega \tag{11.18}$$

Unlike the coefficient matrices of the modal truncation scheme of Equations (11.7) and (11.8), the matrices M, C and K in this case are not necessarily diagonal. Note however, that they are symmetric as long as the operators L_1 and L_2 are self-adjoint. The order, N, of the finite-dimensional approximation, Equation (11.11), is chosen so that $w_n(x, t)$ is as small as possible for the purpose at hand. Note that the difference between the functions $\phi_i(x)$ in Section 11.2 and the $\theta_n(x)$ in this section, is that the $\phi_i(x)$ are eigenfunctions of the stiffness operator. In this section, the trial functions $\theta_n(x)$ are chosen in a somewhat arbitrary fashion. Hence, this method is also called the *assumed mode method.* The bottom line with approximation methods is that the closer the trial function is to the exact eigenfunction (mode shape) the better the estimate is. If, in fact, an exact set of mode shapes is used, and the damping is proportional, the approximation will be exact.

Starting with the Rayleigh Quotient, the Ritz method minimizes the quotient over the coefficients of expansion for $\phi(x)$ and provides an approximation of the system natural frequencies and mode shapes for undamped systems. Let the approximate spatial dependence have the form

$$\phi(x) = \sum_{i=1}^{N} c_i\phi_i(x) \tag{11.19}$$

where the $\phi_i(x)$ are the trial functions and the constants are c_i to be determined. Recall the statement of the operator eigenvalue problem resulting from separation of variables, as given in Equation (8.7). Re-writing Equation (8.7) with the mass density placed on the right-hand side yields

$$L\phi(x) = \lambda\rho\phi(x) \tag{11.20}$$

subject to the appropriate boundary conditions. Multiplying Equation (11.20) by $\phi(x)$ and integrating yields the Rayleigh quotient

$$\lambda = \frac{\int_0^l \phi(x)L\phi(x)dx}{\int_0^l \rho\phi(x)\phi(x)dx} = \frac{N}{D}, \tag{11.21}$$

$$\text{where } N = \int_0^l \phi(x)L\phi(x)dx \text{ and } D = \int_0^l \rho\phi(x)\phi(x)dx$$

The Ritz approximation process is to substitute Equation (11.19) into Equation (11.21) and compute the coefficients c_i that minimize the Rayleigh Quotient given by Equation (11.21).

Differentiating Equation (11.21) with respect to the coefficients c_i yields

$$\frac{\partial\lambda}{\partial c_i} = \frac{D\left(\frac{\partial N}{\partial c_i}\right) - N\left(\frac{\partial D}{\partial c_i}\right)}{D^2} = 0 \Rightarrow \frac{\partial N}{\partial c_i} - \lambda\frac{\partial D}{\partial c_i} = 0 \tag{11.22}$$

since D is never zero. Equation (11.22) computes the values of the expansion coefficients that minimize the Rayleigh quotient and hence allow the approximation of the eigenvalues. Next, consider writing N and D in terms of the constants c_i using Equation (11.19). With a little manipulation it can be shown (Problem 11.11) that Equation (11.22) is the generalized eigenvalue-eigenvector problem

$$K\mathbf{c} = \lambda M\mathbf{c} \tag{11.23}$$

where the column vector $\mathbf{c}$ consists of the expansion coefficients c_i . The elements of the "mass" and "stiffness" matrix are given by

$$k_{ij} = \int_0^l \phi_i(x)L\phi_j(x)dx \text{ and } m_{ij} = \int_0^l \rho\phi_i(x)\phi_j(x)dx \tag{11.24}$$

The solution of the generalized eigenvalue problem of Equation (11.23) yields an approximation of the eigenvalues λ_n and hence the natural frequencies. The eigenvectors $\mathbf{c}$ approximate the system's eigenfunctions and hence the mode shapes. The number of approximated frequencies and mode shapes is N, the number of trial functions used in Equation (11.19).

The power of this approach is in finding approximate solutions when Equation (11.20) cannot be solved analytically, such as for odd boundary conditions and/or for spatially varying coefficients, such as $\rho(x)$ and $EI(x)$. Note that if exact eigenfunctions are used, exact frequencies result. Also, note that the boundary conditions come into play when evaluating the integrals in Equation (11.24).

11.4 Finite Element Method

Probably the most popular method of representing distributed mass structures is the *finite element method* (FEM). This section presents a very brief introduction to the topic. A classic reference for FEM is Hughes (2000). The method divides the structure of interest into subsections of finite size, called *finite elements*. These elements are connected to adjacent elements at various points on their boundaries, called *nodes*. Once this procedure is finished, the distributed mass structure is represented by a finite number of nodes and elements referred to as a finite element *grid*, or *mesh*.

The displacement of each element is approximated by some function of the spatial variables between nodes. The next step in the finite element analysis (FEA) is to calculate the energy in each element as a function of the displacement. The total energy of the structure is then expressed as the sum of the energy in each element. External forces are included by using the principle of virtual work to derive forces per element. Lagrange's equations (Meirovitch, 2001) are then applied to the total energy of the structure, which yields the approximate equations of motion. These equations are finite-dimensional. This procedure is illustrated in the following example.

Example 11.4.1

This example considers the longitudinal vibration of a bar of length l and derives a finite element stiffness matrix of the bar. The bar in Figure 11.1 is configured as one finite element with a node at each end. The axial stiffness is regarded

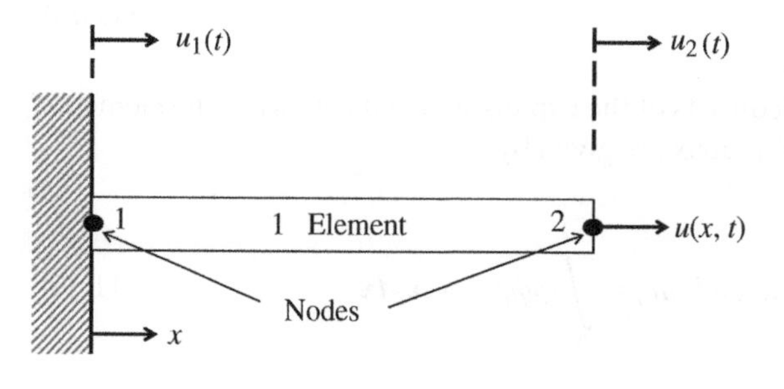

Figure 11.1 A two-node, one-element model of a cantilevered beam.

as time-independent throughout the element, so that the displacement must satisfy

$$EA\frac{d^2u(x)}{dx^2} = 0,\ x \in (0, l) \tag{11.25}$$

Integrating this expression yields

$$u(x) = c_1x + c_2,\ x \in (0, l) \tag{11.26}$$

where c_1 and c_2 are constants of integration. At each node, the value of u is allowed to be a time dependent coordinate denoted by $u_1(t)$, as labeled in Figure 11.1. Using these as boundary conditions, the constants c_1 and c_2 are evaluated as

$$c_2 = u(t) \tag{11.27}$$

$$c_1 = \frac{u_2(t) - u_1(t)}{l} \tag{11.28}$$

so that $u(x, t)$ is approximated by

$$u(x,t) = \left(1 - \frac{x}{l}\right)u_1(t) + \frac{x}{l}u_2(t) \tag{11.29}$$

Next, the nodal forces f_1 and f_2 are related to the displacement $u(x)$ by

$$EAu'(0) = -f_1, EAu'(l) = f_2 \tag{11.30}$$

or

$$EA\frac{u_2 - u_1}{l} = -f_1,\ EA\frac{u_2 - u_1}{l} = f_2 \tag{11.31}$$

where the prime indicates differentiation with respect to x. This last expression can be written in the matrix form

$$K\mathbf{u} = \mathbf{f} \tag{11.32}$$

where $\mathbf{u}(t) = [u_1(t)\ u_2(t)]^T$, $\mathbf{f} = [f_1(t)\, f_2(t)]^T$ and

$$K = \frac{EA}{l}\begin{bmatrix} 1 & -1 \\ -1 & 1 \end{bmatrix} \tag{11.33}$$

Here the vector $\mathbf{u}(t)$ is called the *nodal displacement vector*, the vector $\mathbf{f}(t)$ is called the *nodal force vector*, and the matrix K is the *element stiffness matrix*.

In Example 11.4.1, note that the displacement in the element is written in the form

$$u(x,t) = a_1(x)u_1(t) + a_2(x)u_2(t) = \mathbf{a}^T(x)\mathbf{u}(t) \tag{11.34}$$

where

$$\mathbf{a}(x) = [a_1(x)\ a_2(x)]^T$$

The functions $u_1(t)$ and $u_2(t)$ are the time-dependent nodal displacements, and in Example 11.4.1 they approximate $u(0,t)$ and $u(l,t)$, respectively. The functions $a_1(x)$ and $a_2(x)$ are called *shape functions*, or *interpolation functions*. In the example, $a_1(x) = (1 - x/l)$ and $a_2(x) = (x/l)$. However, the shape functions are not unique in general. They are referred to as interpolation functions because they allow the displacement to be specified, or interpolated, at points along the structure that lie between nodes. As will be illustrated in the following, the solution of the dynamic finite element equations yields only the nodal displacements $u_1(t)$ and $u_2(t)$.

Next a dynamic model is needed. A mass matrix is required that is consistent with the preceding stiffness matrix for the bar element. The mass matrix can be determined from an expression for the kinetic energy of the element, denoted by $T(t)$ and defined by

$$T(t) = \frac{1}{2}\int_0^l \rho(x)u_1(x,t)^2 dx \tag{11.35}$$

Substitution of $u_t(x, t)$ from Equation (11.32) yields

$$T(t) = \frac{1}{2}\int_0^l \rho(x)\dot{\mathbf{u}}^T(t)\mathbf{a}(x)\mathbf{a}^T(x)\dot{\mathbf{u}}(t)dx \tag{11.36}$$

or

$$T(t) = \frac{1}{2}\dot{\mathbf{u}}^T(t)\left[\int_0^l \rho(x)\mathbf{a}(x)\mathbf{a}^T(x)dx\right]\dot{\mathbf{u}}(t) \tag{11.37}$$

The expression in brackets is clearly a matrix that is defined to be the element mass matrix, denoted by M. Examination of Equation (11.37) yields that the mass matrix is given by

$$M = \int_0^l \rho(x)\mathbf{a}(x)\mathbf{a}^T(x)dx \tag{11.38}$$

Since the mass matrix is calculated by using the same shape functions as the stiffness matrix, the resulting mass matrix is called a *consistent mass matrix*. An alternative means of constructing the mass matrix is just to lump the mass of the structure at the various nodes. If this is done, the result is called an *inconsistent mass matrix*.

Note that the stiffness matrix of Equation (11.36) can also be represented in terms of the *shape* functions $\mathbf{a}(t)$. Examination of the potential energy in the system yields (for the bar in Example 11.4.1)

$$K = \int_0^l EA(x)\mathbf{a}(x)\mathbf{a}^T(x)dx \tag{11.39}$$

With K defined by Equation (11.31), the potential energy per element, denoted by $V(t)$, is given by

$$V(t) = \frac{1}{2}\mathbf{u}^T(t)K\mathbf{u}(t) \tag{11.40}$$

Example 11.4.2

Calculate the consistent mass matrix for the bar element of Example 11.4.1. Substituting the shape functions of Equation (11.29) into Equation (11.38) yields

$$M = \rho\int_0^l \left(\begin{bmatrix} \frac{1-x}{l} \\ \frac{x}{l} \end{bmatrix}\begin{bmatrix} \frac{1-x}{l} & \frac{x}{l} \end{bmatrix}\right)dx = \frac{\rho l}{6}\begin{bmatrix} 2 & 1 \\ 1 & 2 \end{bmatrix} \tag{11.41}$$

These definitions of the finite element mass and stiffness matrix can be assembled by using the potential and kinetic energies along with Lagrange's equations to formulate the approximate equations of a distributed parameter structure.

Recall that Lagrange's equations (Sections, 2.2 and 7.2) simply state that the equations of motion of an n degree of freedom structure with coordinates u_i can be calculated from the energy of the structure by

$$\left[\frac{\partial}{\partial t}\left(\frac{\partial T}{\partial \dot{u}_i}\right)\right] - \frac{\partial T}{\partial u_i} + \frac{\partial V}{\partial u_i} = f_i \tag{11.42}$$

where the f_i denote external forces.

With Lagrange's equations, the equations of motion of a structure modeled by one or more finite elements can be derived. This is done by first modeling the structure of interest as several finite elements (like the bar in Examples 11.4.1 and 11.4.2). Next, the total energy of each element is added to produce the total energy of the distributed structure. Then Lagrange's equations are applied to produce the dynamic equations for the structure. The procedure is best illustrated by the following example.

Example 11.4.3

Again, consider the bar element of Examples 11.4.1 and 11.4.2 and use these to model the vibration of a cantilevered bar. In this example, the clamped free bar will be modeled by three (an arbitrary choice) finite elements, and hence four nodes, as depicted in Figure 11.2. Note that because of the clamped boundary condition, $u_1(t) = 0$. Taking this into consideration, the total potential energy, denoted by $V_T(t)$, is the sum of the potential energy in each element

$$V_T(t) = \sum_{i=1}^{3} V_i(t) \tag{11.43}$$

With $l/3$ substituted for l in Equation (11.33) and the appropriate displacement vector $\mathbf{u}$, $V_T(t)$ becomes

$$\begin{aligned} V_T(t) = \frac{3EA}{2l}\begin{bmatrix} 0 \\ u_2 \end{bmatrix}^T \begin{bmatrix} 1 & -1 \\ -1 & 1 \end{bmatrix}\begin{bmatrix} 0 \\ u_3 \end{bmatrix} + \frac{3EA}{2l}\begin{bmatrix} u_2 \\ u_3 \end{bmatrix}^T \begin{bmatrix} 1 & -1 \\ -1 & 1 \end{bmatrix}\begin{bmatrix} u_2 \\ u_3 \end{bmatrix} \\ + \frac{3EA}{2l}\begin{bmatrix} u_3 \\ u_4 \end{bmatrix}^T \begin{bmatrix} 1 & -1 \\ -1 & 1 \end{bmatrix}\begin{bmatrix} u_3 \\ u_4 \end{bmatrix} \end{aligned} \tag{11.44}$$

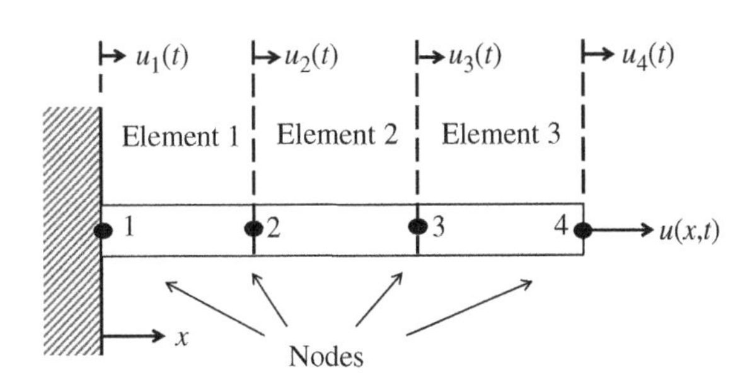

Figure 11.2 A four-node, three-element model of a cantilevered beam.

Calculating the derivatives of V with respect to u_i yields

$$\begin{bmatrix} \dfrac{\partial V_T}{\partial u_2} \\ \dfrac{\partial V_T}{\partial u_3} \\ \dfrac{\partial V_T}{\partial u_4} \end{bmatrix} = \frac{3EA}{l}\begin{bmatrix} 2 & -1 & 0 \\ -1 & 2 & -1 \\ 0 & -1 & 1 \end{bmatrix}\begin{bmatrix} u_2 \\ u_3 \\ u_4 \end{bmatrix} \tag{11.45}$$

where the coefficient of the displacement vector $\mathbf{u} = [u_2\ u_3\ u_4]^T$ is the global stiffness matrix, K, for the entire structure based on a three-element finite approximation.

Calculation of the total kinetic energy $T(t)$ yields

$$T = \frac{1}{2}\frac{\rho Al}{18}\left\{\begin{bmatrix} 0 \\ \dot{u}_2 \end{bmatrix}^T\begin{bmatrix} 2 & 1 \\ 1 & 2 \end{bmatrix}\begin{bmatrix} 0 \\ \dot{u}_2 \end{bmatrix} + \begin{bmatrix} \dot{u}_2 \\ \dot{u}_3 \end{bmatrix}^T\begin{bmatrix} 2 & 1 \\ 1 & 2 \end{bmatrix}\begin{bmatrix} \dot{u}_2 \\ \dot{u}_3 \end{bmatrix} + \begin{bmatrix} \dot{u}_3 \\ \dot{u}_4 \end{bmatrix}^T\begin{bmatrix} 2 & 1 \\ 1 & 2 \end{bmatrix}\begin{bmatrix} \dot{u}_3 \\ \dot{u}_4 \end{bmatrix}\right\} \tag{11.46}$$

Calculation of the various derivatives of T required for Lagrange's equations yields

$$\begin{bmatrix} \dfrac{d}{dt}\left(\dfrac{\partial T}{\partial \dot{u}_2}\right) \\ \dfrac{d}{dt}\left(\dfrac{\partial T}{\partial \dot{u}_3}\right) \\ \dfrac{d}{dt}\left(\dfrac{\partial T}{\partial \dot{u}_4}\right) \end{bmatrix} = \frac{\rho Al}{18}\begin{bmatrix} 4 & 1 & 0 \\ 1 & 4 & 1 \\ 0 & 1 & 2 \end{bmatrix}\ddot{\mathbf{u}} \tag{11.47}$$

where the coefficient of $\mathbf{u}$ is the consistent mass matrix of the three-element finite element approximation.

Substitution of Equations (11.45) and (11.47) into Lagrange's Equation (11.42) yields the three-degree of freedom model of the undamped bar as

$$M\ddot{\mathbf{u}} + K\mathbf{u} = \mathbf{0} \tag{11.48}$$

This last expression can be solved for the vibration response of the undamped bar at the nodal point. The response between nodes can be interpolated by using the shape functions (or interpolation functions), i.e. $u(x, t) = \mathbf{a}^T\mathbf{u}$.

These procedures can be generalized to any type of distributed mass structure or combination of structures. The matrices M and K that result are similar to those that result from the Rayleigh-Ritz method. In fact, the FEM can be thought of as a piecewise version of the Rayleigh-Ritz method. For an accurate representation of a response, 10–20 elements per wavelength of the highest frequency of interest must be used.

11.5 Substructure Analysis

A distributed mass structure often yields a large-order finite element model with hundreds or even thousands of nodes. This is especially true of large, complicated and/or very flexible structures. Substructure analysis is a method of predicting the dynamic behavior of such a complicated large-order system, by first dividing the model up into several parts, called substructures, and analyzing these smaller parts first. The dynamic solution of each substructure is then combined to produce the entire structure's response.

Let the n-dimensional vector $\mathbf{x}$ denote the coordinates of a large finite element model. First divide the structure up into parts according to the modal coordinates via the following scheme

$$\mathbf{x} = \begin{bmatrix} \mathbf{x}_1 \\ \mathbf{x}_2 \end{bmatrix} \tag{11.49}$$

Here $\mathbf{x}_1$ represents those nodes associated with the first substructure and $\mathbf{x}_2$ represents the nodes associated with the second substructure. Let $\mathbf{x}_1$ and $\mathbf{x}_2$ be further partitioned into those coordinates that are unique to substructure 1 and those that are common to $\mathbf{x}_1$ and $\mathbf{x}_2$. Divide $\mathbf{x}_1$ into internal coordinates $\mathbf{x}_{1i}$ and common coordinates $\mathbf{x}_c$, i.e. $\mathbf{x}_1^T = \begin{bmatrix} \mathbf{x}_{1i}^T & \mathbf{x}_c^T \end{bmatrix}^T$. Likewise, the nodal coordinates for the second substructure, $\mathbf{x}_2$, are partitioned as

$$\mathbf{x}_2 = \begin{bmatrix} \mathbf{x}_{2i} \\ \mathbf{x}_c \end{bmatrix} \tag{11.50}$$

The subset of nodes $\mathbf{x}_c$ is the same in both $\mathbf{x}_1$ and $\mathbf{x}_2$.

Next, partition the mass and stiffness matrices for each of the two (could be $N \leq n$) parts according to internal ($\mathbf{x}_{2i}$) and external ($\mathbf{x}_c$) coordinates. Let T_1 and V_1 denote the kinetic energy and potential energy, respectively, in substructure 1. These energies are

$$T_1 = \frac{1}{2}\begin{bmatrix} \dot{\mathbf{x}}_{1i} \\ \dot{\mathbf{x}}_c \end{bmatrix}^T \begin{bmatrix} M_{ii}(1) & M_{ic}(1) \\ M_{ci}(1) & M_{cc}(1) \end{bmatrix} \begin{bmatrix} \dot{\mathbf{x}}_{1i} \\ \dot{\mathbf{x}}_c \end{bmatrix} \tag{11.51}$$

$$V_1 = \frac{1}{2}\begin{bmatrix} \mathbf{x}_{1i} \\ \mathbf{x}_c \end{bmatrix}^T \begin{bmatrix} K_{ii}(1) & K_{ic}(1) \\ K_{ci}(1) & K_{cc}(1) \end{bmatrix} \begin{bmatrix} \mathbf{x}_{1i} \\ \mathbf{x}_c \end{bmatrix} \tag{11.52}$$

Likewise the energy in substructure 2 is

$$T_2 = \frac{1}{2}\begin{bmatrix} \dot{\mathbf{x}}_{2i} \\ \dot{\mathbf{x}}_c \end{bmatrix}^T \begin{bmatrix} M_{ii}(2) & M_{ic}(2) \\ M_{ci}(2) & M_{cc}(2) \end{bmatrix} \begin{bmatrix} \dot{\mathbf{x}}_{2i} \\ \dot{\mathbf{x}}_c \end{bmatrix} \tag{11.53}$$

$$V_2 = \frac{1}{2}\begin{bmatrix} \mathbf{x}_{2i} \\ \mathbf{x}_c \end{bmatrix}^T \begin{bmatrix} K_{ii}(2) & K_{ic}(2) \\ K_{ci}(2) & K_{cc}(2) \end{bmatrix} \begin{bmatrix} \mathbf{x}_{2i} \\ \mathbf{x}_c \end{bmatrix} \tag{11.54}$$

Next, the modes $\mathbf{u}_i$ of each substructure are calculated by assuming that the common coordinates (also called connecting nodes) are free and not really constrained by the rest of the structure, i.e. that the coordinates satisfy the equation of motion

$$\begin{bmatrix} M_{ii}(j) & M_{ic}(j) \\ M_{ci}(j) & M_{cc}(j) \end{bmatrix} \begin{bmatrix} \ddot{\mathbf{x}}_{2i} \\ \ddot{\mathbf{x}}_c \end{bmatrix} + \begin{bmatrix} K_{ii}(j) & K_{ic}(j) \\ K_{ci}(j) & K_{cc}(j) \end{bmatrix} \begin{bmatrix} \mathbf{x}_{2i} \\ \mathbf{x}_c \end{bmatrix} = \mathbf{0} \tag{11.55}$$

for each substructure ($j = 1, 2$). Equation (11.55) is obtained by using the energy expressions of Equations (11.51) to (11.54) substituted into Lagrange's equations. Each of the dynamic substructure Equations (11.55) is next solved for the system eigenvalues and eigenvectors. Let $[\phi_{1i}^T(n)\ \phi_c^T(n)]^T$ denote the nth eigenvector of substructure 1. The modal matrix of substructure 1, denoted by $\phi(1)$, is the square matrix defined by

$$\phi(1) = \begin{bmatrix} \phi_{1i}(1) & \phi_{1i}(2) & \cdots & \phi_{1i}(n) \\ \phi_c(1) & \phi_c(2) & \cdots & \phi_c(n) \end{bmatrix} = \begin{bmatrix} \phi_i(1) \\ \phi_c \end{bmatrix}$$

where $\phi_i(1)$ and ϕ_c are rectangular matrix partitions of $\phi(1)$. These partitions are used to define a new coordinate $\mathbf{q}(1)$ by

$$\mathbf{x}_1 = \begin{bmatrix} \mathbf{x}_{1i} \\ \mathbf{x}_c \end{bmatrix} = \begin{bmatrix} \phi_i(1) \\ \phi_c \end{bmatrix} \mathbf{q}(1) \tag{11.56}$$

This yields

$$\mathbf{x}_{1i} = \phi_i(1)\mathbf{q}(1) \tag{11.57}$$

where it should be noted that $\phi_i(1)$, a rectangular matrix, relates the internal coordinates of substructure 1, $\mathbf{x}_{1i}$, to the new yet-to-be determined coordinate $\mathbf{q}(1)$. This procedure can be repeated using the information from the second substructure to determine $\phi_i(2)$ and to define $\mathbf{q}(2)$. These quantities are related by

$$\mathbf{x}_{2i} = \phi_i(2)\mathbf{q}(2) \tag{11.58}$$

The substitution of Equation (11.57) into the expressions for the energy, Equations (11.51) and (11.52) yields

$$T(1) = \frac{1}{2} \begin{bmatrix} \dot{\mathbf{q}}(1) \\ \dot{\mathbf{x}}_c \end{bmatrix}^T \begin{bmatrix} \phi_i^T(1)M_{ii}(1)\phi_i(1) & \phi_i^T(1)M_{ic}(1) \\ M_{ci}(1)\phi_i(1) & M_{cc}(1) \end{bmatrix} \begin{bmatrix} \dot{\mathbf{q}}(1) \\ \dot{\mathbf{x}}_c \end{bmatrix} \tag{11.59}$$

$$V(1) = \frac{1}{2} \begin{bmatrix} \mathbf{q}(1) \\ \mathbf{x}_c \end{bmatrix}^T \begin{bmatrix} \phi_i^T(1)K_{ii}(1)\phi_i(1) & \phi_i^T(1)K_{ic}(1) \\ K_{ci}(1)\phi_i(1) & M_{cc}(1) \end{bmatrix} \begin{bmatrix} \mathbf{q}(1) \\ \mathbf{x}_c \end{bmatrix} \tag{11.60}$$

Similar expressions are obtained for the energies of the second substructure, $T(2)$ and $V(2)$, by substitution of Equation (11.58) into Equations (11.53) and (11.54). The total energy in the complete structure is now considered to be defined by $[T(1) + T(2)]$ and $[V(1) + V(2)]$.

These energy expressions are substituted into Lagrange's equations to produce the equations of motion in terms of substructure quantities. Lagrange's equations for the system are

$$\begin{bmatrix} \phi_i^T(1)M_{ii}(1)\phi_i(1) & 0 & \phi_i^T(1)M_{ic}(1) \\ 0 & \phi_i(2)M_{ii}(2)\phi_i(2) & \phi_i^T(2)M_{ic}(2) \\ M_{ci}(1)\phi_i(1) & M_{ci}(2)\phi_i(2) & M_{cc}(1)+M_{cc}(2) \end{bmatrix} \begin{bmatrix} \ddot{\mathbf{q}}(1) \\ \ddot{\mathbf{q}}(2) \\ \ddot{\mathbf{x}}_c \end{bmatrix} + \begin{bmatrix} \phi_i^T(1)K_{ii}(1)\phi_i(1) & 0 & \phi_i^T(1)K_{ic}(1) \\ 0 & \phi_i(2)K_{ii}(2)\phi_i(2) & \phi_i^T(2)K_{ic}(2) \\ K_{ci}(1)\phi_i(1) & K_{ci}(2)\phi_i(2) & K_{cc}(1)+K_{cc}(2) \end{bmatrix} \begin{bmatrix} \mathbf{q}(1) \\ \mathbf{q}(2) \\ \mathbf{x}_c \end{bmatrix} = \mathbf{0} \tag{11.61}$$

This last expression constitutes a substructure representation of the original structure. The solution of Equation (11.61) is determined by any of the methods discussed in Chapter 3. The matrix coefficients are determined by analyzing each substructure independently. Equations (11.57) and (11.50) are used to recover the solution in physical coordinates from the solution of the substructure equations given by Equation (11.61). Each of the quantities in Equation (11.61) is determined by solving the two substructures separately. Each of these is of order less than the original system. Equation (11.61) is also of order less than the original structure. In fact, it is of order n minus the order of $\mathbf{x}_c$. Hence the response of the entire structure $\mathbf{x}_n$ can be obtained by analyzing several systems of smaller order.

11.6 Truncation in the Presence of Control

The majority of practical control schemes are implemented by actuators and sensors that are fixed at various points throughout the structure and so behave fundamentally as lumped mass elements rather than as distributed mass elements. In addition, most control algorithms are based on finite-dimensional lumped mass models of small order. Thus, it is natural to use a "truncated" model or other finite-dimensional approximation of distributed mass structures when designing control systems for them.

This section examines the problem of controlling the vibrations of a distributed mass structure by using a finite number of lumped mass actuators and sensors acting at various points on the structure. The approach discussed here is first to cast the structure into an infinite-dimensional matrix equation that is transformed and then truncated. A combination of modal methods and impedance methods are used to solve a simple structural control problem. The goal of this section is to present a simple, representative method of reducing vibration levels in flexible mechanical structures.

Consider a distributed mass structure described by a partial differential equation of the form

$$Ly(x,t) = f(x,t),\ x \in \Omega \tag{11.62}$$

and associated boundary and initial conditions. Here the function $y(x, t)$ and $f(x, t)$ are in $L_2^R(\Omega)$, $y(x,t)$ is the system's output and $f(x, t)$ is the system's input.

This model is an abbreviated formulation of the structures presented in Chapter 7. In terms of the notation of Chapter 7, the operator L is of the form

$$L = \frac{\partial^2}{\partial t^2}(\cdot) + L_1 \frac{\partial}{\partial t}(\cdot) + L_2(\cdot) \tag{11.63}$$

where the output equation is just $y(x, t) = w(x, t)$. If the operator L had an easily calculated inverse, the solution would be given by $y(x, t) = L^{-1}f(x, t)$. To that end, consider taking the Laplace or Fourier transform on the temporal variable of Equation (11.62). This yields

$$Ly(x, s) = f(x, s) \tag{11.64}$$

plus boundary conditions. For ease of notation, no special distinction will be made between y in the time domain and y in the s domain (i.e. between y and its transform), as the remainder of the section deals only with the transformed system.

The control problem of interest here is one that could be implemented by sensors and actuators acting at discrete points along the structure, which may have dynamics of their own. Suppose that the structure is measured at m points along its length, labeled by x_i'. Let $\mathbf{y}(s)$ denote an $m \times 1$ column vector with the ith component defined to be $y(x_i', s)$, i.e. the time-transformed output (displacement) measured at the point x_i'. In addition, r actuators are used to apply time-dependent forces $u_i(t)$, or transformed forces $u_i(s)$, at the r points x_i''. The control action, denoted by $f_c(x, s)$, can be written as

$$f_c(x, s) = \sum_{i=1}^{r} \delta(x - x_i'')u_i(s) = \mathbf{r}^T(x)\mathbf{u}(s) \tag{11.65}$$

Here $\mathbf{r}(x)$ is an $r \times 1$ vector with i^{th} component $\delta(x - x_i'')$, the Dirac delta function, and $\mathbf{u}(s)$ is an $r \times 1$ vector with i^{th} component $u_i(s)$.

Negative feedback is used, so that the total force applied to the structure is given by

$$f(x, s) = f_{ext}(x, s) - f_c(x, s) \tag{11.66}$$

where $f_{ext}(x, s)$ represents an externally applied disturbance force and $f_c(x, s)$ represents the control forces. With the actuator just described, Equation (11.66) becomes

$$f(x, s) = f_{ext}(x, s) - \mathbf{r}^T\mathbf{u}(s) \tag{11.67}$$

To complete the feedback loop, $\mathbf{u}(s)$ must depend on the output, or measured response of the structure. Let $H(s)$ be an $r \times m$ transfer matrix defining the dependence of the control action on the output via the expression

$$\mathbf{u}(s) = H(s)\mathbf{y}(s) \tag{11.68}$$

so that Equation (11.67) becomes

$$f(x, s) = f_{ext} - r^T(x)H(s)\mathbf{y}(s) \tag{11.69}$$

This last expression represents output feedback control. An alternative here would be to use state feedback control, as discussed in Chapter 6.

Next consider casting the problem into modal coordinates. Let $\{\Psi_i(x)\}$ be a set of basis functions in $L_2^C(\Omega)$ and consider $L\Psi_i(x)$, also in $L_2^C(\Omega)$. Then

$$L\Psi_i = \sum_{j=1}^{\infty} \lambda_{ij}(s)\Psi_j(x) \tag{11.70}$$

where $\lambda_{ij}(s)$ is an expansion coefficient. Note that if $\lambda_{ij}(s) = 0$ for $i \neq j$, then Equation (11.70) becomes

$$L\Psi_i(x) = \lambda_i(s)\Psi_i(x) \tag{11.71}$$

so that the expansion coefficient, $\lambda_{ii}(s) = \lambda_i(s)$, is an eigenvalue of the operator L with eigenfunction $\Psi_i(x)$. The $\lambda_{ii}(s)$ are also called modal *impedances* (see Section 10.4 for conditions under which this is true).

Example 11.6.1

For a pinned-pinned uniform beam in transverse vibration of length l with no damping

$$\Psi_n(x) = \frac{\sqrt{2}}{l}\sin(k_n x) \tag{11.72}$$

$$\lambda_n(s) = \left[k_n^4 + \frac{s^2}{c^2k^2}\right]EI \tag{11.73}$$

where

$$k_n = \frac{n\pi}{l}, c^2 = \frac{E}{\rho}, k^2 = \frac{I}{A}.$$

Expanding the functions $y(x, s)$ and $f(x, s)$ in terms of this same set of basis functions, $\{\Psi_i(x)\}$, yields

$$y(x, s) = \sum_{i=1}^{\infty} d_i(s)\Psi_i(x) \tag{11.74}$$

and

$$f(x, s) = \sum_{i=1}^{\infty} c_i(s)\Psi_i(x) \tag{11.75}$$

The expansion coefficients $d_i(s)$ are called the modal response coefficients and the coefficients $c_i(s)$ are called the modal input coefficients.

Next compute (assuming proper convergence, i.e. that Ψ_i is an eigenfunction of L)

$$Ly = \sum_{i=1}^{\infty} d_i(s)L\Psi_i(x) = f(x, s) \tag{11.76}$$

or

$$\sum_{i=1}^{\infty} \lambda_i(s)d_i(s)\Psi_i(x) = \sum_{i=1}^{\infty} c_i(s)\Psi_i(x) \tag{11.77}$$

Note that for Equation (11.77) it is assumed that the $\Psi_i(x)$ are, in fact, the eigenfunctions of L.

Using the orthogonality of the $\{\Psi_i(x)\}$, Equation (11.77) implies that

$$\lambda_i(s)d_i(s) = c_i(s) \tag{11.78}$$

for each index i, so that

$$\lambda_i(s) = \frac{c_i(s)}{d_i(s)} \tag{11.79}$$

This gives rise to the interpretation of $\lambda_i(s)$ as a "modal impedance". Note also that, as before

$$d_i(s) = \int_\Omega y(x,s)\Psi_i(x)d\Omega \tag{11.80}$$

and

$$c_i(s) = \int_\Omega f(x,s)\Psi_i(x)d\Omega \tag{11.81}$$

If L is not self-adjoint and/or the functions $\Psi_i(x)$ are not the system's normal modes, then this procedure can be completed using the orthogonality of the complex set of basis functions. In this coupled case, substitution of Equation (11.70) into Equation (11.76) yields

$$\sum_{i=1}^{\infty} d_i(s)\left[\sum_{j=1}^{\infty} \lambda_{ij}(s)\Psi_j(x)\right] = \sum_{j=1}^{\infty} c_j(s)\Psi_j(x) \tag{11.82}$$

Multiplying Equation (11.82) by $\Psi_k^*(x)$, the conjugate of $\Psi_k(x)$ and integrating over Ω yields that

$$\sum_{i=1}^{\infty} d_i(s)\lambda_{ik}(s) = c_k(s) \tag{11.83}$$

where the summation over the index j has been eliminated by the assumed orthogonality of the set $\{\Psi_k(x)\}$. Equation (11.83) constitutes the equivalent version of Equation (11.78) for the case in which the system does not possess classical normal modes.

Next consider applying linear feedback control in a modal coordinate system defined by the set $\{\Psi_k(x)\}$. Equation (11.78) can be written as a single infinite dimensional matrix equation of the form

$$\Lambda \mathbf{d} = \mathbf{c} \tag{11.84}$$

where Λ is the $\infty \times \infty$ modal impedance matrix with the ij^{th} element defined by $\lambda_{ij}(s)$ and $\mathbf{c}$ and $\mathbf{d}$ are $\infty \times 1$ column matrices defined by $c_i(s)$ and $d_i(s)$, respectively.

Defining $\Psi(x)$ as the $\infty \times 1$ column matrix of eigenfunctions $\Psi_i(x)$, the other relevant terms can be written as

$$f_{ext} = \sum_{i=1}^{\infty} e_i(s)\Psi_i(x) = \mathbf{e}^T(s)\Psi(x) \tag{11.85}$$

$$f(x,s) = \sum_{i=1}^{\infty} c_i(s)\Psi_i(x) = \mathbf{c}^T(s)\Psi(x) \tag{11.86}$$

and

$$y(x,s) = \sum_{i=1}^{\infty} d_i(s)\Psi_i(x) = \mathbf{d}^T(s)\Psi(x) \tag{11.87}$$

where the various column vectors have the obvious definitions. For instance, the vector $\mathbf{e}(s)$ is vector of expansion coefficients for the external disturbance force f_{ext} with components $e_i(s)$, etc.

A measurement matrix, denoted by M, can be defined by

$$M_{ij} = \Psi_j(x_i') \tag{11.88}$$

which is an $m \times \infty$ matrix and relates $\mathbf{d}(s)$ directly to $\mathbf{y}(s)$ by

$$\mathbf{y}(s) = M\mathbf{d}(s) \tag{11.89}$$

Likewise, an $r \times \infty$ modal coefficient matrix, denoted by R, can be defined by

$$R_{ij} = \int_{\Omega} r_i(x)\Psi_j(x)d\Omega \tag{11.90}$$

which relates $\mathbf{r}(x)$ to $\Psi(x)$ by

$$\mathbf{r}(x) = R\Psi(x) \tag{11.91}$$

Using the orthogonality of $\Psi_i(x)$, the inner product

$$\int_{\Omega} \Psi(x)\Psi^{\mathrm{T}}(x)d\Omega = I_{\infty} \tag{11.92}$$

where I_{∞} denotes the $\infty \times \infty$ identity matrix with elements

$$\int_{\Omega} \Psi_i(x)\Psi_{\varphi}(x)d\Omega = \delta_{ij}$$

Note that if $\Psi(x)$ is complex, then the transpose should be interpreted as the conjugate transpose. Multiplying Equation (11.91) by Ψ^T from the right and integrating over Ω yields

$$R = \int_{\Omega} \mathbf{r}(x)\Psi^T(x)d\Omega \tag{11.93}$$

This last expression provides a more useful definition of the modal coefficient matrix R.

A relationship between R, **c** and **e** can be found by substituting Equations (11.85) and (11.86) into Equation (11.67). This yields

$$\mathbf{c}^T\Psi = \mathbf{e}^T\Psi - \mathbf{r}^T\mathbf{u} \tag{11.94}$$

Since $\mathbf{r}^T\mathbf{u}$ is a scalar, this can also be written as

$$\mathbf{c}^T(s)\Psi(x) = \mathbf{e}^T(s)\Psi(x) - \mathbf{u}^T(s)\mathbf{r}(x) \tag{11.95}$$

Multiplication from the right by $\Psi^T(x)$ and integrating over Ω yields

$$\mathbf{c}^T(s)\int_\Omega \Psi(x)\Psi^T(x)d\Omega = \mathbf{e}^T(s)\int_\Omega \Psi(x)\Psi^T(x)d\Omega - \mathbf{u}^T(s)\int_\Omega \mathbf{r}(x)\Psi^T(x)d\Omega \tag{11.96}$$

Using Equations (11.92) and (11.93) then yields

$$\mathbf{c}^T(s) = \mathbf{e}^T(s) - \mathbf{u}^T(s)R \tag{11.97}$$

or

$$\mathbf{c}(s) = \mathbf{e}(s) - R^T\mathbf{u}(s) \tag{11.98}$$

Equation (11.84) now becomes

$$\Lambda\mathbf{d}(s) = \mathbf{e}(s) - R^T\mathbf{u}(s) \tag{11.99}$$

or upon substitution of Equation (11.68) for $\mathbf{u}(s)$

$$\Lambda\mathbf{d}(s) = \mathbf{e}(s) - R^T H(s)\mathbf{y}(s) \tag{11.100}$$

Using Equation (11.89), the last term in Equation (11.100) can be placed in terms of $\mathbf{d}(s)$ to yield

$$\Lambda\mathbf{d}(s) = \mathbf{e}(s) - R^T H(s)M\mathbf{d}(s) \tag{11.101}$$

Assuming that Λ^{-1} exists, this last expression can be manipulated to yield

$$\left[I_\infty + \Lambda^{-1}Q(s)\right]\mathbf{d}(s) = \Lambda^{-1}\mathbf{e}(s) \tag{11.102}$$

where $Q(s) = R^T H(s)M$. Equation (11.102) represents the closed-loop configuration for the output feedback control of a distributed parameter structure in terms of infinite-dimensional matrices.

If the infinite matrix inverse Λ^{-1} exists, if the inverse of the impedance matrix $[I_\infty - \Lambda^{-1}Q]$ can be calculated, and if the functions $\{\Psi_i(x)\}$ are known, Equation (11.102) along with Equation (11.74) yields the response $\mathbf{d}(s)$ in terms of the input, $\mathbf{e}(s)$. Several common examples, such as uniform beams and plates of simple geometry, satisfy these assumptions. Unfortunately, in many practical cases, these assumptions are not satisfied, and the matrix Λ must be truncated in some fashion. Even in cases where Λ^{-1} can be calculated, the control $Q(s)$ may be such that $[I_\infty - \Lambda^{-1}Q]$ is difficult to calculate. In cases where truncation of the model is required, Equation (11.102) provides a convenient formula for studying the effects of truncation in the presence of control.

As was true for the procedure of Section 6.12, the truncation method presented here is based on partitioning the various infinite-dimensional matrices of Equation

(11.102). Let $\Lambda^{-1}_{n\infty}$ denote the matrix formed from the matrix Λ^{-1} by partitioning off the first n rows and all the columns. Using this notation, the matrix Λ^{-1} is partitioned as

$$\Lambda^{-1} = \begin{bmatrix} \Lambda^{-1}_{nm} & \Lambda^{-1}_{n\infty} \\ \Lambda^{-1}_{\infty n} & \Lambda^{-1}_{\infty\infty} \end{bmatrix} \tag{11.103}$$

In a similar fashion, the matrices M, R and Q are partitioned as

$$M = [M_{mn} \quad M_{m\infty}] \tag{11.104}$$

$$R^T = \begin{bmatrix} R^T_{nr} \\ R^T_{\infty r} \end{bmatrix} \tag{11.105}$$

and

$$Q = \begin{bmatrix} Q_{nn} & Q_{n\infty} \\ Q_{\infty n} & Q_{\infty\infty} \end{bmatrix} \tag{11.106}$$

The sub matrices of Q can all be written in terms of R, M and H by

$$Q_{nn} = R^T_{nr} H M_{mn} \tag{11.107}$$

$$Q_{n\infty} = R^T_{nr} H M_{m\infty} \tag{11.108}$$

$$Q_{\infty n} = R^T_{\infty r} H M_{mn} \tag{11.109}$$

and

$$Q_{\infty\infty} = R^T_{\infty r} H M_{m\infty} \tag{11.110}$$

Substitution of these partitioned matrices into Equation (11.102) yields the partitioned system

$$\begin{bmatrix} I_n + \Lambda^{-1}_{mn} Q_{nn} + \Lambda^{-1}_{n\infty} Q_{\infty n} & \Lambda^{-1}_{nn} Q_{n\infty} + \Lambda^{-1}_{n\infty} Q_{\infty\infty} \\ \Lambda^{-1}_{\infty n} Q_{nn} + \Lambda^{-1}_{\infty\infty} Q_{\infty n} & I_\infty + \Lambda^{-1}_{\infty n} Q_{n\infty} + \Lambda^{-1}_{\infty\infty} Q_{\infty\infty} \end{bmatrix} \begin{bmatrix} \mathbf{d}_n \\ \mathbf{d}_\infty \end{bmatrix} = \begin{bmatrix} \Lambda^{-1}_{nn} & \Lambda^{-1}_{n\infty} \\ \Lambda^{-1}_{\infty n} & \Lambda^{-1}_{\infty\infty} \end{bmatrix} \begin{bmatrix} \mathbf{e}_n \\ \mathbf{e}_\infty \end{bmatrix} \tag{11.111}$$

Here the response vector $\mathbf{d}$ and the input vector $\mathbf{e}$ have also been partitioned, dividing these infinite-dimensional vectors into an $n \times 1$ finite-dimensional part and an $\infty \times 1$ infinite-dimensional part.

The various partitions of Equation (11.103) can be used to interpret the effects of truncating the modal description of a structure at n modes in the presence of a control law. Structures are generally thought of as low-pass filters in the sense that $\lambda_n^{-1} \to 0$ as $n \to \infty$. Thus, for structures it is reasonable to assume that the matrix $\Lambda^{-1}_{\infty\infty}$ is zero for some value of n.

Sensors often behave like low-pass filters as well, so that it is also reasonable to assume that $M_{n\infty}$ is the zero matrix. This in turn causes $Q_{n\infty} = Q_{\infty\infty} = 0$. If the actuators are slow enough, it can also be argued that $R^T_{\infty r} = 0$, which causes

$Q_{\infty n} = Q_{\infty\infty} = 0$. With these three assumptions, the system of Equation (11.111) is reduced to

$$\begin{bmatrix} I_n + \Lambda_{nn}^{-1} Q_{nn} & 0 \\ \Lambda_{\infty n}^{-1} Q_{nn} & I_\infty \end{bmatrix} \begin{bmatrix} \mathbf{d}_n \\ \mathbf{d}_\infty \end{bmatrix} = \begin{bmatrix} \Lambda_{nn}^{-1} & \Lambda_{n\infty}^{-1} \\ \Lambda_{\infty n}^{-1} & 0 \end{bmatrix} \begin{bmatrix} \mathbf{e}_n \\ \mathbf{e}_\infty \end{bmatrix} \tag{11.112}$$

This can be written as the two coupled vector equations

$$(I_n + \Lambda_{nn}^{-1} Q_{nn})\mathbf{d}_n = \Lambda_{nn}^{-1}\mathbf{e}_n + \Lambda_{n\infty}\mathbf{e}_\infty \tag{11.113}$$

and

$$\Lambda_{\infty n}^{-1} Q_{nn}\mathbf{d}_n + \mathbf{d}_\infty = \Lambda_{\infty n}^{-1}\mathbf{e}_n \tag{11.114}$$

These last two equations provide a simple explanation of some of the problems encountered in the control of distributed mass structures using truncated models.

First consider the case with $\mathbf{e}_\infty = \mathbf{0}$. This corresponds to a "band limited" input. That is, the external disturbance provides energy only to the first n modes. In this case, Equation (11.105) becomes

$$\mathbf{d}_n = (I_n + \Lambda_{nn}^{-1} Q_{nn})^{-1}\Lambda_{nn}^{-1}\mathbf{e}_n \tag{11.115}$$

Equation (11.115) can now be used to solve the control problem, i.e. to calculate Q_{nn} such that the response $\mathbf{d}_n$ has a desired form. In fact, Equation (11.115) is equivalent to first approximating a distributed parameter system by a finite-dimensional system and then designing a finite-dimensional control system for it. However, this is slightly misleading, as can be seen by considering Equation (11.114).

Rearrangement of Equation (11.114) yields

$$\mathbf{d}_\infty = \Lambda_{\infty n}^{-1}\mathbf{e}_n - \Lambda_{\infty n}^{-1} Q_{nn}\mathbf{d}_n \tag{11.116}$$

This states that unless the structure's dynamics decouple (i.e. $\Lambda_{\infty n}^{-1} = 0$) or unless it can be argued that $\Lambda_{\infty n}^{-1}$ is small, the higher, uncontrolled modes of the response $\mathbf{d}_\infty$ will be excited by the control action, Q_{nn}. Such unwanted excitation is called *control spillover*.

In the case that $\Lambda_{\infty n}^{-1}$ is close to zero, Equation (11.115) provides a good approximation to the control problem for distributed mass structures. In fact, the requirement that $\mathbf{e}_\infty = \mathbf{0}$ provides a criteria for determining the proper order, n, to be chosen for the approximation for a given disturbance. The value of n is chosen so that $\mathbf{e}_\infty$ is approximately zero.

Next consider the case that the sensors are not low-pass filters, i.e. $M_{m\infty} \neq 0$, so that $Q_{n\infty} \neq 0$. In this case, the first partition of Equation (11.111) yields

$$(I_n + \Lambda_{nn}^{-1} Q_{nn})\mathbf{d}_n + \Lambda_{nn}^{-1} Q_{n\infty}\mathbf{d}_\infty = \Lambda_{nn}^{-1}\mathbf{e}_n \tag{11.117}$$

The equation describing $\mathbf{d}_n$ is recoupled to the truncated dynamics constrained in the vector $\mathbf{d}_\infty$. If the term $Q_{n\infty}$ is erroneously neglected and Equation (11.115) is used to design the control system, then the resulting solution $\mathbf{d}_n$ will be in error, and the resulting calculation of R_{nr}^T and H will also be in error. The response will suffer from what is often referred to as *observation spillover*, meaning that the sensors have caused a coupling of the truncated system with the neglected modes, producing error in the closed-loop response.

A similar problem arises if the actuators are "fast", i.e. if $R^T_{\infty r} \neq 0$. In this case, Equations (11.113) and (11.114) become

$$(I_n + \Lambda^{-1}_{mn} Q_{nn} + \Lambda^{-1}_{n\infty} Q_{\infty n})\mathbf{d}_n = \Lambda^{-1}_{nn}\mathbf{e}_n \tag{11.118}$$

and

$$(\Lambda^{-1}_{nn} Q_{nn} + \Lambda^{-1}_{n\infty} Q_{\infty n})\mathbf{d}_\infty = \Lambda^{-1}_{\infty n}\mathbf{e}_n \tag{11.119}$$

Again, the introduction of the term $Q_{\infty n}$, associated with high-speed actuator excitation, couples the equation for the solution $\mathbf{d}_n$ and the truncated, or residual, solution $\mathbf{d}_\infty$. Thus, if $R^T_{\infty n}$ is not actually zero and Equation (11.115) is used to compute the control law, error will result. The interpretation here is that $R^T_{\infty n}$ excites the neglected modes, $\mathbf{d}_\infty$ and hence causes energy to appear in the neglected part of the model. This again causes control spillover.

11.7 Impedance Method of Truncation and Control

The modal description of a structure presented in the previous section lends itself to an interpretation of potential problems encountered when using point actuators and sensors in designing a control system for a distributed mass structure. In this section, an alternative approach is presented that uses the modal Equation (11.102), but rather than truncating the response, it uses an impedance method to calculate the closed-loop response vector $\mathbf{d}$.

The sensor-actuator admittance (inverse of impedance) matrix $Y(s)$ is defined by

$$Y(s) = M\Lambda^{-1}(s)R^T \tag{11.120}$$

and is related to the dynamic stiffness matrix. Note that $Y(s)$ is a finite-dimensional matrix, the elements of which are infinite sums.

Consider again the infinite matrix description of the structural control problem, as formulated in Equation (11.102). This expression can be written as

$$I\mathbf{d} + \Lambda^{-1}R^T HM\mathbf{d} = \Lambda^{-1}\mathbf{e} \tag{11.121}$$

From Equation (11.89), the vector $M\mathbf{d}$ can be replaced by $\mathbf{y}$ to yield

$$I\mathbf{d} + \Lambda^{-1}R^T H\mathbf{y} = \Lambda^{-1}\mathbf{e} \tag{11.122}$$

Multiplying this expression by M yields

$$\mathbf{y} + M\Lambda^{-1}R^T H\mathbf{y} = M\Lambda^{-1}\mathbf{e} \tag{11.123}$$

$$\mathbf{y} + Y(s)H(s)\mathbf{y} = M\Lambda^{-1}\mathbf{e} \tag{11.124}$$

This can be written as

$$[I_m + Y(s)H(s)]\mathbf{y} = M\Lambda^{-1}\mathbf{e} \tag{11.125}$$

where I_m is the $m \times m$ identity matrix. Thus the coefficient of **y** is a finite-dimensional matrix. Assuming the coefficient of **y** has an inverse, Equation (11.125) can be written as

$$\mathbf{y} = [I_m + Y(s)H(s)]^{-1}M\Lambda^{-1}\mathbf{e} \tag{11.126}$$

This expression can be substituted into Equation (11.122) to yield

$$\mathbf{d} = \{I_\infty - \Lambda^{-1}R^TH[I_m + Y(s)H(s)]^{-1}M\}\Lambda^{-1}\mathbf{e} \tag{11.127}$$

which expresses the system response, **d**, in terms of the disturbance input, **e**. Equation (11.127) represents the impedance method of dealing with truncation in the control of distributed mass structures, as developed by Berkman and Karmopp (1969). The open-loop system, as represented by Λ, still needs to be truncated using the low-pass filter argument of the previous section, i.e. $\Lambda^{-1} \cong \Lambda_n^{-1}$. However, the feedback control portion is now finite-dimensional and of low order (i.e. equal to the number of measurement points or sensors). Hence, truncation or partitioning of an inverse matrix is required in order to compute the control. Instead, the elements of the matrix $[I_m + Y(s)H(s)]$, which are all infinite sums, can be first calculated and/or truncated. Then, the exact inverse can be calculated. Thus, the value of the truncation index for the structure and that for the control can be separately chosen. This approach also allows for the inclusion of actuator dynamics due to the presence of the matrix $H(s)$. The following example serves to clarify this method.

Example 11.7.1

Consider the transverse vibrations of a pined-pinned beam with an actuator providing dynamics (or an impedance), $Z_1(s)$ acting at the point x_1. Also consider a single displacement measuring sensor, located at x_1, so that $y(x)$ is the scalar quantity $w(x_1, s)$, i.e. so that the index $m = 1$. Let $\Psi_i(x)$ denote the modes (or eigenfunctions) of the pinned-pinned beam without the actuator attached. From Equation (11.88), the matrix M becomes the $1 \times \infty$ vector

$$M = [\Psi_1(x_1)\ \Psi_2(x_1)\ldots] = \Psi(x_1)$$

Likewise, from Equation (11.82) the matrix R becomes the $1 \times \infty$ vector

$$R = [\Psi_1(x_1)\ \Psi_2(x_1)\ldots] = \Psi(x_1)$$

The matrix $H(s)$ in this case is just the scalar element $H(s) = Z_1(s)$.

The sensor actuator admittance matrix becomes the scalar element

$$Y(s) = M\Lambda^{-1}R^T = \sum_{i=1}^{\infty} \lambda_1^{-1}(s)\Psi_i^2(x_1)$$

where the $\lambda_i(s)$ are the open-loop system eigenvalues, since $\Lambda(s)$ is diagonal in this example (i.e. $\Lambda(s)$ consists simply of the structure's eigenvalues).

The response can now be computed from Equation (11.119) to be

$$\mathbf{d}(s) = \left[\Lambda^{-1}(s) - \frac{[\Lambda^{-1}\Psi^T(x_1)][\Lambda^{-1}\Psi^T(x)]^T}{1 + \sum_{i=1}^{\infty} \lambda_i^{-1}(s)\Psi_i^2(x_1)} \right] \mathbf{e}$$

It is important to note that the response (or more exactly, the Laplace transform of the response) is calculated here by truncating (approximating) the structural dynamics $\Lambda^{-1}(s)$ and $\Lambda^{-1}\Psi(x_1)$ independently of the control. The actuator representation, $Z_1(s)$, is not truncated at all in this example. This is in contrast to the completely modal approach of the previous section.

As the example illustrates, the modal impedance inversion technique described in this section reduces the problem of truncation in the presence of control from one of approximating an infinite-order matrix by a finite-order matrix to that of approximating infinite sums with partial finite summations.

Chapter Notes

This chapter introduces some methods of approximating distributed mass models of structures by lumped mass models more suitable for digital computing. Section 11.2 introduced the obvious and popular method of modal truncation.

Modal methods are common and can be found in most texts. See, for instance, Meirovitch (1980, 2001), for a more complete discussion. The Ritz-Galerkin method of Section 11.3 is again common and is found in most vibration texts at almost every level. The common name of Raleigh-Ritz has always been surrounded with a bit of controversy over who actually first penned the method and this is nicely settled in Leissa (2005). The FEM briefly discussed in Section 11.4 is currently the most often used and written about method. An excellent short introduction to FEMs can be found in Meirovitch (1986). A more comprehensive approach can be found in the excellent book by Shames and Dym (1985) or the more advanced treatment by Hughes (2000).

Meirovitch's (1980) book contains a complete treatment of the substructure methods discussed in Section 11.5, as does the paper by Hale and Meirovitch (1980). The paper by Craig (1987) reviews the related topic of component-mode methods. Sections 11.6 and 11.7, dealing with the topic of truncation and control system design, are taken directly from the paper by Berkman and Karnop (1969), which was one of the first papers written in this area. Many other approaches to this same problem can be found in the literature. The survey paper by Balas (1982) provides a useful introduction to the topic.

References

Balas, M. J. (1982) Trends in large space structure control theory: fondest hopes, wildest dreams. *IEEE Transactions on Automatic Control*, AC-27(3), 522–535.

Berkman, F. and Karnopp, D. (1969) Complete response of distributed systems controlled by a finite number of linear feedback loops. *ASME Journal of Engineering for Industry*, 91, 1062–1068.

Craig, R. R., Jr. (1987) A review of time-domain and frequency-domain component-mode synthesis methods. *The International Journal of Analytical and Experimental Modal Analysis*, 2(2), 59–72.

Hale, A. L. and Meirovitch, L. (1980) A general substructure synthesis method for the dynamic simulation of complex structures, *Journal of Sound and Vibration*, 69(2), 309–326.

Hughes, T. J. R. (2000) *The Finite Element Method: Linear Static and Dynamic Finite Element Analysis*. Dover Publications, Mineola, NY.

Leissa, A. W. (2005) The historical bases of the Rayleigh and Ritz methods. *Journal of Sound and Vibration*, 287(4–5), 961–978.

Meirovitch, L. (1980) *Computational Methods in Structural Dynamics*. Alphen aan den Rijn Sijthoff & Noordhoff International Publishers, The Netherlands.

Meirovitch, L. (1986) *Elements of Vibration Analysis*, 2nd Edition. McGraw Hill, New York.

Meirovitch, L. (2001) *Fundamentals of Vibrations*. McGraw Hill, New York.

Shames, I. H. and Dym, C. L. (1985) *Energy and Finite Element Methods in Structural Mechanics*. Prentice Hall, Englewood Cliffs, NJ.

Problems

11.1 Consider a clamped-clamped string of unit length. Consider the case of $N = 4$ and write out the lumped parameter model suggested in Equation (11.7). Compute the initial conditions given in Equation (11.10) for the case of a zero initial velocity and an initial displacement given by

$$w_0(x) = \begin{cases} x & 0 \leq x < \dfrac{1}{2} \\ 1 - x & \dfrac{1}{2} \leq x \leq 1 \end{cases}$$

11.2 Estimate the amount of energy neglected in a three-mode approximation of a fixed-fixed beam of length l in longitudinal vibration.

11.3 Calculate a three-mode approximation of a clamped square plate using modal truncation.

11.4 Use trigonometric functions and perform a Ritz-Galerkin approximation for a transversely vibrating beam (undamped), which is clamped at one end and attached to a spring with constant k and mass m at the other end. Use three terms. For convenience, assume $EI = A\rho = m = k = 1$. Calculate the natural frequencies and compare them to those obtained by the method in Section 8.4.

11.5 Compare the finite element model of Example 11.4.1 with a three-mode Ritz-Galerkin model of the same structure. How do the eigenvalues compare with those of the distributed parameter model?

11.6 Show that the matrices M and K defined by the FEM of Section 11.4 are both symmetric (in general).

11.7 Derive the finite element matrix for a transversely vibrating beam modeled with three elements.

11.8 Consider a three degree of freedom system with mass matrix $M = I$ and stiffness matrix

$$K = \begin{bmatrix} 3 & -1 & 0 \\ -1 & 1.5 & -.5 \\ 0 & -.5 & .5 \end{bmatrix}$$

that corresponds to three masses connected in series by three springs. Define two substructures by letting substructure 1 be the first two masses and substructure 2 be the remaining mass. Calculate the coefficient matrices of Equation (11.61).

11.9 Calculate $\lambda(s)$ for a cantilevered beam in transverse vibration. Use this information to calculate the matrix Λ^{-1}.

11.10 For Problem (11.8), suppose a disturbance force of sin $2t$ is applied to the structure at the midpoint and calculate the value of the index n such that $\mathbf{e}_\infty$ is negligible. For simplicity, set each of the physical parameter values to unity.

11.11 Recalculate the equations of Example 11.7.1 using two elements as part of the control, i.e. $z_1(s)$ and $z_2(s)$, acting at points x_1 and x_2, respectively.

11.12 Derive Equations (11.23) and (11.24) for the case $N = 2$, by substituting the sum of Equation (11.19) into the Rayleigh Quotient and taking the indicated derivatives (hint, use the fact that L is self-adjoint and write the two derivative equations as one equation in matrix form).

12

Vibration Measurement

12.1 Introduction

This chapter introduces the basic topics associated with dynamic measurement and testing of structures. In the previous chapters, the differential equations, or models, of the system are assumed to be known, and the theory developed consists of calculating and characterizing the response of the system to known inputs. This is called the *forward problem*. In this chapter however, the interest lies in measuring the response of a structure and in some way determining the equations of motion from test data. The problem of determining a system of equations from information about inputs and responses belongs to a class of problems called *inverse problems*. Although measurements of vibrating systems are made for a variety of reasons, this chapter focuses on the use of vibration measurement to identify, or verify, a mathematical model of a test structure. Other uses of vibration measurements include environmental testing, balancing of rotating machinery, prediction of failures, structural health monitoring and machinery diagnostics for maintenance (Inman et al., 2005).

The two main types of modeling are lumped mass modeling (Chapters 1 to 6) and distributive mass modeling (Chapters 7 to 11). Just as numerical solutions of distributed mass systems end up as lumped mass models, the same is true for measurements. That is, measurements of distributed mass structures often result in data that resemble the response of a lumped mass structure. Hence, many test procedures, including those described here, assume the structure being tested can be adequately described by a lumped parameter model. The significance of such an assumption is addressed in Chapter 11 on the finite dimensional modeling of distributed parameter systems.

There are several other assumptions commonly made but not stated (or understated) in vibration testing. The most obvious of these is that the system under test is linear and is driven by the test input only in its linear range. This assumption is essential and should not be neglected. Measurement of nonlinear systems is presented in Virgin (2000).

Vibration testing and measurement for modeling purposes forms a large industry. This field is referred to as *modal testing*, *modal analysis* or *experimental modal analysis* (EMA). Understanding modal testing requires knowledge of

Vibration with Control, Second Edition. Daniel John Inman.

Companion Website: www.wiley.com/go/inmanvibrationcontrol2e

several areas. These include instrumentation, signal processing, parameter estimation and vibration analysis. These topics are introduced in the following sections; however, Ewins (2000) should be consulted for a complete description.

It is worth noting that modal testing activity is centered mainly around determining the modal parameters of a lumped parameter model of the test structure. The process of determining a mathematical model of a system from measured inputs and outputs is called *system identification theory*. System identification and parameter estimation theories have grown rapidly in the discipline of control theory rather than vibrations. System identification theory, in turn, is a subclass of inverse problems, as opposed to the direct problem discussed in earlier chapters. Much of the early work in modal analysis did not take advantage of the disciplines of system identification and parameter estimation, a situation rectified by efforts such as the book by Juang (1994). In the material that follows, a few of the basic approaches are presented.

As mentioned, the first eleven chapters consider only the *forward problem*, i.e. given the matrices M, C and K along with the appropriate initial conditions and forcing functions, determine the solution of Equation (2.19) describing the dynamics of the structure. The *inverse problem*, on the other hand, is to determine the matrices M, C and K from knowledge of the measurements of the responses (position, velocity or acceleration). The modal testing problem, a subclass of inverse problems, is to recover the mode shapes, natural frequencies and damping ratios from response measurements. In general, modal testing methods cannot fully determine the matrices of physical parameters M, C and K. While forward problems have a unique solution for linear systems, inverse problems do not. Model updating, introduced in Section 12.8, is an attempt to justify the analytically derived values of M, C and K with measured modal data.

The fundamental idea behind modal testing is that of resonance first introduced in Section 1.4. If a structure is excited at resonance, its response exhibits two distinct phenomena, as indicated by Equations (1.23) and (1.24). As the driving frequency approaches the natural frequency of the structure, the magnitude at resonance rapidly approaches a sharp maximum value, provided the damping ratio is less than about 0.5. The second, often neglected, phenomena of resonance is that the phase of the response shifts by 180° as the frequency sweeps through resonance, with the value of the phase at resonance being 90°.

The first few sections of this chapter deal with the hardware considerations and digital signal analysis necessary for making a vibration measurement for any purpose. The data acquisition and signal processing hardware has changed considerably over the past decade and continues to change rapidly as the result of advances in solid-state and computer technology. The remaining sections discuss the modal analysis version of model identification and parameter estimation. The last section introduces model updating.

12.2 Measurement Hardware

A vibration measurement generally requires several hardware components. The basic hardware elements required consist of a source of excitation for providing a known or controlled input to the structure, a *transducer* to convert the mechanical

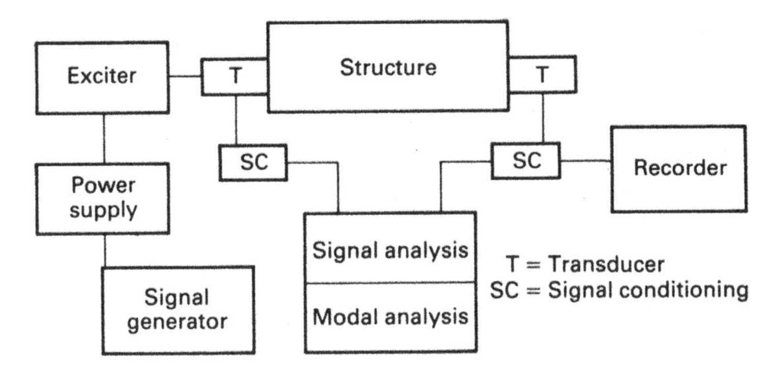

Figure 12.1 Components of a vibration measurement system for modal analysis.

motion of the structure into an electrical signal, a signal conditioning amplifier to match the characteristics of the transducer to the input electronics of the digital data acquisition system, and an analysis system (or analyzer), in which signal processing and modal analysis programs reside. The arrangement is illustrated in Figure 12.1; it includes a power amplifier and a signal generator for the exciter, as well as a transducer to measure, and possibly control, the driving force or other input. Each of these devices and their functions are discussed briefly in this section.

First, consider the excitation system, denoted "exciter" in Figure 12.1. This system provides an input motion or, more commonly, a driving force $f_i(t)$, as in Equation (2.19). The physical device may take several forms, depending on the desired input and the physical properties of the test structure. The two most commonly used exciters in modal testing are the shaker (electromagnetic or electrohydraulic) and the impulse hammer. The preferred device is often the electromagnetic exciter, which has the ability, when properly sized, to provide inputs large enough to result in easily measured responses. Also, the output is easily controlled electronically, sometimes using force feedback. The excitation signal, which can be tailored to match the requirements of the structure being tested, can be a swept sine, random or other appropriate signal. The *electromagnetic shaker* is basically a linear electric motor consisting of coils of wire surrounding a shaft in a magnetic field. An alternating current applied to the coil causes a force to be applied to the shaft which, in turn, transfers force to the structure. The input electrical signal to the shaker is usually a voltage that causes a proportional force to be applied to the test structure.

Since shakers are attached to the test structure and since they have significant mass, care should be taken by the experimenter in choosing the size of shaker and method of attachment to minimize the effect of the shaker on the structure. The shaker and its attachment can add mass to the structure under test (called *mass loading*), as well as otherwise constraining the structure.

Mass loading and other constraints can be minimized by attaching the shaker to the structure through a *stinger*. A stinger consists of a short thin rod (usually made of steel or nylon) running from the driving point of the shaker to a force transducer mounted directly on the structure. The stinger serves to isolate the shaker from the structure, reduces the added mass, and causes the force to be transmitted axially along the stinger, controlling the direction of applied the force.

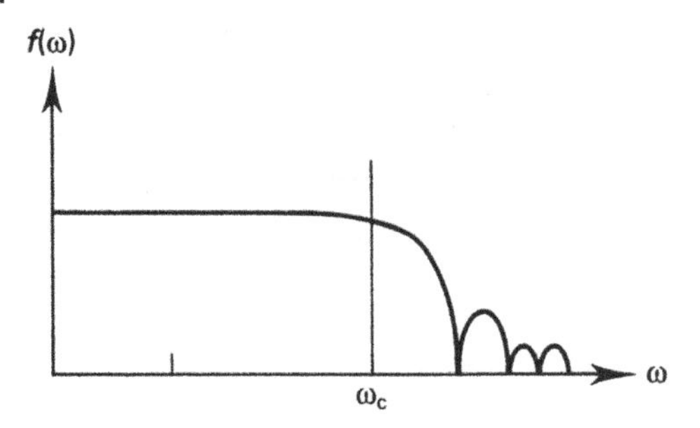

Figure 12.2 Frequency spectrum of a typical hammer impact.

Excitation can also be applied by an *impulse* by using an *impact hammer*. An impact hammer consists of a hammer with a force transducer built into the head of the hammer. The hammer is then used to hit (impact) the test structure and thus excite a broad range of frequencies. The peak impact force is nearly proportional to the hammer head mass and the impact velocity. The upper frequency limit excited by the hammer is decreased by increasing the hammer head mass and increases with increasing stiffness of the tip of the hammer.

As a simple example of how the choice of excitation device is critical, consider Figure 12.2. This plot illustrates the frequency spectrum of a typical force resulting from a hammer impact or hit. It illustrates that the impact force is approximately constant up to a certain value of frequency, denoted by ω_c. This hammer hit is less effective in exciting the modes of the structure with frequencies larger than ω_c than it is for those less than ω_c. For a given hammer mass, ω_c can be lowered by using a softer tip.

The built-in force transducer in impact hammers should be dynamically calibrated for each tip used, as this will affect the sensitivity. Although the impact hammer is simple and does not add mass loading to the structure, it is often incapable of transforming sufficient energy to the structure to obtain adequate response signals in the frequency range of interest. Also, peak impact loads are potentially damaging, and the direction of the applied load is difficult to control. Nonetheless, impact hammers remain a popular and useful excitation device, as they generally are much faster to use than shakers.

Next consider the transducers required to measure the response of the structure as well as the impact force. The most popular and widely used transducers are made from piezoelectric materials. Piezoelectric materials generate electrical charge when strained. By various designs, transducers incorporating these materials can be built to produce signals proportional to either force or local acceleration. *Accelerometers*, as they are called, actually consist of two masses, one of which is attached to the structure, separated by a piezoelectric material, which acts like a very stiff spring. This causes the transducer to have a resonant frequency. The maximum measurable frequency is usually a fraction of the accelerometer's resonance frequency. In fact, the upper frequency limit is usually determined by the so-called mounted resonance, since the connection of the transducer to the structure is always somewhat compliant.

Piezoelectric materials (Section 7.10) also produce a strain when excited with a voltage. Hence, piezoelectric materials are emerging as a reasonable means of vibration excitation for special purpose situations, as described in Cole et al. (1995) and Inman et al. (2005). Piezoelectric wafers used as exciters (actuators) or sensors have the potential for less mass loading than traditional shakers and accelerometers.

The output impedance of many transducers is not well suited for direct input into signal analysis equipment. Hence, *signal conditioners*, which may be charge amplifiers or voltage amplifiers, match and often amplify signals prior to analyzing the signal. It is very important that each set of transducers along with signal conditioning are properly calibrated in terms of both magnitude and phase over the frequency range of interest. While accelerometers are convenient for many applications, they provide weak signals if one is interested in lower-frequency vibrations incurred in terms of velocity of displacement. Even substantial, low-frequency vibration displacements may result in only small accelerations, recalling that for a harmonic displacement of amplitude X, the acceleration amplitude is $-\omega^2 X$. Strain gauges and potentiometers as well as various optical (fiber optics, laser vibrometers), capacitive and inductive transducers are often more suitable than accelerometers for low-frequency motion measurement.

Once the response signal has been properly conditioned, it is routed to an analyzer for signal processing. The standard is called a digital or discrete Fourier analyzer (DFA), also called the Fast Fourier transform (FFT) analyzer; it is introduced briefly here. Basically, the DFA accepts analog voltage signals, which represent the acceleration (force, velocity, displacement or strain) from a signal conditioning amplifier. This signal is filtered and digitized for computation. Discrete frequency spectra of individual signals and cross-spectra between the input and various outputs are computed. The analyzed signals can then be manipulated in a variety of ways to produce such information as natural frequencies, damping ratios and mode shapes in numerical or graphic displays.

While almost all the commercially available analyzers are marketed as turnkey devices, it is important to understand a few details of the signal processing performed by these analysis units in order to carry out valid experiments. This forms the topic of the next two sections.

12.3 Digital Signal Processing

Much of the analysis done in modal testing is performed in the frequency domain, as discussed in Section 5.7. The analyzer's task is to convert analog time domain signals into digital frequency domain information compatible with digital computing and then to perform the required computations with these signals. The method used to change an analog signal, $x(t)$, into frequency domain information is the Fourier transform (similar to the Laplace transform used in Chapter 5), or a *Fourier series*. The Fourier series is used here to introduce the Digital Fourier transform (DFT). See Newland (1985: p. 38) for the exact relationship between a Fourier transform and a Fourier series.

Any periodic in time signal, $x(t)$, of period T can be represented by a Fourier series of the form

$$x(t) = \frac{a_0}{2} + \sum_{i=1}^{\infty} \left[a_i \cos \frac{2\pi i t}{T} + b_i \sin \frac{2\pi i t}{T} \right] \tag{12.1}$$

where the constants a_i and b_i, called the *Fourier coefficients*, or *spectral coefficients*, are defined by

$$a_i = \frac{2}{T}\int_0^T x(t)\cos\frac{2\pi it}{T}dt \tag{12.2}$$

$$b_i = \frac{2}{T}\int_0^T x(t)\sin\frac{2\pi it}{T}dt$$

The expression in Equation (12.1) is referred to as the Fourier series for the periodic function $x(t)$. The spectral coefficients represent frequency domain information about the time signal $x(t)$.

The coefficients a_i and b_i also represent the connection to vibration experiments. The analog output signals of accelerometers and force transducers, represented by $x(t)$, are inputs to the analyzer. The analyzer in turn calculates the spectral coefficients of these signals, thus setting the stage for a frequency domain analysis of the signals. Some signals and their Fourier spectrum are illustrated in Figure 12.3. The analyzer first converts the analog signals into digital records. It samples the signals $x(t)$ at many different equally spaced values and produces a digital record of the signal in the form of a set of numbers $\{x(t_k)\}$. Here, $k = 1,2,\ldots,N$, the digit N denotes the number of samples, and t_k indicates a discrete value of the time.

This process is performed by an analog-to-digital (A/D) converter. This conversion from an A/D signal can be thought of in two ways. First, one can imagine a grating that samples the signal every t_k seconds and passes through the signal $x(t_k)$. One can also consider the A/D converters as multiplying the signal $x(t)$ by a square-wave function, which is zero over alternate values of t_k and has the value of 1 at each t_k for a short time. Some signals and their digital representation are illustrated in Figure 12.4.

In calculating DFTs, one must be careful in choosing the sampling time, i.e. the time elapsed between successive t_k's. A common error introduced in digital signal analysis caused by improper sampling time is called *aliasing*. Aliasing results from A/D conversion and refers to the misrepresentation of the analog signal by the

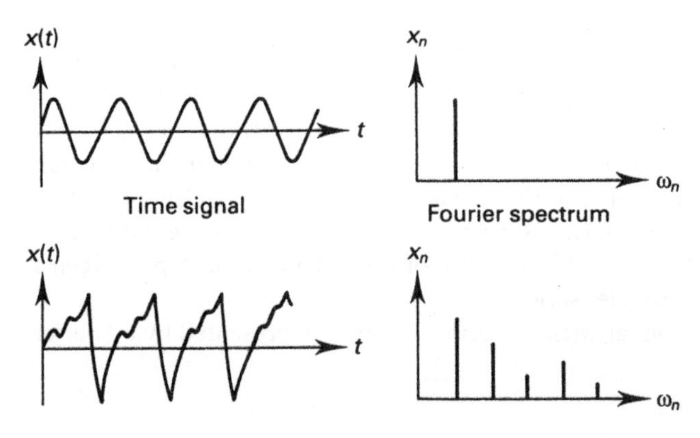

Figure 12.3 Some signals and their Fourier spectrum.

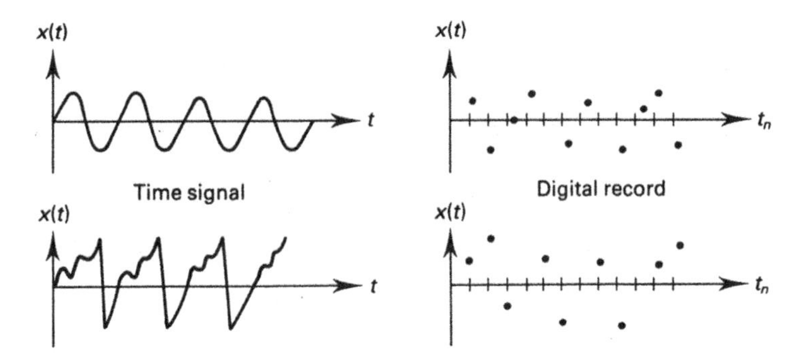

Figure 12.4 Sample time histories and their corresponding digital record.

digital record. Basically, if the sampling rate is too slow to catch the details of the analog signal, the digital representation will cause high frequencies to appear as low frequencies. The following example illustrates two analog signals of different frequency that produce the same digital record.

Example 12.3.1

Consider the signals $x_1(t) = \sin[(\pi/4)t]$ and $x_2(t) = -\sin[(7\pi/4)t]$, and suppose these signals are both sampled at 1-second intervals. The digital record of each signal is given in the following table.

t_k	0	1	2	3	4	5	6	7	ω_i
x_1	0	0.707	1	0.707	0	−0.707	−1	−0.707	$\frac{1}{8}$
x_2	0	0.707	1	0.707	0	−0.707	−1	−0.707	$\frac{7}{8}$

As is easily seen from the table, the digital sample records of x_1 and x_2 are the same, i.e. $x_1(t_k) = x_2(t_k)$ for each value of k. Hence, no matter what analysis is performed on the digital record, x_1 and x_2 will appear the same. Here the sampling frequency, $\Delta\omega$, is 1. Note that $\omega_1 - \Delta\omega = 1/8 - 1 = -7/8$, which is the frequency of $x_2(t)$.

To avoid aliasing, the sampling interval, denoted by Δt, must be chosen small enough to provide at least two samples per cycle of the highest frequency to be calculated. That is, in order to recover a signal from its digital samples, the signal must be sampled at a rate at least twice the highest frequency in the signal. In fact, experience (Otnes and Enachson, 1972) indicates that 2.5 samples per cycle is a better choice. This is called the *sampling theorem*, or *Shannon's sampling theorem*.

Aliasing can be avoided in signals containing many frequencies by subjecting the analog signal $x(t)$ to an *anti-aliasing* filter. An anti-aliasing filter is a low-pass (i.e. only allows low frequencies through) sharp cutoff filter. The filter effectively cuts off frequencies higher than about half the maximum frequency of interest, denoted by ω_{max}, and also called the *Nyquist frequency*. Most digital analyzers provide built-in anti-aliasing filters.

Once the digital record of the signal is available, the discrete version of the Fourier transform is performed. This is accomplished by a DFT or series defined by

$$x_k = x(t_k) = \frac{a_0}{2} + \sum_{i=1}^{N/2} \left[a_i \cos \frac{2\pi i t_k}{T} + b_i \sin \frac{2\pi i t_k}{T} \right], k = 1, 2, \ldots, N, \quad (12.3)$$

where the *digital spectral coefficients* are given by

$$\begin{aligned} a_0 &= \frac{1}{N} \sum_{k=1}^{N} x_k \\ a_i &= \frac{1}{N} \sum_{k=1}^{N} x_k \cos \frac{2\pi i k}{N} \\ b_i &= \frac{1}{N} \sum_{k=1}^{N} x_k \sin \frac{2\pi i k}{N} \end{aligned} \quad (12.4)$$

The task of the analyzer is to solve Equations (12.3) given the digital record $x(t_k)$, also denoted by x_k, for the measured signals. The transform size or number of samples, N, is usually fixed for a given analyzer and is a power of 2.

Writing out Equations (12.3) for each of the N samples yields N linear equations in the N spectral coefficients $(a_1, \ldots, a_{N/2}, b_1, \ldots, b_{N/2})$. These equations can also be written in the form of matrix equations. Equations (12.3) in matrix form become

$$\mathbf{x} = C\mathbf{a} \quad (12.5)$$

where $\mathbf{x}$ is the vector of samples with elements x_k and $\mathbf{a}$ is the vector of spectral coefficients. The solution of Equations (12.5) for the spectral coefficients is then given simply by

$$\mathbf{a} = C^{-1}\mathbf{x} \quad (12.6)$$

The task of the analyzer is to compute C^{-1} and hence the coefficients $\mathbf{a}$. The most widely used method of computing the inverse of this matrix C is called the FFT developed by Cooley and Tukey (1965). Note that C used in this context is not to be confused with the C used to denote a damping matrix.

In order to make digital analysis feasible, the periodic signal must be sampled over a finite time (N must obviously be finite). This can give rise to another problem referred to as *leakage*. To make the signal finite, one could simply cut the signal at any integral multiple of its period. Unfortunately, there is no convenient way to do this for complicated signals containing many different frequencies. Hence, if no further steps are taken, the signal may be cut off mid-period. This causes erroneous frequencies to appear in the digital representation, because the DFT of the finite-length signal assumes that the signal is periodic within the sample record length. Thus, the actual frequency will "leak" into a number of fictitious frequencies. This is illustrated in Figure 12.5.

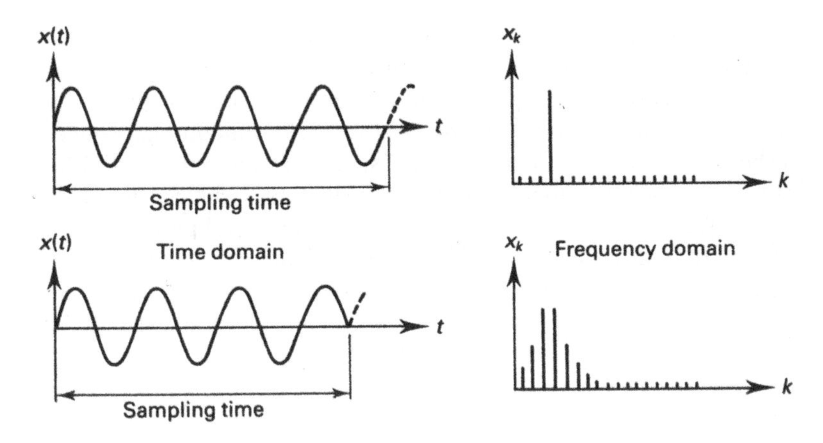

Figure 12.5 An illustration of leakage.

Leakage can be corrected to some degree by the use of a *window function*. Windowing, as it is called, involves multiplying the original analog signal by a weighting function, or window function, $w(t)$, which forces the signal to be zero outside the sampling period.

A common window function, called the *Hanning window*, is illustrated in Figure 12.6, along with the effect it has on a periodic signal. A properly windowed signal will yield a spectral plot with much less leakage. This is also illustrated in the figures.

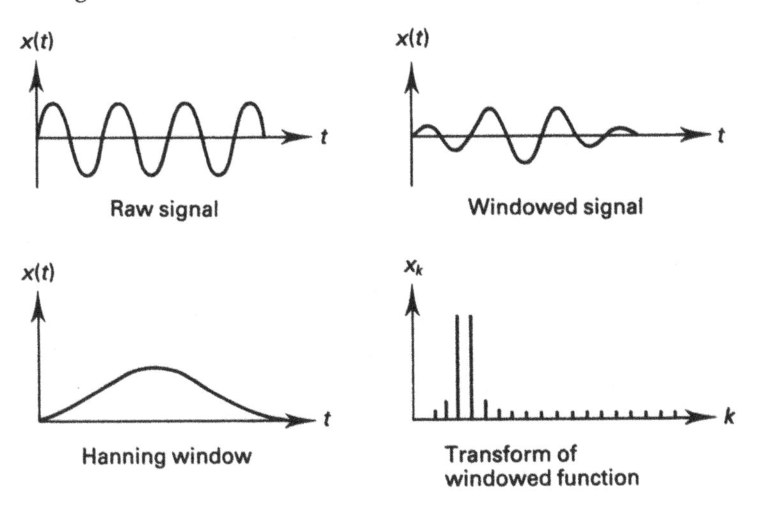

Figure 12.6 A Hanning window and its effect on transform signals.

12.4 Random Signal Analysis

This section introduces the topic of transfer function identification from signals, which is the most commonly used approach to system identification concerned with trying to determine the modal parameters of a linear time-invariant system.

In most test situations, it becomes desirable to average measurements to increase confidence in the measured parameter. This, along with the fact that a random excitation is often used in testing, requires some background in random processes.

First, some concepts and terminology required to describe random signals need to be established. Consider a random signal $x(t)$ such as pictured in Figure 12.7. The first distinction to be made about a random time history is whether or not the signal is stationary. A random signal is *stationary* if its statistical properties (i.e. its root mean square (RMS) value, defined later) are time-independent (i.e. do not change with time). For random signals, it is not possible to focus on the details of the signal, as it is with a pure deterministic signal. Hence, random signals are classified and manipulated in terms of their statistical properties.

Figure 12.7 A sample random signal.

The *average* of the random signal $x(t)$ is defined and denoted by

$$\bar{x}(t) = \lim_{T\to\infty} \frac{1}{T} \int_0^T x(t)dt \tag{12.7}$$

For a digital signal, the definition of average becomes

$$\bar{x} = \lim_{N\to\infty} \frac{1}{N} \sum_{k=1}^{N} x(t_k) \tag{12.8}$$

Often, it is convenient to consider signals with zero average, or *mean*, i.e. $\bar{x}(t) = 0$. Zero mean is not too restrictive an assumption, since if $\bar{x}(t) \neq 0$, a new variable $x' = x(t) - \bar{x}(t)$ can be defined. The new variable $x'(t)$ now has zero mean. In what follows, the signals are assumed to have zero mean.

The *mean square value* of the random variable $x(t)$ is denoted by $\bar{x}^2(t)$ and is defined by

$$\bar{x}^2 = \lim_{T\to\infty} \frac{1}{T} \int_0^T x^2(t)dt \tag{12.9}$$

or, in digital form

$$\bar{x}^2 = \lim_{N\to\infty} \frac{1}{N} \sum_{k=1}^{N} x^2(t_k) \tag{12.10}$$

The mean square value, or *variance* as it is often called, provides a measure of the magnitude of the fluctuations in the signal. A related quantity, called the *root mean square* value, or simply the *RMS* value, of $x(t)$, is just the square root of the variance, i.e.

$$x_{\text{RMS}} = \sqrt{\bar{x}^2} \tag{12.11}$$

Another question of interest for a random variable is that of calculating the probability that the variable $x(t)$ will lie in a given interval. However, neither the RMS value of $x(t)$ nor its probability yield information about how "fast" the values of $x(t)$ change and hence how long it takes to measure enough $x(t_k)$'s to compute statistically meaningful RMS values and probabilities. A measure of the speed with which the random variable $x(t)$ is changing is provided in the time domain by the *autocorrelation* function. For analog signals, the autocorrelation function, denoted by $R(t)$, defined by

$$R(t) = x^2 = \lim_{T\to\infty} \frac{1}{T} \int_0^T x(\tau)x(\tau + t)d\tau \tag{12.12}$$

is used as a measure of how fast the signal $x(t)$ is changing. Note that $R(0)$ is the mean square value of $x(t)$. The autocorrelation function is also useful for detecting the presence of periodic signals contained in, or buried in, random signals. If $x(t)$ is periodic or has a periodic component, then $R(t)$ will be periodic instead of approaching zero as $T \to \infty$, as required for a purely random $x(t)$.

The digital form of the autocorrelation function is

$$R(r, \Delta t) = \frac{1}{N - r} \sum_{k=0}^{N-r} x_k x_{k+r} \tag{12.13}$$

Here, N is the number of samples, Δt the sampling interval and r is an adjustable parameter, called a *lag value*, which controls the number of points used to calculate the autocorrelation function.

In the frequency domain, the *power spectral density* (often denoted PSD) is used to measure the speed with which the random variable $x(t)$ changes. The power spectral density, denoted by $S(\omega)$, is defined to be the Fourier transform of $R(t)$, i.e.

$$S(\omega) = \frac{1}{2\pi} \int_{-\infty}^{\infty} R(\tau)e^{-j\omega\tau} d\tau \tag{12.14}$$

The digital version of the power spectral density is given by

$$S(\Delta\omega) = \frac{|x(\omega)|^2}{N\Delta t} \tag{12.15}$$

where $|x(\omega)|^2$ is the magnitude of the Fourier transform of the sampled data corresponding to $x(t)$.

This definition of the autocorrelation function can also be applied to two different signals to provide a measure of the transfer function between the two signals. The *cross-correlation* function, denoted by $R_{xf}(t)$, for the two signals $x(t)$ and $f(t)$ is defined as

$$R_{xf}(t) = \lim_{T\to\infty} \frac{1}{T} \int_0^T x(\tau)f(\tau + t)d\tau \tag{12.16}$$

Likewise, the *cross-spectral density* is defined as the Fourier transform of the cross-correlation, i.e.

$$S_{xf}(\omega) = \frac{1}{2\pi}\int_{-\infty}^{\infty} R_{xf}(\tau)e^{-j\omega\tau}d\tau \tag{12.17}$$

If the function $f(\tau + t)$ is replaced with $x(\tau + t)$ in Equation (12.16), the power spectral density Equation (12.14), also denoted by S_{xx}, results. These correlation and density functions allow calculation of the transfer functions of test structures. The frequency response function (Section 1.5), $G(j\omega)$, can be shown (Ewins, 2000) to be related to the spectral density functions by the two equations

$$S_{fx}(\omega) = G(j\omega)S_{ff}(\omega) \tag{12.18}$$

and

$$S_{xx}(\omega) = G(j\omega)S_{xf}(\omega) \tag{12.19}$$

These hold if the structure is excited by a random input f resulting in the response x. Here S_{ff} denotes the power spectral density for the function $f(t)$, which is taken to be the excitation force in a vibration test.

The spectrum analyzer calculates (or estimates) the various spectral density functions. Then, using Equation (12.18) or Equation (12.19), the analyzer can calculate the desired frequency response function $G(j\omega)$. Note that Equations (12.18) and (12.19) use different power spectral densities to calculate the same quantity. This can be used to check the value of $G(j\omega)$. The *coherence function*, denoted by γ^2, is defined to be the ratio of the two values of $G(j\omega)$ calculated from Equations (12.18) and (12.19). Let $G'(j\omega) = S_{fx}(\omega)/S_{ff}(\omega)$ denote the value of G obtained from Equation (12.18) from measurements of $S_{fx}(\omega)$ and $S_{ff}(\omega)$. Likewise, let $G''(j\omega) = S_{xx}(\omega)/S_{fx}(\omega)$ denote the frequency response as determined by measurements made for Equation (12.19). Then the coherence function becomes

$$\gamma^2 = \frac{G'(j\omega)}{G''(j\omega)} \tag{12.20}$$

which is always less than or equal to 1. In fact, if the measurements are consistent, $G(j\omega)$ should be the same value, independent of how it is calculated and the coherence should be $1(\gamma^2 = 1)$. In practice, coherence versus frequency is plotted (Figure 12.8) and is taken as an indication of how accurate the measurement process is over a given range of frequencies. Generally, the values of $\gamma^2 = 1$ should occur at values of ω near the structure's resonant frequencies.

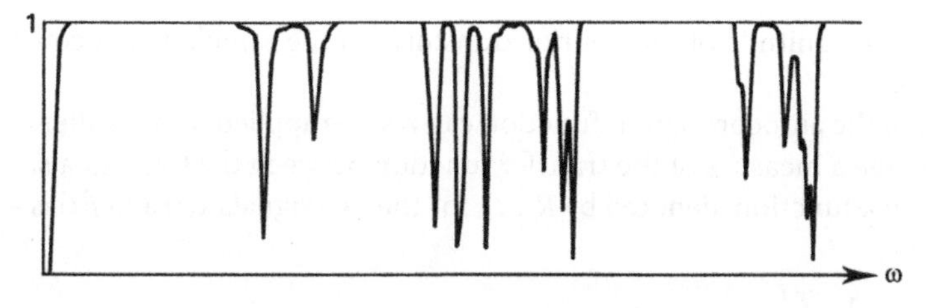

Figure 12.8 A plot of the coherence of a sample signal versus frequency.

12.5 Modal Data Extraction (Frequency Domain)

Once the frequency response of a test structure is calculated from Equation (12.18) or Equation (12.19), the analyzer is used to construct various vibrational information from the processed measurements. This is what is referred to as EMA. In what follows, it is assumed that the frequency response function $G(j\omega)$ has been measured via Equation (12.18) or Equation (12.19), or their equivalents.

The task of interest is to measure the natural frequencies, damping ratios and modal amplitudes associated with each resonant peak of the frequency response function. There are several ways to examine the measured frequency response function to extract the desired modal data. To examine all of them is beyond the scope of this text. To illustrate the basic method, consider the somewhat idealized frequency response function record of Figure 12.9, resulting from measurements taken between two points on a simple structure.

One of the gray areas in modal testing is deciding on the number of degrees of freedom to assign to a test structure. In many cases, simply counting the number of clearly defined peaks or resonances, three in Figure 12.9, determines the order, and the procedure continues with a three-mode model. However, as is illustrated later, this procedure may not be accurate if the structure has closely spaced natural frequencies or high damping values.

The easiest method to use on this data is the so-called single-degree-of-freedom (SDOF) curve fit approach alluded to in Section 2.6. In this approach, the frequency response function is sectioned off into frequency ranges bracketing each successive peak. Each peak is then analyzed by assuming it is the frequency response of an SDOF system. This assumes that in the vicinity of the resonance, the frequency response function is dominated by that single mode.

In other words, in the frequency range around the first resonant peak, it is assumed that the curve is due to the response of a damped SDOF system due to a harmonic input at and near the first natural frequency. Recall from Section 1.4 that the point of resonance corresponds to that value of frequency for which the magnification curve has its maximum or peak value and the phase changes by 180°.

The method is basically the peak-picking method referred to in Section 1.7 and illustrated in Figure 1.18. The approach is to assume that the peak occurs at the

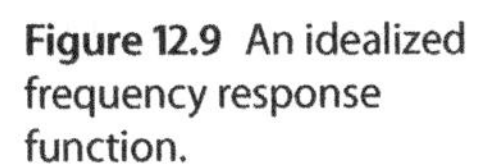
Figure 12.9 An idealized frequency response function.

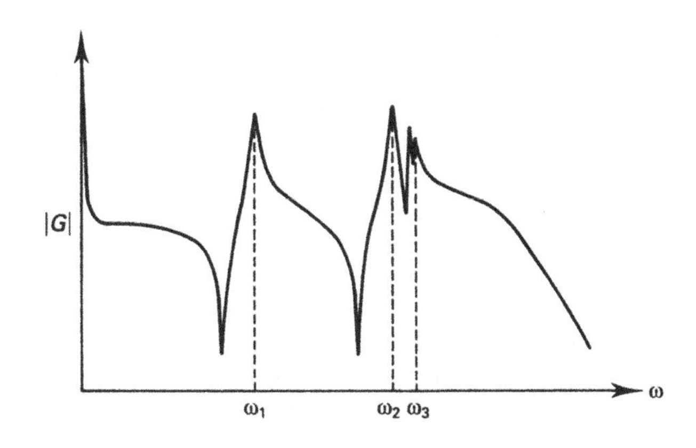

frequency corresponding to the damped natural frequency of that particular mode, denoted by ω_{di} (for the ith mode). The two frequencies on each side of this peak correspond to the points on the curve that are 0.707 of $|G(j\omega_{di})|$ (also called the half-power points). Denoting these two frequencies as ω_{1i} and ω_{2i}, the formula for calculating the damping ratio for the ith peak is (Blevins, 1994: p. 318)

$$\zeta_i = \frac{\omega_{2i} - \omega_{1i}}{2\omega_{di}} \tag{12.21}$$

Next, even though these formulas result from examining an SDOF model, it is recognized that the system is a distributed mass system that is being modeled as a multiple-degree-of-freedom (MDOF) system with three modes. The frequency response function for an MDOF system is discussed in Section 5.7 and leads to the concept of a receptance matrix, $\alpha(\omega)$, which relates the input vector (driving force) to the response (position in this case) via Equation (5.61). For a three-degree-of-freedom system, with the assumption of proportional damping and that the frequency response near resonance ω_{d1} is not influenced by contributions from ω_{d2} and ω_{d3}, the magnitude of the measured frequency response at ω_{d1}, denoted by $G(j\omega_{d1})$, allows the calculation of one of the elements of the receptance matrix.

If the structure is underdamped, the ik^{th} element of the receptance matrix is given by

$$\alpha_{ik} = \sum_{r=1}^{n} \frac{[\mathbf{s}_r \mathbf{s}_r^T]_{ik}}{(\omega_r^2 + 2j\zeta_r\omega_r\omega - \omega^2)} \tag{12.22}$$

where $\mathbf{s}_r$ if the rth eigenvector of the system. The magnitude of $\alpha_{ij}(\omega_r)$ is the measured amplitude of the peak at ω_r. In this case, Equation (12.22) becomes

$$\alpha_{ik}(\omega_r) = \frac{[\mathbf{s}_r \mathbf{s}_r^T]_{ik}}{\omega_r^2 + 2\zeta_r\omega_r^2 j - \omega_r^2} \tag{12.23}$$

Hence

$$\left|[\mathbf{s}_r \mathbf{s}_r^T]_{ik}\right| = \left|G_{ik}(j\omega_r)\right| \left|2\zeta_r\omega_r^2 j\right| = 2\zeta_r\omega_r^2 \left|G_{ik}(j\omega_r)\right| \tag{12.24}$$

Here the measured value of the maximum of the frequency response function at $\omega_r = \omega$ with input at point i and response measured at point k is denoted by $G_{ik}(\omega)$ and is approximated by $|\alpha_{ik}(\omega)|$. With the assumption, stated earlier that the response is due to only one frequency near resonance, Equation (12.24) yields one value for the *modal constant*, which is defined as the magnitude of the ik^{th} element of the matrix $\mathbf{s}_1\mathbf{s}_1^T$. The phase plot yields the relative sign of the ik^{th} element of $\mathbf{s}_1\mathbf{s}_1^T$. Equation (12.24) is the mathematical equivalent of assuming that the first peak in the curve of Figure 12.9 results from only an SDOF system.

The subscripts ik denote the output coordinate and the relative position of the input force. In order words, the quantity $|\alpha_{ik}(\omega_r)|$ represents the magnitude of the transfer function between an input at point i and a measured output at point k.

In this example, an estimate of the eigenvectors, or mode shapes, can be calculated by making a series of measurements at different points, applying the modal constant formula, Equation (12.24), and examining the relative phase shift at ω_{di}. To see this, suppose that the transfer function of Figure 12.9 is between the input f_1

Figure 12.10 Measurement positions marked on a simple plate for performing a modal test.

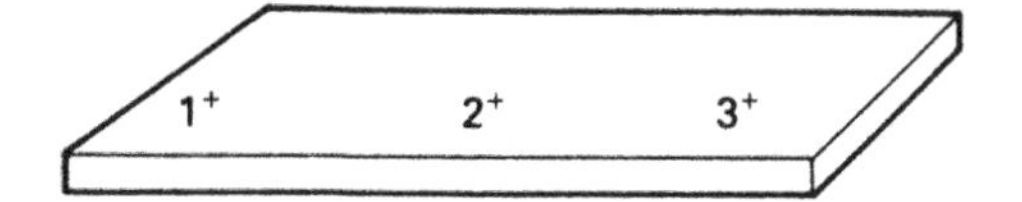

(at point 1) and the measured response x_3 (at point 3 on the structure, as labeled in Figure 12.10). Then, from Equation (12.24) with $i = 1$, $k = 3$ and $r = 1, 2, 3$, the three values of the matrix elements $\left[\mathbf{s}_1\mathbf{s}_1^T\right]_{13}$, $\left[\mathbf{s}_2\mathbf{s}_2^T\right]_{13}$ and $\left[\mathbf{s}_3\mathbf{s}_3^T\right]_{13}$ can be measured. Furthermore, if the preceding experiment is completed two more times with the response measurement (or the force) being moved to the two remaining positions ($k = 2,1$) shown in Figure 12.10, the following matrix elements can also be measured: $\left[\mathbf{s}_1\mathbf{s}_1^T\right]_{12}$, $\left[\mathbf{s}_2\mathbf{s}_2^T\right]_{12}$ and $\left[\mathbf{s}_3\mathbf{s}_3^T\right]_{12}$ as well as $\left[\mathbf{s}_1\mathbf{s}_1^T\right]_{11}$, $\left[\mathbf{s}_2\mathbf{s}_2^T\right]_{11}$ and $\left[\mathbf{s}_3\mathbf{s}_3^T\right]_{11}$. Gathering up these nine elements of the three different outer product matrices $\mathbf{s}_1\mathbf{s}_1^T$, $\mathbf{s}_2\mathbf{s}_2^T$ and $\mathbf{s}_3\mathbf{s}_3^T$, allows the three vectors $\mathbf{s}_1$, $\mathbf{s}_2$ and $\mathbf{s}_3$ to be calculated (Problem 12.5). Note, that because of its special structure, knowing one row or one column of the matrix $\mathbf{s}_i\mathbf{s}_i^T$ is enough to determine the entire matrix and hence the value of $\mathbf{s}_i$ to within a multiplicative constant.

Hence, through measurements taken from the plot in Figure 12.9, along with the above computations and two more measurements at the remaining two measurement points, the receptance matrix, given by Equation (5.71), is completely determined, as are the eigenvalues and eigenvectors of the system. However, in general this does *not* provide enough information to measure the values of the matrices M, D and K, because the eigenvectors may not be scaled (normalized) properly. Also, the mathematical model constructed in this way is dependent upon the choice of measurement points and as such should not be considered to be unique.

Given the modal data, however, the receptance matrix can be used to predict the response of the system as the result of any input, to compare the modal data to theoretical models or, along with other information, to construct physical models of the structure. The question of uniqueness remains, and the physical models developed should always be referenced to the measurement and disturbance position used in performing the test.

The SDOF method is really not based on a principle of parameter identification but rather is an unsophisticated first attempt to "read" the experimental data and to fit the data into a specific underdamped vibration model. An attempt to make the SDOF approach more sophisticated is to plot the real part of the frequency response function versus the imaginary part near the resonant frequency. This yields the Nyquist plot of Figure 1.14. In this domain, the theory predicts a circle; hence, the experimental data around a resonant peak can be "fit" using a least squares circle fit to produce the "best" circle for the given data. Then the frequency and damping values can be taken from the Nyquist plot. This method is referred to as the *circle fit method* and brings the problem a little closer to using the science of parameter estimation (see Ewins (2000) or Zaveri (1984) for a more complete account).

The SDOF method is in obvious error for systems with closely spaced modes and/or highly coupled modes. Hence several other methods have been developed to take into consideration the effects of coupling. These methods are usually referred to as MDOF curve fits and consist largely of adding correction terms to Equation (12.23) to take into consideration mode coupling effects. Because of space limitations, these methods are not discussed here. The reader is referred to the note section at the end of the chapter for references to more advanced methods.

12.6 Modal Data Extraction (Time Domain)

An alternative approach to extracting a structure's modal data from the vibrational response of the structure is based in the time domain. Time domain methods have been more successful in identifying closely spaced or repeated natural frequencies. They also offer a more systematic way of determining the appropriate order (number degrees of freedom) of a test structure and generally identify a larger number of natural frequencies than frequency domain methods do. Time domain methods are also referred to as damped complex exponential response methods (Allemang, 1984, 2013).

The first time domain method to make an impact on the vibration testing community was developed by Ibrahim (1973) and has since become known as the Ibrahim time domain method. The original method is based on the state equations of a dynamic system, Equation (2.27). A simplified version of the method is presented here.

The solution for the dynamic system in physical coordinates in the form of Equation (2.19) is given in Equation (3.68) and repeated here as

$$\mathbf{q}(t) = \sum_{r=1}^{2n} c_r \mathbf{r}_r e^{\lambda_r t} \tag{12.25}$$

where λ_r are the complex latent roots, or eigenvalues, of the system and the $\mathbf{u}_r$ are the system eigenvectors. Here, it is convenient to absorb the constant c_r into the vector $\mathbf{u}_r$ and write Equation (12.25) as

$$\mathbf{q}(t) = \sum_{r=1}^{2n} \mathbf{p}_r e^{\lambda_r t} \tag{12.26}$$

where the vector $\mathbf{p}_r$ has an arbitrary norm. If the response is measured at discrete times, t_i, then Equation (12.26) becomes simply

$$\mathbf{q}(t_i) = \sum_{r=1}^{2n} \mathbf{p}_r e^{\lambda_r t_i} \tag{12.27}$$

For simplicity of explanation, assume that the structure is measured in n places (this assumption can be relaxed) at $2n$ times, where n is the number of degrees of freedom exhibited by the test structure. Writing Equation (12.27) $2n$ times results in the matrix equation

$$X = PE(t_i) \tag{12.28}$$

where X is an $n \times 2n$ matrix with columns consisting of the $2n$ vectors $\mathbf{q}(t_1)$, $\mathbf{q}(t_2)$, …, $\mathbf{q}(t_{2n})$, P is the $n \times 2n$ matrix with columns consisting of the $2n$ vectors $\mathbf{p}_r$, and $E(t_i)$ is the $2n \times 2n$ matrix given by

$$E(t_i) = \begin{bmatrix} e^{\lambda_1 t_1} & e^{\lambda_1 t_2} & \cdots & e^{\lambda_1 t_{2n}} \\ e^{\lambda_2 t_1} & e^{\lambda_2 t_2} & & e^{\lambda_2 t_{2n}} \\ \vdots & \vdots & & \vdots \\ e^{\lambda_{2n} t_1} & e^{\lambda_{2n} t_2} & \cdots & e^{\lambda_{2n} t_{2n}} \end{bmatrix} \tag{12.29}$$

Likewise, if these same responses are measured Δt seconds later, i.e. at time $t_i + \Delta t$, Equation (12.28) becomes

$$Y = PE(t_i + \Delta t) \tag{12.30}$$

where the columns of Y are defined by $\mathbf{y}(t_i) = \mathbf{q}(t_i + \Delta t)$, and $E(t_i + \Delta t)$ is the matrix with ij^{th} element equal to $e^{\lambda_i(t_j+\Delta t)}$. Equation (12.30) can be factored into the form

$$Y = P'E(t_i) \tag{12.31}$$

where the matrix P' has as its ith column the vector $\mathbf{p}_i e^{\lambda_i \Delta t}$. This process can be repeated for another time increment Δt later to provide the equation

$$Z = P''E(t_i) \tag{12.32}$$

where the columns of Z are the vectors $\mathbf{q}(t_i + 2\Delta t)$, etc.

Collecting Equations (12.28) and (12.31) together yields

$$\Phi = \begin{bmatrix} X \\ Y \end{bmatrix} = \begin{bmatrix} P \\ P' \end{bmatrix} E(t_i) = \Psi E(t_i) \tag{12.33}$$

where the $2n \times 2n$ matrices Φ and Ψ have obvious definitions. Likewise, Equations (12.31) and (12.32) can be combined to form

$$\Phi' = \begin{bmatrix} Y \\ Z \end{bmatrix} = \begin{bmatrix} P' \\ P'' \end{bmatrix} E(t_i) = \Psi' E(t_i) \tag{12.34}$$

where $\Phi' = [Y^T\ Z^T]^T$ and $\Psi' = [P'^T\ P''^T]^T$. Note that the matrices Φ, Φ', Ψ and Ψ' are all square $2n \times 2n$ matrices that are assumed to be nonsingular (Ibrahim and Mikulcik, 1976).

Equations (12.33) and (12.34) can be used to calculate a relationship between the vectors that make up the columns of the matrix Ψ and those of Ψ', namely

$$\Phi'\Phi^{-1}\Psi = \Psi' \tag{12.35}$$

so that

$$\Phi'\Phi^{-1}\Psi_i = \Psi'_i, i = 1, 2, \ldots, 2n \tag{12.36}$$

where the subscript i denotes the ith column of the indexed matrix. However, the ith column of P' is $\mathbf{p}_i e^{\lambda_i \Delta t}$, so that

$$\Psi'_i = e^{\lambda_i \Delta t}\Psi_i \tag{12.37}$$

Comparison of Equations (12.37) and (12.36) yields

$$\Phi'\Phi^{-1}\Psi_i = e^{\lambda_i \Delta t}\Psi_i \tag{12.38}$$

Note that this last expression states that the complex scalar $e^{\lambda_i \Delta t} = \beta + \gamma j$ is an eigenvalue of the matrix $\Phi'\Phi^{-1}$ with complex eigenvector Ψ_i. The $2n \times 1$ eigenvector Ψ_i has as its first n elements the eigenvector or mode shape of the structure. That is, $\Psi_i = [\mathbf{p}_i^T \quad \mathbf{p}_i^T e^{\lambda_i \Delta t}]^T$ where $\mathbf{p}_i$ are the system eigenvectors.

The eigenvalues of Equation (12.38) can be used to calculate the eigenvalues of the system and hence the damping ratios and natural frequencies. In particular, since $e^{\lambda_i \Delta t} = \beta + \gamma j$, it follows that

$$\text{Re}\lambda_i = \frac{1}{(\Delta t)} \ln(\gamma^2 + \beta^2) \tag{12.39}$$

and

$$\text{Im}\lambda_i = \frac{1}{(\Delta t)} \tan^{-1}\left(\frac{\gamma}{\beta}\right) \tag{12.40}$$

With proper consideration for sampling time and the interval for the arc tangent in Equation (12.40), Equation (12.38) yields the desired modal data.

To summarize, this method first constructs a matrix of measured time responses, forms the product $\Phi'\Phi^{-1}$, and then numerically calculates the eigenvalue-eigenvector problem for this matrix. The resulting calculation then yields the mode shapes, natural frequencies and damping ratios of the structure. The natural frequencies and damping ratios follow from Equations (12.39) and (12.40) since (Chapter 1)

$$\omega_i = [(\text{Re}\lambda_i)^2 + (\text{Im}\lambda_i)^2]^{1/2} \tag{12.41}$$

and

$$\zeta_i = -\frac{(\text{Re}\lambda_i)}{\omega_i} \tag{12.42}$$

Note that the key difference between the time domain approach to modal analysis and the frequency domain approach is that the time domain approach constructs a matrix from the time response and numerically computes the modal data by solving an eigenvalue problem. On the other hand, the frequency domain method extracts and curve-fits the modal data from the DFT of the measured input and response data.

Two other major approaches in the time domain have been developed. One, called the polyreference method (Vold and Rocklin, 1982) again uses a state space approach and has been developed commercially. The other time domain method is based on realization theory developed by control theorists, as introduced in Section 6.11. This method, advocated by Juang and Pappa (1985), also uses a state space approach and is called the *eigensystem realization algorithm* (ERA). The ERA method introduces several important aspects to modal test analysis and hence is discussed next. The method is based on the realization theory introduced in Section 6.11, and puts modal testing on the firm theoretical foundation of linear systems theory (Juang, 1994).

Unlike the other time domain approaches, the ERA determines a complete state space model of the structure under test. In particular, using the notation of Equations (6.30) and (6.31), the ERA identifies all three of the matrices A, B and C_c. This is an important aspect because it addresses the non-uniqueness of modal analysis by specifying the measurement locations through the matrices B and C_c. In addition, the ERA approach firmly attaches the modal testing problem to the more mature discipline of realization used in control theory (Section 6.11). This approach is computationally more efficient and provides a systematic means of determining the order of the test model. The method is based on a discrete time version of the state space solution given by Equation (5.16) with $f(\tau) = Bu(\tau)$. The solution of Equation (6.30), starting at time t_0, can be written as

$$\mathbf{x}(t) = e^{A(t-t_0)}\mathbf{x}(t_0) + \int_{t_0}^{t} e^{A(t-\tau)}B\mathbf{u}(\tau)d\tau \tag{12.43}$$

This expression can be written at discrete time intervals equally spaced at times 0, Δt, $2\Delta t$, ..., $k\Delta t$,..., by making the substitutions $t = (k+1)\Delta t$ and $t_0 = k\Delta t$. This yields

$$\mathbf{x}((k+1)\Delta t) = e^{A\Delta t}\mathbf{x}(k\Delta t) + \int_{k\Delta t}^{(k+1)\Delta t} e^{A((k+1)\Delta t-\tau)}B\mathbf{u}(\tau)d\tau \tag{12.44}$$

Note that the integration period is now performed over a small increment of time, Δt. If $\mathbf{u}(t)$ is assumed to be constant over the interval Δt and to have the value $\mathbf{u}(k\Delta t)$, this last expression can be simplified to

$$\mathbf{x}(k+1) = e^{A\Delta t}\mathbf{x}(k) + \left[\int_{0}^{\Delta t} e^{A\sigma}d\sigma B\right]\mathbf{u}(k) \tag{12.45}$$

Here, $\mathbf{x}(k+1)$ is used to denote $\mathbf{x}((k+1)\Delta t)$ etc., and the integration variable has been changed to $\sigma = (k+1)\Delta t - \tau$.

Equation (12.45) can be further reduced in form by defining two new matrices A' and B' by

$$A' = e^{A\Delta t}, B' = \int_{0}^{\Delta t} e^{A\sigma}d\sigma B \tag{12.46}$$

Thus, a discrete time version of Equations (6.30) and (6.31) becomes

$$\begin{aligned} \mathbf{x}(k+1) &= A'\mathbf{x}(k) + B'\mathbf{u} \\ \mathbf{y}(k) &= C_c\mathbf{x}(k) \end{aligned} \tag{12.47}$$

The last expression allows the measurement taken at discrete times, $\mathbf{y}(k)$, to be related to the state matrix by the definition of the matrices $A'$and B'.

Consider a system subject to an impulse, $\mathbf{y}(1) = C_c B'_{\mathbf{u}}(1)$, where $\mathbf{u}(1)$ is a vector of zeros and ones, depending upon where the impulse is applied. If the impulse is applied at m different points one at a time, all the response vectors at $k = 1$, $\mathbf{y}(k)$, corresponding to each of these m impulses, can be gathered together into a single $s \times m$ matrix Y, given by

$$Y(1) = C_c B' \tag{12.48}$$

In a similar fashion, it can be shown using Equation (12.47) that for any measurement time k

$$Y(k) = C_c A k^{-1} B' \tag{12.49}$$

The matrices $Y(k)$ defined in this manner are called *Markov parameters.*

The Markov parameters for the test structure can next be used to define a larger matrix of measurements, denoted by $H_{ij}(k)$, for any arbitrary positive integers i and j by

$$H_{ij}(k) = \begin{bmatrix} Y(k) & Y(k+1) & \cdots & Y(k+j) \\ Y(k+1) & Y(k+2) & & \\ \vdots & \vdots & & \vdots \\ Y(k+i) & Y(k+u+1) & & Y(k+i+j) \end{bmatrix} \tag{12.50}$$

This $(i + 1) \times (j + 1)$ block matrix is called a *generalized Hankel matrix.* If the arbitrarily chosen integers $i + 1$ and $j + 1$ are both greater than $2n$, the rank of $H_{ij}(k)$ will be $2n$. This follows from realizing that the Hankel matrix can be factored and written as

$$H_{ij}(k) = O_i A^{k-1} R_j \tag{12.51}$$

where O_i is the observability matrix and R_j is the controllability matrix given by

$$O_i = \begin{bmatrix} C_c \\ C_c A' \\ \vdots \\ C_c (A')^i \end{bmatrix}, R_j = [B' A' B' \cdots (A')^j B'] \tag{12.52}$$

introduced in Section 6.7. Note that the rank of $H_{ij}(k)$ is the order of the matrix A' and, hence, the order of the system state matrix A, as long as i and j are large enough. Equation (12.52) also points out that the smallest-order realization of A will have order equal to the rank of O_i and R_j for a controllable and observable system. This results from the balanced model reduction method presented in Section 6.12.

The ERA is based on a singular-value decomposition (Section 6.12) of the Hankel matrix $H(0)$. Using Equation (6.89), the matrix $H(0)$ can be written as

$$H(0) = U \sum V^T \tag{12.53}$$

where $\sum$ is the diagonal matrix of singular values, σ_i, of $H(0)$ and U and V are defined in Equations (6.89) to (6.92). If enough measurements are taken, so that i and j are large enough, there will be an index r such that $\sigma_r \gg \sigma_r + 1$. Just as in the model reduction problem of Section 6.12, this value of r yields a logical index for defining the order of the structure under test, i.e. $r = 2n$.

Partitioning U and V^T into the first r columns and rows, respectively, and denoting the reduced matrices by U_r and V_r, yields

$$H(0) = U_r \Sigma_r V_r^T \tag{12.54}$$

Here, Σ_r denotes the $r \times r$ diagonal matrix of the first r singular values of $H(0)$. Juang and Pappa (1985) showed that

$$A' = \Sigma_r^{-1/2} U_r^T H(1) V_r \Sigma_r^{-1/2} \tag{12.55}$$

$$B' = \Sigma_r^{1/2} V_r E_p \tag{12.56}$$

$$C_c = E_p^T U_r \Sigma_r^{1/2} \tag{12.57}$$

where E_p is defined as the $p \times r$ matrix

$$E_p^T = [I_p \quad 0_p \cdots 0_p] \tag{12.58}$$

Here I_p is a $p \times p$ identity matrix and 0_p is a $p \times p$ matrix of zeros.

The modal parameters of the test structure are calculated from the numerical eigensolution of the matrix A'. The mode shapes of the structure are taken from the eigenvectors of A' and the natural frequencies and damping ratios are taken from the eigenvalues of A. The eigenvalues of A, the continuous time state matrix, denoted by λ_i, are found from the eigenvalues of A' via the formula (Juang and Pappa, 1985)

$$\lambda_i = \frac{\ln \lambda_i(A')}{\Delta t} \tag{12.59}$$

For underdamped structures, the modal parameters are determined from Equations (12.41) and (12.42).

12.7 Model Identification

A major reason for performing a modal test is for validation and verification of analytical models of the test structure. What is often desired is a mathematical model of the structure under consideration for the purpose of predicting how the structure will behave under a variety of different loadings, to provide the plant in a control design, or to aid in the design process in general. This section discusses three types of models of the structure that result from using modal test data. The three types of models considered are the *modal model*, discussed in the previous section, the *response model* and the *physical model* (also called the *spatial model*).

The modal model is simply the natural frequencies, damping ratios and mode shapes, as given in the previous two sections. The modal model is useful in several ways. First, it can be used to generate both static and dynamic displays of the mode shapes, lending visual insight into the manner in which the structure vibrates. These displays are much more useful to examine in many cases, than reading a list of numbers. An example of such a display is given in Figure 12.11.

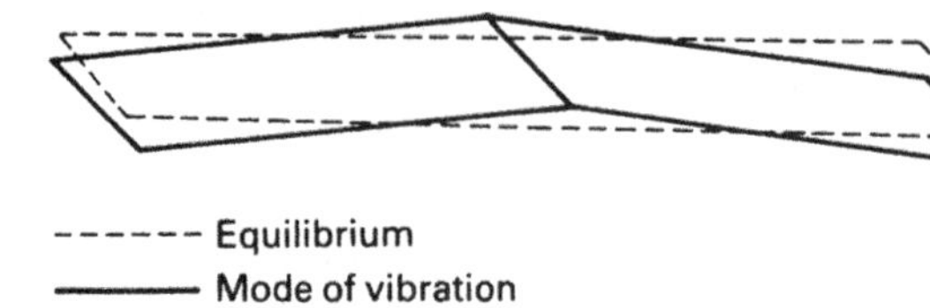

Figure 12.11 A display of the mode shape of a simple test structure.

A second use of modal models is for direct comparison with the predicted frequencies and modal data for an analytical model. Frequently, the analytical model will not predict the measured frequencies. As a result, the analytical model is changed, iteratively, until it produces the measured natural frequencies. The modified analytical model is then considered an improvement over the previous model. This procedure is referred to as model updating and is introduced in the following section.

The frequency response model of the system is given by the receptance matrix, defined in Section 5.7 and used in Section 12.5, to extract the modal parameters from the measured frequency response functions. From Equations (5.70) and (5.67), the measured receptance matrix can be used to calculate the response of the test structure data to any input $\mathbf{f}(t)$. To see this, consider taking the Laplace transform, denoted £[·], of the dynamic Equation (2.19), i.e.

$$(s^2M + sC + K)\pounds[\mathbf{q}] = \pounds[\mathbf{f}] \tag{12.60}$$

so that

$$\pounds[\mathbf{q}(t)] = \frac{1}{s^2M + sC + K}\pounds[\mathbf{f}] \tag{12.61}$$

Letting $s = j\omega$ in Equation (12.61) yields

$$\Im(\mathbf{q}) = \alpha(\omega)\Im(\mathbf{f}) \tag{12.62}$$

where it is noted that the Laplace transform of a function evaluated at $s = j\omega$ yields the exponential Fourier transform, denoted by $\Im[\bullet]$. Hence, Equation (12.62) predicts the response of the structure for any input $\mathbf{f}(t)$.

In some instances, such as control or design applications, it would be productive to have a physical model of the structure in spatial coordinates, i.e. a model of the form

$$M\ddot{\mathbf{q}} + C\dot{\mathbf{q}} + K\mathbf{q} = \mathbf{0} \tag{12.63}$$

where $\mathbf{q}(t)$ denotes a vector of physical positions on the structure. The obvious way to construct this type of model from measured data is to use the orthogonal mass normalized matrix of eigenvectors, S_m, defined in Section 3.5 and to "undo" the theoretical modal analysis by using Equations (3.70) and (3.71). That is, if these equations are pre- and post-multiplied by the inverses of S_m^T and S_m, they yield

$$\begin{aligned} M &= (S_m^T)^{-1}S_m^{-1} \\ C &= (S_m^T)^{-1}\text{diag}[2\zeta_i\omega_i]S_m^{-1} \\ K &= (S_m^T)^{-1}\text{diag}[\omega_i^2]S_m^{-1} \end{aligned} \tag{12.64}$$

where the ω_i, ζ_i and the columns of S_m are the measured modal data. Unfortunately, this formulation requires that the damping mechanism is proportional to velocity and that the measured data for all n modes are available. These two assumptions are very seldom met. Usually, the modal data are too incomplete to form all S_m. In addition, the measured eigenvectors are seldom scaled properly.

The problem of calculating the mass, damping and stiffness matrices from experimental data is very difficult and does not result in a unique solution. It is after all an inverse problem (Lancaster and Maroulas, 1987; Starek and Inman, 1997). Research continues in this area by examining reduced-order models and approaching the problem as one of improving existing analytical models, the topic of the next section.

12.8 Model Updating

Analytical models are only useful if they are verified against an experimental model. Often analysis gets fairly close to predicting experimental data and an approach called *model updating* is used to slightly change or update the analytical model to predict or agree with experimental data. Friswell and Mottershead (1995) present both an introduction to model updating and a good summary of most techniques developed up to that date. One approach to model updating is to use the analytical model of a structure to compute natural frequencies, damping ratios and mode shapes. Then test, or perform an EMA on the structure and examine how well the analytical modal data predicts the measured modal data. If there is some difference, then the analytical model is adjusted or updated until the updated analytical model predicts the measured frequencies and mode shapes. The procedure is illustrated by the following simple example.

Example 12.8.1

Suppose the analytical model of a structure is derived to be the following two-degree of freedom model (Example 3.3.2)

$$I\ddot{\mathbf{q}}(t) + \begin{bmatrix} 3 & -1 \\ -1 & 1 \end{bmatrix} \mathbf{q}(t) = 0$$

This has natural frequencies computed to be

$$\omega_1 = \sqrt{2 - \sqrt{2}}, \text{ and } \omega_2 = \sqrt{2 + \sqrt{2}}$$

Suppose a modal test is performed and that the measured frequencies are $\omega_1 = \sqrt{2}$ and $\omega_2 = \sqrt{3}$. Find adjustments to the analytical model such that the new updated model predicts the measured frequencies.

Let ΔK denote the desired correction in the analytical model. To this end, consider the correction matrix

$$\Delta K = \begin{bmatrix} 0 & 0 \\ \Delta k_1 & \Delta k_2 \end{bmatrix} \tag{12.65}$$

The characteristic equation of the updated model is then

$$\lambda^2 - (4 + \Delta k_1)\lambda + (2 + 3\Delta k_2 + \Delta k_1) = 0 \tag{12.66}$$

The characteristic equation of the system with the experimental frequencies would be

$$\lambda^2 - 5\lambda + 6 = 0 \tag{12.67}$$

Comparing coefficients in Equations (12.66) and (12.67) yields the updated parameters

$$\Delta k_1 = 1 \text{ and } \Delta k_2 = 1$$

The new updated model is

$$I\ddot{\mathbf{q}}(t) + \begin{bmatrix} 3 & -1 \\ 0 & 2 \end{bmatrix} \mathbf{q}(t) = 0$$

which has frequencies $\omega_1 = \sqrt{2}$ and $\omega_2 = \sqrt{3}$, as measured.

While the updated model in Example 12.8.1 does in fact produce the correct frequencies, the stiffness matrix of the updated model is no longer symmetric, nor does it retain the original connectivity of, in this case, the two degree of freedom, spring mass system. Other methods exist which address these issues (Halevi et al., 2004). Model updating, as stated here, is related to pole placement and eigenstructure assignment methods used in Section 6.8. Instead of computing control gains, as done in Section 6.8, the same computation yields an updating procedure if the gain matrices are considered to be the correction matrices (e.g. ΔK). Basically, if measured mode shapes, damping ratios and frequencies are available, then an eigenstructure assignment method may be used to compute updated damping and stiffness matrices. If only damping ratios and frequencies are available, then pole placement can be used to update the damping and stiffness matrices. Friswell and Mottershead (1995) should be consulted for other methods. The literature should be consulted for latest approaches.

12.9 Verification and Validation

EMA results are often used to verify and/or validate a model (Allameng, 2013). Verification of a structural model refers to examining whether or not a given model is able to successfully describe the dynamic response of the structure, assuming all of its various physical parameters (i.e. M, C, K) are correct. This means in particular that there are no missing terms or features in the model. For example, suppose the real structure is nonlinear, yet the assumed model is nonlinear. Then this model could not be verified by any procedure. Correcting the parameters M, C and K in this case, will never lead to an agreement between theory and experiment.

Validation, on the other hand, refers to examining whether or not a given model's parameters are accurate enough to correctly describe the structures vibration response. In the case of a linear model, validation refers to knowing that the parameters in M, C and K are correct, producing an analytical response that agrees with experimental evidence. Clearly if a model is not verified, it cannot be validated. The following examples attempt to illustrate these concepts.

Example 12.9.1

Consider a modal test of a cantilever beam of parameters known from other tests (say static deflection, mass measurements and geometric values). The first four natural frequencies are measured as

$$f_1 = 10.000 \quad f_2 = 62.1875 \quad f_3 = 172.8125 \quad f_4 = 335.5626 \text{ Hz}$$

From the modeling of a cantilever beam given in Example 7.5.1, the frequencies are (Inman, 2014)

$$f_1 = 9.9105 \quad f_2 = 62.1082 \quad f_3 = 173.0946 \quad f_4 = 340.7837 \text{ Hz}$$

Comparing the analytical frequencies with the measured frequency shows that they are fairly close, thus we could say that the model must be both validated and verified.

Example 12.9.2

Consider the Timoshenko beam equation given by Equation (7.85) with pinned-pinned boundary conditions. Assuming a solution of the form

$$w_n(x,\ t) = \sin\frac{n\pi x}{l}\cos\omega_n t$$

compute the equation for the beam natural frequencies for the case of no rotary inertia effect, but just shear deformation and compare this to the same beam under Euler-Bernoulli conditions. Use these expressions to discuss the concept of validation.

Solution: Substitution of the assumed solution form into Equation (7.85) yields

$$EI\left(\frac{n\pi}{l}\right)^4 \sin\frac{n\pi x}{l}\cos\ \omega_n t - \rho I\left(1+\frac{E}{\kappa^2 G}\right)\left(\frac{n\pi}{l}\right)^2 \omega_n^2 \sin\frac{n\pi}{l}\cos\omega_n t$$

$$= -\frac{\rho^2 I}{\kappa^2 G}\omega_n^4 \sin\frac{n\pi x}{l}\cos\omega_n t + \rho A\omega_n^2 \sin\frac{n\pi x}{l}\cos\omega_n t$$

Each term contains the factor $\sin(n\pi x/l)\cos(\omega_n t)$, which can be factored out to reveal the characteristic equation

$$\omega_n^4\frac{\rho r^2}{\kappa^2 G} - \left(1+\frac{n^2\pi^2 r^2}{l^2}+\frac{n^2\pi^2 r^2}{l^2}\frac{E}{\kappa^2 G}\right)\omega_n^2 + \frac{\alpha^2 n^4\pi^4}{l^4} = 0 \tag{12.68}$$

where r and α are defined by

$$\alpha^2 = \frac{EI}{\rho A}, r^2 = \frac{I}{A}$$

The characteristic equation for the frequencies ω_n is quadratic in ω_n^2, and hence easily solved. Here note that of the two roots for each value of n are determined by the frequency equation, the smaller value is associated with bending deformation, and the larger root is associated with shear deformation. The natural frequencies for just the Euler–Bernoulli beam are

$$\omega_n^2 = \frac{\alpha^2 n^4 \pi^4}{l^4} \tag{12.69}$$

The natural frequencies for neglecting the rotary inertia and including the shear deformation are

$$\omega_n^2 = \frac{\alpha^2 n^4 \pi^4}{l^4 \left[1 + \left(n^2\pi^2 r^2/l^2\right) E/\kappa G\right]} \tag{12.70}$$

Next, suppose that one does an experiment to measure the frequencies of a beam, which has significant shear deformation so that the values will be close to those predicted by Equation (12.70). If an analyst were to model the experiment as an Euler-Bernoulli beam, and use the experimental data to validate Equation (12.69), no agreement would be obtained, as in Example 12.9.1. This is because the term in the bracket of the denominator in Equation (12.70) contains the shear coefficient, which is not modeled in the Euler-Bernoulli frequency equation given in Equation (12.70). Thus, the model cannot be validated because it is not verified. The model cannot be verified because of the missing shear term.

Chapter Notes

Modal testing dominates the field of vibration testing. There are, of course, other identification methods and other types of vibration experiments. The main reference for this chapter is the text by Ewins (1984, 2000), which discusses each topic of this chapter, with the exception of Section 8.6 on time domain methods, in much more detail. Ewins (1984) was the first to describe modal testing. Another good general reference for modal testing is the book by Zaveri (1984). Allemang (1984, 2013) has provided a review article on the topic as well as an extended bibliography. Each year since 1981, an International Modal Analysis Conference (IMAC) has been held, indicating the continued activity in this area. McConnell (1995) presents a more general look at vibration testing. Maia and Silva (1997) present an edited volume on modal analysis and testing written by Ewins' former students.

The material in Section 12.3 is a brief introduction to signal processing. More complete treatments along with the required material from Section 12.4 on random signals can be found in Otnes and Enochson (1972) or Doeblin (1980, 1983). An excellent introduction to signal processing and the appropriate transforms is the book by Newland (1985). There are many methods of extracting modal data from frequency response data and only one is covered in Section 12.5. Ewins (2000) and Zaveri (1984) discuss others. Most commercially available modal software packages include details of the various modal data extraction methods used in the package.

The time domain methods are mentioned briefly in Ewins (2000) and by Allemang (1984, 2013). Leuridan et al. (1986) discuss multiple input and multiple output time domain methods and present some comparison of methodology.

However, the ERA is not included in these works. An excellent treatment of identification methods driven by structural dynamics is given in the text by Juang (1994). A good introduction to the realization theory upon which ERA depends can be found in Chen (1970). A least squares regression technique for system identification of discrete time dynamic systems is given by Graupe (1976). The topic of physical modeling using measurements discussed in Section 12.7 is really an inverse problem and/or an identification problem, such as defined in Rajaram and Junkins (1985) and Lancaster and Maroulas (1987). Virgin (2000) provides an excellent introduction to experimental analysis for nonlinear systems. The use of vibration testing in structural health monitoring is reviewed in Doebling et al. (1998). Inman, et al. (2005) provide reviews of health monitoring and machinery diagnostics.

References

Allemang, R. J. (1984) Experimental modal analysis. *Modal Testing and Refinement*, ASME-59, American Society of Mechanical Engineers, 1–129.

Allemang, R. J. (2013) Verification and validation. *Sound & Vibration*, 47(1), 5–7.

Blevins, R. D. (1994) *Flow Induced Vibration*, 2nd Edition. Krieger Publishing, Malabar, FL.

Chen, C. J. (1970) *Introduction to Linear System Theory*. Holt, Rinehart & Winston, New York.

Cole, D. G., Saunders, W. R. and Robertshaw, H. H. (1995) Modal parameter estimation for piezostructures. *ASME Journal of Vibration and Acoustics*, 117(4), 431–438.

Cooley, J. W. and Tukey, J. W. (1965) An algorithm for the machine calculation of complex Fourier series. *Mathematics of Computation*, 19(90), 297–311.

Doeblin, E. O. (1980) *System Modeling and Response: Theoretical and Experimental Approaches*. John Wiley & Sons, New York.

Doeblin, E. O. (1983) *Measurement Systems Application and Design*, 3rd Edition. McGraw-Hill, New York.

Doebling, S. S., Farrar, C.R. and Prime, M. B. (1998) A summary review of vibration-based damage identification methods. *The Shock and Vibration Digest*, 30, 91–105.

Ewins, D. J. (1984) *Modal Testing: Theory and Practice*. Research Studies Press, Hertfordshire, UK.

Ewins, D. J. (2000) *Modal Testing: Theory and Practice*, 2nd Edition. Research Studies Press, Hertfordshire, UK.

Graupe, D. 1976. *Identification of Systems*. R. E. Krieger Publishing Co., Melbourne, FL.

Friswell, M. I. and Motershead, J. E. (1995) *Finite Element Updating in Structural Dynamics*. Kluwer Academic Publications, Dordrecht, The Netherlands.

Ibrahim, S. R. (1973) *A Time Domain Vibration Test Technique*. PhD Thesis, Department of Mechanical Engineering, University of Calgary, Alberta, Canada.

Ibrahim, S. R. and Mikulcik, E. C. (1976) The experimental determination of vibration parameters from time responses. *Shock and Vibration Bulletin*, 46(5), 187–196.

Inman, D. J. (2014) *Engineering Vibration*, 4th Edition. Pearson Education, Upper Saddle River, NJ.

Inman, D. J., Farrar, C. R., Steffen, Jr., V. and Lopes, V. (eds) (2005) *Damage Prognosis.* John Wiley & Sons, New York.

Juang, J. -N. (1994) *Applied System Identification.* Prentice Hall, Upper Saddle River, NJ.

Juang, J. -N. and Pappa, R. (1985) An eigensystem realization algorithm for modal parameter identification and modal reduction. *AIAA Journal of Guidance, Control, and Dynamics*, 8, 620–627.

Lancaster, P. and Maroulas, J. (1987) Inverse eigenvalue problems for damped vibrating systems. *Journal of Mathematical Analysis and Application*, 123(1), 238–261.

Leuridan, J. M., Brown, D. L. and Allemang, R. J. (1986) Time domain parameter identification methods for linear modal analysis: a unifying approach. *ASME Journal of Vibration, Acoustics, Stress, and Reliability in Design*, 108, 1–8.

Maia, N. M. M. and Silva, J. M. M. (1997) *Theoretical and Experimental Modal Analysis.* Research Studies Press, Hertfordshire, UK.

McConnell, K. G. (1995) *Vibration Testing; Theory and Practice.* John Wiley & Sons, New York.

Newland, D. E. (1985) *An Introduction to Random Vibrations and Spectral Analysis*, 2nd Edition. Longman Group Limited, London.

Otnes, R. K. and Enochson, L. (1972) *Digital Time Series Analysis.* John Wiley & Sons, New York.

Rajaram, S. and Junkins, J. L. (1985) Identification of vibrating flexible structures. *AIAA Journal of Guidance Control and Dynamics*, 8(4), 463–470.

Starek, L. and Inman, D. J. (1997) A symmetric inverse vibration problem for non-proportional, underdamped systems. *ASME Journal of Applied Mechanics*, 64(3), 601–605.

Halevi, Y., Tarazaga, P. A. and Inman, D. J. (2004) connectivity constrained reference basis model updating. *Proceedings of the ISMA 2004 International Conference on Noise and Vibration*, Leuven, Belgium, 19–22 September, on CD.

Virgin, L. N. (2000) *Introduction to Experimental Nonlinear Dynamics.* Cambridge University Press, Cambridge, UK.

Vold, H. and Rocklin, T. (1982) The numerical implementation of a multi-input modal estimation algorithm for mini computers. *Proceedings of the 1st International Modal Analysis Conference*, pp. 542–548.

Zaveri, K. (1984) *Modal Analysis of Large Structures-Multiple Exciter Systems.* Bruel & Kjaar Publications, Naerum.

Problems

12.1 Plot the error in measuring the natural frequency of an SDOF system of mass 10 and stiffness 35, if the mass of the exciter is included in the calculation and ranges from 0.4 to 5.0.

12.2 Calculate the Fourier transform of $f(t) = 3\sin 2t + 2\sin t - \cos t$ and plot the spectral coefficients.

12.3 Consider a signal $x(t)$ with maximum frequency of 500 Hz. Discuss the choice of record length and sampling interval.

12.4 The eigenvalues and mode shapes of a structure are given next. Develop a two degree of freedom model of the structure that yields the same modal data if the mass matrix is known to be diag[4 1]

$$\lambda_1 = -0.2134 \pm 1.2890j$$
$$\lambda_2 = -0.0366 \pm 0.5400j$$
$$\mathbf{u}_1 = [0.4142 \quad 1]^T$$
$$\mathbf{u}_2 = [1.000 \quad -0.4142]^T$$

12.5 Consider the vector

$$\mathbf{s}^T = [s_1 \quad s_2 \quad s_3]$$

Write the outer product matrix $\mathbf{ss}^T$ and show that the elements in one row (or one column) of the matrix completely determine the other six elements of the matrix. In particular, calculate the receptance matrix of Section 12.5 for the following measured data.

$$\zeta_1 = 0.01, \quad \zeta_2 = 0.2, \quad \zeta_3 = 0.01$$
$$\omega_{d1} = 2, \quad \omega_{d2} = 10, \quad \omega_{d3} = 12$$
$$G(\omega_{d1}) = 1, \quad G(\omega_{d2}) = -1, \quad G(\omega_3) = 3; \quad \text{(force at position 1 and transducer at position 3.)}$$
$$G(\omega_{d1}) = -3, \quad G(\omega_{d2}) = 2, \quad G(\omega_{d3}) = 4; \quad \text{(force at 1 transducer at 2)}$$
$$G(\omega_{d1}) = 5, \quad G(\omega_{d2}) = 2, \quad G(\omega_{d3}) = -2; \quad \text{(force at 1 transducer at 1)}$$

What are the eigenvectors of the system?

12.6 Calculate the receptance matrix for the two degree of freedom system with

$$M = \begin{bmatrix} 5 & 0 \\ 0 & 10 \end{bmatrix} K = \begin{bmatrix} 4 & -2 \\ -2 & 6 \end{bmatrix} C = \begin{bmatrix} 6 & -4 \\ -4 & 5 \end{bmatrix}$$

12.7

a) Consider the system of Problem 12.6. Using any integration package, numerically solve for the free response of this system to the initial conditions

$$\mathbf{q}(0) = [1 \quad 0]^T \text{ and } \dot{\mathbf{q}}(0) = [0 \quad 0]^T.$$

b) Using the solution to (a), generate the matrices Φ and Φ' of Equations (8.33) and (8.34). Using Φ and Φ', and Equation (8.38), calculate the eigenvalues and eigenvectors of the system. Check to see if they satisfy the eigenvalue problem for this system.

12.8 Repeat Problem 12.7 using ERA. This will require the availability of software performing SVD, and the like, such as MATLAB.

12.9 Solve the problem of Example 12.1.1 again for the case that the measured frequencies are $\omega_1 = \sqrt{2}$, and $\omega_2 = 2$.

12.10 Solve the model-updating problem of Example 12.8.1 using the pole placement method of Section 6.8, to see if it is possible to obtain a symmetric updated stiffness matrix with the required eigenvalues.

A

Comments on Units

This text omits units on quantities wherever convenient, in order to shorten the presentation and to allow the reader to focus on the development of ideas and principles. However, in an application of these principles, the proper use of units becomes critical. Both design and measurement results will be erroneous and could lead to disastrous engineering decisions if they are based on improper consideration of units. This appendix supplies some basic information regarding units, which will allow the reader to apply the techniques presented in the text to real problems.

Beginning physics texts defines three fundamental units of length, mass and time and considers all other units to be derived from these. In the *International System* of units, denoted by SI (from the French Système International), the fundamental units are as in Table A.1.

Table A.1 Fundamental units of length, mass and time.

Unit	Name	Symbol
Length	meter	m
Mass	kilogram	kg
Time	second	s

These units are chosen as fundamental units because they are permanent and reproducible quantities.

The United States has historically used length (in inches), time (in seconds) and force (in pounds) as the fundamental units. The conversion between the US Customary system of units, as it is called, and the SI units is provided as a standard feature of most scientific calculators and in many text and notebook covers. Some simple conversions are:

$$
\begin{aligned}
1 \text{ kg} &= 2.204622622 \text{ pounds mass} \\
4.5 \text{ N} &\cong 1 \text{ pound} \\
2.54 \times 10^{-2} \text{ m} &= 1 \text{ inch}
\end{aligned}
$$

Vibration with Control, Second Edition. Daniel John Inman.

Companion Website: www.wiley.com/go/inmanvibrationcontrol2e

Table A.2 Units of mass, damping and stiffness.

Coefficient	Unit	Name
m	kg	Mass
c	kg/s	Damping
k	kg/s^2	Stiffness

As most equipment and machines in the United States used in vibration design, control and measurement are manufactured in US Customary units, it is important to be able to convert between the two systems. This is discussed in detail, with examples, by Thomson (1988).

In vibration analysis, position is measured in units of length in meters (m), velocity in meters per second (m/s), and acceleration in meters per second squared (m/s^2). Since the basic equation of motion comes from a balance of force, each term in the equation

$$m\ddot{x}(t) + c\dot{x}(t) + kx(t) = f(t) \tag{A.1}$$

must be in units of force. Force is mass times acceleration, kg·m/s^2. This unit is given the special name of Newton, abbreviated to N (i.e. 1 N = 1 kg·m/s^2). The coefficients in Equation (A.1) then must have the following units (Table A.2):

Note that stiffness may also be expressed in terms of N/m, and damping (viscous friction) in terms of N/m/s and these units are normally used. Using other formulas and definitions in the text, the following units and quantities can be derived. Some are given special names because of their usefulness in mechanics (Table A.3).

Table A.3 Units and quantities in mechanics.

Quantity	Unit Name	Common Abbreviation	Definition
Force	Newton	N	kg·m/s^2
Velocity	–	–	m/s
Acceleration	–	–	m/s^2
Frequency	hertz	Hz	1/s
Stress	Pascal	Pa	N/m^2
Work	joule	J	N·m
Power	watt	W	J/s
Area moment of inertia	–	–	m^4
Mass moment of inertia	–	–	kg/m^2
Density	–	–	kg/m^3
Torque	–	–	N/m
Elastic modulus	–	–	Pa

Because vibration measurement instruments and control actuators are often electromechanical transducers, it is useful to recall some electrical quantities. The fundamental electrical unit is often taken as the ampere, denoted by A. Units often encountered with transducer specification are as in Table A.4.

Table A.4 Units encountered with transducer specifications.

Unit	Name	Common Abbreviation	Definition
Electrical potential	volt	V	W/A
Electrical resistance	ohm	Ω	V/A
Capacitance	farad	F	A·s/V
Magnetic flux	weber	Wb	V·s
Inductance	henry	H	V·s/N

The gains used in the control formulation for vibration control problems have the units required to satisfy the units of force when multiplied by the appropriate velocity or displacement term. For instance, the elements of the feedback matrix G_f of Equation (7.18) must have units of stiffness, i.e. N/m.

Often these quantities are too large or too small to be convenient for numerical and computer work. For instance, a Newton, which is about equal to the force exerted in moving an apple, would be inappropriately small if the vibration problem under study is that of a large building. The meter, on the other hand, is too large to be used when discussing the vibration of a compact disc in a stereo system. Hence, it is common to use units with prefixes, such as the millimeter, which is 10^3 m or gigapascal, which is 10^9 Pa. Of course, the fundamental unit kilogram is 10^3 g. For example, a common rating for a capacitor is microfarads, abbreviated as μF. Table A.5 lists some commonly used prefixes.

Table A.5 Commonly used prefixes.

Factor	10^{-12}	10^{-9}	10^{-6}	10^{-3}	10^{-2}	10^{3}	10^{6}	10^{9}
Prefix	pico-	nano-	micro-	milli-	centi-	kilo-	mega-	giga-
Abbreviation	P	n	μ	m	c	k	M	G

Many of the experimental results given in the text are discussed in the frequency domain in terms of magnitude and phase plots. Magnitude plots are often given in logarithmic coordinates. In this situation, the *decibel*, abbreviated db, is used as the unit of measure. The decibel is defined in terms of a power ratio of an electrical signal. Power is proportional to the square of the signal's voltage, so that the decibel can be defined as

$$1 \text{ db} = 20 \log_{10} \frac{V_1}{V_2} \tag{A.2}$$

where V_1 and V_2 represent different values of the voltage signal (i.e. from an accelerometer).

The phase is given in terms of either degrees (°) or radians (rad).

Reference

Thomson, W. T. and Dahleh, M. D. (1998) *Theory of Vibration with Applications*, 5th Edition. Prentice Hall, Englewood Cliffs, NJ.

B

Supplementary Mathematics

B.1 Vector Space

Fundamental to the discipline of matrix theory, as well as the operator theory of functional analysis, is the definition of a *linear space*, also called a *vector space*. A linear space, denoted by V, is a collection of objects (vectors or functions in the cases of interest here) for which the following statements hold for all elements $\mathbf{x}, \mathbf{y}, \mathbf{z} \in V$ (this denotes that the vectors $\mathbf{x}$, $\mathbf{y}$ and $\mathbf{z}$ are all constrained in the set V) and for any real valued scalars α and β:

1) $\mathbf{x} + \mathbf{y} \in V, \alpha\mathbf{x} \in V$
2) $\mathbf{x} + \mathbf{y} = \mathbf{y} + \mathbf{x}$
3) $(\mathbf{x} + \mathbf{y}) + \mathbf{z} = \mathbf{x} + (\mathbf{y} + \mathbf{z})$
4) There exists an element $\mathbf{0} \in V$, such that $0\mathbf{x} = \mathbf{0}$
5) There exists an element $1 \in V$, such that $1\mathbf{x} = \mathbf{x}$
6) $\alpha(\beta\mathbf{x}) = (\alpha\beta)\mathbf{x}$
7) $(\alpha + \beta)\mathbf{x} = \alpha\mathbf{x} + \beta\mathbf{x}$
8) $\alpha(\mathbf{x} + \mathbf{y}) = \alpha\mathbf{x} + \alpha\mathbf{y}$

The examples of linear spaces V used in this text are the set of real vectors of dimension n, the set of complex vectors of dimension n, and the set of functions that are square integrable in the Lebseque sense.

B.2 Rank

An extremely useful concept in matrix analysis is the idea of rank introduced in Section 3.2. Let $R^{m \times n}$ denote the set of all $m \times n$ matrices with m rows and n columns. Consider the matrix $A \in R^{m \times n}$. If the columns of the matrix A are considered as vectors, the number of linearly independent columns is defined to be the *column rank* of the matrix A. Likewise, the number of linearly independent rows of the matrix A is called the *row rank* of A. The row rank of a matrix and the column rank of the matrix are equal, and this integer is called the *rank* of the matrix A. The concept of rank is useful in solving equations, as well as checking stability of a system (Chapter 4) or the controllability and observability of a system (Chapter 6). Perhaps the best way to determine the rank of a matrix is to calculate

Vibration with Control, Second Edition. Daniel John Inman.

Companion Website: www.wiley.com/go/inmanvibrationcontrol2e

the singular values of the matrix of interest (see Section 6.12 and the following comments). The rank of a matrix can be shown to be equal to the number of nonzero singular values of the matrix. The singular values also provide a very precise way of investigating the numerical difficulties frequently encountered in situations where the rank of the matrix is near the desired value. This shows up as very small but nonzero singular values.

A simple procedure to calculate the singular values of a matrix A, and hence determine its rank, is provided by calculating the eigenvalues of the symmetric matrix

$$\tilde{A} = \begin{bmatrix} 0 & A^T \\ A & 0 \end{bmatrix}$$

If $A \in R^{m \times n}$ of rank r, the first r eigenvalues of $\tilde{A}$ are equal to the singular values of A, the next r eigenvalues are equal to the negative of the singular values of A, and the remaining eigenvalues of $\tilde{A}$ are zero. The rank of A is thus the number of positive eigenvalues of the symmetric matrix $\tilde{A}$.

B.3 Inverses

For $A \in R^{m \times n}$, the linear equation

$$A\mathbf{x} = \mathbf{b}$$

with $\det A \neq 0$ has the solution $\mathbf{x} = A^{-1}\mathbf{b}$, where A^{-1} denotes the unique inverse of the matrix A. The matrix A^{-1} is the matrix that satisfies

$$A^{-1}A = AA^{-1} = I_n$$

Next consider $A \in R^{m \times n}$. If $m \geq n$ and if the rank of A is n, then there exists an $n \times m$ matrix A_L of rank n such that

$$A_L A = I_n$$

where I_n denotes the $n \times n$ identity matrix. Let A and $B \in R^{m \times n}$, then

$$(AB)^T = B^T A^T$$

The matrix A_L is called the left inverse of A. If, on the other hand, $n \geq m$ and the rank of A is m, then there exists an $n \times m$ matrix A_R of rank m, called a right inverse of A such that

$$AA_R = I_m$$

where I_m denotes the $m \times m$ identity matrix. If $m = n = \text{rank}\, A$, then A is nonsingular and

$$A_R = A_L = A^{-1}$$

Consider the matrix $A^T A$ and note that it is an $n \times n$ symmetric matrix. If A is of rank n (this requires that $m \geq n$), then $A^T A$ is nonsingular. A solution of

$$A\mathbf{x} = \mathbf{b}$$

for $A \in R^{m \times n}$ can then be calculated by multiplying both sides of this last expression by $(A^TA)^{-1}A^T$, which yields

$$\mathbf{x} = (A^TA)^{-1}A^T\mathbf{b}$$

The quantity $(A^TA)^{-1}A^T$ is called the *generalized inverse* of A, denoted by $A^\dagger$.

The matrix $A^\dagger$ is also called a *pseudo-inverse* or *Moore-Penrose inverse* and can be expressed in terms of a singular-value decomposition (Section 7.7) of the matrix A. In the notation of Section 7.7, any matrix $A \in R^{m \times n}$ can be expressed in terms of its singular-value factors as

$$A = U\Sigma V^T$$

where Σ denotes that the diagonal matrix of singular values of A, U and V are orthogonal. For the case that $m \geq n$, if the rank of A is r, then the last $n - r$ (or $m - r$ if $m \leq n$) singular values are zero, so that Σ has the partitioned form

$$\Sigma = \begin{bmatrix} \Sigma_r & 0 \\ 0 & 0 \end{bmatrix}$$

where the zeros indicate matrices of zeros of the appropriate size and Σ_r is an $r \times r$ diagonal matrix of the nonzero singular values of A. Define the matrix Σ' by

$$\Sigma' = \begin{bmatrix} \Sigma_r^{-1} & 0 \\ 0 & 0 \end{bmatrix}$$

The matrix $A^\dagger$ can be shown to be

$$A^\dagger = V\Sigma U^T$$

which is the singular value decomposition of the generalized inverse. This last expression constitutes a more numerically stable way of calculating the generalized inverse than using the definition

$$(A^TA^{-1})A^T$$

The following Moore-Penrose conditions can be stated for the pseudo-inverse. If $A \in R^{m \times n}$ has singular value decomposition $A = U\Sigma V^T$, then $A^\dagger = V\Sigma' U^T$ satisfies

$$\begin{array}{ll} 1)\ \ AA^\dagger A = A & 2)\ \ A^\dagger AA^\dagger = A^\dagger \\ 3)\ \ (AA^\dagger)^T = AA^\dagger & 4)\ \ (A^\dagger A)^T = A^\dagger A \end{array}$$

The matrix $A^\dagger$ satisfying all four of these conditions is unique. If A has full rank, then $A^\dagger$ is identical to the left (and right) inverse just discussed.

Finally, note that the least squares solution of the general equation

$$A\mathbf{x} = \mathbf{b}$$

calculated by using the generalized inverse of A is *not* a solution in the sense that $\mathbf{x} = A^{-1}\mathbf{b}$ is a solution in the nonsingular case, but is rather a vector $\mathbf{x}$ that minimizes the quantity

$$||A\mathbf{x} - \mathbf{b}||$$

The preceding is a quick summary of material contained in most modern texts on linear algebra and matrix theory, such as the excellent text by Ortega (2013). Computational issues and algorithms are discussed in the text by Golub and Van Loan (2012), which also mentions several convenient software packages. In most cases, the matrix computations required in the vibration analysis covered in this text can be performed by using standard software packages, most of which are in the public domain.

References

Golub, G. H. and Van Loan, C. F. (2012) *Matrix Computations*, vol. 3. JHU Press.

Ortega, J. M. (2013) *Matrix Theory: A Second Course*. Springer Science & Business Media.

Index

Vibration with Control, Second Edition. Daniel John Inman.

Companion Website: www.wiley.com/go/inmanvibrationcontrol2e

Printed and bound by CPI Group (UK) Ltd, Croydon, CR0 4YY
16/04/2025
14658562-0003